CW01369842

Variability of blazars

Blazars (also known as BL Lac objects), first discovered in 1978, are unusually energetic objects in the extragalactic universe. About 200 are known or suspected. They are radio sources with highly variable optical and radio emission, as well as high polarisation, and their optical spectra normally have no distinguishable features. It is generally accepted that they belong to that class of galaxies with active galactic nuclei, which are presumed to be driven by infall of matter to a supermassive black hole. In this book researchers give a complete summary of the observations of blazars and the theoretical interpretation. A comprehensive listing of confirmed and candidate objects is included. Mechanisms in which the variability can arise from shocks and relativistic jets are discussed. There are at least four different answers given to the question: what is a blazar? This book is a complete overview of the violent activity observed in these extreme active galactic nuclei.

Variability of Blazars

Proceedings of a Conference in honour of
the 100th anniversary of the birth of Academician Yrjö Väisälä,
held at Tuorla Observatory, University of Turku, Finland,
January 6-10, 1991

Edited by

Esko Valtaoja
Metsähovi Radio Research Station, Helsinki University of Technology

and

Mauri Valtonen
Tuorla Observatory, University of Turku

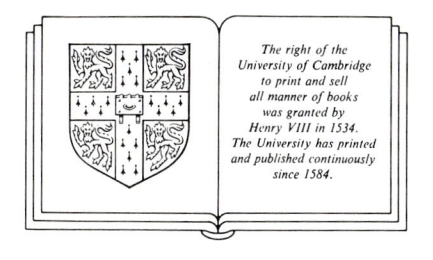

CAMBRIDGE UNIVERSITY PRESS

Cambridge

New York   Port Chester

Melbourne   Sydney

Published by the Press Syndicate of the University of Cambridge
The Pitt Building, Trumpington Street, Cambridge CB2 1RP
40 West 20th Street, New York, NY 10022–4211, USA
10 Stamford Road, Oakleigh, Victoria 3166, Australia

© Cambridge University Press 1992

First published 1992

Printed in Great Britain at the University Press, Cambridge

British Library cataloguing in publication data available

Library of Congress cataloguing in publication data available

ISBN 0 521 41351 6 hardback

Contents

Acknowledgements	x
Preface	xi
Participants	xii
Opening Address Pekka Teerikorpi	1
BL Lac Objects and Rapidly Variable QSOs – an Overview G. Burbidge and A. Hewitt	4
Many Black Hole Nuclei in Blazars Mauri Valtonen	39
Optically Violently Variable Active Galactic Nuclei as Supermassive Binary Systems M. K. Babadzhanyants, Yu. V. Baryshev, and E. T. Belokon	45
Pulsation of a Supermassive Star in Tensor Field Gravitation Theory Yu. V. Baryshev	52
Blazars: Faster than Light? C. D. Impey	55
Two Classes of BL Lac Objects? E. Valtaoja, H. Teräsranta, M. Lainela, and P. Teerikorpi	70
Are Highly Evolved Radio Quiet QSOs the "Parent" Population of Lacertae-type Objects? B. V. Komberg	79
Blazars within an Evolutionary Unified Scheme F. Vagnetti, E. Giallongo, and A. Cavaliere	82
Variability of Nonthermal Continuum Emission in Blazars Alan P. Marscher, Walter K. Gear, and John P. Travis	85
Variability of 3C 273 and Shock Models M. Lainela, E. Valtaoja, and M. Tornikoski	102
Multifrequency Observations of the Flaring Characteristics of Blazars Ian Robson	111
The Cm-Wavelength Flux Behavior of AGNs M. F. Aller, H. D. Aller, P. A. Hughes, and G. E. Latimer	126

Centimeter-Wavelength Linear Polarization Observations of Active   134
Galactic Nuclei
    H. D. Aller, M. F. Aller, and P. A. Hughes

Modeling the UMRAO Database: Status and Application to Other Sources   142
    P. A. Hughes, H. D. Aller, and M. F. Aller

Observations of Southern Blazars   151
    D. Bramwell and G. D. Nicolson

Ten Years Monitoring of Blazars at Metsähovi   159
    H. Teräsranta, M. Tornikoski, K. Karlamaa, E. Valtaoja, S. Urpo,
    M. Lainela, J. Kotilainen, S. Wiren, S. Laine, K. Nilsson,
    A. Lähteenmäki, R. Korpi, and M. Valtonen

A Study of Long Term Variability of Blazars in Multifrequency Spectral   167
Monitoring Program
    A. B. Berlin, Yu. A. Kovalev, Yu. Yu. Kovalev, G. M. Larionov,
    N. A. Nidgelski, and V. A. Soglasnov

A Study of Quasars, BL Lacertae Objects and Active Galactic Nuclei   170
at 22 and 43 GHz
    L. C. L. Botti and Z. Abraham

SEST Observations of Southern AGN   175
    M. Tornikoski, H. Teräsranta, E. Valtaoja, J. Kotilainen,
    M. Lainela, and L. C. L. Botti

Northern Hemisphere Survey of Quasar Variability   184
    S. Wirén, H. Teräsranta, E. Valtaoja, and J. Kotilainen

VLBI Monitoring of Blazars   187
    A. Zensus

High Resolution Images of Blazar Cores   196
    L. B. Bååth

Structural Variability of Active Galactic Nuclei at 43 GHz   205
    T. P. Krichbaum and A. Witzel

Monitoring of the Milliarcsecond-Structure of S5-1928+738: Apparent   221
Superluminal Motion along a Fixed Path?
    C. J. Schalinski, A. Witzel, C. A. Hummel, T. P. Krichbaum,
    A. Quirrenbach, and K. J. Johnston

Structural Variability of Blazars from the Complete S5-VLBI-Sample   225
    C. J. Schalinski, A. Witzel, T. P. Krichbaum, C. A. Hummel,
    A. Quirrenbach, and K. J. Johnston

Varying Gamma in 3C 273   229
    L. B. Bååth

What Is Responsible for the New Outburst of 0735+178 *F. J. Zhang, L. B. Bååth, and H. S. Chu*	234
Milliarcsecond Polarization Structure and the Classification of "Blazars" *D. C. Gabuzda and T. V. Cawthorne*	238
Radio Structures of Selected Blazars at Sub-arcsec Resolutions *Chidi E. Akujor*	251
Jets in a Sample of Superluminal Active Galactic Nuclei *Irene Cruz-Gonzalez and Rene Carrillo*	256
Radio Structure of Variable AGN and Cosmological Tests *L. I. Gurvits, N. S. Kardashev, and A. P. Lobanov*	259
Blazar Models *C.-I. Björnsson*	264
Polarized Synchrotron Emission from Shock Accelerated Particles *Klaus-Dieter Fritz*	278
The Connection Between Growing Radio Shocks and Frequency Dependent Optical Polarization in Blazars and Low Polarization Quasars *L. Valtaoja, E. Valtaoja, N. M. Shakhovskoy, and Yu. S. Efimov*	284
Photometry and Polarimetry of OJ 287 and Mrk 421 in 1980-90 *S. Kikuchi*	289
First Detection of Nightly Variations in the Frequency Dependent Polarization of OJ 287 *L. O. Takalo, A. Sillanpää, M. Kidger, and J. A. de Diego*	294
Polarimetric Monitoring of Blazars at the Nordic Optical Telescope *L. O. Takalo*	300
Statistical Analysis on the Frequency Dependent Polarization in Blazars *K. Nilsson, L. O. Takalo, and A. Sillanpää*	303
Rotation of Position Angles in Blazars *A. Sillanpää and L. O. Takalo*	306
Accretion Disk Models for Microvariability *Paul J. Wiita, H. Richard Miller, Navarun Gupta, and Sandip K. Chakrabarti*	311
Blazar Microvariability: A Case Study of PKS 2155-304 *H. R. Miller, M. T. Carini, J. C. Noble, J. R. Webb, and P. J. Wiita*	320

On the Possible Cause of Rapid Flux Variability in Active 327
Galactic Nuclei
    V. G. Gorbatsky

Intraday Variability of Compact Extragalactic Radio Sources 331
    T. P. Krichbaum, A. Quirrenbach, and A. Witzel

Rapid Variability in the BL Lac Object S5 0716+714 346
    Stefan J. Wagner

Fast Radio Variability of Blazars in Observations with RATAN-600 352
    S. A. Pustil'nik and K. D. Aliakberov

Rapid Light Variations of BL Lacertae 358
    V. T. Doroshenko, V. M. Lyuty, A. Sillanpää, and E. Valtaoja

Detection of Microvariability in Mrk 421 361
    U. C. Joshi and M. R. Deshpande

Rapid Variability of Mkn 501 367
    J. A. de Diego, M. R. Kidger, and L. O. Takalo

Blazar Type Variability in a Seyfert Nucleus 372
    D. Dultzin-Hacyan, W. Schuster, J. Peña, L. Parrao, R. Peniche,
    E. Benitez, and R. Costero

Periodicity in OJ 287? 377
    M. R. Kidger, J. A. de Diego, L. O. Takalo, A. Sillanpää,
    and K. Nilsson

Long-Term Periodicity in 3C 345? 382
    M. R. Kidger, J. A. de Diego, L. O. Takalo, A. Sillanpää,
    and K. Nilsson

100 Years of Observations of 3C 273: New Evidence for Possible 384
Periodic Behaviour
    M. K. Babadzhanyants and E. T. Belokon

Rapid 1.3cm Variability of Blazars 391
    H. J. Lehto

Multi Frequency Observations and Variability of Blazars 399
    Thierry J.-L. Courvoisier

The Energetics and Morphologies of Blazar Outbursts 414
    James R. Webb

Optical Spectral Energy Distributions of Variable Sources in Blazars 427
    V. A. Hagen-Thorn, S. G. Marchenko, and O. V. Mikolaichuk

Redshift-Dependent Optical Variability of Quasars: an Interpretation 438
in Terms of Spectral Changes
    E. Giallongo, D. Trevese, and F. Vagnetti

Emission-Line Variability in the Seyfert 1 Galaxy NGC 3516 441
    Ignaz Wanders and Ernst van Groningen

Near-Infrared Observations of Hard X-Ray Selected AGN 444
    Jari Kotilainen

UV and X-Ray Emission of Blazars 447
    Laura Maraschi

Index of Objects 459

Index of Subjects 463

Acknowledgements

The conference was sponsored by the University of Turku, the Turku University Foundation, the Ministry of Education, the Vilho, Yrjö and Kalle Väisälä Foundation, and the optical manufacturing company Opteon. The latter sponsored a visit to Tuorla Observatory for a demonstration of the manufacturing of telescope mirrors which was started by Yrjö Väisälä nearly 70 years ago in Turku.

The Scientific Organizing Committee included, in addition to the editors, the following scientists: Roy Booth (Onsala), Thierry Courvoisier (Geneva), Sen Kikuchi (Tokyo), Martin Rees (Cambridge), Ian Robson (Lancashire), and Roland Svensson (Nordita). In the Local Organizing Committee Aimo Niemi, Aimo Sillanpää, Mats-Olof Snåre, and Leena Valtaoja carried out much of the responsibilities. We would like to thank all the committee members for their valuable contributions which made the meeting a success.

We would also like to extend special thanks to Anne Lähteenmäki and Sirpa Reinikainen for help beyond the call of duty. Ms. Reinikainen also carefully retyped many of the manuscripts.

Preface

The conference "Variability of Blazars" was held 6-10 January 1991 at the Tuorla Observatory of the University of Turku, Finland, in honour of the 100th anniversary of the birth of Yrjö Väisälä, the founder and long time director of the Observatory. Variability of blazars could hardly have been of major interest to Väisälä; indeed this class of objects was discovered after his death in 1971. However, Väisälä has left his mark also in this field through the development of the wide field telescopes which are often used in survey work, and through his extensive collection of photographic plates which have been used as a historical archive of AGN variability. A description of Väisälä's primary scientific interests and achievements is found in the Opening Address.

The conference also coincided with the promotion of the Tuorla Observatory to the status of a national research institute, the first of its kind in Finland.

The previous international astronomical conference in Turku had ended with snow on the ground on Midsummer Eve; despite this, over sixty astronomers from around the world braved January in Finland to attend the meeting which concentrated on both the theory and the observations of variability in blazars. Tragically, the winter proved to be among the mildest in living memory, and the organizers were unable to provide the traditional hole in the ice for a refreshing swim after the Conference Sauna, the reason being that the sea ice was too thin for the operation.

Having convened, the participants promptly agreed that there are no such things as blazars (although no two people seemed to agree on anything else concerning the classification of active galaxies). However, the blazar *phenomenon* was found to be alive and well, its study continuing to provide major insights into the variability mechanisms of AGN. As several contributions attest, the shock-in-jet models have begun to shown their power in explaining the main aspects of nonthermal variability in extragalactic radio sources; many participants expressed their belief that such unexpected and puzzling new phenomena as microvariability and deviant polarization behaviour, both prominently featured in these proceedings, would eventually also be explained within the framework of shocks and relativistic jets. Yet one need not fear that the field is becoming dull; the reader of this book will find at least four different answers even to the simplest of questions: what are the BL Lac objects?

This volume contains the invited review talks as well as most of the oral contributions and posters presented in the conference. We hope that the book provides an up-to-date overview of our knowledge of the most extreme and violent manifestation of AGN activity, the blazar phenomenon.

Esko Valtaoja   *Tuorla*
Mauri Valtonen   *July 1991*

Participants

Akujor, Chidi E., University of Manchester, UK
Aller, Hugh D., Department of Astronomy, University of Michigan, US
Aller, Margo F., Department of Astronomy, University of Michigan, US
Babadzhanyants, M., Astronomical Observatory, Leningrad University, USSR
Baryshev, Yuri, Astronomical Observatory, Leningrad University, USSR
Björnsson, Claes-Ingvar, Stockholm Observatory, Sweden
Botti, Luiz C.L., CRAAE, Universidade de Sao Paulo, Brazil
Bramwell, Don, Hartebeesthoek Radio Astronomy Observatory, S. Africa
Burbidge, Geoffrey, University of California, San Diego, US
Bååth, Lars, Onsala Space Observatory, Sweden
Clavel, Jean, ESA IUE Observatory, Madrid, Spain
Courvoisier, Thierry, Geneva Observatory, Switzerland
De Diego, José A., Instituto de Astrofísica de Canarias, Spain
Donner, Karl Johan, Tuorla Observatory, University of Turku, Finland
Dultzin-Hacyan, Deborah, Instituto de Astronomía, UNAM, México
Fritz, Klaus-Dieter, MPI für Kernphysik, Heidelberg, Germany
Gabuzda, Denise C., JPL, California Institute of Technology, US
Gurvits, Leonid, Astro Space Center, Lebedev Physical Institute, USSR
Hagen-Thorn, V.A., Dept. of Astronomy, Leningrad State University, USSR
Hughes, Philip, Department of Astronomy, University of Michigan, US
Impey, Chris, Steward Observatory, University of Arizona, US
Inoue, Makoto, Nobeyama Radio Observatory, Japan
Jaakkola, Toivo, Tuorla Observatory, University of Turku, Finland
Joshi, U.C., Physical Research Laboratory, Ahmedabad, India
Kikuchi, Sen, National Astronomical Observatory, Mitaka, Japan
Komberg, Boris, Institute of Cosmical Research, Moscow, USSR
Korhonen, Tapio, Tuorla Observatory, University of Turku, Finland
Kotilainen, Jari, Institute of Astronomy, Cambridge, UK
Kovalev, Yuri A., Astro Space Center, Lebedev Physical Institute, USSR
Krichbaum, Thomas, MPI für Radioastronomie, Bonn, Germany
Lainela, Markku, Tuorla Observatory, University of Turku, Finland
Laurikainen, Eija, Tuorla Observatory, University of Turku, Finland
Lehto, Harry, University of Southampton, UK
Leppänen, Kari, Metsähovi Radio Research Station, Finland
Lähteenmäki, Anne, Tuorla Observatory, University of Turku, Finland
Maraschi, Laura, Dipartimento di Fisica, Università di Milano, Italy
Marscher, Alan, Boston University, US
Mikkola, Seppo, Tuorla Observatory, University of Turku, Finland
Nesterov, Nikolai S., Crimean Astrophysical Observatory, USSR
Niemi, Aimo, Tuorla Observatory, University of Turku, Finland
Pustil'nik, S.A., Special Astrophysical Observatory, USSR
Robson, Ian, Lancashire Polytechnic, UK
Romeo, Alessandro, Tuorla Observatory, University of Turku, Finland
Salo, Heikki, Department of Astronomy, University of Oulu, Finland
Shakhovskoy, Nikolai M., Crimean Astrophysical Observatory, USSR
Sillanpää, Aimo, Tuorla Observatory, University of Turku, Finland
Snåre, Mats-Olof, Tuorla Observatory, University of Turku, Finland
Svensson, Roland, Stockholm Observatory, Sweden
Teerikorpi, Pekka, Tuorla Observatory, University of Turku, Finland
Teräsranta, Harri, Metsähovi Radio Research Station, Finland
Tornikoski, Merja, Metsähovi Radio Research Station, Finland
Trevese, Dario, Università di Roma "La Sapienza", Italy
Urpo, Seppo, Metsähovi Radio Research Station, Finland
Valtaoja, Esko, Metsähovi Radio Research Station, Finland
Valtaoja, Leena, Tuorla Observatory, University of Turku, Finland
Valtonen, Mauri, Tuorla Observatory, University of Turku, Finland
Vagnetti, Fausto, II Università di Roma, Italy
Vestergaard, Marianne, Copenhagen University Observatory, Denmark
Wagner, Stefan, Landessternwarte, Königstuhl, Heidelberg, Germany
Wanders, Ignaz, Astronomiska observatoriet, Uppsala, Sweden
Webb, James R., Dept. of Physics, Florida International University, US
Wiren, Seppo, Tuorla Observatory, University of Turku, Finland
Wiik, Kaj, Metsähovi Radio Research Station, Finland
Wiita, Paul J., Georgia State University, Atlanta, US
Zensus, Anton, NRAO, Socorro, US
Zhang, F.J. University of Manchester, UK
Zheng, Jia-Qing, Tuorla Observatory, University of Turku, Finland

KEY TO PHOTO

Front row (seated, left to right): Maraschi, Wiren, De Diego, Romeo, Sillanpää, Teerikorpi, Burbidge, Wagner, Krichbaum, Urpo. *Second row*: Kikuchi, Botti, Robson, Tornikoski, Teräsranta, Björnsson, Wanders, Vestergaard, Svensson, Niemi. *Third row* : Valtonen, Salo, Zheng, Zhang, Jaakkola, Inoue, Impey, Hughes, Gurvits, Gabuzda. *Fourth row*: Akujor, Webb, Dultzin-Hacyan, Joshi, Clavel, Babadzhanyants, Baryshev, Kotilainen, Lainela, H. Aller, Lähteenmäki. *Standing*: L. Valtaoja, Kornberg, Zensus, Bååth, Trevese, Vagnetti, Marscher, Lehto, Bramwell, Nesterov, Pustil'nik, Mikkola, Leppänen, Wiik, E. Valtaoja, Kovalev, M. Aller and daughter. *Hunting reindeers*: Courvoisier, Donner, Fritz, Hagen-Thorn, Korhonen, Laurikainen, Shakhovskoy, Snåre, Wiita.

Opening Address

PEKKA TEERIKORPI

Tuorla Observatory, University of Turku, SF-21500 Piikkiö, Finland

The Local Organizing Committee has kindly asked me to say a few words for the beginning of this meeting. Perhaps it is appropriate to tell something concerning our Observatory – the host institute of the meeting. Especially so because the beginning new year has special significance for the Observatory.

First, in this year we are celebrating Yrjö Väisälä, the founder of the Tuorla Observatory. He was born one hundred years ago, in September of 1891, and he died in 1971, not quite reaching the age of 80 years. Yrjä Väisälä was a remarkable scientist and personality with wide interests in geodesy, astronomical optics and astronomy. He studied in the University of Helsinki where he obtained his doctor's degree in 1923. His thesis was concerned with the investigation of optical surfaces with interferometric methods, and one can see here the beginning of the road which led to one of the main activities in the present day Observatory, i.e. investigation and manufacture of astronomical optics.

At that time the University of Turku had just been founded, and in 1924 Yrjö Väisälä was elected to the first permanent physics chair of this young university. In addition, from 1927 he held the professorship in astronomy as a side occupation. After having been appointed to the membership of the Academy of Finland in the year 1951 he retired, and he could concentrate his efforts to his dream: establishment of an astronomical-optical research institute in Tuorla, about 12 km from Turku. This institute was founded in 1952, with Yrjö Väisälä as its first director to the end of his life. The first large construction, the underground optical laboratory, was completed in 1954. At the present time, Tuorla's main astronomical instruments are the one meter telescope and the 70 cm Schmidt camera.

When Turku University became a state university in 1975, the Astronomical-Optical Institute lost its independent status and became a part of the Physics Department. During this period of "non-independency", however, the observatory and its activities have greatly evolved towards a modern, versatile astrophysical centre, thanks especially to the efforts of Professor Mauri Valtonen. Finally, last year it was decided that the observatory will regain its status as a separate research institute, and this new national institute, Tuorla Observatory, started its operation in the beginning of this year, about one week ago. This is another reason why we, the astronomers of Turku, should be happy with the new year.

In fact, the new institute was especially created in order to support the Finnish research work with the Nordic Optical Telescope at La Palma.

At the present time the main fields of research at Tuorla Observatory are

- observational studies of active galactic nuclei,
- theoretical work on stellar systems, quasars and cosmology,
- investigation and manufacture of telescope optics.

The last topic in this list continues the tradition established by Yrjö Väisälä. Thus far the culmination was the 2.5 m telescope mirror for the Nordic Optical Telescope. This mirror of an exceptionally high quality was made by a small group of only four people under the direction of Väisälä's pupil, Dr. Tapio Korhonen.

We, of course, regard it as a richness that the three fundamental stones for astronomy: telescope making, observational study, and theoretical investigation, can fruitfully coexist even in such a rather small institute as ours.

Finally, let me say still something about Yrjö Väisälä as an astronomer and maker of telescopes. He made about 25 anastigmatic telescopes, and one of these, completed in 1934, with its opening of 50 cm was for a few years the biggest Schmidt telescope in the world. Actually it was Väisälä's variant of the Schmidt telescope, with a truely planar focal plane. With this instrument about 10000 plates were photographed and about 800 minor planets and 6 comets were discovered. This collection of plates can still be used for complementary studies of the historical light curves of blazars and quasars, not yet fully utilized for this purpose.

Yrjö Väisälä's interest in wide-field telescopes can be traced to year 1924. In a notebook from this year one finds a sketch of the Schmidt telescope, several years before Bernhardt Schmidt published his own discovery. Meanwhile Väisälä did not develop his independent idea, possibly because of his heavy engagement with other works at that time (his interferometric method for measuring long geodetic baselines), or possibly because he regarded the curved focal plane impractical. Väisälä was ready to acknowledge, of course, that Schmidt was the inventor of the Schmidt telescope. However, he sometimes regretted that he did not publish his idea, not even in a local newspaper, as he added with his typical humour.

The present meeting has been dedicated to the memory of Academician Yrjö Väisälä. It is good to remind of an answer by him, when it was inquired why he devoted his time to many seemingly useless things, such as measuring the distance between two points with an utmost accuracy. He said that it is a lot of fun and it is not wrong, not immoral. He was not ashamed after having done it. I think that we also feel that there is nothing wrong in investigating the variations of blazars. In fact, many of us think that these points of radiation in the sky are one important key to understanding the Universe, and there is nothing wrong in trying to turn this key. I hope that these days in Turku will also offer a lot of fun to you, as Väisälä might have put it, and that you will have pleasant memories from our country.

Yrjö Väisälä 1891 – 1971

BL Lac Objects and Rapidly Variable QSOs
– an Overview

G. Burbidge and A. Hewitt

Center for Astrophysics & Space Sciences
University of California, San Diego
La Jolla, California, 92093-0111

ABSTRACT

The definitions of BL Lac objects and blazars etc. are discussed. A list of 215 objects which have been called BL Lac objects, blazars, optically violent variables (OVV), high polarization quasars (HPQ) and superluminal sources (SLS) is given in Table 1. As far as BL Lac objects are concerned, it is suggested that in future they be classified as follows:

a. When first identified, objects with apparently no spectral features be designated as BL Lac <u>candidates</u>.

b. Candidate BL Lac objects which are then shown to have spectra typical of bright elliptical galaxies containing non-thermal sources should be called BL Lac objects.

c. Candidate BL Lac objects which are shown to have QSO-like spectra should be redefined as QSOs and removed from the list of BL Lac objects.

Evidence is then presented which shows that the majority of the QSOs lie at distances ~100 Mpc or less. It is then shown that if this limit is placed on the objects in Table 1 (other than the genuine BL Lac objects as defined above) the motions observed by the radio astronomers previously called superluminal, are reduced in general to values less than or equal to 0.5c. Also rapid variability no longer requires that highly relativistic motions be invoked. Finally it is suggested that coherent radiation processes may be of importance in explaining the variability, and that more work should be done on such processes.

1. THE PHENOMENA TO BE EXPLAINED

The first questions to ask are which objects are we talking about and what are the observational properties which determine them? It seems to us that a general discussion of this whole question is warranted. Let us start with what we believe is the classical definition of a BL Lac object. It is an object presumed to be extragalactic which is a radio source, shows variability in its optical and radio flux, may have high polarization in its optical and radio flux which also may be variable, may be an x-ray source, and has at least at first sight an optical spectrum in which no features can be distinguished. The prototype is BL Lac, originally thought to be a variable star.

This was the class of objects which was the topic of the first conference held in Pittsburgh in 1978 (Wolfe 1978). At that conference the detailed properties of a number of these objects were described. At the dinner at the end of that meeting, Spiegel coined the term "Blazar" a pictorial term which he proposed be applied to rapidly variable objects some of which, but not all, can also be classified as BL Lac objects.

The variability of blazars is the subject of this conference.

To add to the complication, in 1980 Angel and Stockman (1980) produced a list of what they showed were objects with high and (usually variable) polarization and these were also called blazars, since with changes in the direction and the magnitude of polarization come changes in total flux. Further observations on objects of this type (HPQ) were carried out by Moore and Stockman (1984). Furthermore, the term optically violent variable (OVV) was also introduced (cf Angel and Stockman, 1980; Ledden and O'Dell, 1985.)

Further complications have arisen over the years as the original BL Lac objects have been studied in detail. From the earliest days it was clear that in some cases weak lines could be seen in the spectra. This was the case for MKN 501 and MKN 421 which were shown by Ulrich et al. (1975) to be elliptical galaxies. This was also shown to be true of BL Lac itself (Miller, French and Hawley 1978).

The number of objects classified as BL Lac objects according to the classical definition has remained small. Our most recent lists (Burbidge and Hewitt 1987, 1989) contain less than 90 objects. In recent years there have been systematic attempts to find more from complete samples of radio and x-ray sources. However as we show in Table 1 the numbers have not increased by a large factor. The spectroscopy has remained very difficult. In many of the BL Lac objects in Table 1 no absorption or emission lines have been detected. However, where lines are detected, and except for one or two ambiguous cases, they fall into one of two categories:

   a. They are broad but sometimes very weak emission lines similar (except for their strength) to those seen in normal quasi-stellar objects. In a few cases absorption features of the kind seen in high-redshift QSOs are seen, and in one or two cases absorption of this kind is seen and not emission.
   b. They are absorption lines typically coming from starlight in elliptical galaxies, and/or weak narrow emission lines of [OII] 3727, Hα, etc. coming from ionized gas in such galaxies.

If the BL Lac objects are separated into the two groups depending on their spectra, then as has been shown by us (Burbidge and Hewitt 1987, 1989) and by Ulrich (1989), those with spectra of galactic origin show a good Hubble relation indicating that they are indeed nothing more than highly visible non-thermal sources in the center of highly luminous elliptical galaxies. While it is not classified in this way, M87 in the Virgo cluster fits well into this category.

These are all objects with $z \lesssim 0.6$. In the other category are the BL Lac objects which have much larger redshifts and whose Hubble diagram shows the tremendous scatter known for QSOs for more than 20 years.

In Table 1 we have attempted to give an updated catalogue of BL Lac objects, QSOs and galaxies, which based on measured properties have been designated blazars, HPV, or OVV. In each case we have given the appropriate references with sources listed at the end of the table.

In our earlier lists we attempted to restrict the objects to those strictly following the original definition of a BL Lac object. However, in view of the fact that the major topic to be discussed here in Turku is the variability of blazars, and that the term blazar is used (in our opinion) for many different types of object, we have attempted in Table 1 to compile a list of all of the objects which have been identified as BL Lac objects, blazars, optically violent variables (OVV) high polarization variables (HPQ) and superluminal sources (SLS), the idea being that they have some physics (associated with a non-thermal source) in common.

It can be seen from Table 1 that there is still a good deal of confusion in the literature. For example, there are many objects classified as BL Lac objects by us and by others which have also been called QSOs. It is clear that for some authors it does not matter, since they believe that the underlying phenomena in all of these objects are identical in their physics.

We should mention here that the definition of a BL Lac object used by the x-ray astronomers in their identification of BL Lac objects is not the classical definition. As was pointed out in the introduction a classical BL lac object must show flux variation. Those who attempt to identify x-ray BL Lacs do not require that this condition be satisfied. Thus, while they have gone to great pains to show that all of the x-ray BL Lac are also radio emitters, many of these objects are not known flux variables. Stocke et al. (1989) define an x-ray BL Lac as a point-like x-ray source with an optical spectrum which is featureless in the sense that it has no emission lines with equivalent width more than 5 Å, or if the CaII H and the break is present the break is less than it would be in a normal galaxy thus implying the presence of a substantial non-thermal component. This definition is used in the identification of all of the x-ray BL Lac objects listed in Table 1, with the exception of one or two objects which have been studied in more detail and show a redshift and/or variable flux.

There are more than 200 objects listed in Table 1. All of the objects following the original definition used by us (Burbidge and Hewitt 1987, 1989) are marked with an asterisk. A considerable number of new BL Lac objects come from the radio and x-ray survey papers referenced in Table 1. In addition we have added all of the objects not originally called BL

Lac objects, but put in the blazar or related category by authors also listed in Table 1.

The contents of the Table are as follows:

The first column contains objects which have been described as BL Lac objects and an asterisk indicates objects listed in the 1987 Burbidge-Hewitt list. In the second, we list those objects which are either BL Lac candidates with QSO-like spectra, or variable QSOs which have now been put in the BL Lac category. Clearly it follows that those objects which which appear in both columns were originally identified as BL Lacs and those which only appear in the second column were defined as OVV, HPQs, Blazars, etc.

In the 3rd column we give the name of the object.

In the 4th column an apparent magnitude is given, and since the objects are highly variable, these magnitudes are an inhomogeneous set. In cases where a range of magnitudes have been reported we give an average.

In columns 5, 6, and 7 we give redshifts as follows: column 5, emission redshift for QSO-type objects; column 6, absorption redshifts similar to those found in high-redshift QSOs; column 7, redshift determined from absorption lines and/or weak emission lines which are characteristic of elliptical galaxies.

In column 8 we list HPQ, OVV, BZR designations as found in the literature and in column 9 source references. Finally in column 10 we note SLS (= superluminal source), followed by the reference number; X (= objects originally discovered as x-ray sources); and XD (= objects that have been detected as x-ray sources).

Study of this table suggests that the objects fall into three basic categories.

1. BL Lac candidates. These are objects identified in either radio or x-ray surveys, or found serendipitously, which appear to show continuous spectra with no features.

2. BL Lac objects. These are BL Lac candidates which have been shown by spectroscopy to be non-thermal sources lying in bright elliptical galaxies whose prototype is BL Lac. These are the objects in Table 1 with redshifts listed under heading z_{galaxy}.

3. QSOs. Those are (a) BL Lac candidates which have been shown by spectroscopy to have broad emission-line spectra or in a few cases absorption spectra of QSO type, and (b) rapidly variable QSOs already in the blazar category.

If these definitions are adopted we shall be tacitly accepting the following general hypothesis. Very powerful variable non-thermal sources reside in, or may be ejected from bright elliptical galaxies. Similar sources sometimes dominate the spectra of QSOs leading to a sub-group of QSOs which are rapidly variable. Whether these QSOs lie in galaxies or not, is not known, but the physics of the variable sources both in galaxies and in QSOs is assumed to be the same.

2. GENERAL PROPERTIES OF THE OBJECTS IN TABLE 1

There are 215 objects in Table 1. This is to be compared with 87 BL Lac objects (according to the strict definition) which were listed in 1987 (Burbidge and Hewitt 1987). They break down as follows: Forty BL Lac candidates have redshifts indicating an underlying galaxy, and 39 show QSO spectra (emission and/or absorption). There are 71 BL Lac candidates with no evidence for any spectral features, 64 QSOs which have been described as OVV, HPQ, Blazars, etc., and 28 out of all of the objects in the table in which evidence for superluminal motion has been reported. The table also contains three radio galaxies, 3C 84 (NGC 1275), 3C 120 and 3C 390.3, which some authors have chosen to call (erroneously in our view) blazars or BL Lac objects. The majority of the BL Lac objects (with galaxy spectra or QSO spectra) or BL Lac candidates were first detected as radio sources and a fraction were first detected as x-ray sources.

The mean redshift for the BL Lac objects with galaxy spectra is 0.191, and the mean redshift for BL Lac objects with QSO spectra is 0.786. In Fig 1 we show the Hubble diagram for the BL Lac objects in galaxies. Clearly the correlation expected if we are looking at a parent population of bright elliptical galaxies is present thus supporting the original hypothesis based on only about 20 objects. The Hubble relation should give a slope of 5 if the absolute magnitudes of the elliptical galaxies are all the same, and if the non-thermal components only make minor contributions to the apparent magnitudes plotted. In our original plot (Burbidge & Hewitt 1987) we used the minimum magnitudes for the objects where they were available. While it was clear that a line of slope 5 fitted the data points quite well, the actually least-squares slope was closer to 6. In Fig 1 here the slope is even steeper – about 7.2. We attribute this to two factors. For many of the objects we have no data on variability and the optical magnitudes may in many cases include significant contributions from the variable non-thermal components. Also since progressively more distant objects have been identified the brighter ones will tend to be selected (the so-called Malmquist effect) and these will tend to have significant non-thermal contributions when they are detected. These effects will tend to steepen the slope.

In Fig 2 we show a similar diagram for the 38 BL Lac objects with QSO spectra. For one object we have no apparent magnitude. As expected the scatter is large, as it is in the Hubble diagram for the QSOs in general.

There has been a considerable amount of discussion in the literature about the evolutionary properties of BL Lac objects. (cf Giommi et al. 1989; Stickel et al. 1991). The log N – log S studies are all based on a very small complete sample of objects both x-ray detected and radio detected contained in Table 1.

It appears to us that since line spectra obtained for BL Lac candidates either turn out to

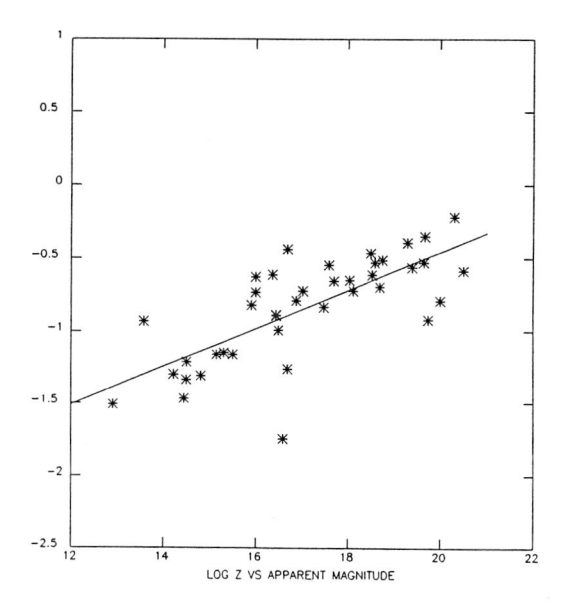

Fig 1. Hubble diagram for 40 BL Lac objects with galaxy-like spectra taken from Table 1. The least squares solution is plotted.

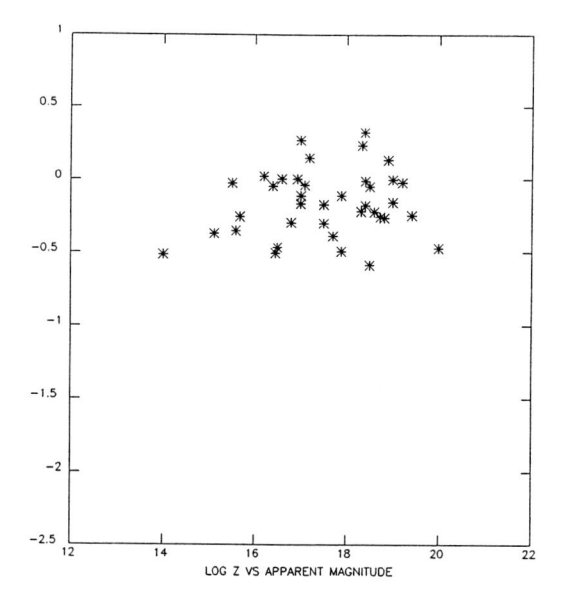

Fig 2. Hubble diagram for 38 BL Lac objects with QSO-like spectra taken from Table 1. It can be seen that there is no correlation of magnitude with redshift.

be associated with bright elliptical galaxies, or rapidly variable QSOs, it necessarily follows that a $\log N - \log S$ plot will contain a mixture of these two distinct types of object. Now we know that when investigated separately they show very different $\log N - \log S$ slopes. The QSOs on the assumption of cosmological redshifts show strong evidence for evolution (cf Schmidt 1968 and many subsequent investigations), while over the range of redshifts observed in the elliptical galaxies there is no significant evolution.

Thus a mix of the two components should lead to a slope for the $\log N - \log S$ relation not as flat as that obtained from the QSOs but flatter then the Euclidean slope. This is indeed what is found.

3. THE PHYSICAL PROPERTIES OF THESE OBJECTS

The major feature which distinguishes these objects from other classes of extragalactic objects is their rapid variability.

It is important to trace historically the reasoning that has led to the current models. The discovery of rapid optical flux variability of objects with large redshifts led to the statement of the Compton paradox (Hoyle, Burbidge and Sargent 1966) and the conclusion that the problem could be solved by bringing the objects closer than was implied by their redshifts, or by invoking relativistic expansion, or in some situations by invoking coherent radiation processes (Hoyle et al. 1966; Woltjer 1966; Rees 1967). The discovery of motions of very compact radio structures which led to velocities apparently well in excess of the velocity of light if cosmological redshifts are assumed led to the proposal that we are seeing relativistic jets, which meant in fact bulk relativistic motion of plasma clouds with values of $\gamma = (1-v^2/c^2)^{-1/2} \gg 1$. (Blandford and Rees 1978). The extensive use of, and belief in, this model has led to the idea that the flux from these sources is emitted in narrow jets so that we only see such sources when they are pointed nearly towards us. This, in turn, has led to a search for the parent population of these sources - the objects which form the bulk of the population and are seen when the rapidly variable objects are not pointed towards us.

When we look at all of the evidence, how well does this chain of reasoning hold up? In our view the answer is "not very well". There are several reasons for this. We discuss these in the following sections. They are (1) the very serious question of whether or not for the QSOs the redshifts are measures of distance; (2) The real question of whether or not the observed variability gives scales which are so small that the Compton paradox is present, and (3) the question of whether or not incoherent synchrotron radiation is the basic radiating mechanism for the highly compact components.

4. THE DISTANCE OF THE OBJECTS IN TABLE 1 AND SUPERLUMINAL VELOCITIES

Almost since the discovery of the QSOs the question of whether or not their distances are those given by their redshifts has been debated, and we shall not give an account of that discussion here. Arguments in favor of cosmological redshifts are the continuity idea first proposed by Sandage (1973) in which it is supposed that all QSOs lie in the nuclei of galaxies, the weaker of them being Seyfert nuclei, and the fact that many low redshift QSOs lie close to galaxies with approximately the same redshifts. Another argument given different weight by different people has been that the lack of understanding of the redshift phenomenon if it is not associated with the expansion of the universe, is itself an argument against the non-cosmological redshift hypothesis.

The evidence for the idea that QSOs do not lie at cosmological distances has been over the years principally the evidence that many low redshift galaxies and high redshift QSOs are physically associated. Early statistical evidence of this for the 3CR QSOs was given in 1971 by Burbidge et al. (1971) and a luminous connection between NGC 4319 and MKN 205 was found by Arp (1971). Arp has provided many more examples of high redshift QSOs lying close to low redshift galaxies (cf Arp 1987).

A comprehensive study has recently been completed (Burbidge et al. 1990) in which all QSOs lying within 10' of bright galaxies have been catalogued and also in which all faint galaxies which have been noted to lie close to QSOs are listed.

Statistical analyses of the pairs in this catalog strongly confirm the earlier work which always suggested that there were far more close pairs (high redshift QSO and low redshift galaxy) than were expected by chance.

We show in Figs (3, 4, 5, & 6) some of this evidence. Several hundred pairs were used in this analysis. The conclusion from these results together with the evidence that QSOs and galaxies with the same redshifts tend to occur in weak clusters leads to the conclusion that QSOs tend to lie in the vicinity of galaxies much more often than would be expected by chance, whether or not the redshifts are the same. The magnitude of the effect is large. To give an example. There are at present about 5000 QSOs with measured redshifts. Based on the known numbers of bright galaxies we would expect about 6 to be found to lie within 2' of the QSOs for this whole sample if the configuration were accidental. In spite of the fact that the regions around most bright galaxies have never been searched for QSOs, Burbidge et al. have already found 38 cases in their table. Given these results, is there any way to save the cosmological hypothesis? The only possibility is to invoke gravitational lensing of QSOs due to discrete point masses in the halos of the galaxies. However this is ruled out quantitively from the observed form of the luminosity function of QSOs (see references in Burbidge et al., and Arp 1989, Ostriker 1989). Also, gravitational microlensing cannot explain the correlations in Figs 5 and 6.

We must therefore conclude that the majority of the QSOs that we are studying are no further away than the galaxies with which they are associated. If we suppose that most of these galaxies are bright with an average $cz < 3000$ km sec^{-1} corresponding to $D < 60$ Mpc, $(H_o = 50$ km sec$^{-1})$ Mpc^{-1}, this distance can be treated as a limit to the distances of most of the QSOs which have been claimed to exhibit superluminal motion. Therefore the picture immediately changes. Let us take as an example 1928+738 which according to Zensus (1989) has a measured motion of its radio components of 0.6 mas/year.

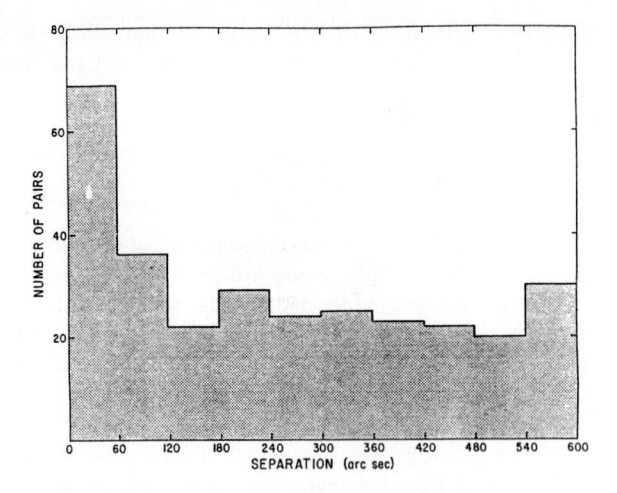

Fig 3. This figure is taken from the paper by Burbidge et al. (1990). It is a histogram of the distribution of separations of 300 QSO-galaxy pairs for separations ≤600″ taken from the table in Burbidge et al. It clearly shows the increase in the number of small separations. For the analysis see Burbidge, et al.

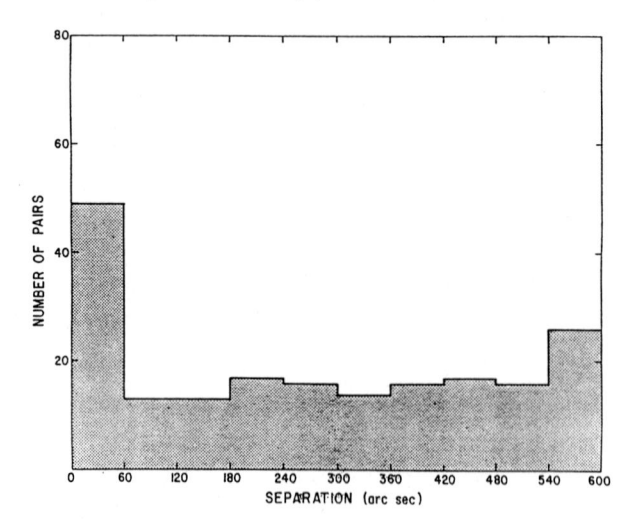

Fig 4. This figure is also taken from Burbidge et al. (1990) and it shows the histogram containing only 197 pairs with separations ≤600″. The difference between Fig 4 and Fig 3 is that in this case pairs that were found by deliberate searches around galaxies are omitted.

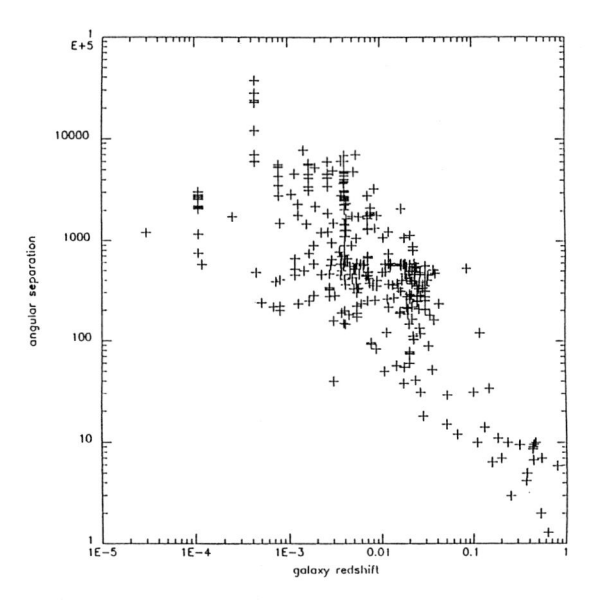

Fig 5. This figure is also taken from Burbidge et al. (1990). It shows the correlation of z_{galaxy} with angular separation for 392 pairs of galaxies and QSOs. This plot strongly suggests that there is a genetic connection between the two components in each case.

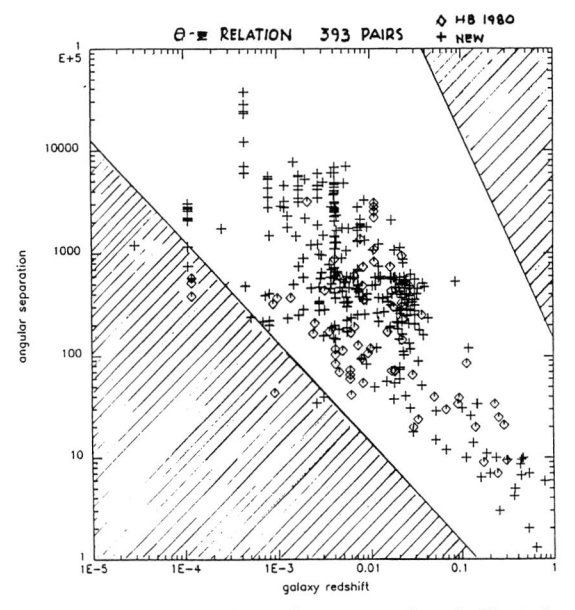

Fig 6. The points plotted are the same as those in Fig 5, but the shaded areas delineate regions where selections effects are operating. It can be seen that the bulk of the points lie well clear of these regions.

At 60 Mpc, $1'' = 300$ pc, so that 0.6 mas = 0.55 l.y.

Thus the true speed is 0.55c, and this is no longer highly relativistic. Practically all of the motions measured in the objects marked in Table 1 as superluminal sources are less than 1 mas/yr which corresponds to 0.9c at 60 Mpc. Thus the superluminal hypothesis largely vanishes. The only objects with motion that are larger than 1 mas/yr include 3C 273 and 3C 120 according to the compilation of Zensus (1989).

However, as has been pointed out by Arp and others (cf Arp and Burbidge 1990) a good case can be made for arguing that 3C 273 really lies in the Virgo cluster at a distance of no more than about 20 Mpc. A sketch showing the alignments in the Virgo cluster which suggest ejection is shown in Fig 7. On this basis the highest velocity measured in this source is about 0.35c. The only other sources with larger motions are 3C 120 and BL Lac, neither of them QSOs. The host galaxy of the nonthermal component of BL Lac has $z = 0.07$ suggesting a distance of 420 Mpc, and the motion is 0.7 mas/yr (Zensus 1989), and 3C 120 has $z = 0.033$ giving a distance of 200 Mpc. Based on the largest angular motions measured of $\sim$2.5 mas/yr for 3C 120 and 0.7 mas/yr for BL Lac, if we suppose that the velocity is $c/2$ the distances of these sources must be about 25 Mpc (for BL Lac) and 13 Mpc (3C 120) respectively. Thus we still have a severe problem with these sources which I do not wish to under-estimate. I have suggested elsewhere that the solution may lie in the fact that in these cases the nonthermal sources have been ejected from the galaxies towards us almost along the line of sight. This is not impossible because they are very rare.

Before we conclude this part of the discussion, some important points need to be made. We have never believed in bulk motion at highly relativistic speeds (remember that the average value of γ used by Zensus corresponds to bulk motion with $\beta > 0.98$), because we have never understood the bulk acceleration process that is required, and more important it is easy to show that if the medium through which the object is moving has any appreciable density at all, the deceleration is very rapid, and it does not seem possible to maintain such speeds for many years as is implied by the radio results. Even if we are looking at highly relativistic shocks as is now being suggested in many models, we do not believe that very large values of $\gamma (>10)$ are realistic.

Bringing the QSOs closer solves this problem. However there is still every reason to believe that beaming is still important but with values of $\beta \leq 0.5$. This still will give some of the selection effects discussed by Browne and others.

However it is important to remember here that some QSOs have galaxies with which they are associated which do have appreciable values of z (cf. the Table in Burbidge et al. 1990). It would be very important to measure the motion of a radio source in any one of these QSOs to see whether at the distance of those galaxies (up to $z \approx 0.4$) the motions appear to be superluminal).

It is also important to mention that the close proximity between some BL Lac objects or blazars and galaxies with smaller redshifts has been used to try and explain what we see.

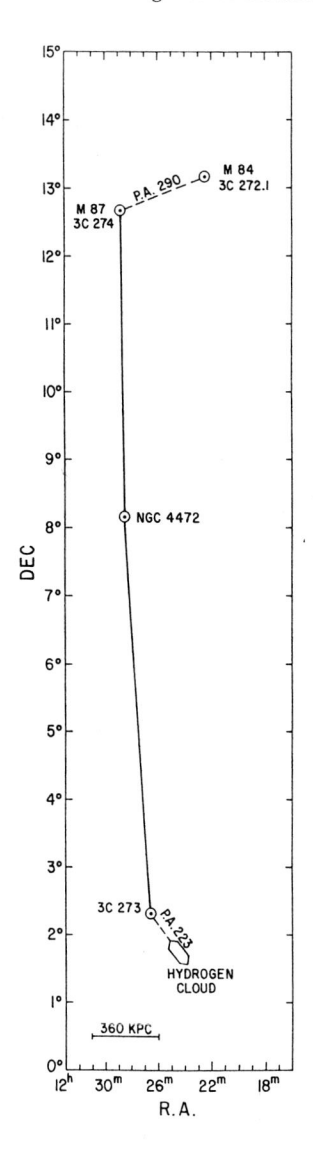

Fig 7. This shows the geometry associated with the alignments involving 3C 273 and M87 in the Virgo cluster. This geometry suggests that ejection processes have occurred.

For example, Ostriker and Vietri (1985, 1990) have proposed that BL Lac objects consist of a background QSO continuum source whose line of sight passes through a foreground galaxy so that the source is gravitationally microlensed, and Kuhr et al. (1990) have argued that in the specific cases of 0235+164 and 0846+513 a similar effect is taking place.

The problem that we have with this hypothesis is that it implies that the apparent association of these objects with normal galaxies with lower redshifts nearby are all accidental configurations. In that case it would be reasonable to argue that the objects with the larger redshifts are more distant. However, the work discussed at the beginning of this section (Burbidge et al. 1990) makes that very unlikely.

We turn finally and very briefly to the problem of rapid flux variations.

5. RAPID VARIABILITY

As was stated earlier, historically, it was the discovery of large flux variations over short time scales in optical wavelengths which first led to the discussion of the Compton paradox and the ways of avoiding it. It was supposed that the radiation was incoherent synchrotron emission. The alternatives were:

a. Relativistic expansion of the source or,

b. Reducing the distance so that the luminosity and hence the photon density is reduced so that the Compton paradox disappears. A mechanism to explain variability is still required.

c. Generation of the radiation by coherent mechanisms.

In recent years a fourth explanation has been put forward;

d. that the flux variations are the result of high amplification events (HAE) which can take place if the source is subject to gravitational microlensing. (Pacynski 1980, Kayser et al. 1986; Chang and Refsdal 1979, 1984).

The evidence presented in the last section strongly suggests that many of the rapidly variable QSOs are no further away than about 100 Mpc. As we now show, in general this conclusion removes the need to invoke relativistic expansion if the radiation is due to the incoherent synchrotron process. If the size of the radiation region is R, the condition is that the energy density of photons in the source $L(\pi R^2 c)^{-1}$ must be less than the energy density in the magnetic field $B^2(8\pi)^{-1}$, or $L < B^2 R^2 c/8$. For a "typical" value of $B \approx 0.1$ gauss this reduces to $L < 4 \times 10^7 R^2$.

If we suppose that a flux change $\triangle L \simeq fL$ takes place in time $\triangle t$ years, then the condition reduces to

$$L < 3 \times 10^{43} \, (\triangle t)^2$$
$$\text{or } \triangle L = L/f < 3 \times 10^{43} \, f^{-1} \, (\triangle t)^2$$

Most of the QSOs in Table 1 have apparent magnitudes between 15 and 20. If we use the arguments of the last section to put a limit on their distances of 100 Mpc, their luminosities will lie in the range $\leq 5 \times 10^{43}$ erg sec^{-1} to $< 5 \times 10^{41}$ erg sec^{-1}. Without knowing specific distances and specific values of $\triangle L$ and $\triangle t$ we cannot be sure that they will all obey condition (1). However the use of representative parameters suggests that there are not likely to be many (if any) exceptions. If this result holds up quantitatively, the incoherent synchrotron process without relativistic expansion effects can still explain the flux and its variability.

We turn next to (d). The explanation of rapid variations in terms of gravitational microlensing is undoubtedly an attractive idea. However it requires that in each case the light from the QSO pass through the halo of a foreground galaxy, and those who have invoked this explanation (cf Borgeest et al. 1990) have seized on the fact that in many cases a galaxy is seen to lie next to the QSO. But it is exactly this phenomenon that we discussed in Section IV where we summarized the evidence that there are far too many pairs (QSO and galaxy) for these to be accidental juxtapositions with the QSO lying far behind the galaxy. Our conclusion is that these pairs are physically associated. What this means is that the QSOs lie in and not behind the halos of these galaxies. Since some QSOs show absorption in their spectra at the redshift of the galaxy, and some do not, this means that some lie towards the rear of the halo (as seen by the observer) and some towards the front. If all galaxy halos were identical, for equal separations we would expect to find 50% with absorption and 50% without. Clearly if this is the case, the gravitational microlensing effect will be significantly modified.

Finally a few words about (c) coherent radiation processes. The term "blazar" was undoubtedly invented to describe variations on a rapid enough timescale such that they were detected serendipitously in studies which involved observing the objects many times over a period of observing covering one or two observing sessions, i.e. we are talking about variations on timescales of $\sim$ 1 year or less. Variations on timescales much shorter than this have been reported at optical, x-ray and radio frequencies.

Thus it has been generally agreed that the variable energy sources are contained within volumes with sizes no bigger than 10^{15}–10^{17}cm. In contrast to the volumes which give rise to the extended radio sources, the volumes here are very small and the density of gas and relativistic particles is expected to be very high when compared with that expected in radio lobes. On these general grounds alone it can be argued that collective plasma processes are likely to be important as we know is the case in regions of comparable size about which we have much more information, i.e. planetary magnetospheres.

Studies of such mechanisms which may explain some of the radio variability (cf. Baker et al 1988) should be followed up.

Also, it is not impossible that coherent plasma processes may explain some of the emission through much of the electromagnetic spectrum. For example, Krishan and Wiita (1990) have agreed that stimulated Raman scattering processes can produce a continuum spectrum all the way from x-rays to the infrared.

Our research is supported in part by NASA Grant NAGW-1737.

REFERENCES

Angel, J. R. P., and Stockman, H. S. 1980, ARA&A, 18, 321

Arp, H. C. 1971, Astrophys Lett, 9, 1

Arp, H. C. 1987, Quasars, Redshifts and Controversies (Berkeley: Interstellar Media)

Arp, H. C. 1989, BL Lac Objects, eds L. Maraschi, T. Maccacaro, M.-H. Ulrich (Springer-Verlag), 475

Arp, H. C. and Burbidge, G. 1990, ApJL, 355, L1

Baker, D., Borovsky, J., Benford, G., and Eilek, J. 1988, ApJ, 326, 110

Blandford, R. and Rees, M. 1978, Pittsburgh Conference on BL Lac Objects, ed A. M. Wolfe, (U Pittsburgh), 328

Borgeest, U., Kayser, R., Refsdal, S., Schramm, J., and Schramm, T. 1990, Proc. Variability of Active Galaxies, Heidelberg

Bozyan, E. P., Hemenway, P. D., and Argue, A. N. 1990, AJ, 99, 1421

Branduardi-Raymont, G., Mason, K. O., Mittay, J. P. D., Murdin, P. G., Allington-Smith, J. R., Giommi, P., Tagliaferri, G., and Augelini, L. 1989, BL Lac Objects, eds L. . Maraschi, T. Maccacaro, M.-H. Ulrich (Springer-Verlag), 261

Bregman, J. N., et al. 1986, ApJ, 301, 708

Brissenden, R. J. V., Remillard, R. A., Tuohy, I. R., Schwartz, D. A. and Hertz, P. L. 1990, ApJ, 350, 578

Burbidge, E. M., Burbidge, G. R., Solomon, P. M., and Strittmatter, P. A. 1972, ApJ, 175, 601

Burbidge, G., and Hewitt, A. 1987, AJ, 92, 1

Burbidge, G. and Hewitt, A. 1989, BL Lac Objects, eds A. Maraschi, T. Maccacaro, M.-H. Ulrich, (NY: Springer Verlag), 412

Burbidge, G., Hewitt, A., Narlikar, J. V., and Das Gupta, P. 1990, ApJS, 74, 679

Chang, V. and Refsdal, S. 1979, Nature, 282, 561

Chang, V. and Refsdal, S. 1984, A&A, 132, 168

Condon, J. J., Hicks, P. D., and Jauncey, D. L. 1977, AJ 82, 692

Falomo, R. 1990, ApJ, 353, 114

Falomo, R. and Treves, A. 1990, PASP, 102, 1120

Fugmann, W. and Meisenheimer, K. 1988, A&AS, 76, 145

Giommi, P., Tagliaferri, K., Beuermann, K., Branduardi-Raymont, G., Brissenden, R., Graser, U., Mason, K. O., Murdin, P., Pooley, G., Thomas, H.-C., and Tuohy, I. 1989, BL Lac Objects eds L. Maraschi, T. Maccacaro, M.-H. Ulrich (Springer-Verlag), 231

Griffiths, R. E., Wilson, A. S., Ward, M. J., Tapia, S. and Ulvestad, J. S. 1989, MNRAS, 240, 33

Halpern, J. P., Chen, V. S., Madejski, G. M. and Chanan, G. A. 1990, preprint.

Hawkins, M. R. S., Veron, P., Hunstead, R. W., and Burgess, A. M. 1991, preprint

Hoyle, F., Burbidge, G., and Sargent, W. L. W. 1966, Nature, 209, 751

Impey, C. D. 1987, Superluminal Radio Sources, eds J. A. Zensus, T. J. Pearson (Cambridge University), 231

Impey,C.D. and Tapia,S. 1988, Preprint

Kayser, R., Refsdal, S., and Stabell, R. 1986, A&A, 166, 36

Krishan, V., and Wiita, P. 1990, MNRAS, 246, 597

Kuhr, H. and Schmidt, G. D. 1990, AJ, 99, 1

Ledden, J. E., and O'Dell, S. L. 1985, ApJ, 298, 630

Maccacaro, T. Gioia, I. M., Schild, R. E., Walter, A., Morris, S. L., and Stocke, J. T. 1989, BL Lac Objects, eds L. Maraschi, T., Maccacaro, M.-H. Ulrich (Springer-Verlag), 222

Maccagni, D., Garilli, B., Barr, P., Giommi, P. and Pollack, A. 1989, BL Lac Objects, eds L. Maraschi, T. Maccacaro, M.-H. Ulrich (Springer-Verlag), 281

Mead, A. R. G., Ballard, K. R., Brand, P. W. J. L., Hough, J. H., Brindle, C. and Bailey, J. A. 1990, A&AS, 83, 183

Miller, J. S., French, H. B., and Hawley, H. S. 1978, ApJL, 219, L85

Moore, R. L., and Stockman, H. S. 1981, ApJ, 243, 60

Moore, R. L., and Stockman, H. S. 1984, ApJ, 279, 465

Mutel, R. L., 1989, Parsec Scale Radio Jet Workshop, Socorro

Ostriker, J. P. 1989, BL Lac Objects, eds L. Maraschi, T. Maccacaro, M.-H. Ulrich (Springer-Verlag), 477.

Ostriker, J. P., and Vietri, M. 1985, Nature, 318, 446

Ostriker, J. P., and Vietri, M. 1990, Nature, 344, 45

Pacynski, B. 1986, ApJ, 301, 503

Pica, A. J., Smith, A. G., Webb, J. R., Leacock, R. J., Clements, S., and Gombola, P. P. 1988, AJ, 96, 1215

Porcas, R. W. 1987, Superluminal Radio Sources, eds. J. A. Zensus, T. J. Pearson (Cambridge University), 12

Rees, M. 1967, MNRAS, 135, 345

Remillard, R. A., Toshy, I. R., Brissenden, R. J. V., Buckley, D. A. H., Schwartz, D. A., Feigelson, E. D., and Tapia, S. 1989, ApJ, 345, 140

Sandage, A. R. 1973, ApJ, 180, 687

Sargent, W. L. W., Steidel, C. C. and Boksenberg, A. 1988, preprint

Schwartz, D. A., Brissenden, R. J. V., Tuohy, I. R., Feigelson, E. D., Hertz, P. L. and Remillard, R. A. 1989, BL Lac Objects eds L. Maraschi, T. Maccacaro, M.-H. Ulrich (Springer-Verlag), 209

Stickel, M., Fried, J. W. and Kuhr, H. 1989, A&AS, 89, 103

Stickel, M., Padovani, P., Urry, C. M., Fried, J. W. and Kuhr, H. 1991 ApJ in press

Stocke, J. T., Morris, S. L., Gioia, I., Maccacaro, T., Schild, R. E. and Wolter, A. 1990, ApJ, 348, 141

Stocke, J. T., Morris, S. L., Gioia, I., Maccacaro, T., Schild, R. E. and Wolter, A. 1989, BL Lac Objects, eds L. Maraschi, T. Maccacaro, M.-H. Ulrich (Springer-Verlag), 242

Stocke, J. T., Schneider, P., Morris, S. L., Gioia, I. M., Maccacaro, T. and Schild, R. E. 1987, ApJL, 315, L11

Ulrich, M.-H. 1989, BL Lac Objects, eds L. Maraschi, T. Maccacaro, M.-H. Ulrich (Springer-Verlag), 45

Ulrich, M.-H., Kinman, T. D., Lynds, C. R., Rieke, G. H., and Ekers, R. D. 1975, ApJ, 198, 261

Walsh, D., Beckers, J. M., Carswell, R. F. and Weymann, R. J. 1984, MNRAS, 211, 1984

Webb, J. R., Smith, A. G., Leacock, R. J., Fitzgibbons, G. L., Gombola, P. P. and Shepherd, D. W. 1988, AJ, 95, 374

Wehrle, A. E., Cohen, M. H., and Unwin, S. C. 1990, ApJL, 351, L1

Witzel, A., Schalinski, C. J., Johnston, K. J., Biermann, P. L., Krichbaum, T. P., Hummel, C. A. and Eckart, A. 1988, A&A, 206, 245

Wolfe, A. M. ed 1978, Pittsburgh Conference on BL Lac Objects (U Pittsburgh)

Wolter, A., Gioia, I. M., Maccacaro, T., Morris, S. L., and Stocke, J. T. 1990, preprint

Woltjer, L. 1966, ApJ, 146, 597

Wright, A. E., Ables, J. G., and Allen, D. A. 1983, MNRAS, 205, 793

Zensus, J. A. 1989, BL Lac Objects, eds L. Maraschi, T. Maccacaro, M.-H. Ulrich (Springer-Verlag), 3

TABLE 1
BL LAC OBJECTS AND RAPIDLY VARIABLE QSOs

Object BL Lac	Object QSO	Name	m	z_{em}	z_{abs}	z_{gal}	Names in Literature	REF	Comments
0003-066		PKS	18.5			0.347		5,31	
0017+200			19.6				BZR	5	
0019+058*		PKS	19.2					1,2	
0045+395*		5C3.178	18.5					1,2	
0048-097*		PKS	16.27				BZR	1,2,3,6	XD
0106+013	0106+013	PKS	18.39	2.107			BZR HPQ	5,23	SLS,7;XD
0109+224*		GC	16.41				BZR	1,2,6	XD
0118-272	0118-272	PKS	15.67		0.557?		BZR	3,4,23	XD
0119+115	0119+115	PKS	19.4	0.570			BZR	5,26	
0122+090	0122+090	1E	19.98	0.339				19,28	X
0138-098*	0138-098	PKS	17.5		0.501		BZR	1,2,3,4	
0158+003		1E	17.96					19	X
0205+3519		1E	19.24					19	X
0208-512	0208-512	PKS	16.93	1.003			BZR HPQ	4	

Object BL Lac	Object QSO	Name	m	z_{em}	z_{abs}	z_{gal}	Names in Literature	REF	Comments
	0212+735	S5	19	2.367			BZR	23	SLS,37;XD
0215+015*	0215+015	PKS	18.33	1.715	1.2547 1.3437 1.4909 1.5475 1.6441 1.6492 1.6855 1.719		BZR OVV	1,2,6 13,23	XD
0219+428*	0219+428	3C 66A	15.58	(0.444)			BZR	1,2,6	XD
0219-164*	0219-164	PKS	19.0	0.698			BZR	1,2,5 11	
	0229+341	3CR 68.1	19	1.238			BZR HPQ	23	
0235+164*	0235+164	AO	15.5	0.94 0.524	0.524 0.852		BZR OVV	1,2,3,6 13	SLS,9;XD
0256+075	0256+075	PKS	18.5	0.893				12,26	
0257+3449		1E	18.53			(0.245)		19,28	X
0300+470*		4C 47.08	17.21				BZR	1,2,6	
0301-243			15.50				BZR	4	
0306+102*		PKS	18				BZR OVV	1,2,13 23	XD
		NGC 1275[1]	11.85	0.0172					motion detected by radio <0.2C

[1] NGC 1275 = 3CR 84 = 0316+413 called blazar in ref. 6 and 23.

Object BL Lac	Object QSO	Name	m	z_{em}	z_{abs}	z_{gal}	Names in Literature	REF	Comments
0317+185*		1E	18.12			0.190		1,2,19,29	X
0323+022*		H	17.48			0.147	BZR	1,2,11,19	X
	0330-262	EXO	21.0	0.67			OVV	30	
0331-364		1E						20	X
	0332-403	PKS	18.5	1.455			BZR HPQ	4	
	0333+321	NRAO 140	17.10	1.258					SLS,8;XD
	0336-019	PKS	17.60	0.852			BZR HPQ	10,11,23	XD
0338-214		PKS	17.5				BZR	11,23	XD
0350-372			18.5					19	X
	0403-132	PKS	17.17	0.571			BZR HPQ	6,10,23	XD
0406+121*		PKS	20.5				BZR	1,2,5	XD
0414+009*		1H	17.59			0.287	BZR	2,4,11 21,27	X
0419+197		1E	20.26					19	X

| Object | | Name | m | z_{em} | z_{abs} | z_{gal} | Names in Literature | REF | Comments |
BL Lac	QSO								
	0420-014	PKS	17.76	0.915			BZR HPQ OVV	6,10,13 23	XD
0422+004*		PKS	16.05				BZR OVV	1,2,6 17	XD
0426-380	0426-380	PKS			1.030?			3	
		3C 120[2]	15.05	0.0331			OVV	13	SLS,8
	0438-436	PKS	18.8	2.852			BZR HPQ	4	XD
	0440-003	PKS	17.	0.844			BZR OVV	5,13	XD
0454-234	0454-234	PKS	16.6	1.003	0.606 0.630 0.752 0.891		BZR HPQ	4,12,23	
0454+844*		S5	16.5				BZR	1,2,3,23	XD
	0458-020	PKS	19.23	2.286	2.0398		BZR HPQ	4,23	XD
0503-043*		A	18.5					1,2	
0521-365*		PKS	14.5			0.061	BZR	1,2,6,23	XD
0537-441	0537-441	PKS	16.40	0.896			BZR HPQ	3,4,23	XD

[2] 3C 120 = 0430+052, a radio galaxy

Object							Names in		
BL Lac	QSO	Name	m	z_{em}	z_{abs}	z_{gal}	Literature	REF	Comments
0548-322*		PKS	15.5			0.069	BZR	1,2,6,21	XD
	0605-085	OH-010	18.5	0.870			BZR HPQ	4	XD
0607+711		1E	18.5					19	X
0622-529		1E	19					19	X
0627-199			18.7				BZR	5,19	X
0646+60			19.2				BZR	5	
0716+714*		S5	15.5				BZR	1,2,3,23	XD
	0723+679	3C 179	18.0	0.846					SLS,8
0735+178*	0735+178	PKS	15.10		0.4246		BZR OVV	1,2,3,6 13	SLS,8;XD
	0736+017	PKS	16.05	0.191			BZR HPQ	6,10,23	XD
0737+746		1E	16.89					19,29	X
0745+24	0745+24	B2	17.7	0.409				12,26	
	0752+258	OI 287	18.41	0.446			BZR HPQ	6,10,23	
0754+100*		PKS	15.71				BZR	1,2,6	XD

Object BL Lac	Object QSO	Name	m	z_{cm}	z_{abs}	z_{gal}	Names in Literature	REF	Comments
0808+019*		PKS	17.5				BZR	1,2,6,23	XD
0814+425	0814+425	OJ 425	18.5	(0.258)				3	
0818-128*		PKS	17.01				BZR OVV	1,2,6,17	
0820+225	0820+225	PKS	19.2	0.951				3	
0823+033	0823+033	PKS	16.8	0.506			BZR	3,4	
0823-223	0823-223	PKS	17.09		0.910		BZR	4,32	
0828+493*	0828+493	OJ 448	18.82	0.548				1,2,3	
0829+046*		PKS	16.5				BZR OVV	1,2,6,17	XD
	0836+710	4C 71.07	16.5	2.17					SLS,8
0836+182*			16.8				BZR	1,2,23	
0846+513*	0846+513	WI	17	1.86			BZR HPQ	1,2,10 23	
	0850+581	4C 58.17	18	1.322					SLS,8
0851+202*	0851+202	OJ 287	14	0.306			BZR OVV	1,2,3 6,8,13	SLS,37;XD
	0855+143	3CR 212	19.06	1.048			BZR HPQ	23	

Object BL Lac	Object QSO	Name	m	z_{cm}	z_{abs}	z_{gal}	Names in Literature	REF	Comments
	0906+430	3CR 216	18.48	0.669			BZR HPQ	6,10,23	SLS,8
	0906+015	PKS	17.79	1.018			BZR HPQ OVV	11,13,23	XD
0912+297*		B2	16				BZR	1,2,6	
0922+749		1E	19.74			(0.120)		19,28	X
	0923+392	4C 39.25	17.86	0.699					SLS,37
0950+494		1E	19.30			(0.41)		19,28	X
0954+658*		S4	16.7			0.368		1,2,3	
0957+227*		4C 22.25	18				BZR	1,2,6,23	
0958+210		1E						20	X
1011+496*			16.15					1,2	
1034-293*	1034-293	PKS	16.46	0.312			BZR HPQ	1,2,4,12 23,33	XD
	1040+123	3CR 245	16.45	1.029					SLS,8;XD
	1055+018	PKS	18.28	0.888			BZR HPQ	4	

Object BL Lac	Object QSO	Name	m	z_{em}	z_{abs}	z_{gal}	Names in Literature	REF	Comments
1057+100*		MC	17.76				BZR	1,2,6	
1101+384*		MKN 421	12.91			0.031	BZR	1,2,6,21	XD
1101-232		1HEAO	16			0.186	BZR	4,22,34	X
1133-704*		MKN 180 S5	14.49			0.046	BZR	1,2,6,21	XD
1133+163		1E	20.04					19	X
	1137+660	3CR 263	16.32	0.646					SLS,8
1144-379*	1144-379	PKS	16.2	(1.048)			BZR	1,2,3,4 12, 23, 35	
1147+245*		B2	16.66				BZR	1,2,3 6,23	XD
	1150+812	S5	18.5	1.25					SLS,8
	1150+497	4C 49.22	17.50	0.334			BZR HPQ	6,23	
	1156+295	4C 29.45	14.41	0.729			BZR HPQ OVV	6,10,11 13,23	
1207+397*		W4;1E	20.3			0.59		1,2,19	XD
1210+121*		MC2	17.8					1,2	

Object BL Lac	Object QSO	Name	m	z_{em}	z_{abs}	z_{gal}	Names in Literature	REF	Comments
1215+303*		ON 325	15.73				BZR OVV	1,2,6,17	XD
1217+348*		GV 136	17.09					1,2	
1218+304*		RS 4	16.45			0.13	BZR	1,2 21,23	XD
1219+285*		W COM	16.5			0.102	BZR OVV	1,2,6,13	XD
1221+248		1E	17.65					19	X
1222+102*		WDM 6	17.6					1,2	
1225+206*		4C 20.29	18				BZR	1,2,23	XD
	1226+023	3CR 273	12.63	0.158					SLS,8
1229+645		1E	16.89			0.163		19,28 29, 36	X
1235+632*		1E	18.59			0.297		1,2,19	X
	1244-255	PKS	17.41	0.638			BZR	4	
	1253-055	3C 279	16.84	0.538			BZR HPQ OVV	6,10,11 17,23	SLS,8;XD
1256+018		1E	20.0			(0.162)		19	X

Object BL Lac	Object QSO	Name	m	z_{em}	z_{abs}	z_{gal}	Names in Literature	REF	Comments
1258+640		1E	19.5					19	X
1307+121*		4C 12.46	18.5				BZR	1,2,23	XD
1308+326*	1308+326	B2	19	0.996	0.879		BZR HPQ OVV	1,2,6,10 13,23	XD
1309-216*	1309-216	PKS	18.9		1.361 1.489		BZR	1,2,4,23	
1312+423			17.0					19	X
	1334-127	PKS	17.2	(0.541)			BZR	4	XD
1349-439		PKS	18.12 16.37				BZR	4	
1400+162*		4C 16.39	16.37			0.245	BZR	1,2,6	XD
1402+042*		1E	17.5				BZR	1,2,19 23	X
1407+022*		PKS	18.65					1,2	
1407+599		1E	19.67			(0.452)		19,28	X
1413+135*		PKS	20.5			0.26	BZR	1,2,11	XD
1415+259*		1E	16.			0.237		1,2,4,21	X

Object BL Lac	Object QSO	Name	m	z_{em}	z_{abs}	z_{gal}	Names in Literature	REF	Comments
1418+546*		OQ 530	15.91			0.152	BZR	1,2,3,6	XD
1424+240			15.10				BZR	4,11,23	XD
	1424-418	PKS	17.7	1.522			BZR HPQ	4	
1426+428		1H	16.45			0.129		21,34	X
1443+638		1E	19.65			(0.298)		19	X
1451+172*		MC3	17.9					1,2	
1458+228		1E	16.79					19	X
	1502+106	MC2	18.56	1.839			BZR HPQ	4	
	1504-166	MC	18.5	0.876			BZR HPQ	4	XD
	1510-089	PKS	16.74	0.361	0.351		BZR HPQ	10,11 23	XD
1514+197*		PKS	18.5				BZR	1,2,6	XD
1514-241*		AP LIB	14.80			0.049	BZR OVV	1,2,3,6 13	XD
1517+656		1H	15.5					22	X
1519-273		PKS	17.90				BZR	3,4	

Object BL Lac	Object QSO	Name	m	z_{em}	z_{abs}	z_{gal}	Names in Literature	REF	Comments
	1522+155	MC3	17.5	0.628			BZR HPQ	6,23	XD
	1532+016	PKS	18.11	1.435			BZR HPQ	4	
1534+018		1E	18.70			(0.202)		19,28	X
1538+149*	1538+149	4C 14.60	18.31	0.605			BZR	1,2,3,6	XD
	1546+027	PKS	16.83	0.413			BZR HPQ	10,23	XD
	1548+056	PKS	17.7	1.422			BZR HPQ	4	
1552+203		1E	17.70			0.222		19,28,29	X
1553+113		PG	15.04					25	
1604+159*		4C 15.54	18.5				BZR	1,2,23	
	1610-771	PKS	19	1.71			BZR HPQ	4	
1620+103*		MC2	17.8					1,2	
	1622+238	3CR 336	17.47	0.927			BZR HPQ	23	
	1638+398	NRAO 512	18.5	1.666			HPQ OVV	13,26	

Object BL Lac	Object QSO	Name	m	z_{em}	z_{abs}	z_{gal}	Names in Literature	REF	Comments
	1641+399	3CR 345	15.20	0.595			BZR HPQ OVV	6,10,11 13,18,23	SLS,37;XD
	1642+690	4C 69.21	19.2	0.751					SLS,8
1652+398*		MKN 501	14.44			0.034	BZR	1,2,3 6, 21	XD
1704+607*		1E	19.4			(0.275)	BZR	1,2,19 23, 28	X
1717+178*		PKS	18.5				BZR	1,2,6	
	1721+343	4C 34.47	16.5	0.206					SLS,8
1722+119		4U	16.6			(0.018)	HPB	15,16	X
1727+502*		OT 546 1 ZW 187	16.70			0.055	BZR	1,2,6,21	XD,
	1730-130	NRAO 530	18.5	0.902			OVV	13	
1732+389	1732+389	OT 355	18.4	0.976				12,26	
1738+476*		OT 465	18.5					1,2	
	1741-038	PKS	18.6	1.054			HPQ	26	
1749+701*	1749+701	WI	17.01	(0.770)			BZR	1,2,3,23	XD,SLS,40

Object							Names in		
BL Lac	QSO	Name	m	z_{em}	z_{abs}	z_{gal}	Literature	REF	Comments
1749+096*	1749+096	4C 09.57	17.88	0.322			BZR	1,2,3,6 23	XD
1757+705		1E	18.27					19	X
1803+784*	1803+784	S5	17	0.68			BZR	2,23	XD
1807+698*		3CR 371	14.22			0.050	BZR OVV	1,2,3,6 13	XD,SLS,40
1823+568	1823+568	4C 56.37	18.4	(0.664)				3,39	SLS,40
		3CR 390.3[3]	14.37	0.0561			BZR HPQ	6,23	SLS,37
	1901+319	3C 395	17.42	0.635					SLS,8
	1921-293	OV-236	16.82	0.352			BZR HPQ OVV	11,23,17	XD
	1928+738	4C 73.18	16.5	0.302					SLS,8;XD
	1936-155	PKS	19.4	1.657			BZR	5	
	1951+498		17.5	0.466					SLS,8
	1954-388	PKS	17.07	0.626			BZR HPQ	4	
2005-489*		PKS	15.3			0.071		1,2,3	XD

[3] 3CR 390.3 = 1845+797, a radio galaxy, called a blazar in ref. 6 and high-polarization quasar in ref. 23.

Object BL Lac	Object QSO	Name	m	z_{em}	z_{abs}	z_{gal}	Names in Literature	REF	Comments
2007-777*	2007+777	S5	16.5	0.342			BZR	1,2,3,12 23	SLS,14;XD
2032+107*	2032+107	PKS	18.6	0.601			BZR	1,2,6,11	
	2121+053	PKS	17.5	1.878			BZR	5	XD
2131-021*	2131-021	4C-02.81	18.73	(0.557)			BZR	1,2,3,23	XD
2133-449		B24	19					41	
2136-428		B7	17.3					41	
2143+070		1E	18.04			0.225		19,28,29	X
	2144+092	PKS	18.9	1.113			OVV	13	
2150+17			17.9					26	
2155-152*	2155-152	PKS	17.5	0.672			BZR	1,2,12,23	XD
2155-304*	2155-304	PKS	13.58		0.083	0.117	BZR	1,2,6,11 21	XD
2200+420*		BL LAC	15.14			0.069	BZR OVV	1,2,3,6 11,13	SLS,8
	2201+171	MC3	18.8	1.075			BZR HPQ	6,23	XD
2206-260			17.74				BZR	4	

Object BL Lac	Object QSO	Name	m	z_{em}	z_{abs}	z_{gal}	Names in Literature	REF	Comments
2207+020*		4C 02.54	19					1,2	
	2208-137	PKS	17	0.392			BZR HPQ	6,10, 11,23	
2223-114			19.6				BZR	5	
2223-052*	2223-052	3C 446	17.19	1.404	0.847		BZR HPQ OVV	1,2,6,10 11,13,23	XD;SLS 40
	2225-055	PHL 5200	17.7	1.981			HPQ	10,38	
	2230+114	CTA 102	17.66	1.037			BZR HPQ	6,10,23	SLS,8;XD
2233-148		PKS	19.0				BZR	24,23	XD
	2234+282	B2	19	0.795			BZR HPQ	23	XD
2240-260	2240-260	PKS	17.88	0.774			BZR	3,4	
	2243-123	PKS	16.45	0.63			HPQ	4	
	2251+158	3CR 454.3	16.04	0.859			BZR HPQ OVV	6,10,13 23	SLS,8;XD
2254+074*		PKS	17.03			0.19	BZR OVV	1,2,3,6 17	XD
2254-204			17.85				BZR	4	
2335+031*		4C 03.59	18.76			0.31	BZR	1,2,23	

Object		Name	m	z_{em}	z_{abs}	z_{gal}	Names in Literature	REF	Comments
BL Lac	QSO								
2336+052		1E	20.3					19	X
2342-155		1E	19.22					19	X
	2345-167	PKS	17.5	0.576			BZR HPQ OVV	6,10,13 23	XD
2347+194		1E	20.78					19	X
	2355-534	PKS	17.8	1.006			BZR HPQ	4,6	
2356-309		1H	17					22	X

REFERENCES.- (1) Burbidge and Hewitt 1987; (2) Burbidge and Hewitt 1989; (3) Stickel et al. 1990; (4) Impey and Tapia 1988; (5) Fugmann and Meisenheimer 1988; (6) Angel and Stockman 1980; (7) Wehrle, Cohen, and Unwin 1990; (8) Zensus 1989; (9) Impey 1987; (10) Moore and Stockman 1981; (11) Mead et al. 1990; (12) Stickel, Fried, and Kuhr 1989; (13) Webb et al. 1988; (14) Witzel et al. 1988; (15) Griffiths et al. 1989; (16) Brissenden et al. 1990; (17) Pica et al. 1988; (18) Bregman et al. 1986; (19) Stocke et al. 1990; (20) Wolter et al. 1990; (21) Maccagni et al. 1989; (22) Schwartz et al. 1989; (23) Ledden and O'Dell 1985; (24) Condon, Hicks, and Jauncey 1977; (25) Falomo and Treves 1990; (26) Kuhr and Schmidt 1990; (27) Halpern et al. 1990; (28) Maccacaro et al. 1989; (29) Stocke et al. 1989; (30) Branduardi-Raymont et al. 1989; (31) Wright, Ables, and Allen 1983; (32) Falomo 1990; (33) Sargent, Steidel, and Boksenberg 1988; (34) Remillard et al. 1989; (35) Bozyan, Hemenway, and Argue 1990; (36) Stocke et al. 1987; (37) Porcas 1987; (38) Moore and Stockman 1984; (39) Walsh et al. 1984; (40) Mutel 1989; (41) Hawkins et al. 1991

Many black hole nuclei in blazars

MAURI VALTONEN

Tuorla Observatory
University of Turku

1 INTRODUCTION

It is nowadays commonly assumed that the power of a blazar ultimately comes from processes near a supermassive black hole. The black hole is thought to be surrounded by an accretion disk (e.g. Ulrich 1989) which feeds matter partly into the black hole, partly into a relativistic outflow which is confined to the form of a jet. In this paper we follow these common assumptions.

The additional factor which we want to discuss is the possibility that more than one black hole is involved in the process. There are several arguments which may be put forward in favor of multiple black hole models.

1) It has been the general lesson of astrophysical studies in this century that multiple systems are important. For example, in case of stars and galaxies the efforts were concentrated for a long time in understanding isolated systems only. Later it was realized that multiple systems are common and important especially in understanding the most active phases of the objects. Blazars are generally recognized as the most active galactic nuclei, and as such strong candidates to be multiple systems.

2) The standard theory of disk accretion, developed first for the stellar scale, works best in binary systems. The angular momentum of the matter in the accretion disk which is convected inwards has to be transferred somewhere and could be most easily transferred to the another star. Even though other possibilities have been discussed in case of supermassive black holes, the extension of the stellar case would be the most simple one. See e.g. Shapiro and Teukolsky (1983) and Begelman et al. (1989) for a discussion of this point.

3) Apart from the question whether the accretion models work at all without a binary component, there is at least a strong feeling that the efficiency of the accretion process is enhanced in a binary system (Lin and Pringle 1976, Lin and Papaloizou 1979 a, b). The companion black hole may cause a strong spiral density wave in the

accretion disk of the primary black hole and drive matter efficiently inwards in the disk (Sillanpää et al.1988, Spruit 1989, Rózyczka and Spruit 1989, Matsuda et al. 1989, Szuszkiewicz 1991).

4) The evidence for periodic flux variations in at least one blazar, OJ 287, is most easily interpreted in terms of a binary model (Sillanpää et al. 1988, Figure 1).

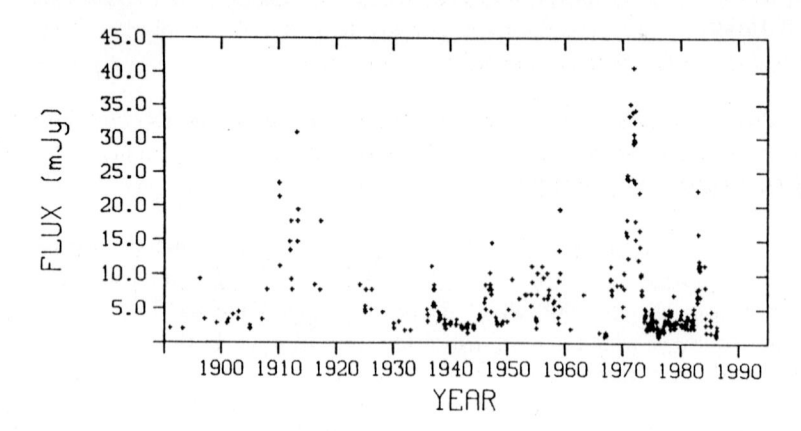

Figure 1. Optical brightness variations in OJ 287 (Sillanpää et al. 1988).

This model can be tested by a continuous flux monitoring programme which should extend at least over the next expected major optical outburst in 1994.

2 ORIGIN OF MULTIPLE SYSTEMS

There are good reasons to expect that supermassive binary black hole systems are common in the nuclei of major galaxies. This is

1) because supermassive black holes are probably common in the centers of even small galaxies (Filippenko and Sargent 1986, Tonry 1987, Dressler and Richstone 1988).
2) because mergers between galaxies are probably common in dense galaxy environments (Aarseth and Fall 1980, Roos 1981a, Navarro, Mosconi and Garcia Lambas 1987);
3) and because supermassive black hole binaries which are created in mergers of galaxies are long lived (Begelman, Blandford and Rees 1980, Roos 1981b, Valtaoja, Valtonen and Byrd 1989).

Mergers of galaxies which themselves possess binary black hole systems may also be important. This leads to the formation of three- or four-black hole systems in the centers of some multiple merger galaxies. Probable sites for such events are the dominant central galaxies in groups and clusters or other galaxies in dense galaxy environments.

The few black hole systems evolve like few body system in general: the black holes have close two-body encounters until the system breaks up through ejections and escapes (Valtonen 1988). The mean separation of the black holes should be initially of the order of one parsec and orbital speeds of the order of 1000 km s^{-1}.

When very close encounters take place, gravitational radiation may become important. Due to a close encounter at distance a_0, gravitational radiation energy losses may lead to the formation of a temporary binary. If

$$a_0 < 0.64 \cdot 10^{-2} \left[\frac{m_1 m_2 \, (m_1+m_2)}{(10^9 \, M_\odot)^3} \right]^{1/4} \left(\frac{\Sigma m}{10^9 M_\odot} \right)^{1/16} \text{pc}, \quad (1)$$

the lifetime of the binary due to gravitational energy losses is less than the crossing time in the original few-body system. Here m_1 and m_2 are the masses of the binary components and Σm is the total mass of the few-black hole system. The binary may have a close encounter with another black hole, with the consequence of a very energetic escape of all three black holes from the galactic nucleus (a slingshot event; Mikkola and Valtonen 1990), or the binary may collapse before any further few-body dynamics take place. In this way the total number of black holes in the galactic nucleus is reduced either by 3 units (slingshot) or by 1 unit (binary merger).

One may describe the state of a galactic nucleus by the occupation number: the state is defined by the number of supermassive black holes in the galactic nucleus at any one time. Figure 2 illustrates the states of a nucleus in the manner which is analogous to the energy level diagrams of the atoms

Mergers of galaxies drive the nucleus towards higher occupation numbers, while slingshot events and mergers of black hole binaries cause downward transitions. The relevance of this diagram to the various forms of activity in galactic nuclei is still under investigation, but it could be an important classification scheme in addition to the widely discussed viewing angle effects. As a working hypothesis one might take, for reasons which have been outlined in the previous section, the view that the most extreme forms of activity in galactic nuclei, such as the blazar phase, are associated with the downward transitions in the occupation number diagram. During the downward transitions binaries with their accretion disks are pushed together and spectacular mass transfer rates towards the black holes could result.

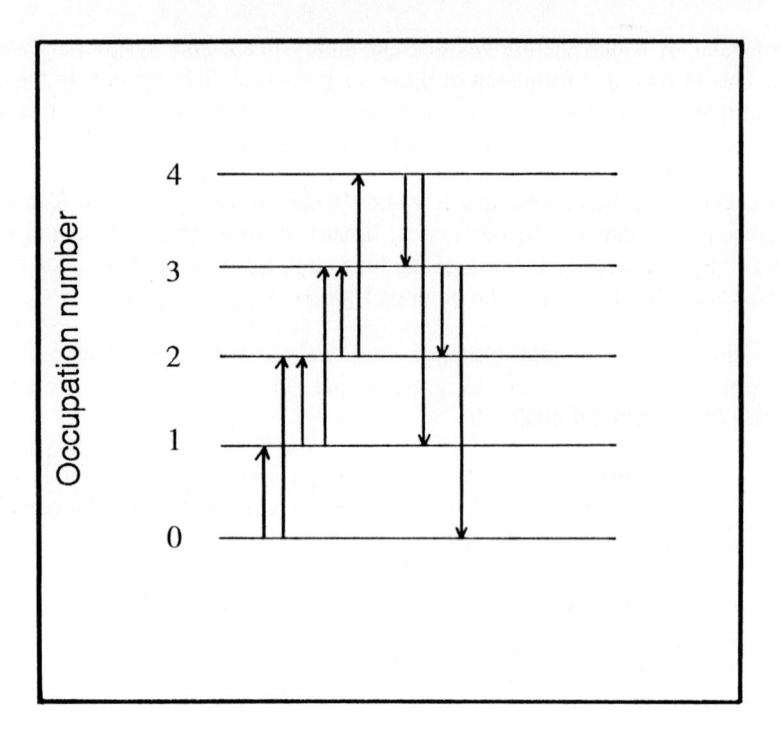

Figure 2. The occupation number of galactic nuclei. The occupation number tells the number of supermassive black holes in a galactic nucleus. Arrows indicate transitions: mergers of galaxies produce upward transitions and mergers of black holes as well as ejections of black holes cause downward transitions.

Roos (1988) has proposed a scenario which would prevent the occupation number ever going higher than two. He visualizes a third black hole approaching the binary gently in nearly circular orbit and at the same time channeling stars to the binary. When the binary of mass $m_1 + m_2$ ejects stars of total mass dm, its energy E changes by

$$\frac{dE}{E} = \frac{\frac{1}{2} dm\, V_{esc}^2}{\frac{1}{2} \frac{m_1 m_2}{m_1+m_2} V_{orb}^2} = \frac{(m_1+m_2)^2}{m_1 m_2} \frac{dm}{m_1+m_2} \left(\frac{V_{esc}}{V_{orb}}\right)^2 = A \frac{dm}{m_1+m_2}. \tag{2}$$

In three-body ejections the ratio of the escape speed to the binary orbital speed $V_{esc}/V_{orb} \cong 1/3$ (Valtonen 1988). Thus the coefficient A is bounded by $0.4 < A < 0.7$ when $m_2 \leq m_1 \leq 5\, m_2$. For larger mass ratios m_1/m_2 the ratio V_{esc}/V_{orb} will go down, and thus the coefficient A is unlikely to be far from 0.5. However, Roos (1988) uses $A = 2.0$. If his assumption were true, and $dm >> m_1 + m_2$, so much energy would be carried away by the escaping stars that the binary would shrink with the same rate as the third body approaches it, and finally the binary would collapse before an efficient few black hole interaction has time to take place.

There are two reasons why the process of Roos (1988) is unlikely to occur in nature. First, the approach of a third black hole after the merger of two galaxies does not happen in a nearly circular orbit (Valtaoja, Valtonen and Byrd 1989), which means that $dm \cong m_1 + m_2$ only, and secondly, the coefficient A is in fact too small for the process to be efficient. While the process of Roos (1988) may not unimportant, the binary radius is probably not reduced enough to prevent the few black hole interactions.

3 SATELLITE BLACK HOLES

In quasar models we usually consider black holes of mass around 10^9 $M_\odot$. However, Eq. 1 shows that the critical encounter distance a_0 is greatly reduced if one of the black holes is much smaller in mass than 10^9 $M_\odot$, say $m_2 \cong 10^{-4}$ m_1. Such "intermediate mass" black holes have been proposed as constituents of dark matter in halos of galaxies and could be as numerous as millions per galaxy (Lacey and Ostriker 1985; Carr 1990 and references therein). Many of the intermediate mass black holes would collect in centers of galaxies and would form stable satellite systems around the supermassive black holes. The satellites could number in thousands per central black hole.

When mergers of galaxies take place and the occupation number increases (we don't count the possible intermediate mass black holes in Fig. 2), the satellite systems become dynamically unstable. Large numbers of intermediate mass black holes will be thrown out of the merged galaxy. The tail end of the escape velocity distribution goes up to relativistic velocities. There are various selection effects which strongly bias us to observe only a subpopulation of the ejected black holes, those which move with high velocity away from us (Valtonen and Basu 1991). Assuming that the ejected black holes appear as quasars, this could explain at least some of the quasar – galaxy associations of very different redshift which are mentioned in the previous article by Burbidge and Hewitt.

4 DISCUSSION

In recent years there has been a tendency to try to explain the different forms of activity in galactic nuclei as due to effects of viewing the same "beast" from different directions. Even though the viewing angle must be important in some instances, this approach is a probably an oversimplification. Among other things, it neglects the multiplicity of supermassive black holes in galactic nuclei. Multiple black hole nuclei may not only be common but they may actually be the ones which are preferentially selected in observed samples. This could be especially true for blazars.

References

Aarseth, S.J. and Fall, S.M. (1980) *Astrophys.J.* **236**, 43.
Begelman, M.C., Blandford, R.D. and Rees, M.J. (1980) *Nature* **287**, 307.
Begelman, M.C., Frank, J. and Shlosman, I. (1989). In *Theory of Accretion Disks*, eds. F. Meyer et al., Kluwer: Dordrecht, p. 373.
Carr, B.J. (1990) *Comments Astrophys.* **14**, 257.
Dressler, A. and Richstone, D.O. (1988) *Astrophys.J.* **324**, 701.
Filippenko, A.V. and Sargent, W.L.W. (1986) in *Structure and Evolution of Active Nuclei*, ed. G. Giuricin et al., Reidel:Dordrecht, p. 21.
Lacey, C.G. and Ostriker, J.P. (1985) *Astrophys.J.* **299**, 633.
Lin, D.N.C. and Papaloizou, J. (1979a) *Mon. Not. R. astr. Soc.* **186**,799.
Lin, D.N.C. and Papaloizou, J. (1979b) *Mon. Not. R. astr. Soc.* **188**, 191.
Lin, D.N.C. and Pringle, J.E. (1976) in *Structure and Evolution of Close Binary Systems*, eds. P. Eggleton et al., Reidel: Dordrecht, p. 237.
Matsuda, T., Sekino, N., Shima, E., Sawada, K. and Spruit, H. (1989). In *Theory of Accretion Disks*, eds. F. Meyer et al., Kluwer: Dordrecht, p. 355.
Mikkola, S. and Valtonen, M.J. (1990) *Astrophys.J.* **348**, 412.
Navarro, J.F., Mosconi, M.B. and Garcia Lambas, D. (1987) *Mon. Not. R. astr. Soc.* **228**, 501.
Roos, N. (1981a) *Astron.Astrophys.* **95**, 349.
Roos, N. (1981b) *Astron.Astrophys.* **104**, 218.
Roos, N. (1988) *Astrophys.J.* **334**, 95.
Rózyczka, M. and Spruit, H.C. (1989). In *Theory of Accretion disks*, eds. F. Meyer et al., Kluwer: Dordrecht, p. 341.
Shapiro, S.L. and Teukolsky, S.A. (1983) *Black Holes, White Dwarfs and Neutron Stars. The Physics of Compact Objects*. Wiley: New York.
Sillanpää, A., Haarala, S., Valtonen, M.J., Sundelius, B. and Byrd, G.G. (1988) *Astrophys.J.* **325**, 628.
Spruit, H.C. (1989). In *Theory of Accretion Disks*, eds. F. Meyer et al., Kluwer: Dordrecht, p. 325.
Szuszkiewicz, E. (1991). In *IAU Colloq. no. 129, Structure and Emission Properties of Accretion Disks*, ed. S. Collin-Souffrin, Kluwer: Dordrecht, in press.
Tonry, J.L. (1987), *Astrophys.J.* **332**, 632.
Ulrich, M.-H. (1989). In *Theory of Accretion Disks*, eds. F. Meyer et al., Kluwer: Dordrecht, p. 3.
Valtaoja, L., Valtonen, M.J. and Byrd, G.G. (1989) *Astrophys.J.* **343**, 47.
Valtonen, M.J. (1988) *Vistas Astron.* **32**, 23.
Valtonen, M.J. and Basu, D. (1991) *J.Astrophys.Astr.*, June 1991.

Optically violently variable active galactic nuclei as supermassive binary systems

M.K. BABADZHANYANTS, YU.V. BARYSHEV AND
E.T. BELOKON'

Astronomical Observatory of Leningrad University

ABSTRACT

Evidence has been found for periodic light variations in four OVV AGN's with characteristic times of 10 – 20 years. It is suggested to be a common property of OVV objects caused by their binary nature. Assuming that the optical emission of the objects originates in a relativistic jet, a hypothesis is proposed that observed optical variability in time scales mentioned is caused, at least partly, by gravitational bending of the jet varying in amount and direction due to orbital motion in the system.

1 OBSERVATIONAL DATA

Prolonged photometric monitoring (20 – 25 yr) in optical region of a number of OVV extragalactic objects, combined with archival data increasing the observational interval to 50 – 100 years, now gives a possibility of revealing some common properties of their light curves. Fig. 1 presents light curves (B-band) for four AGN: the quasar 3C 273, the BL Lac object OJ 287, and the Seyfert galaxies 3C 120 and 3C 390.3. Data sets contain all available observations published by 1990 and some unpublished data (Babadzhanyants and Belokon' 1991a, 1991c; Belokon' 1987; Babadzhanyants, Belokon' and Gamm 1991). The observations obtained during one night were averaged. Some systematic differences between different data sets were eliminated also.

These light curves show several components with different time-scales:
a) $\sim 0\overset{m}{.}5 - 3^m$ flares lasting for several years and arising with the frequency $\sim 1/(10\text{-}20)$ [years^{-1}]; they are either roughly sinusoidal in shape (3C 273 and 3C 120) or sharp outbursts (OJ 287, 3C 390.3);
b) a component with time-scales of several months and of $\sim 0\overset{m}{.}2 - 1^m$ amplitudes;
c) "fast" light variations with the time-scales from weeks to one day.

Similar components were revealed for a number of other OVV's also.

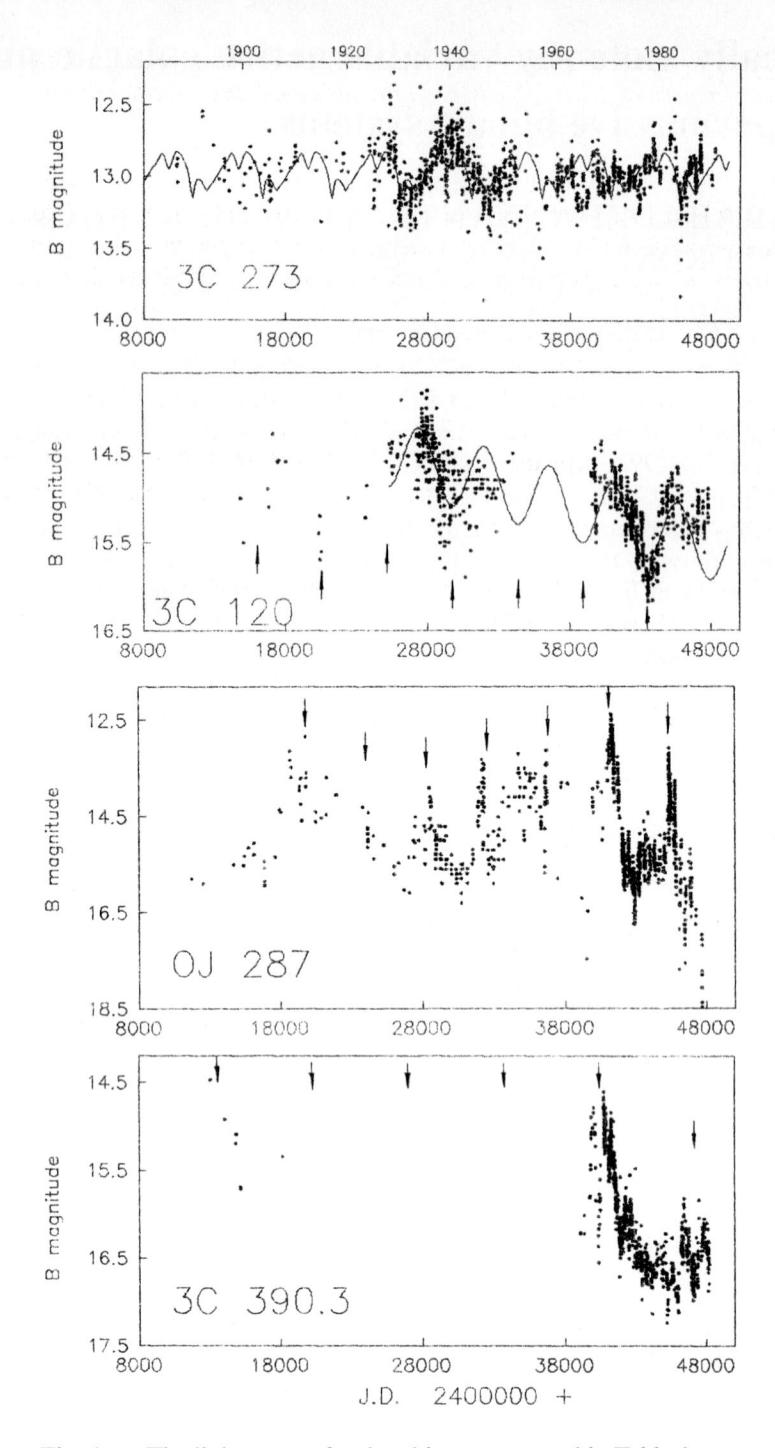

Fig. 1   The light curves for the objects presented in Table 1.

Optical variability of the four above mentioned AGN's has been analyzed for periodicity on different time-scales. There is evidence for periodical behavior at least on a limited time intervals for both variability components (a) and (b).

An important common property for objects mentioned above is strong variability not only in optical but in all observable spectral regions from radio to X-rays. Besides, they are well-known superluminal sources showing apparent superluminal motions in their milliarcsecond radio jets (Unwin et al. 1985; Gabuzda et al. 1989; Walker et al. 1984; Alef et al. 1988).

Four superluminal sources: 3C 345, 3C 120, 3C 273 and OJ 287 show strong evidence for a connection between optical flares and epochs of jet component ejection from the core (Babadzhanyants and Belokon' 1985; Belokon' 1987, 1991; Babadzhanyants and Belokon' 1991a, 1991c). In the case of 3C 273 and 3C 120 such a connection is suggested between the variability components (a) and (b) and the milliarcsecond and ten milliarcsecond scales, respectively.

The general data concerning this correlation is shown in Table I. Columns: (1) – object name; (2) – redshift; (3) – value of "large" periods for optical flares (component (a)); (6) – value of "small" period for optical flares (component (b)); the values of the periods in the rest frame are given in brackets; (4), (7) – amplitudes of optical flares for both of the variability components; (5), (8) – references to period determinations; (9) – "++" means the existence of the connection between VLBI structure and periodic optical flares of both types; (10) – references where these connections were suggested.

Table I. Optical-VLBI connection

Name 1	z 2	P_1(yr) 3	ΔB_1 4	Ref 5	P_2(day) 6	ΔB_2 7	Ref 8	VLB 9	Ref 10
3C 273	0.158	13.4 (11.6)	$0\overset{m}{.}3$	1	307 (265)	$0\overset{m}{.}2$	7	++	1,7
3C 120	0.033	12.5 (12.1)	$0\overset{m}{.}8$	2,3 4	303 (293)	$0\overset{m}{.}6$	3	++	3
OJ 287	0.306	11.65 (8.9)	2^m-3^m	5	184? (141)	$0\overset{m}{.}5$	8	+?	9
3C390.3	0.057	18? (17)	1^m-2^m	6	270 (255)	$0\overset{m}{.}5$	6	??	

References: 1. Babadzhanyants and Belokon' 1991a. 2. Jurkevich, Ucher and Shen 1971. 3. Belokon' 1978. 4. Webb 1990. 5. Sillanpää et al. 1988. 6. Babadzhanyants and Belokon' 1991b. 7. Belokon' 1991. 8. Hagen-Torn et al. 1977. 9. Babadzhanyants and Belokon' 1991c.

In the case of OJ 287 the separation between the jet components which may correspond to $P_2 \sim 200$ days would be ~0.1 mass (apparent proper motion ~0.2 mass yr^{-1}, Gabuzda et al. 1989). For 3C 390.3 VLBI maps are too infrequent to identify components with certainty (Alef et al. 1988).

Another feature of the light curves shown is the presence of sharp brightness drops by $0^m4 - 1^m0$ which last for $\sim 1 - 4$ yr. For OJ 287 such "downfalls" are positioned between the main outbursts and have nearly the same period. In the light curve of 3C 273 at least four "downfalls" are clearly seen exactly at minima or maxima of the 13-yr flares. Besides, both objects have less pronounced additional drops of brightness. By analogy we interpret the double structure of the large flares of 3C 390.3 as similar "downfalls" with occurred at maxima of these flares with proposed period of about 18 years. The data available on 3C 120 do not exhibit such evident drops.

The well investigated quasar 3C 273 shows additional evidence for correlation between optical variations and the superluminal jet which we discuss elsewhere (Babadzhanyants and Belokon', this conference).

2 A MODEL FOR OVV AGN

Further we shall try to show that the observational facts presented above may be explained by a model with a supermassive binary system, where one component produces a relativistic jet directed nearly along the line of sight.

This model is based on an analogy with the Galactic object SS 433 which is a close binary system with a subrelativistic jet originating in the compact object of smaller mass.

According to Begelman, Blandford and Rees (1980) and Whitmire and Matese (1981) supermassive binary systems (SBS) in active galactic nuclei may have masses of the order of $10^7 - 10^{10}$ M$_\odot$. The orbital period of SBS is given by

$$P_{orb} \approx 10 \text{yr} \cdot \left[\frac{\alpha}{7 \cdot 10^{16}}\right]^{3/2} \cdot \left[\frac{10^9}{M_1 + M_2}\right]^{1/2}$$

where α is the semimajor axis of the orbit in cm, and M_1, M_2 are the masses of the two bodies in M$_\odot$.

The lifetime of SBS is restricted by gravitational radiation:

$$T_{grav} \approx 8 \cdot 10^7 \text{yr} \left[\frac{\alpha}{7 \cdot 10^{16}}\right]^4 \cdot \left[\frac{10^7}{M_1}\right] \left[\frac{10^9}{M_2}\right] \cdot \left[\frac{10^9}{M_1+M_2}\right]$$

The value is sufficient to satisfy observational restrictions on duration of the lifetime of the active phase of galactic nuclei.

Optical emission of OVV objects is possibly of synchrotron nature and originates in a relativistic jet (see Babadzhanyants, Baryshev and Belokon' 1990). By analogy with SS 433 let us consider in our case the main component to have mass $M_1 \sim 10^9\ M_\odot$ and its companion to be a compact supermassive object with smaller mass $M_2 \sim 10^7\ M_\odot$ and with radius close to the gravitational one, $R_{g2} \sim 2GM_2/c^2 \sim 3\cdot 10^{12}$ cm. The secondary component is surrounded by a thick accretion disc due to mass outflow from the main component and is the source of a relativistic jet.

The essential property of such a model is "gravitational stagger" of the jet with orbital period – see Fig. 2. Due to the gravitational deflection the angle θ between the line of sight and direction of jet motion will vary and so will vary the projected apparent velocity of the jet component, $\beta_{app} = \beta\cdot\sin\theta/(1-\beta\cos\theta)$, the Doppler factor of component bulk motion, $\delta = (1-\beta^2)^{-1/2}(1-\beta\cos\theta)^{-1}$, and the apparent optical flux from the moving component, $F_{obs} = F_o\cdot\delta^{3+\alpha}$.

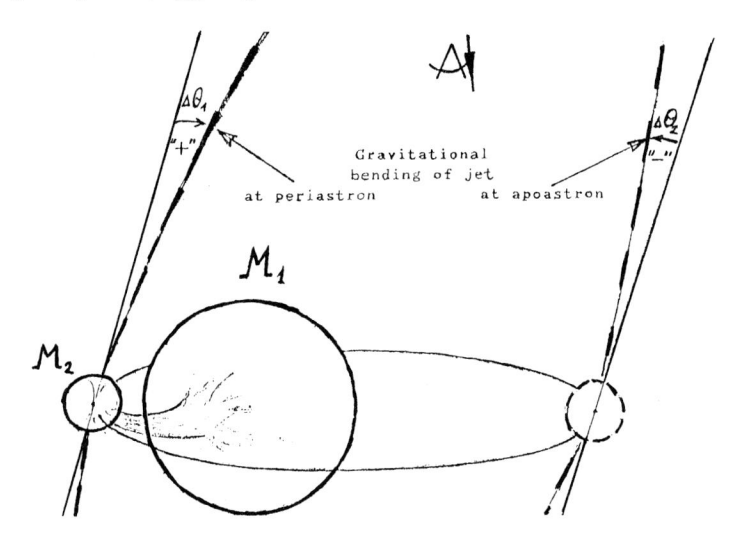

Fig. 2. A "gravitational stagger" model for the relativistic jet in the supermassive binary system. M_1 is the main companion. M_2 is the secondary one, $\Delta\theta$ is the gravitational deflection angle of the jet relative to the unperturbed jet axis.

The characteristic deflection of the jet may be estimated from the relation

$$\Delta\theta \approx \pm 0\overset{\prime\prime}{.}24 \cdot \left[\frac{M_1}{10^9}\right] \cdot \left[\frac{7\cdot 10^{16}}{\alpha}\right],$$

where ± is the sign of the deflection relatively to the unperturbed jet axis.

Depending on the orientation angles of the system (inclination of the jet axis to the orbital plane and its direction relative to the line of sight) one can obtain different types of light curves, from nearly sinusoidal to peak-like, when $\cos\theta \approx \beta$.

Within the "gravitational stagger" model correlated periodic changes in β_{app}, optical flux and characteristic changes in position angles of superluminal components ejection found for 3C 273 (Krichbaum et al. 1990) are naturally understood. Note that according to recent VLBI observations the jet trajectories are not purely ballistic on scales of about 1 pc. Particularly, different kinds of MHD instabilities can amplify initial gravitational oscillations of jet.

Following the analogy with SS 433 one can suppose that the main companion is a classic supermassive star (SMS). Its stability depends essentially on adopted relativistic gravitation theory. In General Relativity SMS is unstable unlike in Tensor Field Gravitation Theory (TFT), in which it is stable in the Post-Newtonian approximation (Baryshev 1991). Moreover, in TFT there are compact massive objects having surface at the gravitational radius (asymptotic saturation of gravity in TFT).

Possibly, pulsations of SMS and/or of accretion discs around the compact object lead to periodic modulation of the mass outflow in a relativistic jet (excitation of shock waves in it) that would appear as periodic ejection of superluminal knots with characteristic times of 200 – 300 days.

The existence of a rather extended SMS in the binary system may lead (with certain inclinations of the jet axis) to a physical interaction of the jet with SMS atmosphere as well as with the accretion disc. Such a direct interaction may be the reason for observed sharp drops in the light curves of OVV AGN's.

In conclusion it should be emphasized that the VLBI resolution now available (about 50 µas at 3 mm) is sufficient for testing our SBS hypothesis in the case of the Seyfert galaxy 3C 120. In this case we have orbital periodic P_1 = 12.5 years and if, for example, mass of the system is $M = 10^8 M_\odot$, then the semimajor axis will be $\alpha = 3.6 \cdot 10^{16}$ cm. An angular resolution of 50 µas corresponds to a linear resolution of $5.8 \cdot 10^{16} \cdot h_{100}^{-1}$ cm for 3C 120. This value lies very close to the predicted orbital diameter $2\alpha = 7.2 \cdot 10^{16}$ cm. In this connection it would be very interesting to make special VLBI observations of 3C 120 similar to those made for 3C 345 with precise astrometry of the "core" (Bartel et al. 1986). These observations will allow to "weigh" the "energy machine" or to set a limit on the Hubble constant if the mass of the SBS is known.

We thank S.V. Sudakov for his contribution to this work.

REFERENCES

Alef M., Gotz M.M.A., Preuss E. and Kellermann K. 1988, *Astron.Astrophys.* **192**, 53.
Babadzhanyants M.K., Baryshev Ju.V. and Belokon' E.T. 1990, *Commun.SAO AS USSR* **64**, 47.
Babadzhanyants M.K., Belokon' E.T. 1985, *Astrofizika* **23**, 459. English translation: 1986, *Astrophysics* **23**, 639.
Babadzhanyants M.K. and Belokon' E.T. 1991a, *Pis'ma Astron.Zh.* (submitted).
Babadzhanyants M.K. and Belokon' E.T. 1991b (in preparation).
Babadzhanyants M.K. and Belokon' E.T. 1991c (in preparation).
Babadzhanyants M.K., Belokon' E.T. and Gamm N.N. 1991 (unpublished).
Bartel N., Herring T.A., Ratner M.I., Shapiro I.I. and Corey B.E. 1986, *Nature* **319**, 733.
Baryshev Yu.V. 1991, this conference.
Begelman M.C., Blandford R.D. and Rees M.J. 1980, *Nature* **287**, 307.
Belokon' E.T. 1987, *Astrofizika* **27**, 429. English translation: *Astrophysics* **27**, 588.
Belokon' E.T. 1991, *Astron.Zh.* **68**, 1.
Benson J.M., Walker R.C., Unwin S.C., Muxlow T.W.B., Wilkinson P.N., Booth R.C., Pilbratt G. and Simon R.C. 1988, *Astrophys.J.* **334**, 560.
Gabuzda D.C., Wardle J.F.C. and Roberts D.H. 1989, *Astrophys.J.* **336**, L59.
Hagen-Torn V.A., Perewozchikova A.I., Ershtadt S.G. and Jakovleva V.A. 1977, *Pis'ma Astron. Zh.* **3**, 54.
Jurkevich I., Ucher P.D. and Shen B.S. 1971, *Astrophys. Space Sci.* **10**, 402.
Krichbaum T.P., Booth R.S., Kus A.J., Rönnäng B.O., Witzel A., Graham D.A., Pauliny-Toth I.I.K., Quirrenbach A., Hummel C.A., Alberdi A., Zensus J.A., Johnston K.J., Spencer J.H., Rogers A.E.E., Lawrence C.R., Readhead A.C.S., Hirabayashi H., Inoue M., Morimoto M., Dhawan V., Bartel N., Shapiro I.I., Burke B.F. and Marcaide J.M. 1990, *Astron. Astrophys.* **237**, 3.
Sillanpää A., Haarala S., Valtonen M.J., Sundelius B. and Byrd G.G. 1988, *Astrophys. J.* **325**, 628.
Unwin S.C., Cohen M.H., Biretta J.A., Pearson T.J., Seielstad G.A., Walker R.C., Simon R.S. and Linfield R.P. 1985, *Astrophys. J.* **289**, 109.
Walker R.C., Benson J.M., Seielstad G.A. and Unwin S.C. 1984, in *IAU Symp. 110, VLBI and Compact Radio Sources*, eds. R. Fanti, K. Kellermann and G. Setti (Dordrecht:Reidel), p. 121.
Webb J.R. 1990, *Astron. J.* **99**, 49.
Whitmire D.P. and Matese J.J. 1981, *Nature* **203**, 722.
Zensus J.A., Bååth L.B., Cohen M.H. and Nicolson G.D. 1988, *Nature* **334**, 410.

Pulsation of a Supermassive Star in tensor Field Gravitation Theory

YURI V. BARYSHEV

Astronomical Observatory of Leningrad University

It is shown that unlike General Relativity, the Tensor Field Gravitation Theory leads to the Post-Newtonian stability of a supermassive star. Its gravitational binding energy can be a more effective power source than nuclear reactions. Within the Field Gravitation Theory framework there might exist supermassive magnetoids, synchro-Compton caldrons, and plasma-turbulent reactors as power sources in Active Galactic Nuclei. Pulsations of these objects could appear in the observed variability of their emission.

1 INTRODUCTION

Hoyle and Fowler [1963] suggested the possibility that a mass of the order of 10^8 $M_\odot$ has condensed in the galactic nucleus into a supermassive star in which nuclear-energy generation takes place. However, a year later Fowler [1964] showed that in General Relativity (GR) the supermassive star (SMS) is destroyed before nuclear reactions begin. The lifetime of SMS in GR is $\tau \sim 10(M_8)^{-1}$ yr, and then SMS will collapse into a black hole. Hence, only black holes can be the primary power sources in AGN.

In the present work we consider the Post-Newtonian (PN) stability of SMS within the framework of the Field Gravitation Theory (FGT). It will be shown that supermassive stars are stable in FGT.

2 SMS STABILITY ANALYSIS

Basic principles and main equations of FGT are discussed in detail in Baryshev [1991].

Internal structure and stability of an equilibrium relativistic star in the PN approximation is defined: 1) by the gravitational field equations, 2) by the hydrostatic equilibrium equation, and 3) by the total energy of the system in equilibrium.

Following Fowler [1966], consider PN hydrostatic equilibrium and small adiabatic pulsations in a slowly rotating SMS taking into account differences between FGT and GR.

Let us write the expression for the total equilibrium energy of SMS (excluding the constant term $\int \rho_0 c^2 dV = M_0 c^2$) in the form:

$$E^{(eq)}_{(SMS)} = \int \left(e - 3p - \frac{1}{2}\rho_0 v^2 \right) dV , \qquad (1)$$

which is in essence of the relativistic virial theorem.

Here e is the thermal energy density, p is pressure, and $\frac{1}{2}\rho_0 v^2$ is the density of kinetic rotation energy. The first two terms in (1) can be expressed through the Newtonian potential energy plus a relativistic correction. Thus we have

$$E^{(eq)}_{(SMS)} = \frac{\beta}{2} \Omega_0 - \Delta E_{rel} - \Delta E_{rot}$$

where β is the ratio of gas pressure to total pressure ($\beta \ll 1$, because radiation pressure prevails), Ω_0 is the Newtonian potential energy of SMS, and ΔE_{rel} is the relativistic correction:

$$\Delta E_{rel} = K_4 \frac{G^2}{c^2} M_0^{7/3} \rho_{0(c)}^{2/3} \qquad (2)$$

where M_0 is the SMS rest mass, $\rho_{0(c)}$ is the SMS central density and K_4 is a constant coefficient depending on the relativistic theory adopted and on the internal structure of SMS. For n = 3 polytrope $K_4 = +1.7349$ in FGT and $K_4 = -0.9183$ in GR. Considering small radial adiabatic pulsations of SMS of the main mode we obtain the expression for the frequency of the oscillations:

$$\omega_0^2 = \frac{1}{I}\left(-\frac{\beta}{2}\Omega_0 + 2\Delta E_{rel} + 2E_{rot} \right) \qquad (3)$$

where $I = \int r^2 \rho_0 dV$ is the moment of inertia of SMS.

Assuming that the internal structure of SMS is well described by an n=3 polytrope, we finally have from (2) and (3) for equilibrium energy and oscillation period of SMS (in FGT):

$$\frac{E^{(eq)}_{(SMS)}}{M_0 c^2} = -\frac{3}{8}\beta \frac{R_g}{R} - \frac{1}{2}\left(\frac{\Phi}{ckM_0 R} \right)^2 - 2.39 \left(\frac{R_g}{R} \right)^2 ,$$

$$P_0 = \frac{2\pi}{\omega_0} = 2.11 \; Rc^{-1} \left\{ \frac{3}{8} \beta \frac{R_g}{R} + \left(\frac{\Phi}{ckM_0 R} \right)^2 + 4.78 \left(\frac{R_g}{R} \right) \right\}^{-1/2}$$

where R, R_g, M_0, and Φ are the radius, gravitational radius, mass and angular moment of SMS, respectively, and k is a structural constant of the order of unity.

3 MAIN CONCLUSIONS

Within the Field Gravitation Theory framework, the PN correction to the equilibrium energy of the system derived from the relativistic hydrostatic equilibrium equation, has the same sign as the term connected with rotation of SMS.

In the GR case the terms proportional to $(R_g/R)^2$ have opposite signs which leads to PN instability of SMS.

The PN stability of a supermassive star in FGT radically changes our concepts of SMS evolution derived earlier using GR. In particular, at last stages of evolution the main energy source will be not nuclear reactions with energy output nearly 1% of $M_\odot c^2$ but the gravitational binding energy of SMS which is of the order of $M_\odot c^2$. The gravitational binding energy might be the power reservoir that keeps "working" synchro-Compton caldrons, plasma-turbulent reactors or supermassive spinars having been proposed earlier as models for activity in galactic nuclei and which has been rejected because of their short lifetimes due to PN instability in GR.

REFERENCES

Baryshev, Yu.V. 1991: Gravitational field theory: foundations and astrophysical consequences (in press). In Russian.
Fowler, W. 1964: *Rev.Mod.Phys.* **36**, 545.
Fowler, W. 1966: *Astrophys.J.* **144**, 180.
Hoyle, F. and Fowler, W. 1963: *Mon.Not.Roy.Astron.Soc.* **125**, 169.

Blazars : Faster than Light?

C.D. IMPEY

Steward Observatory, University of Arizona

1 TAXONOMY OF AGN

Those who study Active Galactic Nuclei (AGN) have long been faced with a bewildering taxonomy. The term "blazar" was coined, half in jest, by Ed Speigel at the first conference on BL Lac objects in Pittsburgh over a dozen years ago. It implies a combination of the properties of quasars and BL Lacertae objects, and gives a sense of the characteristic variability of a nonthermal continuum (Angel and Stockman 1980). Close relatives of blazars include the highly polarized quasars (HPQs) and the optically violent variable quasars (OVVs). There is much overlap among sources in these different categories. All of the properties of blazars reflect the dominance of a nonthermal continuum : compact radio emission, high polarization, rapid variability, and weak emission lines.

A complex observational path is used to classify optical and radio-selected samples of AGN. Satellites have enabled additional but smaller samples to be selected at infrared and X-ray wavelengths. Every vertex of this decision-space is a function of the details of the observational setup, and of the physical properties of the source. For example, the distinction between a BL Lac object and a highly polarized quasar may depend on the luminosity of the host galaxy, and its contrast with respect to the nonthermal continuum. It also depends on the angular resolution and depth of the images used for classification. The detection of significant polarization ($p > 3\%$, $\sigma(p)/p > 3$ in general for synchrotron emission) depends on the intrinsic nonthermal polarization, and the amount of dilution by a stellar component. It also depends on the precision of the polarimeter. The claim of weak emission lines clearly depends on the wavelength range and signal-to-noise of the spectrum used for classification.

The hope is that taxonomy will illuminate the connections between different types of AGN (Lawrence 1987). We should not be shy about looking for insights in the classification process. Gould (1989) has argued persuasively that taxonomy is an incisive tool of the scientific method, using paleontology as an example. However, the classification of fossils rests on the powerful underlying paradigm of natural selection. The theory of AGN also has an underlying paradigm : a gravitational engine, fuelled by accretion, which powers relativistic outflow (Rees 1984). There have been few critical observational tests of this paradigm. One fruitful approach is to work on complete and carefully selected samples of AGN. In this way, it is hoped that the vaguaries of observational selection will not obscure the physical relationships being sought.

2 COMPLETE SAMPLES OF RADIO SOURCES

The primary sample used in this work is the largest available with comprehensive VLBI data, optical spectrscopy and optical polarimetry. The 65 sources in the "VLBI sample" were selected according to the following criteria : $\delta > 35$ deg, $b^{II} > 10$ deg, $S_{5GHz} > 1.3$ Jy (Pearson and Readhead 1981, 1988). In certain parts of the analysis, this sample is augmented by another set of radio sources. The 90 quasars in the "2 Jy sample" were selected according to the following criteria : $b^{II} > 10$ deg, $S_{5GHz} > 2$ Jy (Impey and Tapia 1988). Redshifts are available for $\sim 90\%$ of the VLBI sample and $\sim 80\%$ of the 2 Jy sample. Milliarcsecond radio structures and optical polarimetry exist for over 90% of the combined sample. Polarimetry is only complete for 67% of the brightest galaxies in the VLBI sample. Overall, the luminosity, polarization and radio structural information is complete enough to be virtually immune from selection effects.

3 NONTHERMAL PROPERTIES

Much of the new data described here consists of optical polarimetry. High and variable linear polarization is a primary characteristic of optically thin synchrotron emission. The analysis uses the first measured value of polarization, which has the benefit of giving unbiased statistics for a group of objects. However, synchrotron polarization has a duty cycle of high and low values, and the first measured value may not be representative of the behavior of an individual object. Estimates of the duty cycle of high and low polarization states have been made by Fugmann and Meisenheimer (1988), Impey and Tapia (1990), and Kühr and Schmidt (1990). All polarizations above 3% are assumed to be nonthermal (in most cases this is confirmed by variability), a level designed to discriminate against the low polarization imprinted by dust grains aligned in the magnetic field of our own galaxy.

3.1 Optical Polarization
The polarimetry described here was obtained between March 1984 and April 1987 using the MINIPOL polarimeter. The telescopes used were the Du Pont 100-inch

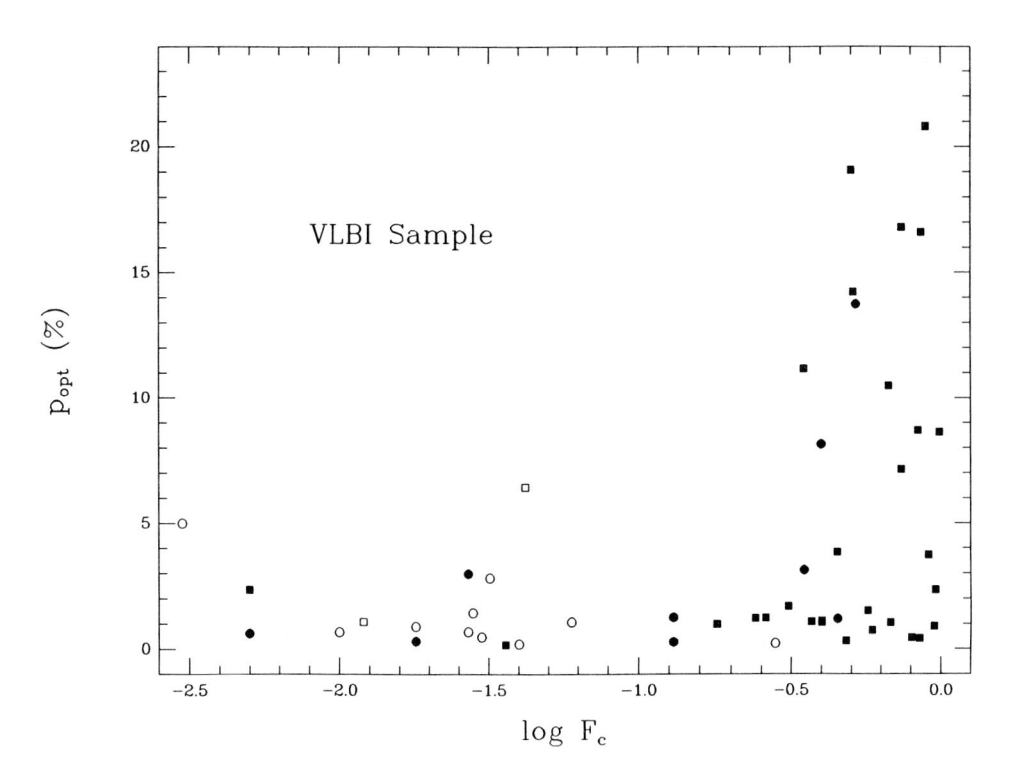

Fig. 1 — First measured value of optical polarization for VLBI sample vs. fraction of 5 GHz flux density in a VLBI core. Circles are radio galaxies and squares are quasars; open symbols indicate upper limits on radio compactness.

at Las Campanas in Chile and the 200-inch at Mount Palomar. The measurements were unfiltered, with a bandpass of 3200-8800Å, and an effective wavelength of about 5700Å. MINIPOL has an instrumental polarization of under 0.01%, and polarization errors on faint radio sources are limited only by photon statistics, and the brightness and polarization of the night sky. The degree of polarization has been corrected for low signal-to-noise ratio bias, according to the prescription of Simmons and Stewart (1985). Impey, Malkan and Tapia (1989) have fully desribed the calibration procedures applied to this data.

Figure 1 shows the data for the VLBI sample. The first measured value of optical polarization is plotted against the fraction of the total radio flux density in a milliarcsecond core. Both quasars and radio galaxies are included, and a clear trend is seen. High polarization is rarely observed in extended radio sources, but is common when more than half the total flux density is unresolved on VLBI scales. The correlation between high polarization and radio compactness (F_c) is significant at the 95% level for the VLBI sample, increasing to 99.5% if the quasars from the 2 Jy sample are added.

3.2 Quasars and Radio Galaxies

As discussed previously, the distinction between radio galaxies and quasars may be somewhat arbitrary. The 65 optical counterparts of the VLBI sample include 29 "galaxies" and 36 "quasi-stellar objects", classified according to their appearance on sky survey plates or CCD images. Figure 1 illustrates the well known fact that quasars have more compact radio emission than radio galaxies. The complication is that nuclear polarization in radio galaxies is diluted by starlight from the host galaxy. As a result, it is difficult determine whether the small polarization seen in most radio galaxies has a thermal or nonthermal origin (Rudy *et al.* 1983; Antonucci 1984).

The VLBI sample contains both radio galaxies and quasars. The hypothesis under consideration is that *all* radio sources contain polarized synchrotron cores, with strength proportional to the strength of the core radio emission. The visibility of high polarization in radio galaxies then depends on the continuum polarization, the luminosity of the host galaxy, the redshift, and the size of the aperture used for the polarimetry. An average radio-optical spectral index of $\alpha_{ro} = -0.73$ is used to predict the optical core strength, the same value found in blazars (Impey and Tapia 1988). The hypothesis that polarized optical luminosity is proportional to radio core luminosity is supported at the 98% level. The detectability of such cores in both quasars and radio galaxies appears to depend primarily on F_c, the compactness of the radio emission.

3.3 Emission Line Strength

It seems that every radio source with emission lines of low equivalent width is, sooner or later, called a BL Lac object. It is preferable to reserve this term for objects related to the prototype, BL Lacertae, which have both low radio luminosity and low emission line luminosity. In this paper, these will be referred to as classical BL Lac objects. There are also sources at high redshift with high polarization and weak emission lines, and these sources may or may not be related to the prototype. High quality spectroscopy of the VLBI sample, with good wavelength coverage, has been reported by Lawrence *et al.* (1986, 1987).

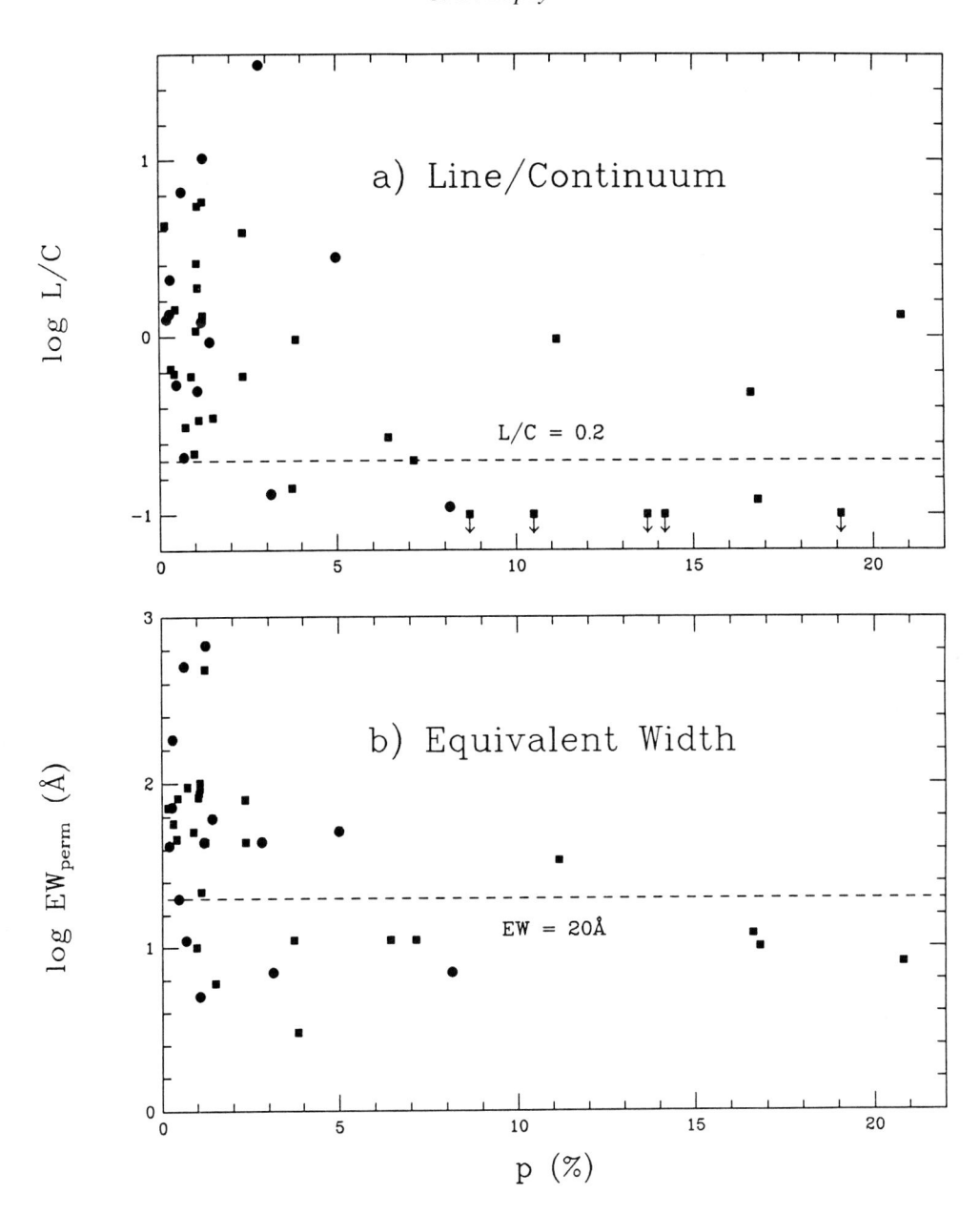

Fig. 2 — First measure of optical polarization plotted against a) line to continuum ratio and b) equivalent width of the strongest permitted line. Squares are quasars, circles are radio galaxies.

Emission line properties are plotted against optical polariztion for quasars and radio galaxies in Figure 2. The correlation betweeen weak emission lines and high polarization is very strong for the quasars, a confidence level of over 99.9% for both line-to-continuum ratio and equivalent width of the strongest permitted line. Combining the VLBI and 2 Jy samples, and defining high polarization as $p > 3\%$ and weak lines as $L/C < 0.2$, 22 out of 23 (96%) quasars with weak emission lines have high polarization, as opposed to 44 out of 133 (33%) of the quasars with strong lines. As expected, emission line strength is anticorrelated with radio compactness.

3.4 Optical Variability

The connection in blazars between high polarization and large amplitude flux variability was first noted by Moore and Stockman (1984). It implies that the variable flux component in quasars carries the high polarization. The amplitude of flux variability on a month timescale can be estimated from the monitoring programs of Pica et al. (1988) and Webb et al. (1988). In the VLBI sample, only 1 out of 13 (8%) of the sources with $F_c < 0.1$ have $\Delta m > 1$ magnitude, whereas 32 out of 57 (56%) of the sources with $F_c > 0.1$ have $\Delta m > 1$ magnitude. The behaviors of polarization and flux variability as a function of radio compactness are identical.

3.5 Radio Polarization

Compact radio sources often have one synchrotron component that dominates from 10^{11} to 10^{15} Hz (Landau et al. 1986, Impey and Neugebauer 1988). At a sufficiently high radio frequency, where the opacity is low, a relationship might be expected between radio and optical polarization. In this comparison, radio polarimetry is taken from Rusk (1988), based on a homogenous set of VLA maps. The degrees of polarization at 2cm and optical wavelengths are correlated at the 99% significance level for the VLBI sample. This correlation is absent at the longer wavelengths of 6cm, 18cm, and 22cm. In addition, sources with high polarization show alignment between the position angles of radio and optical polarization. The radio and optical data is not simultaneous, but nonsimultaneity and source variability will only act to *weaken* these correlations.

To summarize, there are clear relationships between radio and optical properties in samples of compact radio sources. All of the diagnostics of an optical synchrotron source (high polarization, rapid variability, weak emission lines and a prominent power law) correlate with a single radio parameter, the compactness on parsec scales. This is true for both quasars and radio galaxies.

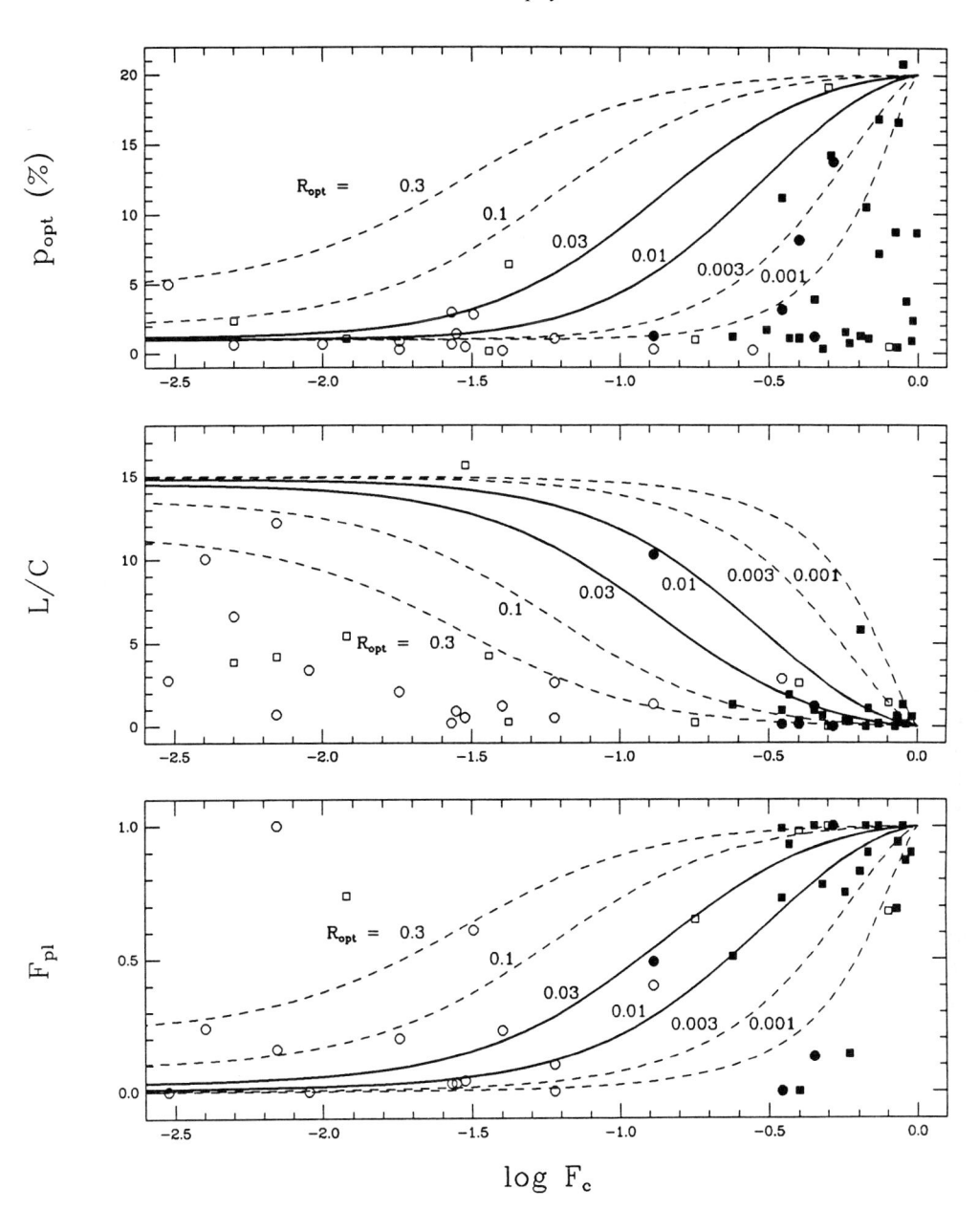

Fig. 3 — Radio core fraction vs. a) optical polarization, b) line to continuum ratio, and c) fraction of the optical continuum which is a power law. The curves represent a model with the same beaming cone used for the optical and radio emission. Symbols as in Figure 1.

4 RELATIVISTIC MOTION

The "standard model" used to explain the apparently superluminal motion of radio source components is based on an idea first put forward by Rees (1967). Radiating blobs or plasmons move at velocities close to c, and an observer whose line-of-sight is nearly aligned with the motion sees large transverse velocities and radiation that is amplified by relativistic aberration (Pearson and Zensus 1987). Additional indirect arguments for relativistic motion in compact radio sources include a lack of Compton-scattered X-rays (Marscher *et al.* 1979), infrared luminosities higher than the Eddington limit, and direct VLBI measurements of brightness temperature (Linfield *et al.* 1989).

Beaming models have been pursued in two ways. First, simple ballistic models have been modified to correspond to what may in reality be a complex hydrodynamic flow (Lind and Blandford 1985). Complications include an opening angle of the beamed emission that may not be simply related to the Lorentz factor, a velocity of the radiating material that may not be the same as the velocity inferred from superluminal motion, and the use of a distribution of Lorentz factors for the flow. Second, it is clear that compact or core-dominated sources are a small percentage of all known active galactic nuclei. An attractive generalization of the beaming model proposes that all high luminosity radio sources have relativistic jets, and that the observed differences between powerful radio galaxies and quasars, and between flat and steep spectrum radio sources are caused by Doppler boosting and projection effects (Orr and Browne 1982, Barthel 1989).

4.1 Apparent Superluminal Motion

Apparent superluminal motion is common among core dominated quasars (Cohen 1989), and superluminal motion and blazar properties are often related (Impey 1987). Thirteen sources in the VLBI sample are superluminal, and nine of these show high optical polarization (two of the remainder have only been measured once). Two other effects are seen. Galaxies have slower values of v/c than quasars, at the 99% confidence level; and classical BL Lac objects have lower values of v/c than quasars, at the 95% confidence level.

4.2 Beamed Optical Emission

Spectral deconvolution has been used to determine the optical power law fraction (F_{pl}) in sources in the VLBI sample. Every case of high polarization is accompanied by a high power law fraction, except for Mk 501 which has a power law fraction which varies with epoch. It is expected that low polarization will become rarer at high redshift, since observations are at short rest wavelengths where red power law

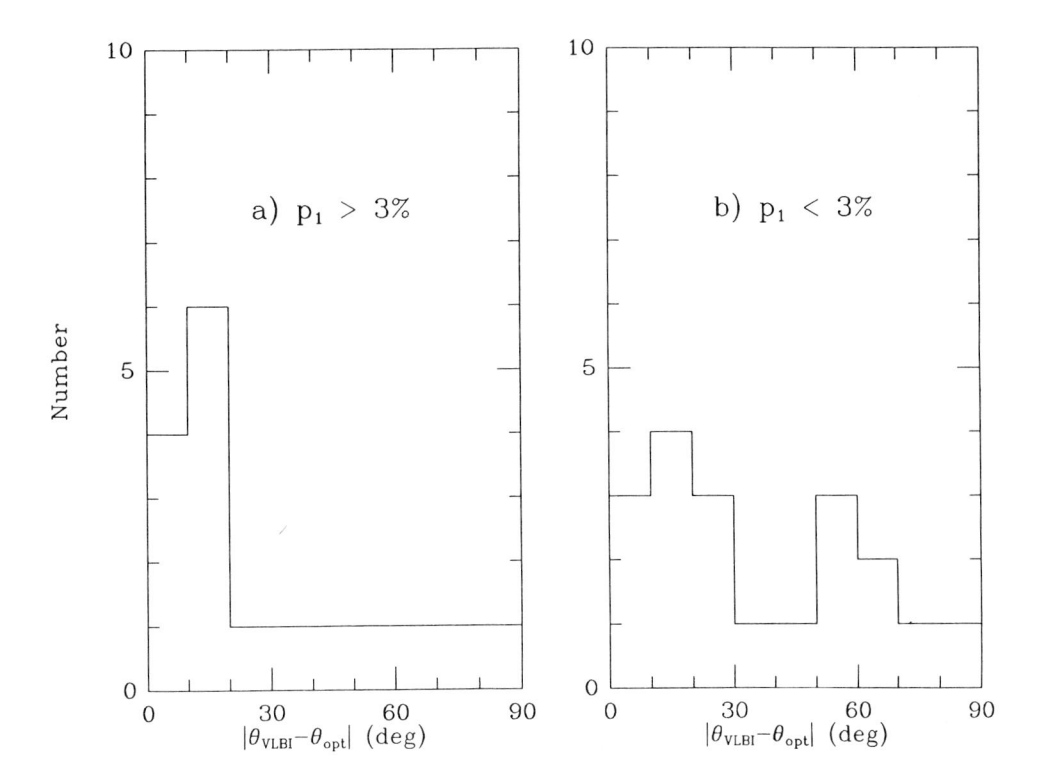

Fig. 4 — The difference between the position angle of the VLBI structure axis and the position angle of optical polarization divided according to a) high and b) low optical polarization.

components are weak, and hot thermal components are strong. In the VLBI plus 2 Jy samples combined, the fraction of quasars with high polarization falls from 0.43±0.09 ($0 < z < 1$) to 0.36±0.10 ($1 < z < 2$) to 0.22±0.16 ($2 < z < 3$). This is consistent with the spectral decomposition of Wills (1989). Weak-lined radio sources, including the classical BL Lac objects, have systematically lower redshifts than low polarization quasars at the 97% confidence level.

Figure 3 shows the relationship between three optical indicators of synchrotron emission and the VLBI compactness F_c. There is a sudden appearance of high polarization, high power law fractions, and weak emission lines, all occuring at a value of $F_c \sim 0.3$. The curves represent the predictions of a beaming model with an isotropic component and a relativistically moving component. Following Orr and

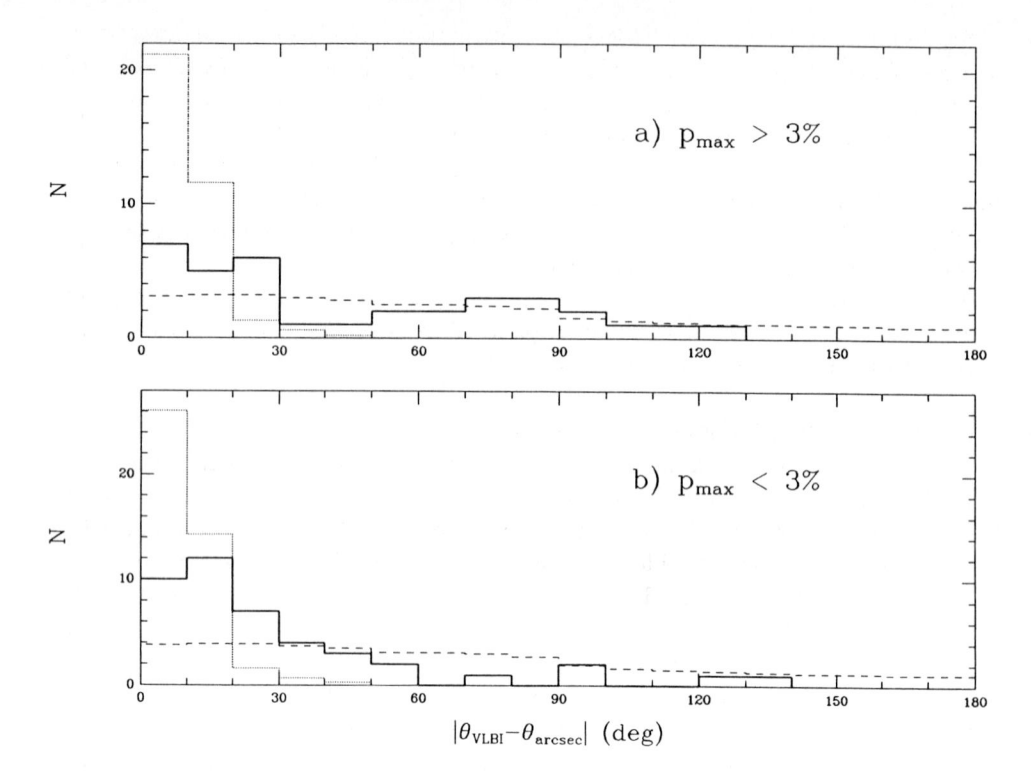

Fig. 5 — Histogram of the position angle difference between arcsecond and milliarcsecond radio structure divided by polarization level. The dotted curve shows a model with no relativistic motion and an intrinsic misalignment of 10 degrees; the dashed curve shows a model with a Lorentz factor of 10.

Browne (1982), we assume that the ratio of the transverse radio jet to the isotropic radio component is 0.01 (and that $\alpha_{rad} = 0.0$, $\alpha_{opt} = -0.7$). The curves are labelled with the only other free parameter, the ratio of the transverse optical jet to the isotropic optical component. A power law polarization of 20% and a ratio of line to unbeamed continuum of 15 is assumed. The model gives the upper envelope of three parameters, and is a reasonable fit to the trends in the data. Amplification by a factor of 10 to 100 is required before the beamed synchrotron component dominates the isotropic (unpolarized, quiescent) component.

4.3 Position Angle Alignments

A striking alignment has been observed between the position angle of compact radio structure and the position angle of optical polarization (Rusk and Seaquist

1985, Impey 1987). Figure 4 shows the distributions of $|\theta_{\text{VLBI}} - \theta_{\text{opt}}|$ for various subsets of this complete sample. VLBI structure angles measured at 5GHz are used where possible. The first measured position angle of optical polarization is used, so variability can only act to smear out any peaks found. The peak at $|\theta_{\text{VLBI}} - \theta_{\text{opt}}| = 0$ for the entire sample is highly significant, the probability from a χ^2 test that it is uniformly distributed is 0.0025. The deviation from a uniform distribution comes mostly from the high polarization sources; $|\theta_{\text{VLBI}} - \theta_{\text{opt}}|$ for low polarization sources is consistent with a uniform distribution. Classical BL Lac objects show a position angle alignment; six out of ten have $|\theta_{\text{VLBI}} - \theta_{\text{opt}}| < 20\,\text{deg}$. The most important comparison will be between the optical polarization and the growing set of VLBI polarization measurements (Gabuzda et al. 1989).

4.4 Position Angle Misalignments

An intriguing relationship is emerging between the degree of optical polarization and the misalignment between radio stuctures on pacsec and kiloparsec scales. Figure 5 shows the data for the VLBI sample and a larger but inhomogeneous set of radio sources from Rusk (1988), divided by optical polarization. Maximum polarization is used rather than first measured value to facilitate the comparison. The probability that misalignment and p_{max} are not related is only 2×10^{-4}. It is striking that none of the sources with $p_{max} > 22\%$ has $|\theta_{\text{VLBI}} - \theta_{\text{arcsec}}| < 50\,\text{deg}$. On the other hand, the distribution of misalignments for low polarization sources is broadly peaked around zero. This effect can be interpreted in terms of relativistic abberation of the highly polarized population, and is yet more evidence that they emit a Doppler-boosted continuum close to the line of sight.

4.5 Blazars, BL Lacs, and Beaming

There is growing evidence that the set of radio sources with "blazar" properties is not homogeneous. In particular, there are significant differences between the continuum properties of classical BL Lac objects and highly polarized quasars. Some of these differences reflect the generally lower redshifts of BL Lac objects, which result in lower emission line luminosities and extended radio powers. However, BL Lac objects have flatter continuum slopes at both optical and X-ray wavelengths than highly polarized quasars (Impey and Neugebauer 1988, Worrall 1989). Finally, the properties of the beamed radio emission appear to be different. All six BL Lac objects with measured VLBI kinematics have $v/c < 4$, whereas highly polarized quasars are faster superluminal sources, 6 out of 11 have $v/c > 8$.

This last result suggests that weak-lined objects either have relatively small values of γ, or have large values of γ but are oriented very close to the line of sight. Suppose that $\gamma_{\text{max}} \approx 4$. Transverse speed reaches its maximum of around γ when

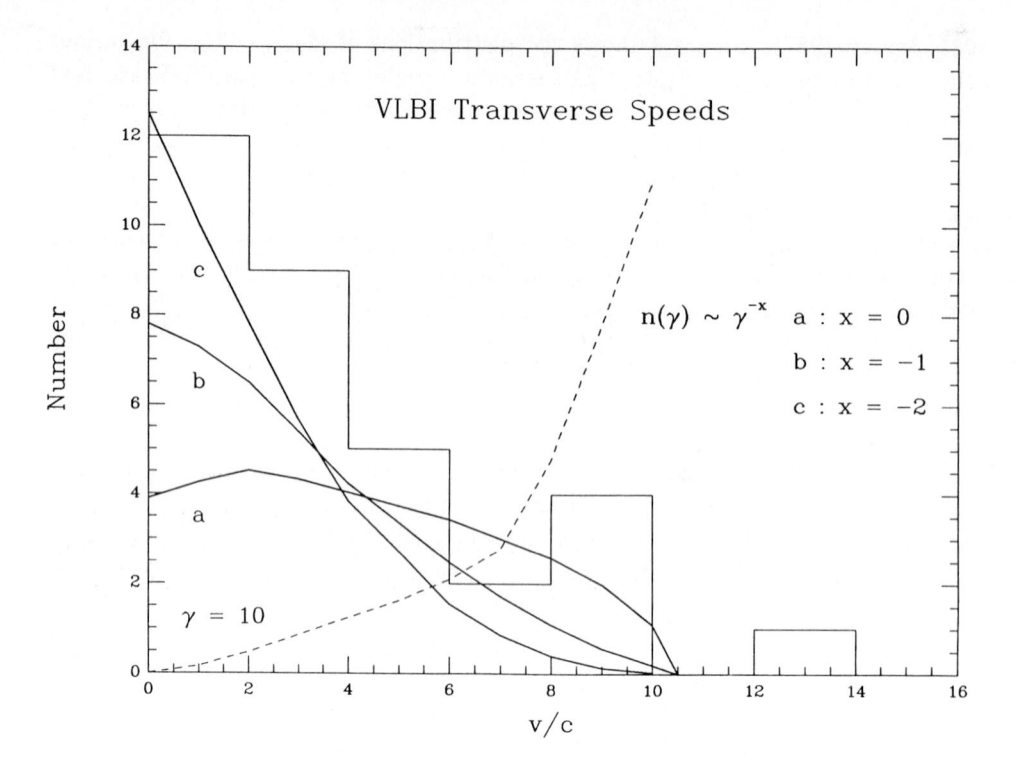

Fig. 6 — Histogram of observed values of apparent velocity from Cohen (1989). The dashed curve is for a single Lorentz factor; the solid curves represent distributions of Lorentz factors.

$\theta \approx 1/\gamma$, so the fastest weak-lined objects would have $\theta \approx 14\,\mathrm{deg}$. Such a modest Doppler boosting would easily swamp the low luminosity emission lines typical of low powered radio galaxies. About 6% of an orientation unbiased sample would have $\theta < 14\,\mathrm{deg}$, consistent with the fraction of BL Lac objects among B2 radio galaxies (Colla et al. 1975). On the other hand, the fastest quasars have $\gamma \approx 10$, so objects with $v/c = 4$ would have $\theta \approx 1\,\mathrm{deg}$. The implied rarity of heavily Doppler boosted sources may in fact be consistent with the space density of luminous, weak-lined sources such as AO 0235+164. The interpretation of the luminous sources is complicated by the possibility of gravitational lensing effects (Stickel et al. 1989). The data presented here supports the idea that classical BL Lac objects observed in elliptical galaxies at low redshift are physically distinct from weak-lined, luminous sources at high redshift.

The correlations derived for the VLBI and 2 Jy samples constitute strong (if indirect) evidence that compact radio sources have beamed optical radiation. A similar beaming geometry explains both the radio and optical data. The most logical modification of the ballistic beaming model is to assume a distribution of Lorentz factors. This is in fact required by the VLBI data (Cohen 1989). Figure 6 shows the distribution of apparent v/c, which is predicted to peak at high values for one-sided jets with any reasonable luminosity function. The distribution $n(\gamma) \sim \gamma^{-2}$ fits the data well. It also leads to agreement with the Orr and Browne (1982) model, where steep spectrum quasars are the parent population of flat spectrum quasars (Peacock 1987).

5 SUMMARY

Blazars occupy a small niche in the taxonomy of AGN. Most AGN are not radio sources, do not vary widely, and do not have high polarization. Yet blazars form the subset which have yielded the best understanding of the emission mechanism and the geometry of the nuclear regions. Considerable progress can be made with the sytematic, multi-wavelength study of complete samples. The one presented here is flux limited at 5GHz. Optical polarization, power law fraction, line to continuum ratio and variability amplitude are all strongly correlated with VLBI measures of compactness. The distributions of these optical properties are well matched by a model where the radio and optical radiation have the same beaming geometry. Optical beaming by a factor of 10 to 100 is required before a radio source shows blazar properties. Positional angle alignments between optical polarization and VLBI structure axis confirm the physical relationship between the two wavelength regimes. BL Lac objects differ significantly from highly polarized quasars in terms of luminosity, continuum shape, and apparent v/c of the milliarcsecond radio components. If both types of AGN are beamed, there is some evidence that the parent populations are distinct, and it may not be useful to lump them together as "blazars". A variety of observations can be used to successfully constrain the distribution of Lorentz factors in a simple beaming model.

I happily acknowledge many insightful conversations with my collaborators in this project, Charlie Lawrence and Santiago Tapia. Our progress was facilitated by the excellent day crews and telescope operators at Palomar and Las Campanas Observatories. This research was supported by Caltech through a Weingart Fellowship, by NSF grant AST-8700741, and by NASA/JPL contract 958028.

6 REFERENCES

Angel, J. R. P., and Stockman, H. S. 1980, *Ann. Rev. Astr. Ap.*, **18**, 321.
Antonucci, R. R. J. 1984, *Ap. J.*, **278**, 499.
Barthel, P. D. 1989, *Ap. J.*, **336**, 606.
Cohen, M. H. 1989, in *BL Lac Objects*, ed. L. Maraschi, T. Maccacaro, and M.-H. Ulrich (Berlin: Springer-Verlag), p. 13.
Colla, G. *et al.* 1975, *A.A.Supp.*, **20**, 1.
Fugmann, W., and Meisenheimer, K. 1988, *Astr. Ap.*, **207**, 211.
Gabuzda, D. C., Wardle, J. F. C., and Roberts, D. H. 1989, *Ap. J. Lett.*, **336**, L59.
Gould, S. J. 1989, *Wonderful Life*; (New York: Norton and Co).
Impey, C. D. 1987, in *Superluminal Radio Sources*, ed. J. A. Zensus and T. J. Pearson (Cambridge: Cambridge University Press), p. 233.
Impey, C. D., and Neugebauer, G. 1988, *A. J.*, **95**, 307.
Impey, C. D., and Tapia, S. 1988, *Ap. J.*, **333**, 666.
Impey, Malkan, M. A., and Tapia, S. 1989, *Ap. J.*, **347**, 145.
Impey, C. D., and Tapia, S. 1990, *Ap. J.*, **354**, 124.
Kühr, H., and Schmidt, G. D. 1989, *A. J.*, **99**, 1.
Landau, R. *et al.* 1986, *Ap. J.*, **308**, 78.
Lawrence, A. 1987, *P.A.S.P.*, **99**, 309.
Lawrence, C. R., Pearson, T. J., Readhead, A. C. S., and Unwin, S. C. 1986, *A. J.*, **91**, 494.
Lawrence, C. R., Readhead, A. C. S., Pearson, T. J., and Unwin, S. C. 1987, in *Superluminal Radio Sources*, ed. J. A. Zensus and T. J. Pearson (Cambridge: Cambridge University Press), p. 260.
Lind, K., and Blandford, R. D. 1985, *Ap. J.*, **295**, 358.
Linfield, R. P., *et al.* 1989, *Ap. J.*, **336**, 1105.
Marscher, A. P., Marshall, R. E., Mushotzsky, R. F., Dent, W. A., Balonek, T. A., and Hartman, M. F. 1979, *Ap. J.*, **233**, 498.
Moore, R. L., and Stockman, H. S. 1984, *Ap. J.*, **279**, 465.
Orr, M. J. L., and Browne, I. W. A. 1982, *M.N.R.A.S.*, **200**, 1067.
Peacock, J. A. 1987, in *Astrophysical Jets and Their Engines*, ed. W. Kundt (Dordrecht: Reidel), p. 185.
Pearson, T. J., and Zensus, J. A. 1987, in *Superluminal Radio Sources*, ed. J. A. Zensus and T. J. Pearson (Cambridge: Cambridge University Press), p. 1.
Pearson, T. J., and Readhead, A. C. S. 1981, *Ap. J.*, **248**, 61.
Pearson, T. J., and Readhead, A. C. S. 1988, *Ap. J.*, **328**, 114.
Pica, A. J., Smith, A. G., Webb, J. R., Leacock, R. J., Clements, S., and Gombola, P. P. 1988, *A. J.*, **96**, 1215.

Rees, M. J. 1967, *M.N.R.A.S.*, **135**, 345.
Rees, M. J. 1984, *Ann.Rev.Ast.Ap.*, **22**, 471.
Rudy, R. J., Schmidt, G. D., Stockman, H. S., and Moore, R. L. 1983, *Ap. J.*, **271**, 59.
Rusk, R. E., and Seaquist, E. R. 1985, *A. J.*, **90**, 30.
Rusk, R. E. 1988, Ph.D. thesis, University of Toronto.
Simmons, J. F. L., and Stewart, B. G. 1985, *Astr. Ap.*, **142**, 100.
Stickel, M., Fried, J. W., and Kühr, H. 1989, in *BL Lac Objects: Ten Years After*, ed. L. Maraschi, T. Maccacaro, and M.-H. Ulrich (Berlin: Springer-Verlag), p. 64.
Webb, J.R., Smith, A. G., Leacock, R. J., Fitzgibbons, G. L., Gombola, P. P., and Shepherd, D. W. 1988, *A. J.*, **95**, 374.
Worrall, D. M. 1989, in *BL Lac Objects*, ed. L. Maraschi, T. Maccacaro, and M.-H. Ulrich (Berlin: Springer-Verlag), p. 305.
Wills, B. J. 1989, in *BL Lac Objects*, ed. L. Maraschi, T. Maccacaro, and M.-H. Ulrich (Berlin: Springer-Verlag), p. 109.

Two Classes of BL Lac Objects?

E. Valtaoja[1,2], H. Teräsranta[1], M. Lainela[2] and P. Teerikorpi[2]

1) Metsähovi Radio Research Station, Helsinki University of Technology, Otakaari 5A, SF-01250 Espoo, Finland
2) Tuorla Observatory, University of Turku, SF-21500 Piikkiö, Finland

Abstract. We propose that the dichotomy in the properties of BL Lac objects – their variability properties, luminosities, space distribution etc. – indicates that there are two classes of "BL Lac objects" with different properties and different parent populations. Nearby BL Lacs are FR I radio galaxies beamed towards us, characterized by moderate radio variability. Distant, strongly variable BL Lacs are the extreme end of the FR II – quasar population, HPQ/OVV quasars with jets pointing almost directly towards us.

1. Introduction

In BL Lac objects synchrotron radiation from the core overwhelms all the other components. The generally accepted explanation is that the observed continuum has been amplified relative to the other components. One, although not the only, way to achieve this is to have the synchrotron radiation originating in a jet pointed towards us, while the other components radiate more or less isotropically. However, just *what* kind of sources BL Lacs are is less clear – how would they appear if seen from other directions? In other words, what is the parent population of BL Lac objects?

BL Lacs were originally proposed to be the extreme end of the quasar population with jets seen end on (Blandford and Rees 1978; Blandford and Königl 1979). However, there is an excess of nearby BL Lacs, and their space density is too great for them to be just a specially selected subset of quasars. An alternative which has gained wide acceptance is that BL Lacs are not quasars but moderately strong (Fanaroff-Riley type I) radio galaxies beamed towards us (e.g., Browne 1983; Padovani and Urry 1990, 1991). According to this view, there are two "unification schemes". Strong (Fanaroff-Riley type II) radio galaxies are seen as themselves, normal quasars or blazar-type (highly polarized/OVV) quasars depending on the viewing angle (Barthel 1989); the weaker FR I radio galaxies appear as BL Lacs when the viewing angle is sufficiently small. Another possibility which has recently gained favor is that BL Lacs are microlensed OVV quasars (e.g., Stickel et al. 1989; Ostriker and Vietri 1990), although such models seem to create more problems than they solve (Gear 1991).

There are at least two major problems in having FR I radio galaxies as the parent population of BL Lacs. The first is the close similarity between BL Lacs and HPQ/OVV quasars, also recognized in their common "blazar" classification. The second is the two-peaked redshift distribution of BL Lacs in flux-limited samples, with maxima at $z \approx 0$ and at $z \approx 1$ (Browne 1989). In this paper we consider these questions using new variability data from the Metsähovi monitoring program (Teräsranta et al., these proceedings). The solution we propose is in many ways obvious, although (as far as we know) it has not been explicitly formulated before: there are not one, but two, different parent populations of BL Lac objects, FR I and FR II radio galaxies. A single class of "BL Lac objects" does not exist.

2. Radio properties of the BL Lac objects in the Metsähovi sample

We have recently completed an extensive investigation of the spectral and variability properties of the sources monitored in Metsähovi between 1980 and 1986, using much larger high radio frequency data set than has been possible previously (see Salonen et al., 1987 for the data and Valtaoja et al. 1988, 1991a, 1991b for the details). We found an overwhelming similarity in the spectra, the evolution of outbursts, and in all the other variability characteristics in all compact radio sources, including BL Lacs. The outbursts in BL Lacs follow the same pattern, typical for shocks (Marscher and Gear 1985), and show different burst amplitudes and time delays between neighboring frequencies just as shock models predict. Their radio variations also connect to IR and shorter wavelengths, just as in quasars, and a separate origin for optical and radio variations is not a viable alternative (cf. Gear 1991). Thus, there seems to be no reason to assume a different origin for the variability in BL Lacs: if their variations really were due to gravitational microlensing, it would be hard to understand why two totally different mechanisms produce so similar variations (cf. also Valtaoja 1990; Gear 1991).

For quasars we found the correlation between spectral flatness and the rapidity of variations predicted by orientation dependent models (Eckart et al. 1989; Valtaoja et al. 1991b). We also found that highly polarized quasars (HPQs) have, on the average, flatter spectra than ordinary quasars (LPQs), and that HPQs are more variable than LPQs. Thus, quasars seem to fit in Barthel's unified scenario (Barthel 1989), and the differences between the radio properties of LPQs and HPQs can be understood as resulting from different viewing angles. However, BL Lacs do not fit in. The only agreement is that they do have flatter spectra than HPQs as would be expected for sources more closely aligned, but there is no correlation between spectral flatness and variability in BL Lacs, nor are they on the average more variable than quasars.

One possible reason for the general lack of correlations is the wide range of variability of BL Lacs in our sample. Some BL Lacs are among the least variable in our whole sample of flat-spectrum sources. Others are by any indicator among the most active and rapidly variable of all known radio sources. However, when considered as a class BL Lac objects are not more variable than other radio sources, due to the large fraction of nonvariable BL Lacs (cf. Altschuler 1982, 1983).

3. OJ 287: small viewing angles and large Lorentz factors

OJ 287 is the most rapidly variable of all known bright radio sources, with outbursts so closely spaced and overlapping that their nature or evolution is not easily studied. During the first half of 1989 there was a lull in the activity, during which OJ 287 reached unprecedented low flux levels before the onset of the next flare. In the beginning of May 1989 the fluxes started to rise rapidly, growing by a factor of 4 in less than 1.5 months at 37 GHz (Figure 1). Comparison of our 22 and 37 GHz lightcurves with 4.8, 8 and 14.5 GHz lightcurves from the University of Michigan monitoring (Aller, Aller and Hughes, private communication) shows that the variations were due to the decay of an earlier outburst and the rise of a new one: the multifrequency lightcurves exhibited the canonical behavior expected from an expanding outburst, with smaller amplitudes and growing time delays towards lower frequencies. The time interval between the outbursts was unusually long, and the flares could therefore be separated.

We have calculated the variability timescales and associated brightness temperatures for the outburst. At 37 and 22 GHz only lower limits can be estimated due to gaps in the data: $T_b(37) > 1.6\ 10^{14}$ K and $T_b(22) > 2.5\ 10^{14}$ K, requiring $D > 6$. Even these lower limits for T_b are among the highest found in any source during our Metsähovi monitoring. From best fits to the University of Michigan data we can derive corresponding values for the lower frequencies: $D(14.5) \approx 7 - 11$ and $D(8) \approx 12 - 14$. $D > 10$ means that the viewing angle *must* be less than 6° independent of the intrinsic Lorentz factor of the flow (e.g., Cohen and Unwin 1984). If we further accept the measured superluminal speeds in other outbursts of OJ 287, 2.4-3.2 (Gabuzda et al. 1989) as representative for this the source, the viewing angle is less than 3° and the Lorentz factor $\gamma > 6$ (Cohen and Unwin 1984). Thus, at least one BL Lac object, OJ 287, is beamed closely towards us, and has a large intrinsic Lorentz factor comparable to those typically found in quasars.

This causes problems for the FR I parent scenario. Where are all the BL Lacs similar to OJ 287 (i.e., with $\gamma > 6$ and similar intrinsic luminosity), but with larger viewing angles? Calculations are somewhat model-dependent, but in a flux-limited sample one should find at least 10 OJ's with larger viewing angles (Cohen 1989), easily identifiable by their respectable Doppler factors (5 – 10) and superluminal speeds ($v/c > 5$). Although there may be one or two such BL Lacs, 0235+164 being one candidate, these numbers are more characteristic of OVV/HPQ quasars than BL Lacs (cf. Mutel 1990).

OJ 287 is a source with a highly relativistic jet aligned to within a few degrees with our line of sight; it seems to represent the extreme end of FR II / quasar population. Do all BL Lacs, then, have small viewing angles? The answer seems to be a definite no.

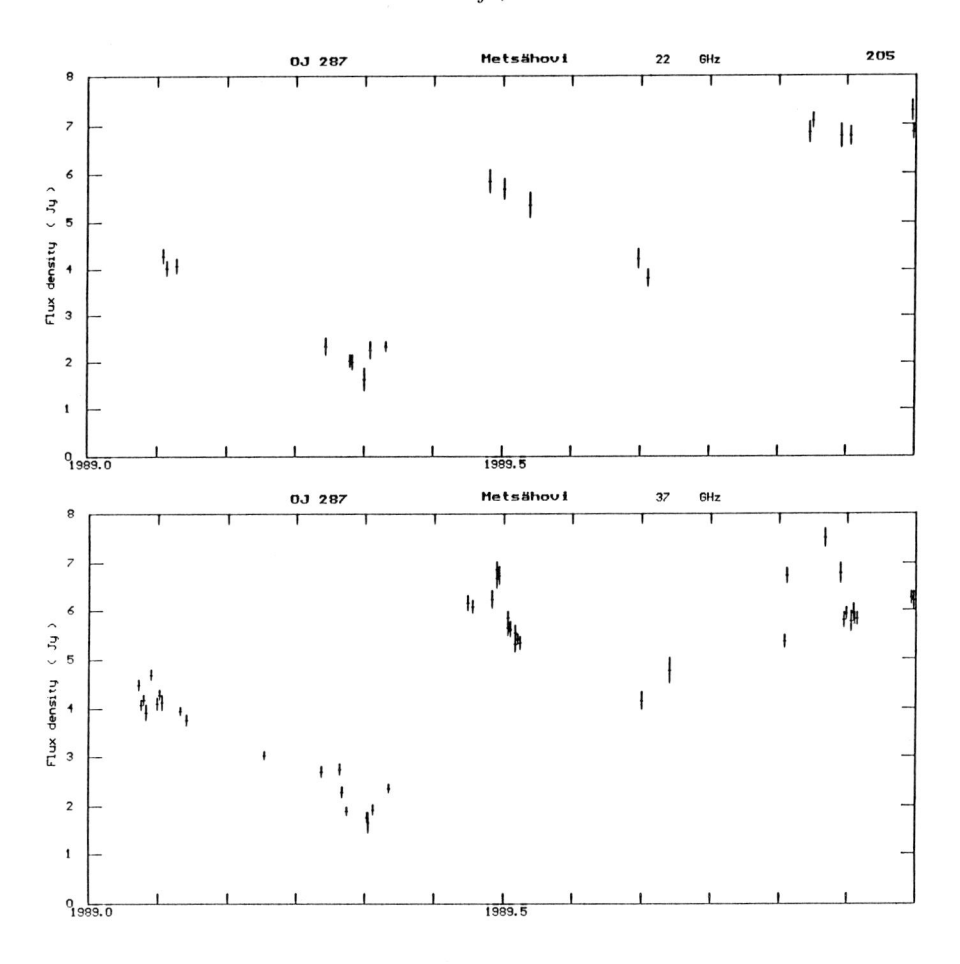

Figure 1. The 22 and 37 GHz flux variations of OJ 287 during 1989, measured with the Metsähovi radio telescope.

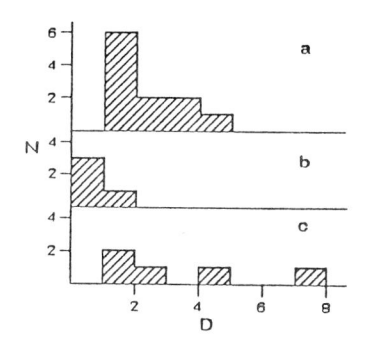

Figure 2. The distribution of the Doppler boosting factor $D = (T_b/10^{12}K)^{1/3}$ for quasars (*a*), nearby z < 0.3 BL Lacs (*b*) and distant z > 0.3 BL Lacs (*c*).

4. Large viewing angles and small Lorentz factors

Several BL Lacs are among the least variable sources in the Metsähovi monitoring sample, a fact hard to reconcile with supposedly closely aligned sources and consequent relativistic enhancement of variability. Calculating the brightness temperatures for the outbursts in these sources we find $T_b \approx 10^{12}$ K or even less, with at best marginal need for relativistic boosting to explain the variations. For example, even though BL Lac, the prototype itself, does have spectacular outbursts, the associated $D \approx 1$: the outbursts are only mildly relativistic.

Larger viewing angles and/or smaller intrinsic Lorentz factors than in quasars are also indicated by the general lack of extra radio variability in BL Lacs as compared to quasars (Altschuler 1982, 1983; Valtaoja et al. 1991b), and by their smaller superluminal component speeds (Mutel 1990). Finally, detailed model fits to the centimeter flux and polarization data of 1749+096 and BL Lac itself also require viewing angles of about 40° (Hughes et al. 1989, 1991). There seems to be no avoiding the fact that some BL Lacs are *not* oriented particularly close to the line of sight.

5. Two parent populations for BL Lac objects

There is strong evidence that at least some BL Lacs are highly relativistic sources with jets aligned to within a few degrees with the line of sight, and strong evidence that at least some are not so terribly relativistic sources seen at moderate viewing angles. Moreover, their variability properties span the whole observed range in AGN, and correlations between spectral and variability properties are lacking. How does one explain these contradictory results with a model for BL Lac objects?

We propose that one does not. *The whole BL Lac classification is a misnomer: the sources classified as "BL Lac objects" are a mixture of two physically different populations with different parent objects.* One population consists of sources which are closely aligned with our line of sight, and consequently also are strongly and rapidly variable; they are the extreme end of the quasar population, HPQ/OVVs with the smallest viewing angles. They thus belong to Barthel's unification scheme, and their parent population is the strong, distant FR II radio galaxies. The other population consists of sources which are also beamed towards us, but not as closely as the "distant" BL Lacs: they are moderately strong FR I radio galaxies, favorably oriented.

Burbidge and Hewitt (1989) have divided BL Lacs into two classes: "BL Lacs in galaxies", mainly nearby sources with z < 0.3, and "BL Lacs with QSO spectra", all with z > 0.3. In this conference they have further suggested that the sources in the second category should be reclassified as QSOs. This division, reached from different considerations, corresponds to our proposed reclassification, although we here use the neutral terms "nearby" and "distant" BL Lacs.

If BL Lacs really are a mixture of two populations, the wide range of variability characteristics and lack of correlations mentioned earlier is easier to understand. The dichotomy should also be directly visible in a number of properties. Being quasars, the "distant" BL Lacs with QSO spectra should be distributed in space in a similar manner as quasars, have similar (unbeamed) luminosities, parent galaxies, intrinsic Lorentz factors etc. as other quasars. The "nearby" BL Lacs in galaxies should have a quite different space distribution, smaller unbeamed luminosities and surrounding FR I parent galaxies. It would also be reasonable to expect that they have smaller Lorentz factors, since the parent galaxies are less energetic, and that they are not as closely beamed towards us as the very heavily selected subclass of distant BL Lacs. We consider some of these points.

a) Space distribution. The two-peaked redshift distribution of BL Lacs in the Kühr 1 Jy catalogue (Browne 1989; Stickel et al. 1991) results from the mixture of the two classes of BL Lacs. The nearby Lacs are distributed similarly to the FR I radio galaxies in the same flux-limited sample, and the distant BL Lacs have similar z distribution as flat spectrum quasars (cf. Figure 1 in Browne 1989). The two-peaked distribution is very hard to explain with a single population of objects, but is a natural consequence of the two populations hypothesis.

b) Radio variability, Lorentz factors, and boosting. The dichotomy in variability is best seen in the Doppler boosting factors, which we have derived from the observed variability timescales in our Metsähovi monitoring program. We have calculated D for the most rapid observed outbursts in quasars, nearby $z < 0.3$ BL Lacs and distant $z > 0.3$ BL Lacs. The results are shown in Figure 2. Nearby BL Lacs are hardly Doppler boosted at all: at most a little relativistic enhancement is needed to explain the observed variability. Thus, they do not need to be closely aligned with the line of sight, nor do their γ's need to be large. Distant BL Lacs have boosting factors as large or larger than quasars: the implication is that they have similar Lorentz factors as quasars, but still smaller viewing angles. (If their larger boosting factors were instead due to larger Lorentz factors, they should exhibit larger superluminal speeds than quasars, which is not the case.)

c) Unbeamed luminosity. Line luminosities are (one hopes) unbeamed. For nearby BL Lacs, the O III line luminosities are much smaller than those of quasars, but comparable to those of radio galaxies (Stickel et al. 1991). For distant BL Lacs the O III line cannot be observed, and Mg II line strengths must be used instead. Now one finds considerable overlap between the Mg II line luminosities of distant BL Lacs and quasars (Stickel et al. 1991). One must either invoke some unknown selection effects or accept that the line luminosities of nearby BL Lacs are similar to those of radio galaxies and the line luminosities of distant BL Lacs are similar to those of quasars. This agrees perfectly with the two populations hypothesis.

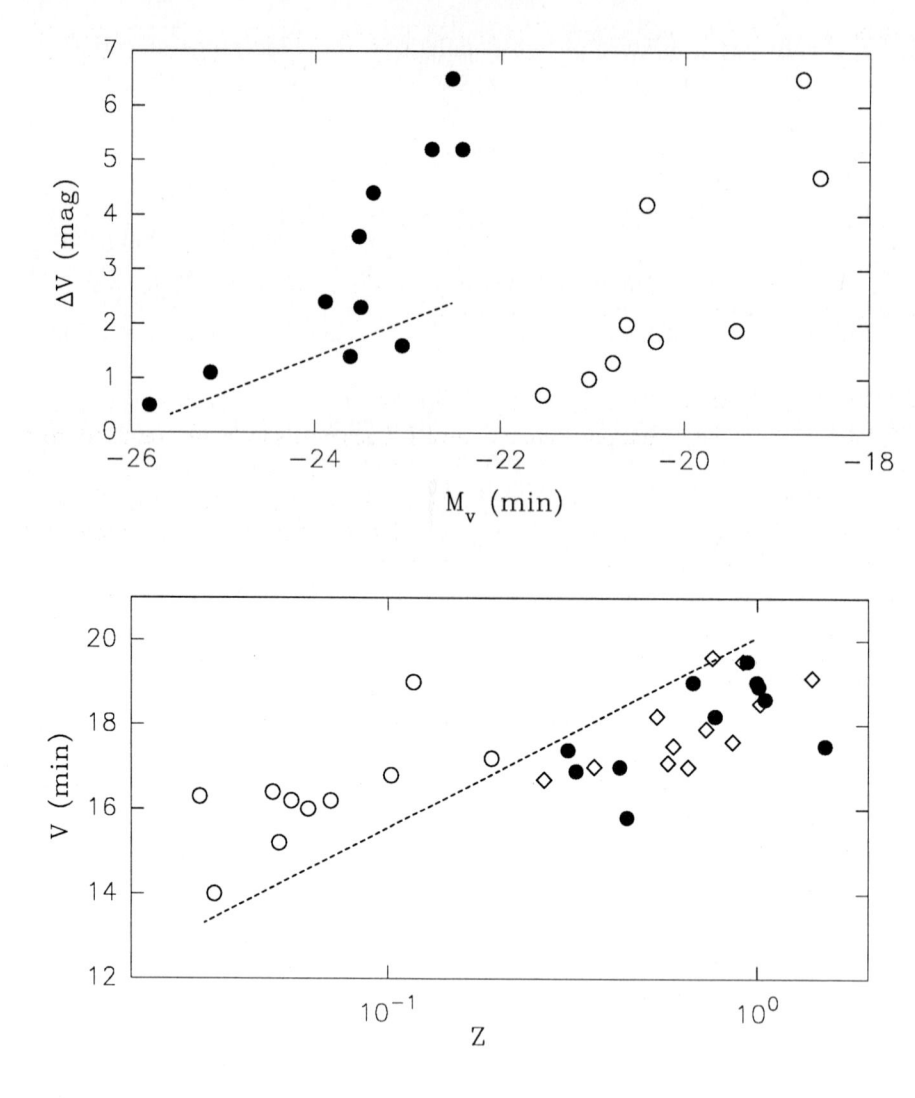

Figure 3.(upper) The observed variation amplitude ΔV (in magnitudes) versus the absolute magnitude M_V at minimum brightness for nearby ($z < 0.3$) BL Lacs (*open circles*) and for distant ($z > 0.3$) BL Lacs (*filled circles*). Data on ΔV and M_V collected from the literature (Teerikorpi 1991, in preparation). The dashed line shows the observed dependence for quasars. (*lower*) The observed minimum optical magnitude V versus redshift for nearby BL Lacs (*open circles*), distant BL Lacs (*filled circles*) and OVV quasars (*diamonds*). The dashed line shows the magnitude-redshift relation for first-ranked cluster galaxies. Data from Teerikorpi (1991).

d) Optical luminosity and variability. The Hubble diagrams for nearby and distant BL Lacs are very different (cf. Burbidge and Hewitt, these proceedings). If we plot the observed *minimum* optical magnitude vs. redshift, we find that distant BL Lacs approach the luminosities of first-ranked cluster galaxies, just as OVV quasars (Figure 3). In contrast, the nearby BL Lacs are one to two magnitudes dimmer. Another approach is to consider the observed variation amplitude ΔV vs. absolute magnitude at minimum brightness (Figure 3). For quasars there exists a dependence between these two quantities, although the reason for such dependence is as yet unknown (Teerikorpi 1991, in preparation). The same dependence seems to be the lower limit for the variation amplitude of distant BL Lacs. Once again, the nearby BL Lacs form a clearly separate group.

e) Number counts, luminosity functions and all that. Much of the work done to establish that FR I galaxies are the parent population of all BL Lacs needs to be re-examined from the two population hypothesis point of view. Practically all available data on the host galaxies, their absolute magnitudes, radio powers etc. relates only to the nearby BL Lacs. Number counts, derived luminosity functions and other statistics, which have been used to establish the parenthood of FR I's, typically have so large uncertainties that an extra factor of 2 or so (if only half of the BL Lacs belong to the FR I scenario) does not cause much trouble. X-ray selected samples seem to preferentially contain lower-z BL Lacs (cf. Padovani and Urry 1990) than radio selected samples; preliminary work (Padovani and Urry 1991) suggests that for the X-ray selected BL Lacs $\gamma \approx 3$, and for the radio selected BL Lacs $\gamma \approx 8$. An alternative to the assumption that the X-rays and the radio come from different parts of the jet (in which case the close coupling between X-rays and radio is difficult to understand) with different Lorentz factors is that two different populations have been selected. For nearby BL Lacs $\gamma \approx 3$ is in accordance with the variability data, and the average $\gamma \approx 8$ for distant BL Lacs is similar to the values derived for quasars.

6. Conclusions

The spectral and variability properties of BL Lacs are so similar to those of other AGN that a common intrinsic origin for their variability seems certain and gravitational lensing scenarios can be ruled out for *classes* of objects, although some *individual* sources may be lensed. While the properties of LPQ and HPQ quasars can be understood in terms of orientation effects, as proposed by Barthel (1989), BL Lacs cannot be easily fitted into a single scenario. We propose instead that the dichotomy in their properties can best be explained by assuming the existence of two different classes of "BL Lac" objects with different parent populations. Some BL Lacs, the distant, strongly variable sources, are quasars with jets seen directly end on, and would appear as FR II radio galaxies if viewed from the side; they should be reclassified as belonging to blazar-type quasars, with still more extreme properties due to still smaller viewing angles. Other BL Lacs are relatively nearby FR I radio galaxies beamed towards us; their viewing angles are larger and Lorentz factors smaller, and consequently they appear less variable. In both classes of sources the synchrotron component is enhanced by orientation effects, and they have therefore mistakenly been grouped together under the single name of BL Lacs.

References
Altschuler, D.R.: 1982, *Astron. J.* **87**, 387
Altschuler, D.R.: 1983, *Astron. J.* **88**, 16
Barthel, P.D.: 1989, *Astrophys. J.* **336**, 606
Blandford, R.D., Königl, A.: 1979, *Astrophys. J.* **232**, 34
Blandford, R.D., Rees, M.J.: 1978, in: *Pittsburgh Conference on BL Lac Objects*, ed. A.M. Wolfe, Pittsburgh, p. 328
Browne, I.W.A.: 1983, *M.N.R.A.S.* **204**, 23p
Browne, I.W.A.: 1989, in *BL Lac Objects*, eds. L. Maraschi, T. Maccacaro, M.-H. Ulrich, Springer, Berlin, p. 401
Burbidge, G., Hewitt, A.: 1989, in *BL Lac Objects*, eds. L. Maraschi, T. Maccacaro, M.-H. Ulrich, Springer, Berlin, p. 412
Cohen, M.H., Unwin, S.C.: 1984, in *VLBI and Compact Radio Sources*, eds. R. Fanti, K. Kellerman, G. Setti, Reidel, Dordrecht, p. 95
Cohen, M.H.: 1989, in *BL Lac Objects*, eds. L. Maraschi, T. Maccacaro, M.-H. Ulrich, Springer, Berlin, p. 13
Eckart, A., Hummel, C.A., Witzel, A.: 1989, *M.N.R.A.S.* **239**, 381
Gabuzda, D.C., Wardle, J.F.C., Roberts, D.H.: 1989, *Astrophys. J.* **336**, L59
Gear, W.K.: 1991, *Nature* **349**, 676
Hughes, P.A., Aller, H.D., Aller, M.F.: 1989, *Astrophys. J.* **341**, 68
Hughes, P.A., Aller, H.D., Aller, M.F.: 1991 (preprint)
Marscher, A.P., Gear, W.K.: 1985, *Astrophys. J.* **298**, 114
Mutel, R.L.: 1990, in *Parsec-scale Radio Jets*, eds. J.A. Zensus, T.J. Pearson, Cambridge University Press, Cambridge, p. 98
Ostriker, J.P., Vietri, M.: 1990, *Nature* **344**, 45
Padovani, P., Urry, C.M.: 1990, *Astrophys. J.* **356**, 75
Padovani, P., Urry, C.M.: 1991, *Astrophys. J.* (in press)
Salonen, E. *et al.*: 1987, *Astron. Astrophys. Suppl.* **70**, 409
Stickel, M., Fried, J.W., Kühr, H.: 1989, in *BL Lac Objects*, eds. L. Maraschi, T. Maccacaro, M.-H. Ulrich, Springer, Berlin, p. 64
Stickel, M. *et al.*: 1991 (preprint)
Teerikorpi, P: 1991 (in preparation)
Valtaoja, E.: 1990, in *Variability of Active Galaxies*, ed. S. Wagner, Springer, Berlin (in press)
Valtaoja, E. *et al.*: 1988, *Astron. Astrophys.* **203**, 1
Valtaoja, E. *et al.*: 1991a (preprint)
Valtaoja, E. *et al.*: 1991b (preprint)

Are highly evolved radio quiet QSO the "Parent" population of Lacertae-type objects?

B.V. KOMBERG

Institute of Cosmical Research, Moscow

1 INTRODUCTION

1. In some papers (see the references) it was suggested that QSO's may be formed only in the central regions of gas rich massive "mergers" which are formed in the process of merging of low-mass members in groups or in clusters of galaxies. In this case the ignition moment of the QSO's is determined by the time necessary for the formation of massive "mergers" in these systems. It is clear that for systems of larger angular momentum the "merger" would happen later and the QSO's would be observed at lower z. The value of the angular momentum of the system has to be related to the morphology of its members: for larger fraction of disk galaxies the angular momentum is larger. This leads to later formation of "mergers" with active nuclei, which have higher activity in radio region, because the probability of generation of stronger magnetic fields in the process of turbulent "dynamo" is higher.

2. Based on comparison of the observed properties of different types of AGN one can conclude that the parent population of the Lacertae-type objects may be the highly evolved radio quiet QSO's which are observed as double compact steep spectrum radio sources (db CSS RS), identified with E-galaxies. Younger radio quiet QSO's may be parent population for relatively quiet radio objects of the type HPQ (QSO's of high degree of optical polarization). Thus in this scenario (see Figure 1) the objects of the type BL Lac (LAC) and radio quiet HPQ are different phases of evolution of radio quiet QSO's with small angle between the line of sight and the radio jet axis. Radio loud HPQ's and compact radio sources with steep spectra (CSS RS) may be different evolution phases of radio loud QSO's with the jet axis directed to the observer (see Figure 1).

In accordance with our suggestions the following relations between volume densities of different types of AGN should be valid:

$$\rho_{HPQ}^{\text{radio quiet}} / \rho_{QSG} \approx \rho_{HPQ}^{\text{radio loud}} / \rho_{QSS}$$

$$\rho_{LAC}/\rho_{db\ CSS\ RS} \approx \rho_{CSS\ RS} / \rho_{db\ SS\ RS} \ ;$$

$$\rho_{HPQ}^{\text{radio quiet}} / \rho_{LAC} \approx \rho_{QSG} / \rho_{db\ CSS\ RS} \ ;$$

$$\rho_{HPQ}^{\text{radio loud}} / \rho_{CSS\ RS} \approx \rho_{QSS} / \rho_{db\ SS\ RS}$$

(db SS RS: steep spectrum extended double radio sources).

3. The observed weak cosmological evolution of LACs as compared with the strong evolution of QSO's could be understood in this scheme under assumption that more massive thick near nuclear discs are formed in the earlier epochs which are responsible for more collimated emission from active nuclei. Therefore actually the same cosmological evolution of LACs and QSOs would be masked in case of LACs by a more narrow radiation pattern at higher redshifts. This could be correct also for HPQ.

The high spatial density of LACs, as derived from X-ray surveys, could be connected to using too high a limit of the QSO luminosity ($M_V < -23$) in the calculation of the spatial density of optical QSOs. Decreasing of this limit by 1 magnitude increases more than 10 times the local spatial density of QSOs. It is possible also that QSO's have a wider emission pattern in X-rays than in optical. This naturally increases the probability that a line of sight falls inside a beam which, following our suggestions, leads to discovering LACs more often .

References

Byrd G.G., Sundelius B., Valtonen M., *Astron. Astrophys.* **171**, 16, 1987.
Heckman T.M., Smith E.P., Baum S.A., Van Breugel W.J.M., Miley G.K., Illingworth G.D., Bothun G.D., Balick B., *Astrophys. J.* **311**, 526, 1986.
Komberg B.V., *Astrofizika* (USSR) **20**, 73, 1984.
Komberg B.V., *Commun. Spec. Astrophys. Obs.* (USSR) **61**, 134, 1989.

$T < 10^7$ years $T > 10^8$ years

HPQ (radio quiet)   LAC
⟵   ⟵
$\Theta < 20°$ $\Theta < 20°$

$\Theta > 60°$ ↓ $\Theta > 60°$ ↓
radio quiet QSO   db CSS RS

HPQ (radio loud)   CSS RS
⟵   ⟵
$\Theta < 20°$ $\Theta < 20°$

$\Theta > 60°$ ↓ $\Theta > 60°$ ↓
radio loud QSO   db SS RS

evolution ⟶

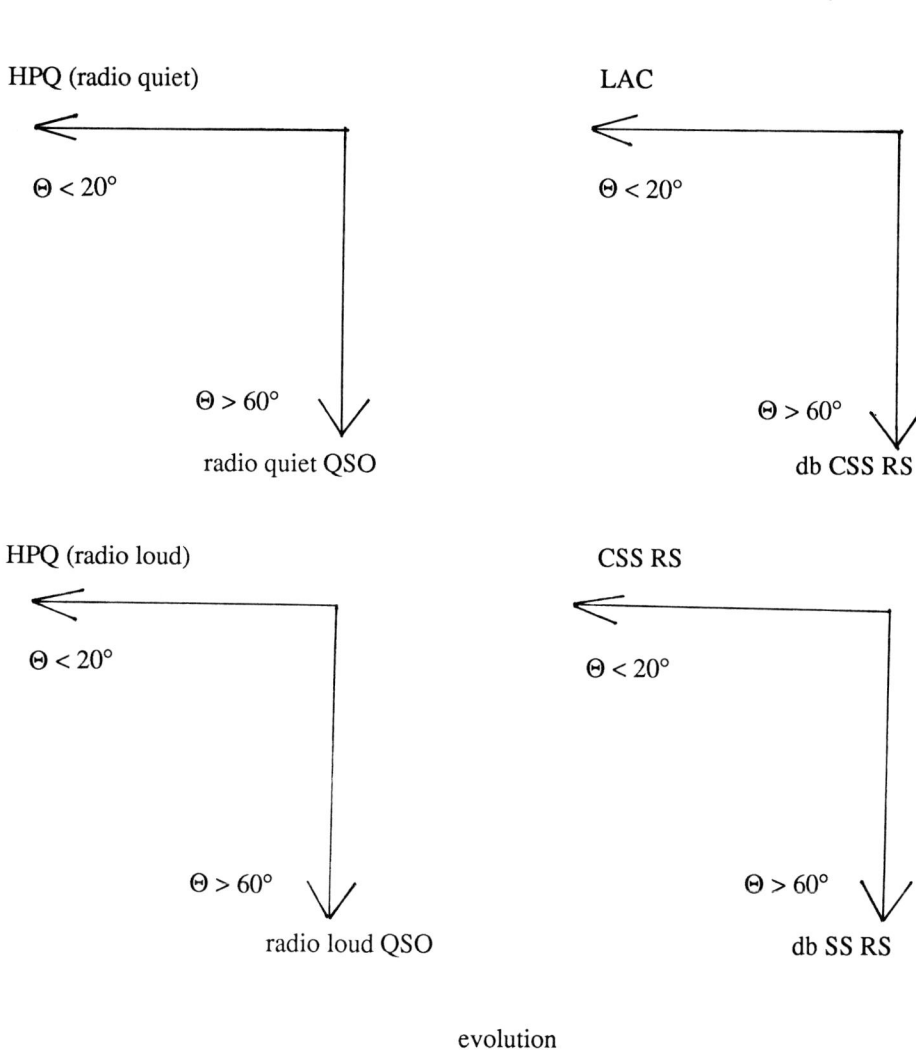

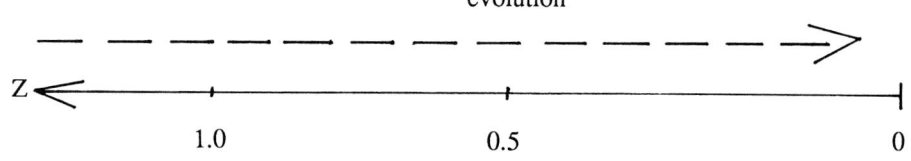

Z ⟵——|————|————————|
     1.0   0.5   0

Figure 1.

Blazars within an Evolutionary Unified Scheme

F. VAGNETTI[1], E. GIALLONGO[2], A. CAVALIERE[1]

[1] Astrofisica, Dipartimento di Fisica, II Università di Roma, I-00173 Roma, Italy
[2] Osservatorio Astronomico di Roma, I-00040 Monteporzio, Italy

1. INTRODUCTION

The set of blazars includes the BL Lac objects and many (virtually all, after Fugmann 1988) radio-loud flat-spectrum quasars. The two subsets differ mainly for the strength of the emission lines and for the apparent distribution in redshift. In fact, at variance with the quasars, BL Lac objects are classified as "lineless" active galactic nuclei and are mostly selected at low redshifts on the basis of the absorption lines of the underlying elliptical galaxies (Burbidge and Hewitt 1987). Several *transitional* objects show considerable variability in the equivalent width (EW) of the emission lines and are classified sometimes as (high redshift) BL Lac objects and sometimes as quasars (Antonucci et al. 1987).

Within the *unified scheme* of Barthel (1989), the blazars are objects relativistically beamed toward the observer, while misaligned objects would be observed as radio galaxies and/or steep-spectrum quasars. We try a connection between high and low redshift blazars in terms of an *evolutionary unified scheme* that adds the time dimension to the geometrical scheme (Vagnetti, Giallongo and Cavaliere 1991, VGC). As for the EW of the emission lines, we propose that in cosmological time strong-lined change into weak-lined objects, based on the following consideration. In the framework of luminosity evolution, two components of different behavior are envisaged for the optical band: the isotropic "thermal" component L_i typical of radio-quiet and steep-spectrum radio-loud quasars, that mainly excites the lines; and a beamed component L_b which dims more slowly to remain dominant at low redshifts, possibly swamping some broad emission lines in BL Lac Objects.

2. ANALYSIS

To disentangle the evolution of the beamed component, we first test the dependence of the optical to radio ratio as a function of the radio power L_R and of the look-back time $T(z)$, separately for flat-spectrum and steep-spectrum quasars. We perform a linear regression analysis, trying the simple form: $\log(L_{opt}/L_R) = A_L \log L_R + A_z T + A$.

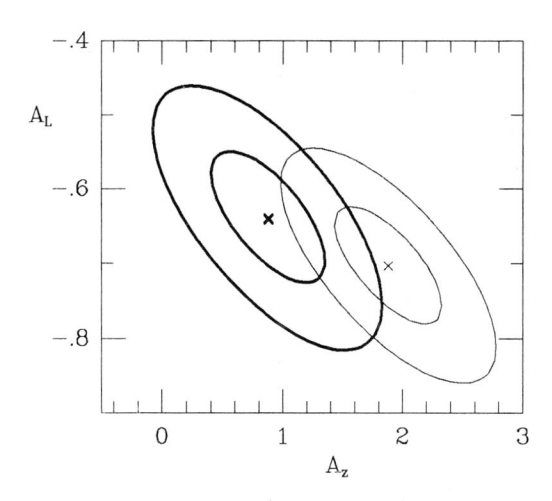

Fig. 1.
Confidence regions at 1 and 2 σ levels for each of the two interesting correlation parameters A_L and A_z, with luminosities in units of 10^{33} erg s^{-1} Hz^{-1} at 2.7 GHz and at 5550Å and $T(z)$ in units of H_o^{-1}; $\Omega_o = 0.2$, $H_o = 50$ Km s^{-1} Mpc^{-1}. Thin contours and best fits: steep-spectrum quasars ($\alpha \geq 0.2$); thick contours and best fits: flat-spectrum quasars ($\alpha < 0.2$). The separation is as sharp as could be expected considering the basic continuity of the two spectral sets (cf. VGC).

The data base and the results of the analysis are fully reported in VGC. The confidence regions of the two interesting parameters A_L and A_z are shown in Fig. 1 for a particular case. The distributions of $\langle L_{opt}/L_R \rangle$ of flat-spectrum and steep-spectrum quasars differ: the flat set is shifted toward slower evolution while the steep one is shifted toward faster evolution, with the best fits differing by $\gtrsim 2\sigma$.

We use these different correlations for the two radio spectral sets to deduce the evolutionary properties of the two luminosity functions in the optical band. Assuming pure luminosity evolutions both in the radio and optical bands, i.e. $L_R \propto \exp(k_R T)$ and $L_{opt} \propto \exp(k_{opt} T)$, the evolutionary parameters are related by: $k_{opt} = k_R(A_L + 1) + A_z \ln 10$; in turn, k_R is estimated through a maximum likelihood analysis of the radio luminosity function. We find $k_R = 5.2$ and $k_{opt} = 3.9$ for the flat-spectrum quasars, $k_R = 6.3$ and $k_{opt} = 6.2$ for the steep-spectrum quasars.

3. DISCUSSION

The slower evolution of flat-spectrum quasars in the optical can be understood within the *evolutionary unified scheme*, in terms of a slower evolution of the beamed component compared to the isotropic one. The observed evolution may be caused by the secular change of the intrinsic jet power L_j and/or of the Lorentz factor Γ: $L_b(z) \propto L_j(z)\Gamma^{2+\alpha}(z)$ (see e.g. Urry and Shafer 1984). Since the evolution in the radio band is stronger than in the optical, a spectral evolution is implied. This may be explained within the inhomogeneous synchrotron model of Marscher (1980), whose spectrum is broken in the IR at a frequency $\nu_2 \propto \Gamma$. An increase of Γ by a factor $\simeq 1.5$ from $z \simeq 2$ to $z \simeq 0$ causes a drift of the spectral break toward higher frequencies, accounting for the difference of k_R and k_{opt} given above. Circumstantial evidence for such increase of Γ in the cosmic time is discussed in VGC.

The peculiar characteristics of BL Lac Objects are understood as follows. The observed EW of the emission lines ought to be reduced by a factor $L_i/[L_i + L_b]$. Considering the evolution of the flat-spectrum quasars as an upper limit to the evolution of the beamed component L_b, the evolutions of the two components of the optical luminosity ought to differ by an amount $\Delta k \gtrsim 2.3$. This implies a reduction in EW by a factor $\sim 1/5$ from $z = 2$ to $z = 0$, enough to explain the absence of emission lines of intermediate strength such as OIII ($\lambda 5007$) and Hβ. The stronger Hα is more difficult to swamp, and in fact is often observed even in local BL Lac objects (VGC). Moreover, since the average $\langle L_{opt}/L_R \rangle$ and the density of local BL Lac objects agree with the extrapolations at the appropriate L_R and z found for flat-spectrum quasars from the analysis above, typical BL Lac objects can be considered as the low z end of the spectral and statistical evolution of flat-spectrum quasars.

The evolutionary unified scheme implies also a connection between the parent populations, namely between the radiogalaxies of different powers and morphologies. This might be understood in terms of an evolution towards an increasing richness of the environment (Prestage and Peacock 1988), and agrees with the observations of weak nuclear activity in some low redshift radio galaxies (De Robertis and Yee 1990).

REFERENCES

Antonucci, R. R. J., et al. 1987, AJ, 93, 785

Barthel, P. D. 1989, ApJ, 336, 606

Burbidge J., and Hewitt, A. 1987, AJ, 92, 1

De Robertis, M.M., and Yee, H.K.C. 1990, AJ, 100, 84

Fugmann, W. 1988, A&A, 205, 86

Marscher, A. P. 1980, ApJ, 235, 386

Prestage, R. M., and Peacock, J. A. 1988, MNRAS, 230, 131

Urry, C.M., and Shafer, R.A. 1984, ApJ, 280, 569

Vagnetti, F., Giallongo, E., and Cavaliere, A. 1991, ApJ, 368, 366 (VGC)

Variability of Nonthermal Continuum Emission in Blazars

ALAN P. MARSCHER

Boston University and Harvard-Smithsonian Center for Astrophysics

WALTER K. GEAR

Royal Observatory Edinburgh

JOHN P. TRAVIS

Boston University

Abstract Shock waves propagating down relativistic jets reproduce quite well the general characteristics of variability of the nonthermal continuum in blazars. Here we review the physical basis of the model and its application to variability observations. Plausible additions to the basic model, such as bending of the jet and turbulence, can explain a diverse range of flare properties within a single, unifying theory.

1 INTRODUCTION

Explaining variability in quasars and active galactic nuclei has proven to be quite a challenging task for theorists. Work on variability of the nonthermal continuum emission has focussed on two areas: reconciliation of the observations with general theoretical constraints and modeling of the spectral "light curves." The former category includes avoidance of the Compton problem and satisfaction of the energy requirements, while the latter includes polarization variability and the spectral evolution of individual components as imaged using VLBI. The past 13 years have seen the evolution of a unifying physical scheme that seems capable of solving all these difficulties: the relativistic jet model.

The basic outline of the relativistic jet model was first presented by Blandford and Rees (1978) and expanded upon by Blandford and Königl (1979) and Marscher (1980). The jet is assumed to be generated at some point R_0, beyond which it flows at a constant Lorentz factor Γ_j (speed $\beta_j c$), confined to a cone of constant opening half-angle ϕ. The radial coordinate R is measured from the apex of this cone. Although it is not necessary for the cone to have a constant opening angle, this simple geometry conforms quite well to the available VLBI observations. If one further assumes that the jet is free (confined solely by its own inertia), it resembles a conic section of a spherical wind. The density therefore falls off as R^{-2}, while the component of the magnetic field parallel to the jet $B_\parallel \propto R^{-2}$ and the component transverse to the jet $B_\perp \propto R^{-1}$. The region near $R = R_0$ contains the highest density and magnetic field strength of the conic portion of the jet, and therefore appears as a stationary, bright feature even though plasma streams through it at a relativistic speed. In the model, this region is identified with the VLBI core. Since it is not clear whether the portion of the jet at radii $R < R_0$ is directly observable at any waveband, one can at present only speculate as to the origin and source of collimation of the jet.

The relativistic jet model relieves the major theoretical difficulties associated with blazars. Relativistic beaming of the emission toward the observer causes one to miscalculate the luminosity and photon density of the source. In order to derive these quantities, one needs to multiply the observed integrated flux F by $4\pi D_\ell^2$ to obtain the luminosity at the source, where D_ℓ is the luminosity distance. However, if relativistic beaming occurs, the radiation is not emitted isotropically, and the timescales are contracted in the observer's frame. Therefore, one must multiply F by a factor δ^{-4} for a variable emission region moving within the jet, where $\delta \equiv [\Gamma(1 - \beta\cos\theta)]^{-1}$ and θ is the angle between the jet axis and the line of sight. The Doppler factor δ ranges from $(1+\beta)\Gamma$ for $\theta = 0$, to Γ for $\theta = \sin^{-1}(1/\Gamma)$, to Γ^{-1} for $\theta = \pi/2$. Any given flux-limited survey tends to be dominated by beamed sources, so one has no problem with statistics as long as there is a sufficient supply of "parent" objects that have low apparent luminosities because most of their radiation is beamed away from us. Since the actual luminosities of sources with jets beamed toward us are much lower than previously inferred, the enormous energy requirements and "Compton catastrophe" are greatly diminished.

Since we have yet to understand how a relativistic jet is focussed and accelerated, we cannot yet obtain a firm theoretical upper limit to the Lorentz factor of the jet. However, if the jet is accelerated near the putative massive black hole, Compton drag should limit the Lorentz factor to $\Gamma_j \lesssim 10$ (Phinney 1987). There are other arguments that lead to a similar constraint (see also Phinney 1987). One can use the measured angular size, turnover frequency, and flux density to derive the magnetic field and energy density in relativistic electrons of a source. When this is done for a typical compact component in a blazar without correcting for relativistic beaming, one finds that the energetics are dominated by the relativistic electrons (see Burbidge, Jones, and O'Dell 1974; Marscher and Broderick 1985). If one assumes that the component is moving relativistically with a Doppler factor δ, this imbalance is lessened and the total energy requirement is correspondingly diminished. But once the Lorentz factor exceeds about 10, the magnetic field begins to dominate and the energy requirements become extremely high. The other consequence of very high Lorentz factors is that the more highly beamed blazars are, the greater is the required number of unfavorably beamed parent objects. This also restricts the *typical* value of Γ to be $\lesssim 10$, although *a few* extraordinary objects could have higher values. The "critical" value $\Gamma \sim 10$ is a rough one, but the arguments are general.

Blandford and Königl (1979) proposed that disturbances in the jet flow would cause

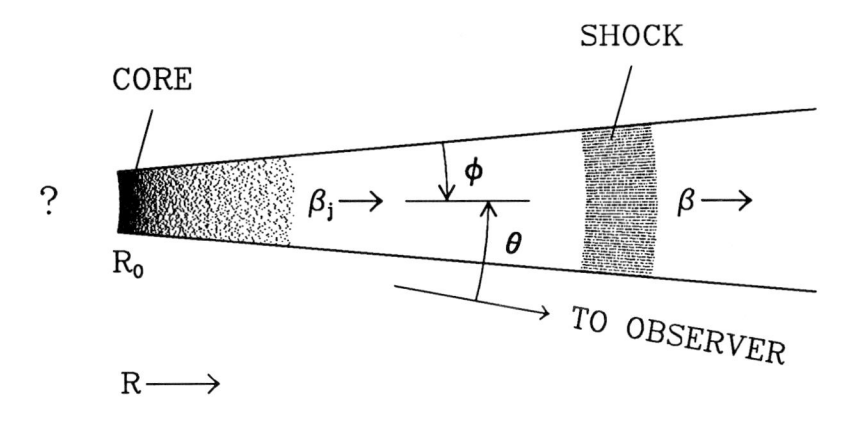

Figure 1 Basic geometry of the relativistic jet model as viewed from a direction transverse to the axis. The vertical scale is expanded by a factor of a few.

shocks to form and propagate down the jet. They associated these shocks with the knots observed to move at apparent superluminal speeds in VLBI images of compact extragalactic jets. Jones (1982), Marscher and Gear (1985), and Hughes, Aller, and Aller (1985) subsequently showed that such shocks could reproduce quite well many of the detailed characteristics of the structural, spectral, and polarization variability of blazars. Figure 1 shows the basic geometry of the model. In many ways, the shock-in-jet model is a modified, physical version of the early, simple adiabatic expansion model (van der Laan 1966; Pauliny-Toth and Kellermann 1966). Indeed, the shock expands homologously as it propagates down the jet, but the evolution of the shock is similar to that of a fluid cell in a wind rather than that of a uniformly expanding ball of gas.

Probably the greatest success of the detailed shock-in-jet models is their reproduction of complex spectral and polarization variability observations while obeying the constraints of a specific physical scheme. This success gives us courage to advance the model to the next step of complexity: numerical modeling and the addition of physically reasonable complications. Here we summarize the general characteristics of spectral variability in compact sources, review the basic shock-in-jet model as proposed by Marscher and Gear (1985, hereafter abbreviated as M&G), discuss the effects of bending of the jet and turbulence in the jet plasma, and provide a guide to the range of observed behavior that one might expect from variable sources.

2 GENERAL PROPERTIES OF BLAZAR VARIABILITY

The vast assortment of observations of spectral variability of blazars have revealed a diverse range in characteristic behavior. There are many similarities and differences both among sources and among flares within the same source. In the process of developing a model to explain such a diversity, one needs to oversimplify in order to make progress. We therefore list the most salient properties of spectral variability that we feel best exemplify the nature of the phenomenon, while recognizing that there are exceptions to these patterns. Reproducing these characteristics should be the goal of any respectable theory.

(i) Flares usually begin at high frequencies, then propagate to low frequencies. Some flares, however, develop simultaneously over a wide range of frequencies before propagating to lower frequencies, and still others do not seem to move to lower frequencies at all. Some events begin at optical-infrared wavelengths, while others begin at submillimeter, millimeter, or even centimeter wavelengths.

(ii) The flaring emission usually rises abruptly, propagates to lower frequencies while maintaining essentially a constant amplitude, and then dies, sometimes gradually (as in AO 0235+164; O'Dell et al. 1988) and sometimes rapidly (as in 1156+295; McHardy et al. 1990).

(iii) Superposed on the flares, and sometimes on the "quiescent state," are mini-flares and/or rapid flickering; viz., the light curves are not smooth.

(iv) The timescale of variability is shorter at the higher frequencies. The polarization behavior also tends to be much more erratic at these frequencies. An exception to this rule is the rapid flickering observed at the few percent level at centimeter wavelengths by the German group (Quirrenbach et al. 1989; see also the contributions to these proceedings by Krichbaum and Wagner), which occurs on the same timescales as found in the optical.

(v) The polarization of the flares and VLBI knots tends to be parallel to the jet, indicating that the magnetic field associated with the flare is transverse to the jet axis. The VLBI result holds for BL Lac objects but not quasars, perhaps because polarization in the quasars suffers from strong opacity effects at the wavelength of observation (6 cm) (as suggested by H. Aller at the conference).

(vi) To first order, the multi-waveband spectra connect smoothly from the radio to the infrared. However, when observed with sufficient frequency coverage, the spectra often show double peaks (Brown et al. 1989).

3 THE BASIC SHOCK-IN-JET MODEL

We sketched out the generic properties of the relativistic jet model in section 1 above. In order to model particular flares in a given source, it is necessary to specify the details somewhat further. This is necessary to reduce the number of free parameters.

The equation of state of the jet plasma must be chosen. The most obvious is to assume that the plasma is completely relativistic, so that the adiabatic index is 4/3 and the pressure $p = e/3$, where e is the energy density. However, an ultrarelativistic gas will convert its internal energy into bulk motion, thereby accelerating the flow. While this may not conflict with observations of regions very close to the core (see the contribution to these proceedings by Zensus), the velocity of a given knot is found to be constant once it is well separated from the core. In our version of the model, we assume that the protons are essentially nonrelativistic while the electrons are ultrarelativistic (although both species may have the same mean internal energy as long as it is much less than the proton rest-mass energy). The corresponding "semirelativistic" adiabatic index is 13/9. Unfortunately, the gas dynamic equations are then much more complex.

One must also decide what to do about the acceleration of relativistic electrons. Königl (1981) assumes that the electrons are constantly reaccelerated so as to compensate for adiabatic losses. M&G and Hughes *et al.* (1985) assume that there is no compensation for these losses. In dealing with shocks, the latter two groups assume that the electrons are heated solely by compression as they pass through the shock front. As Björnsson (these proceedings) has emphasized, a relativistic shock with a magnetic field nearly parallel to the shock front may not be able to accelerate electrons through the first-order Fermi process. Hence, compressional heating, which is the most basic acceleration mechanism, may in fact dominate over other processes. We hasten to mention that no detailed models have demonstrated a need for particle acceleration other than through compression, save for the obvious requirement that the electrons be initially relativistic (with the observed power-law energy distribution) at the core radius R_0.

Shocks are created when a significant disturbance occurs in the otherwise steady jet flow. (See Marscher 1990 for a definition of "significant disturbance.") The basic shock structure (pressure, density, or velocity vs. distance R along the jet) is a shock pair, with a forward-facing shock sweeping up the previously undisturbed jet medium and a reverse shock (moving backward relative to the forward shock but moving forward relative to the observer) that decelerates the enhanced "driver gas" flow propagating from inner radii. The strength of the reverse shock depends on how strong and prolonged the disturbance is, with more impulsive disturbances resulting in weaker reverse shocks. A contact discontinuity separating the shocked ambient jet plasma from the enhanced flow is situated between the forward and reverse shocks. For non-prolonged disturbances, a rarefaction occurs to the low-R side of the reverse shock. Figure 2 shows the radial dependence of the density, pressure, and velocity of a strongly driven shock. The numerical calculations that generated this figure were based on a nonrelativistic code. We are currently working on the development of a relativistic code. In the absence of actual relativistic gas dynamical calculations, we use square-wave disturbances to approximate the density, magnetic field, and velocity structure of a shock. This has the advantage of leading to analytic results that can be used as first-order guides in interpreting the observations.

Under the square-wave approximation, there are a number of free parameters (e.g., shock Lorentz factor and compression ratio) that one can use to match the model spectra and polarization at any chosen epoch. The time evolution of the source, however, is highly constrained if one adopts a particular model for the underlying jet. The assumption that the jet is adiabatic leads to the following relativistic electron distribution (taken to be a power law in energy E):

$$N(E, R) = K(R)E^{-s}, \qquad [E_1(R) < E < E_2(R)], \qquad (1)$$

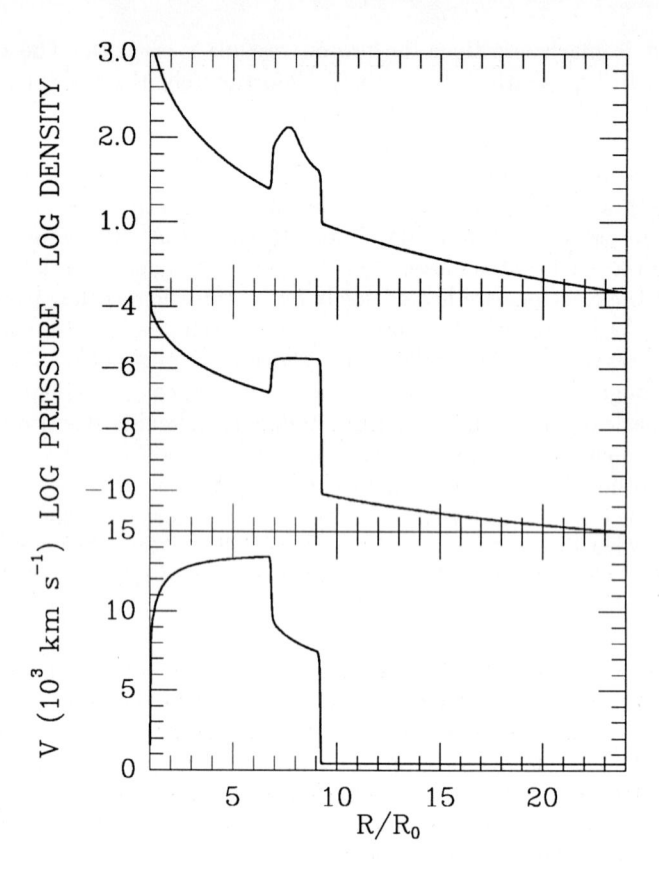

Figure 2 Density, pressure, and velocity profiles of a nonrelativistic shock wave in a conical jet. The shock is driven by a strong, prolonged disturbance. The forward shock is evident at $R/R_0 \approx 9.2$ and the reverse shock at $R/R_0 \approx 6.8$. From Balser and Marscher, research in progress.

with $K(R) \propto R^{-2(s+2)/3}$ and $E_i(R) \propto R^{-2/3}$. The magnetic field falls off as $B(R) \propto R^{-a}$, with $a = 1$ behind the shock front unless the ambient jet field is strongly parallel to the jet axis, in which case $a = 2$. Since the shock amplifies the component of the field that is parallel to the shock front by a factor equal to the compression ratio, one expects the polarization electric vector of the emission from the shock to be perpendicular to the shock front, *i.e.* parallel to the jet axis if the shock is a transverse one.

After the electrons are heated at the shock front, they suffer synchrotron and, if the photon density is high, inverse Compton losses. The highest energy electrons, which radiate at the highest frequencies, lose most of their energy close to the shock front, while those that radiate at lower frequencies survive much longer. During the early development of a major flare, the high-frequency emission is therefore confined to a

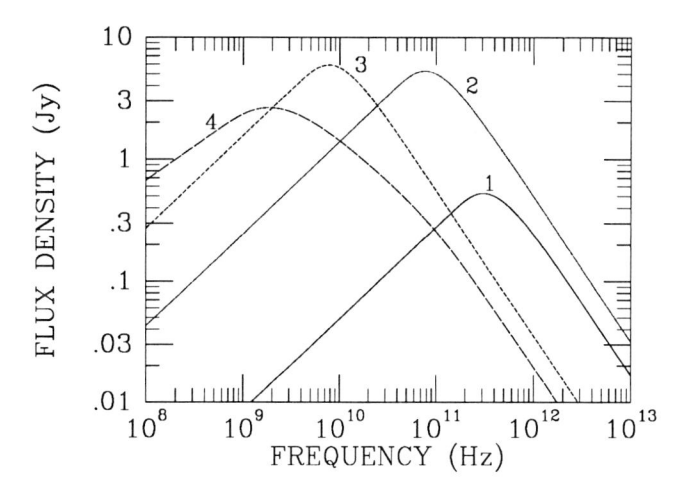

Figure 3 Schematic spectral evolution of a flare arising from a shock propagating down a jet. The outburst progresses from the Compton-loss stage (1 to 2) to the synchrotron-loss stage (2 to 3) to the adiabatic-loss stage (3 to 4).

thin region near the shock front, with the thickness of this region increasing toward lower frequencies. As the shock evolves, the photon density and magnetic field (which causes radiative losses) decrease, and the thickness of the emission region increases. The emission region has a maximum effective thickness given by the extent of the entire shock structure. Therefore, at sufficiently low frequencies or at later times at the higher frequencies, radiative losses are unimportant, the effective thickness equals the maximum, and the shock emission evolves adiabatically. The work of Hughes *et al.* (1985) concentrates on flares that are in this adiabatic-loss phase.

The flare begins with a sharp rise in the spectrum, with the turnover frequency ν_m decreasing slightly as the flux density rises simultaneously at all optically thin frequencies. This abrupt rise corresponds to either the initial formation of the shock or the progressive weakening of Compton losses as the shock expands. Synchrotron losses soon dominate, and the entire spectrum shifts toward lower frequencies while the flux density at the turnover remains roughly constant. Once the adiabatic-loss phase begins, the flare subsides such that the low-frequency side ($\nu < \nu_m$) "slides" downward in the sense that the flux density S_ν remains roughly constant until the turnover passes by. The entire evolution is given in Figure 3.

Quantitatively, the evolution proceeds according to the formulas of M&G [see also Marscher 1990; note that the minus sign in the exponent of expression (8a) of that paper should be deleted]. As was pointed out by Björnsson at the conference, one cannot ignore second and possibly higher order Compton losses during the Compton-loss stage. The rise of the spectrum will therefore be even more abrupt than specified by M&G.

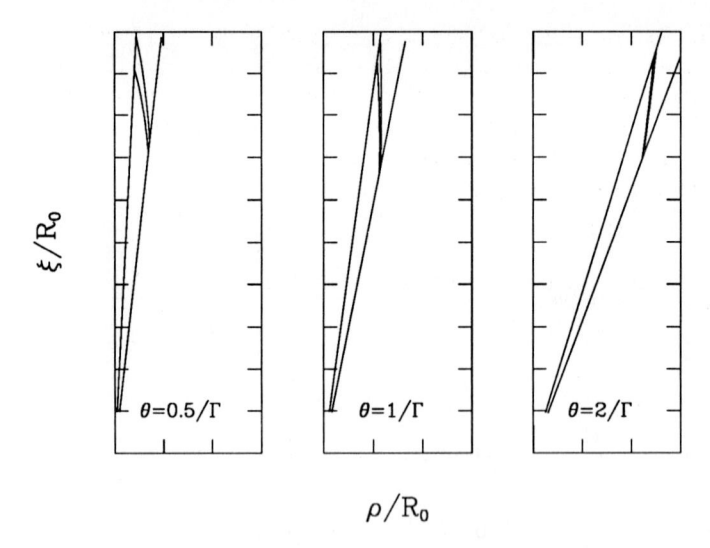

Figure 4 Cross-section of a shock moving down a relativistic jet for three angles of inclination of the jet axis relative to the line of sight. The Lorentz factor of the shock is $\Gamma = 7$. The figure is drawn to scale, with ξ as the coordinate along the line-of-sight direction and ρ as the coordinate transverse to the line of sight. The observer lies at $\xi = \infty$. The intrinsic opening half-angle of the jet is $1°.5$.

In order to obtain analytic results, M&G performed their calculations for a plane shock viewed head-on. The proportionalities they derive are, however, much more general than this. Figure 4 illustrates the geometry of a cross-section of the shock in observer's coordinates for three angles of inclination θ of the line-of-sight relative to the jet axis. There are three ranges of viewing angles that lead to qualitatively different observed orientations of the shock: (i) $\theta < \sin^{-1}(1/\Gamma) - \phi$, (ii) $\sin^{-1}(1/\Gamma) - \phi < \theta < \sin^{-1}(1/\Gamma) + \phi$, and (iii) $\theta > \sin^{-1}(1/\Gamma) + \phi$. For case (i), the line of sight penetrates first the front, then the back of the shock. For case (iii), the reverse is true! For case (ii) most of the shock is viewed nearly sideways. These peculiar properties reflect the well known rotation of an object moving relativistically at an angle to the line of sight. The boundary of the shock has its greatest velocity along the $\hat{\rho}$ direction (transverse to the line of sight) at the angle $\sin^{-1}(1/\Gamma)$ or, if this lies outside the boundaries of the jet, the angle closest to this. This fact leads directly to the above conclusions regarding the orientation of the shock in the observer's frame. Except for case (ii), the line-of-sight depth of the shock is directly proportional to its thickness, as assumed in the calculations of M&G. For those sources observed near the angle of maximum superluminal motion (case ii), the shock is viewed sideways, which alters the expression for the self-absorption turnover frequency ν_m.

Probably the most readily observed property of an outburst is the shape and evolution of its spectrum as characterized by the turnover frequency ν_m and the flux density at turnover S_m. The evolution during the first phase (either the Compton-loss stage or the initial formation of the shock) cannot at present be calculated without detailed numerical modeling. Nevertheless, we can state that the spectrum will rise sharply without much evolution of the turnover frequency. For the phases when synchrotron losses and, later, adiabatic expansion determine the thickness of the emission region at any given frequency, we repeat the results of M&G for viewing angle ranges (i) and (iii) and present new expressions for case (ii).

Synchrotron-loss stage:

$$S_m \propto \nu_m^{[(2s-5)(2+3a)]/[4(s+2)+3a(s-1)]} \quad \text{(cases i \& iii)} \quad (2a)$$
$$\propto \nu_m^{-[4(3a+2)-5s]/[2(2s+1)+3a(s+2)]} \quad \text{(case ii)} \quad (2b)$$

Adiabatic-loss stage:

$$S_m \propto \nu_m^{[(4s-19)+3a(2s+3)]/[2(2s+1)+3a(s+2)]} \quad (3)$$

We parametrize the evolution of the spectrum by the expression $S_m \propto \nu_m^y$. During the synchrotron-loss phase, y ranges from -0.65 for $s = 1.5$ to 0.19 for $s = 3$ (with $y = 0$ for $s = 2.5$) over the viewing angle ranges (i) and (iii), and from -0.68 for $s = 1.5$ to -0.17 for $s = 3$ over viewing angle range (ii). Here we consider only shocks whose dominant magnetic field direction is parallel to the shock front, hence $a = 1$. The optically thin (high frequency) spectrum during the Compton- and synchrotron-loss stages has a slope of $-s/2$, steeper by $1/2$ than the usual $-(s-1)/2$. Since the slopes of high frequency spectra of variable sources tend to fall between -1 and -1.5, s typically lies between 2 and 3. Over this range, the spectral evolution is very flat (small y) for cases (i) and (iii), while the spectrum rises as it moves to lower frequencies for case (ii). It is therefore possible, at least in principle, to identify sources with viewing angles near $\sin^{-1}(1/\Gamma)$ by measuring the high-frequency spectral index to determine s and the evolution of the spectrum to determine y.

The spectrum of the emission from a single shock during the synchrotron-loss stage is essentially that of a uniform self-absorbed synchrotron source. The spectrum broadens out near the turnover as synchrotron losses subside and the transition is made to the adiabatic-loss stage. In addition, if a reverse shock forms, the flare spectrum is a composite of the forward and reverse shock spectra. In the observer's frame, the reverse shock is at a considerably smaller radius R than is the forward shock. It also, however, has a lower Lorentz factor, and hence its emission is less beamed. The precise shape of the composite spectrum therefore depends on the separation between the forward and reverse shocks and on the observer's viewing angle. As an illustration, Figure 5 shows the spectrum of a flare with $a = 1$, $s = 2.5$, and viewing angle $\theta \approx 0.5 \sin^{-1}(1/\Gamma)$. The distance x_s between the forward and reverse shocks is set equal to $0.15\Gamma^{-1}R$, which is roughly the value observed in strong solar wind shock pairs, multiplied by the relativistic length contraction factor. The compression ratio η of the reverse shock is assumed to be 1.5 times that of the forward shock. (We use the definition of η as the factor by which the density is enhanced by

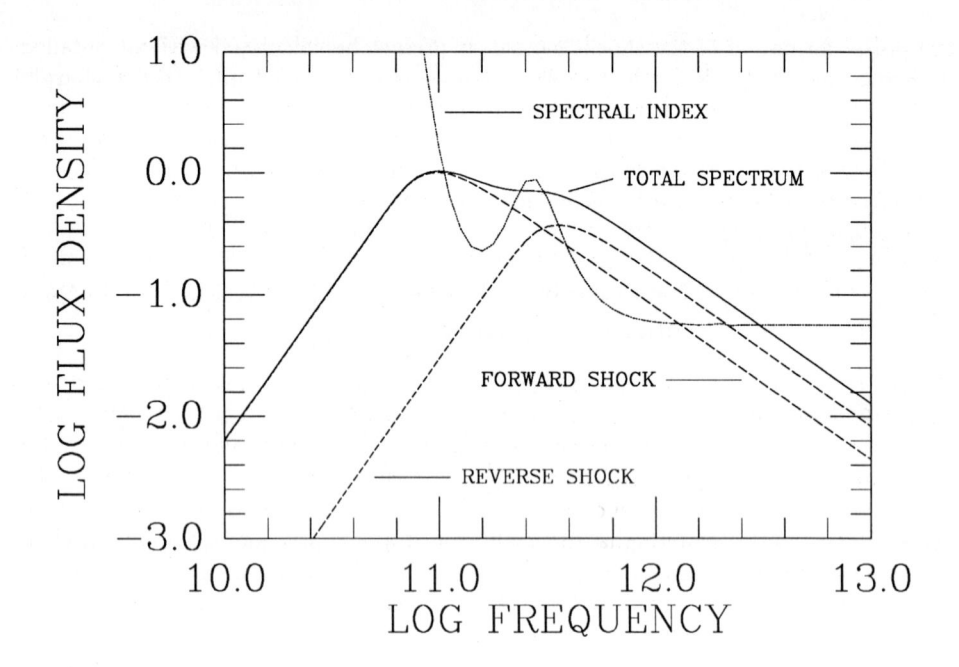

Figure 5 Spectrum of a shock pair caused by a prolonged disturbance in a relativistic jet during the synchrotron-loss stage, as described in the text. Although the units of flux density and frequency in the plot are arbitrary, the numbers chosen correspond roughly to Janskys and Hertz.

the shock over its undisturbed value, which is the inverse of the definition employed by Hughes, Aller, and Aller.) This value of η is chosen so as to give a flat spectrum above the turnover, as observed (Valtaoja et al. 1988). Of course, one would like the values of x_s and η to be determined by gas dynamics; this is in our future plans.

One could possibly explain the double-humped multiwavelength spectra observed by Brown et al. (1989) as the superposition of forward and reverse shocks. However, since the spectra were taken during more quiescent phases, another explanation seems likely: the higher frequency peak represents the spectral turnover of the steady "core" emission at the narrow end of the jet, while the lower frequency peak corresponds to a decaying flare that is propagating toward centimeter wavelengths.

4 BENDING RELATIVISTIC JETS

VLBI and VLA observations of jets show that they often bend considerably. For relativistic jets, the small angle of the jet axis to the line-of-sight amplifies the bending angle. Even so, bends of at least several degrees are required by the observations.

Although it is not necessarily true that the jet axis itself bends, this is the most straightforward interpretation of the observations. Hardee (1990) has shown that macroscopic fluid instabilities can cause a jet to bend. Curvature in a relativistic jet causes variations in Doppler boosting and observed motion transverse to the line of sight. These effects lead to systematic changes in the observed brightness and apparent superluminal motion of a compact source.

The observational consequences of variations in the inclination angle of the jet θ are described by Marscher (1990). The Doppler boosting of the flux density can change by a factor of 2 or so even if the jet bends by only a few degrees. This is accompanied by a noticeable change in the apparent superluminal speed if θ changes from $\lesssim 15°/\Gamma$ to larger angles (or *vice versa*), or by only a modest change in v_{app} if θ remains in the $30°/\Gamma$ to $110°/\Gamma$ range.

It is straightforward to calculate (to first order) the spectral evolution of a thin shock that follows the bend in a jet. [In order to avoid the complication of significantly different Doppler factors for the near and far side of the shock emission region, the thickness of the emission region must obey the relation $x \ll \mathcal{R}(1-\beta\cos\theta)$, where $\mathcal{R}$ is the distance over which the bend occurs. Otherwise, one must perform the calculation numerically.] Let us assume that the change in Doppler factor with distance R along the jet can be approximated as a power law: $\delta \propto R^d$. Then, for example, in the synchrotron-loss phase (see expressions 2), one can derive:

$$\nu_m \propto R^{[3d(s+3)-3a(s-1)-4(s+2)]/[3(s+5)]} \tag{4a}$$
$$\propto R^{-0.3} \quad (d=1, s=2.5, a=1) \tag{4b}$$

$$S_m \propto \nu_m^{[3d(3s+10)-(3a+2)(2s-5)]/[3d(s+3)-3a(s-1)-4(s+2)]} \tag{5a}$$
$$\propto \nu_m^{1.9} \quad (d=-2, s=2.5, a=1) \tag{5b}$$
$$\propto \nu_m^{1.3} \quad (d=-1, s=2.5, a=1) \tag{5c}$$
$$\propto \nu_m^{0} \quad (d=0, s=2.5, a=1) \tag{5d}$$
$$\propto \nu_m^{-8.8} \quad (d=1, s=2.5, a=1) \tag{5e}$$
$$\propto \nu_m^{10} \quad (d=2, s=2.5, a=1) \tag{5f}$$

Marscher(1990) gives similar expressions for the adiabatic-loss stage. The exponent approaches the limit $(3s+10)/(s+3)$ as $d \to \infty$. Expressions (4-5) show that a wide range of spectral evolution is possible if the jet is bent. Note that, as in the Compton-loss stage of a shock in a straight jet, it is possible for the turnover frequency to decrease as the flux density increases. This behavior is difficult for other expanding source models to reproduce (see M&G for a discussion), but is indeed observed in some sources. The bending jet model can explain this phenomenon even if radiative losses of the relativistic electrons are negligible. This may be relevant to centimeter-wave outbursts such as that observed in 1979 in the source 1921-29 (Dent and Balonek 1980).

Marscher *et al.* (1991) have interpreted the peculiar superluminal quasar 4C 39.25 in terms of a jet that bends by about 4° relative to its preferred direction $\theta \approx 8°$.

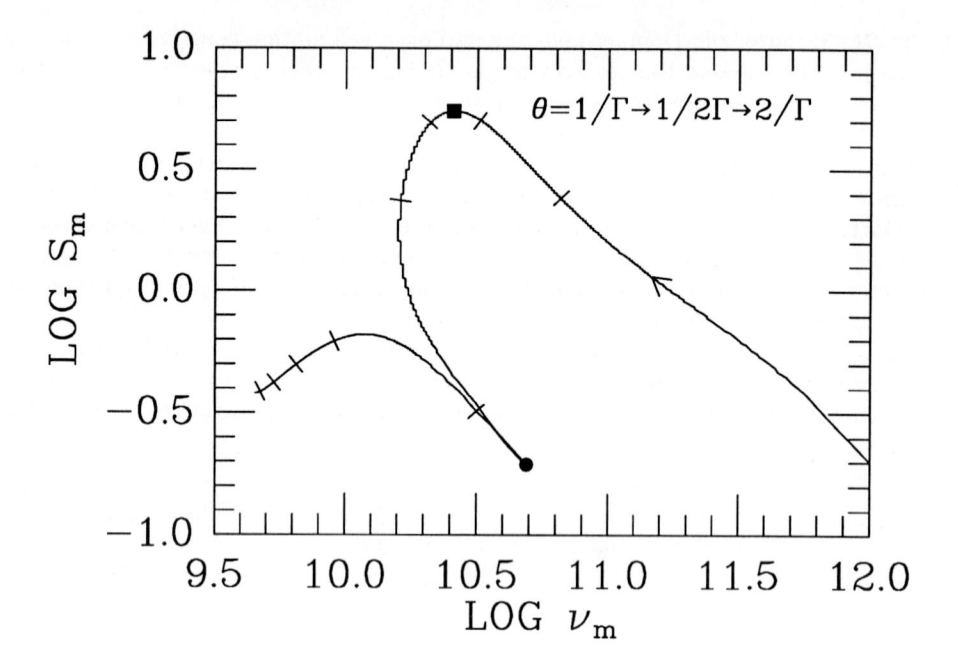

Figure 6 Spectral evolution of a shock propagating down a curved jet during the synchrotron-loss stage. The arrow indicates increasing time, with the hash marks along the curve marking equally spaced time intervals. In this simulation, the orientation angle θ varies sinusoidally with distance R along the jet. The filled square denotes the minimum orientation angle ($\theta = 0.5/\Gamma$), while the beginning of the curve and the filled circle both correspond to $\theta = 1/\Gamma$. Although the units of flux density and frequency in the plot are arbitrary, the numbers chosen correspond roughly to Janskys and Hertz.

The model interprets the stationary features as points in the jet where $\theta \approx 4°$, at which point the Doppler beaming is enhanced but the apparent transverse motion is greatly diminished. The evolution of the spectrum follows that expected, roughly in accord with expressions (4) and (5) above.

In order to provide a concrete example of the evolution of the spectrum of a shock in a bent jet, we present in Figure 6 the numerically calculated S_m vs. ν_m curve for a jet whose orientation relative to the line of sight θ follows a sinusoidal curve (through half a cycle). The spectrum rises sharply while moving to lower frequencies, but the decline is even more rapid. The movement toward a higher turnover frequency as θ approaches $1/\Gamma$ is caused by the higher opacity as the shock is observed from the side (cf. Fig. 4). The evolution of the spectrum in this particular simulation is quite similar to that of the 1984-86 flare of 1156+295 analyzed by McHardy et al. (1990).

5 TURBULENCE

We expect the Reynolds number in a relativistic jet to be extremely high, so the plasma in the jet should be quite turbulent. The effect that this has on the time variability depends on the amplitude and power spectrum of the turbulence as well as the thickness of the shock emission region. We have performed numerical simulations of a square-wave disturbance propagating down a jet whose plasma is hydromagnetically turbulent. Artificial compressible turbulence is generated through Fourier transform techniques. As the disturbance propagates down the jet, it brightens at sites where it encounters density and/or magnetic field enhancements and fades where it encounters diminishments. It is mainly the magnetic field fluctuations that amplify or reduce the flux density at a given location in the shock. Especially effective are changes in the direction of the magnetic vector, since the observed synchrotron radiation depends strongly on the orientation of the magnetic field relative to the observer.

The disturbance is allowed to have a frequency-dependent dimension x along the jet axis, as is the case if the electrons suffer significant radiative losses. The main effect occurs on a timescale determined by the decorrelation length scale of the turbulence. However, minor fluctuations superposed on the more pronounced flux density variations can occur over shorter timescales as the shock encounters eddies with sizes of order x. The variability timescale is found to be as short as $\sim x/(\gamma c)$, so very rapid variability, much faster than $\sim R/c$, is possible if the shock is very thin ($x \ll R\phi$) The timescales of these minor variations are shorter and the amplitudes larger at the higher frequencies corresponding to smaller values of x. The power spectrum of the fluctuations has a slope in the range ~ -1.0 to -1.8, but is quite uncertain owing to our ignorance of the detailed physics of compressible hydromagnetic turbulence in relativistic plasmas.

We present in Figure 7 the results of one of our simulations. The nature of the time variations is that expected for any object whose magnetic field and density and energy of relativistic electrons is enhanced: the turnover frequency rises as the flux density increases. There can be a slight time delay between the peaks in the light curves at high frequencies if x is frequency dependent, even if the source is optically thin. The dominant effect of the turbulence is to cause random minor outbursts as the shock propagates down the jet. The simulated flickering agrees quite well with optical observations of blazars (see, e.g., Bregman *et al.* 1990).

The turbulent fluctuations can occur far downstream of the compact core of the source. Shocks in turbulent jets are therefore good candidates for explaining much of the "flickering" of radio sources (Heeschen *et al.* 1987; Quirrenbach *et al.* 1989). However, it is not clear how shocks thin enough to cause variations on timescales ~ 1 day can occur at centimeter wavelengths, since radiative losses should be negligible. If such thin shocks are possible, the flickering radio emission could be synchrotron radiation from shocks moving through turbulence, while the similar optical variations

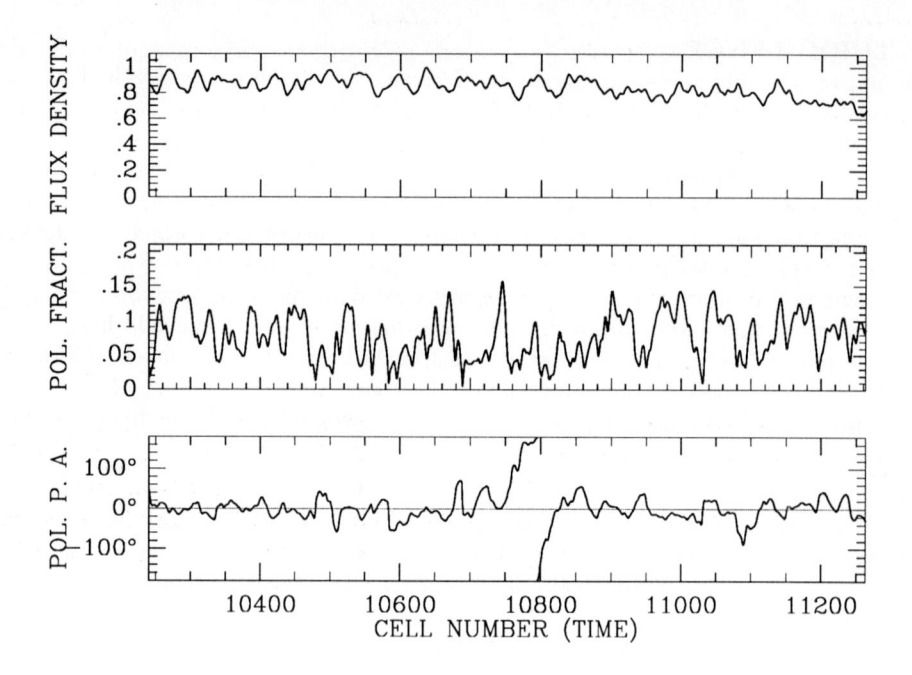

Figure 7 Total flux density (normalized), polarization fraction, and polarization position angle "light curves" corresponding to a numerical simulation of a thin shock propagating through a turbulent jet. The gradual decline in flux density is caused by the overall evolution of the shock emission, while the rapid flickering occurs as the shock encounters turbulent fluctuations in magnetic field and density. The mean magnetic field is parallel to the jet axis (perpendicular to the shock front); superposed on this is a random component with magnitude equal to 80% of the mean field. The density fluctuations correspond to 10% of the mean value.

observed in 0716+714 (see the contributions to these proceedings by Krichbaum *et al.* and Wagner *et al.*) could be inverse Compton scattered radiation generated inside the flickering radio synchrotron component. There are too few constraints at present to develop a detailed model, however.

The simulations demonstrate that the polarization variability of a shock moving through a turbulent jet is quite pronounced. As discussed by Jones (1988) and noted repeatedly at the conference by H. Aller, even apparent systematic rotations of the polarization vector can be observed (cf. Fig. 7). Observers are therefore cautioned that apparent polarization rotations should be considered evidence for helical field configurations only if the "rotations" persist for more than about one full cycle.

6 INTERSTELLAR SCATTERING

Interstellar refractive scintillation (Rickett, Coles, and Bourgois 1984) can cause variations in compact radio sources at low frequencies. In the source NRAO 140 the changes in the VLBI image at 18 cm are dramatic on a timescale of ~ 1 year. The quasar lies behind a complex region of Galactic molecular, atomic, and ionized gas, as well as a nonthermal loop (Fiedler et al. 1991), so the interpretation of the variationsin terms of scattering seems rather solid. Based on the first-order theory, one would not have expected the effect to be so strong at this short a wavelength, so it is clear that we have much to learn about this phenomenon. In general, one expects shorter timescales at shorter wavelengths, so refractive scintillation is a natural explanation for flickering, although the radio-optical correlations and polarization variability discussed by Krichbaum et al. and Wagner et al. (these proceedings) strongly imply that these variations at least are intrinsic to the source.

7 HIGH ENERGIES

The radio-infrared synchrotron flare caused by a shock in a jet should be accompanied by a self-Compton X-ray and γ-ray flare. Since the self-Compton flux density is proportional to the synchrotron photon density as well as the density of relativistic electrons, both of which increase during an outburst, the high-energy flare should be more pronounced than the synchrotron flare, after subtraction of the "quiescent" emission. If the shock goes through a Compton-loss stage, the synchrotron flare is suppressed at the start, so the increase in high-energy flux precedes the rise of the infrared emission, which peaks some time after the maximum in X-ray emission. The spectrum of the X-ray emission from the shock has a slope $-(s-1)/2$ as opposed to the value $-s/2$ for the infrared spectrum during the synchrotron-loss stage (see M&G for details). The self-Compton X-ray emission from a component in a bending jet should vary simultaneously and with about the same fractional amplitude as the optically thin radio synchrotron emission from an isolated component. This is because the Doppler factor δ boosts both types of emission identically.

If the jet is turbulent, essentially simultaneous synchrotron and self-Compton flickering will occur as the shock encounters inhomogeneities. However, those synchrotron variations that are primarily caused by changes in magnetic field *direction* will cause little, if any, high-energy self-Compton fluctuations.

The multifrequency flare in 3C 279 observed in 1988 by Makino et al. (1989) qualitatively follows the expectations of the model as stated above, but the time/frequency time coverage was too poor to compare the data with the details of the theory.

8 SUMMARY

The shock-in-jet model appears capable of explaining not only the general trends of variable source behavior, but also some of the idiosyncrasies displayed by particular flares in certain blazars. While this is accomplished at the expense of adding free parameters, the geometry of the model provides a fundamental constraint on the possible evolutionary paths that a flare may follow.

(i) A shock moving down a straight jet produces a flare that rises suddenly, then propagates to lower frequencies at nearly constant amplitude before fading. The decline is either gradual if allowed to decay adiabatically, or more abrupt if the only relativistic electrons encountered by the shock already suffered heavy synchrotron losses before they flowed out of the core region. Some disturbances may not steepen into shocks until they are well down the jet; in this case, the flare begins at centimeter-millimeter wavelengths rather that in the submillimeter-infrared as expected for more dramatic disturbances. The accompanying high-energy self-Compton flare peaks before or at the same time as the maximum in the synchrotron emission.

(ii) A shock in a jet that bends toward the observer produces a flare whose spectrum rises abruptly at all frequencies, with little change (perhaps even a decrease) in turnover frequency. The VLBI structure should appear twisted, as in the case of 1156+295 (McHardy *et al.* 1990). The X-ray flare caused by bending is simultaneous with, and has the same amplitude as, the synchrotron outburst.

(iii) A shock in a jet that bends away from the observer decays rapidly at all frequencies simultaneously.

(iv) In all cases, rapid fluctuations occur, especially at high frequencies, as the shock encounters turbulence in the jet. The polarization can vary erratically, with apparent rotations in positions angle up to about a full cycle.

The basic evolution of the flare [items (i) and (iv) above] does not require the flow velocity of the jet to be relativistic. Twists in the jet will not, however, alter the emission from a nonrelativistic shock.

Given the successes of the shock-in-jet models, it seems appropriate to use them as the primary working hypotheses against which observations of blazar variability are compared. With the further development of millimeter-wave VLBI, completion of the VLBA, and better monitoring at submillimeter, infrared, and X-ray frequencies over the next several years, the predictions of the model will be thoroughly tested. If theoretical work on the dynamics of jets and acceleration of relativistic particles keeps pace, we should understand compact nonthermal sources in active galaxies much better than we do now.

The work discussed in this paper has been supported in part by National Science Foundation grant AST-8815848 and NASA grants NAGW-1068 and NGT-50314.

9 REFERENCES

Blandford, R.D., and Königl, A. 1979, ApJ, 232, 34

Blandford, R.D., and Rees, M.J. 1978, in *BL Lac Objects*, ed. A.M. Wolfe (Univ. of Pittsburgh), p. 328

Bregman, J.D., *et al.* 1990, ApJ, 352, 574

Brown, L.M.J., *et al.* 1989, ApJ, 340, 129

Burbidge, G.R., Jones, T.W., and O'Dell, S.L. 1974, ApJ, 193, 43

Dent, W.A., and Balonek, T.J. 1980, Nature, 283, 747

Fiedler, R.L., Pauls, T., Johnston, K.J., and Dennison, B. 1991, ApJ, submitted

Hardee, P.E. 1990, in *Parsec-Scale Radio Jets*, ed. J.A. Zensus and T.J. Pearson (Cambridge Univ. Press), p. 266

Heeschen, D.S., Krichbaum, T., Schalinski, C.J., and Witzel, A. 1987, AJ, 94, 1493

Hughes, P.A., Aller, H.D., and Aller, M.F. 1985, ApJ 298, 301

Jones, T.W. 1982, BAAS, 14, 963 (Abstract)

Jones, T.W. 1988, ApJ, 332, 678

Königl, A. 1981, ApJ, 243, 700

Makino, F., *et al.* 1989, ApJ, 347, L9

Marscher, A.P. 1980, ApJ, 235, 386

Marscher, A.P. 1990, in *Parsec-Scale Radio Jets*, ed. J.A. Zensus and T.J. Pearson (Cambridge Univ. Press), p. 236

Marscher, A.P., and Broderick, J.J. 1985, ApJ, 290, 735

Marscher, A.P., and Gear, W.K. 1985, ApJ, 298, 114

Marscher, A.P., Zhang, Y.F., Shaffer, D.B., Aller, H.D., and Aller, M.F. 1991, ApJ, 371, in press

McHardy, I.M., Marscher, A.P., Gear, W.K., Muxlow, T., Lehto, H.J., and Abraham, R.G. 1990, MNRAS, 246, 305

O'Dell, S.L., *et al.* 1988, ApJ, 326, 668

Pauliny-Toth, I.I.K., and Kellermann, K.I. 1966, ApJ, 146, 634

Phinney, E.S. 1987, in *Superluminal Radio Sources*, ed. J.A. Zensus and T.J. Pearson (Cambridge Univ. Press), p. 301

Quirrenbach, A., Witzel, A., Krichbaum, T., Hummel, C.A., Alberdi, A., and Schalinski, C. 1989, Nature, 337, 442

Rickett, B.J., Coles, W.A., and Bourgois, G. 1984, A&A, 134, 390

van der Laan, J. 1966, Nature, 211, 1131

Valtaoja, E., *et al.* 1988, A&A, 203, 1

Variability of 3C 273 and Shock Models

M. LAINELA[1], E. VALTAOJA[1,2] AND M. TORNIKOSKI[2]

1) Tuorla Observatory, University of Turku, SF-21500 Piikkiö, Finland
2) Metsähovi Radio Research Station, Helsinki University of Technology, Otakaari 5A, SF-02150 Espoo, Finland

ABSTRACT. Variable extragalactic radio sources, associated with the nuclei of galaxies and quasars, are interpreted in terms of a supersonic relativistic jet. It is proposed that radio emission originates both from the quasi-steady jet itself and from behind strong shock waves which propagate in the jet. No complete and satisfactory physical models for the shock evolution can as yet be formulated. To provide a framework for comparison with observations, we present instead a generalized shock model, which incorporates the growth and the decay stages of the shock in a simple parametrized form. In order to compare the model predictions with data, we present a discussion of the outbursts of the variable source 3C 273 that occurred 1987 and 1989.

1 INTRODUCTION

The first quantitative attempt to understand the main features of multifrequency lightcurves – their shapes, maximum amplitudes of outbursts and time delays between maxima at different frequencies – was the adiabatic, spherically expanding source model developed by Shklovsky (1965), van der Laan (1966) and Pauliny-Toth and Kellerman (1966). The model could explain qualitatively many observed features, but it was clear from the beginning that the model could not be extended to the earliest stages of the outbursts, since the turnover flux was a monotonously increasing function of the turnover frequency. In addition to variability studies, several other lines evidence have led to the general agreement that shock structures evolving in a relativistic jet provide a better approximation of the true situation than spherical symmetry. Valtaoja et al. (1988) established that such shock models can, in general, explain both the observed shape of the radio outburst spectrum and its time evolution. Independent other work has demonstrated that quite detailed modeling of the flux variations (Marscher and Gear 1985) or even both the flux and polarization variations (Hughes et al. 1989a,b, 1991) is possible using the concept of growing and decaying shocks in a relativistic jet. We therefore begin by investigating what general consequences shock models have for multifrequency lightcurves, and then proceed to compare the predictions with the data of the outbursts occurred 1987 and 1988 in 3C 273.

2 A GENERALIZED SHOCK MODEL

We are still far from a realistic physical theory of the formation and evolution of shocks in the relativistic jets of active galactic nuclei. Detailed models will require observational data we do not yet possess; however, it seems to be possible to give an overall description which is applicable to shock models in general, and which can be used as a convenient framework in discussing the evolution of outbursts, in defining the relationships between the observed phenomena and the underlying models, and in comparing predictions of individual models with the data.

Here we present a short summary of the generalized shock model. A fuller version of the model is presented in the paper of Valtaoja et al. (1991). Speaking in as general terms as possible, all outbursts will have a growth stage, between the initial impetus causing the formation of the shock and the maximal development of the shock, and a decay stage when expansion and other losses extinguish the shock. Between the growth and the decay stages there may be a plateau, when energy losses and energy gains are temporarily balanced. The decay stage is the easiest to describe, since one can start with an already existing synchrotron component and just allow it to expand or otherwise lose energy in some model-specific manner. The self-absorption turnover peak ν_m will move to lower frequencies, and the peak flux S_m will diminish. The evolution of the earlier growth stages depends critically on the as yet unknown physics of the shock and the underlying jet. Both observations and rather general theoretical arguments indicate that during the growth stage the self-absorption peak moves to lower frequencies while the peak flux increases (cf. Marscher and Gear 1985).

Consequently, we can construct a "generalized shock model" which incorporates the main characteristics of the three stages of shock evolution, but deliberately ignores all the other details of the physics of the shock. Figure 1 gives a sketch of the model, showing the evolution of the shock spectrum. The thick curve traces the motion of the turnover peak of the shock spectrum through flux-frequency-space. The actual slopes and the detailed shape of the $\nu_m(S,\nu)$-track depend on the details of the adopted model, but for the following discussion only the overall morphology is needed. We define the three stages as simply as possible as follows:

1) the growth stage: $S_m \propto \nu_m^a$,
2) the plateau stage: $S_m \approx$ constant, and
3) the decay stage: $S_m \propto \nu_m^b$,

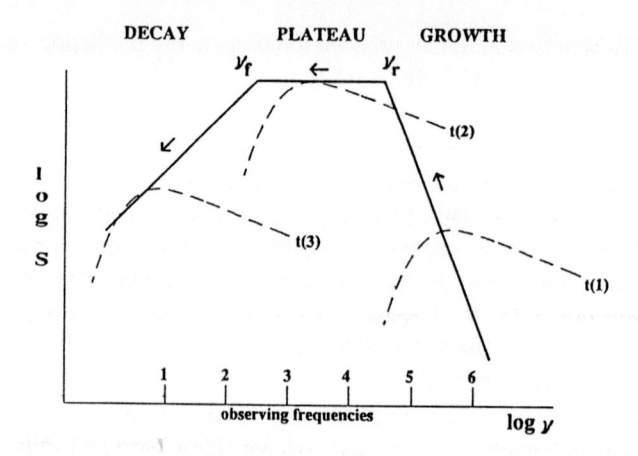

Figure 1. Spectral evolution in the generalized shock model. The appearance of the shock to an observer depends on whether the monitoring frequencies are below ν_f (a high-peaking flare, $\nu_{1,2}$), between ν_f and ν_r ($\nu_{3,4}$), or above ν_r (a low-peaking flare, $\nu_{5,6}$).

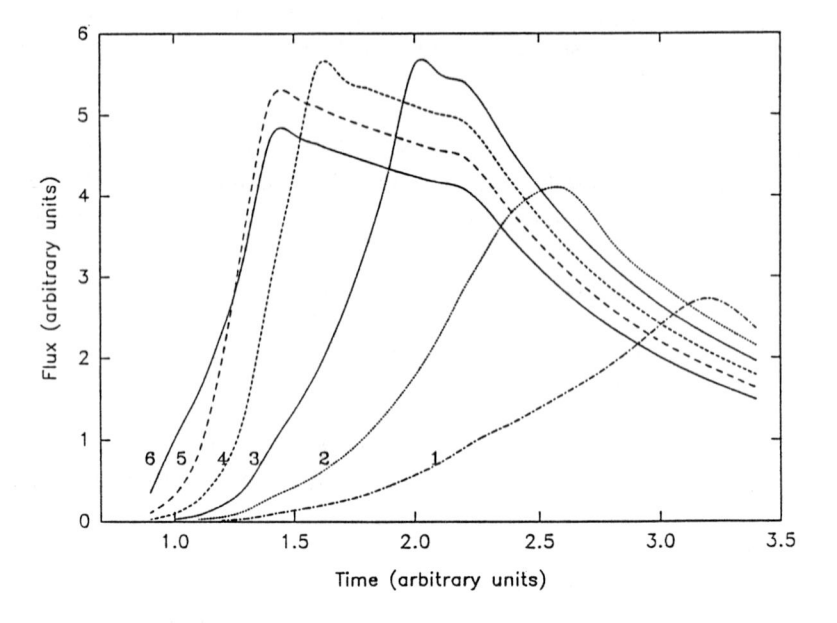

Figure 2. Computed model lightcurves in the generalized shock model. The six logarithmically equally spaced frequencies $\nu_1 \ldots \nu_6$ are as in Figure 1.

where S_m is the turnover flux, ν_m the turnover frequency, and a and b model-dependent parameters.

Figure 1 does not describe a true physical model, but a scenario for defining the concepts of shock models and for comparison with observations: a framework intended to replace the old expanding source model. The only really relevant and necessary part of the description is the concept of growth and decay stages; all the rest should be considered as a parametrized description, and the aim of observations should be to determine whether such a parametrization is at all reasonable and what the numerical values of the parameters are.

As a first approximation we assume that all shocks follow a similar evolutionary track. Even if all the shocks evolve in a similar manner, the observed behaviour of a burst depends on the frequency ν_r, where the flare reaches its maximum development, on ν_f, where the decay stage starts, and on the observing frequency ν_{obs}. The flux curve at given observing frequency will depend on whether the frequency is smaller than ν_f, larger than ν_r, or somewhere in between. In what follows we use the term high-peaking flare for outbursts which reach their maximum development at high frequencies relative to the observing frequency (i.e., $\nu_{obs} < \nu_f$), and the term low-peaking flare for outbursts which reach their maximum development at low frequencies relative to the observing frequency ($\nu_{obs} > \nu_r$).

The general shape of the lightcurves is not sensitive to chosen exponents a and b. We calculate model lightcurves for six observing frequencies with equal logarithmic spacings. The two lowest frequencies, ν_1 and ν_2 are below ν_f, and according to our terminology the outburst appears as a high-peaking one to the observer at these frequencies. The intermediate frequencies ν_3 and ν_4 lie on the plateau where the shock reaches it maximal development (in the sense that the turnover flux is higher than at any other time or frequency). Finally, ν_5 and ν_6 are higher than ν_r, and observer sees a low-peaking flare. The computed model lightcurves are shown in figure 2.

3 A FEW CONSEQUENCES OF THE GENERALIZED SHOCK MODEL
Not surprisingly, since the observed flux variations occur during the decay stage of the burst, the lightcurves at the two lowest frequencies show similar behaviour as the old canonical model. The maximum is reached when the turnover passes the observing frequency (when the source becomes optically thin), and the relationship between observed maximum amplitude ΔS_{max} and the frequency is determined by the slope b ($\approx +1$) of the decay track: $\Delta S_{max}(\nu) = S_m(\nu) \propto \nu^b$, just as in the canonical model. There is also a time delay between the higher and the lower frequencies. For the two frequencies situated on the plateau the maxima are also reached when the turnover passes the observing frequency, and there is a time delay between neighbouring frequencies.

However, the maximum amplitude is now the same for both frequencies. For the two highest frequencies the shock is still growing when the turnover passes the observing frequencies. The shock becomes optically thin, but the flux continues to grow until the turnover peak reaches the plateau at v_r. Since the shock has a flat optically thin spectrum, the fluxes decrease significantly first when the final decay stage ($v_m < v_f$) is reached. For all the frequencies above v_f, the difference between the maximum amplitudes does not depend on $S_m(v)$, but is now instead a function of α_{thin} only: $\Delta S_{max} \propto v^{\alpha(thin)}$, and the fluxes vary in unison. Figure 3a shows graphically the dependence between ΔS_{max} and v over the whole frequency range.

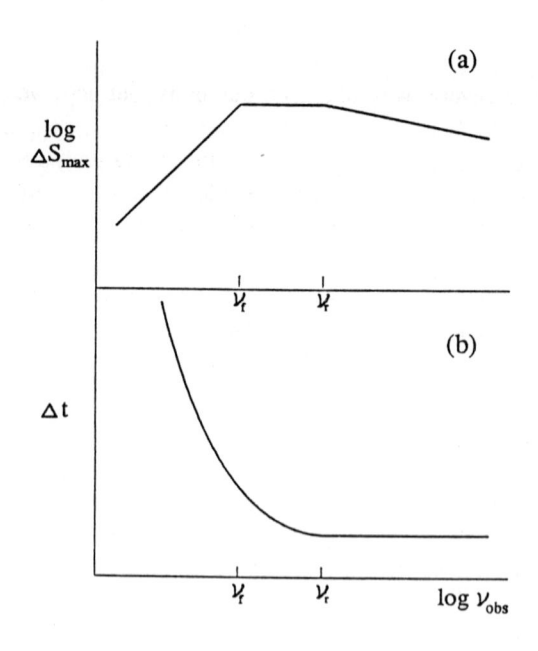

Figure 3. (a) The maximum amplitude of the outburst, ΔS_{max}, vs. the observing frequency v_{obs}, as predicted by the generalized shock model. (b) The predicted time delay between burst maxima, Δt, vs. frequency v_{obs}.

For the high-peaking flares the behaviour of the time delay is similar to the old canonical model, with lower frequencies trailing the higher. For low-peaking flares the situation is very different. Most of the observed flux variations occur when the burst peak has already passed the observing frequencies, and lightcurves on nearby frequencies will track each other closely. The maxima at neighbouring frequencies occur simultaneously, at the time when the turnover peak reaches the plateau at v_r. Thus, for a given outburst we expect to find increasing time delays for all frequencies lower than v_r. Above v_r the variations are virtually simultaneous with no time delays between frequencies. Figure 3b shows the expected Δt vs v schematically.

4 COMPARISON WITH THE MONITORING DATA OF 3C 273

Measurements of extragalactic sources have been done on a regular basis with the Metsähovi radio telescope since 1980. Most of the observations are done at frequencies 22 and 37 GHz. Since 1985 also 87 GHz observations have been done (Teräsranta et al.). 3C 273 is one of the most frequently observed sources in our monitoring program and so it is a good candidate for testing the generalized shock model.

In 3C 273 there has been several well observed outbursts, but here we concentrate only on the outbursts in 1987 and 1988. The data which we use here is from different sources: the 22, 37 and 87 GHz data is from the Metsähovi monitoring program. There are also a few 90 GHz data points from SEST telescope. The 4.8, 8.0 and 14.5 GHz data is from the monitoring program of the University of Michigan. The data is presented in figure 4.

We already mentioned in chapter 3 that in some sources outbursts are high-peaking and in others low-peaking. In 3C 273 we have both cases: the outburst in 1987 is low-peaking and the later one in 1988 is high-peaking. The first outburst appeared simultaneously at 22, 37 and 87 GHz. In contrast, the second outburst behaved in the canonical manner: the peak appeared first at high frequencies and moved then towards lower frequencies.

In figure 5 we have plotted the burst amplitudes ΔS in different frequencies. It seems that the behaviour is similar in both outbursts, except that maximum frequency is shifted. First the outburst peak moves towards lower frequencies and the burst amplitude grows rapidly. Then there is possibly a short plateau and after that the burst amplitude starts to decay. This is in good agreement with the generalized shock model. The outburst in 1987 reaches its maximum amplitude around 20 GHz and the later outburst around 200 GHz.

In figure 6 we have plotted the time delays between different frequencies. The behaviour is again similar in both the 1987 and the 1988 outbursts, except shifted frequencies. At high frequencies there is no time delay, but when the outburst move towards lower frequencies, at some point the time delay starts to grow. In the model this corresponds to the time when the outburst peak reaches v_r.

Thus, there need to be nothing fundamentally different in the physics of these two outbursts: the shock mechanism is the same, only the frequency domain is different. There must, of course, be some physical reason for different v_r in different bursts; this may, for example, be related to the distance from the core at which the shock forms: if

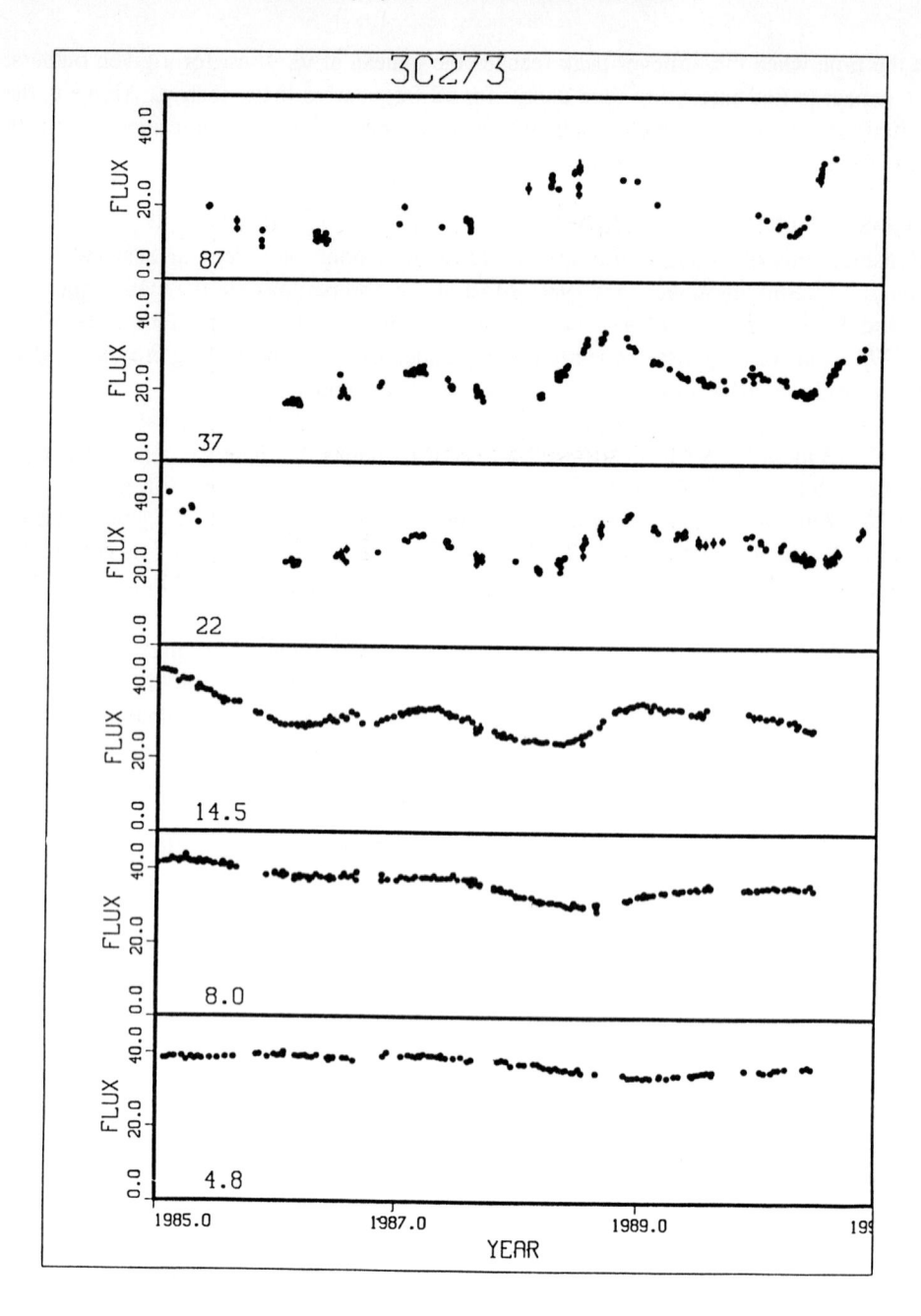

Figure 4. The flux variations of 3C 273 during 1985-1990. The 22, 37 and 87 GHz measured with Metsähovi radio telescope. The 4.8, 8.0 and 14.5 GHz data is from University of Michigan (Hughes, Aller and Aller, private communication).

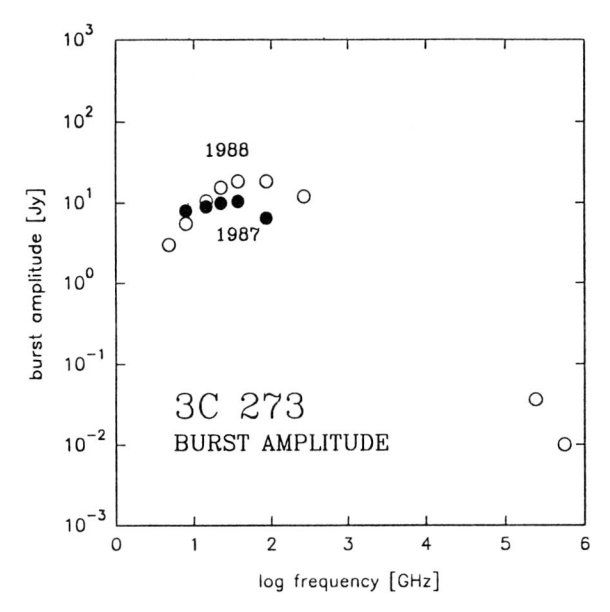

Figure 5. The observed maximum burst amplitudes of 3C 273 in 1987 and 1988.

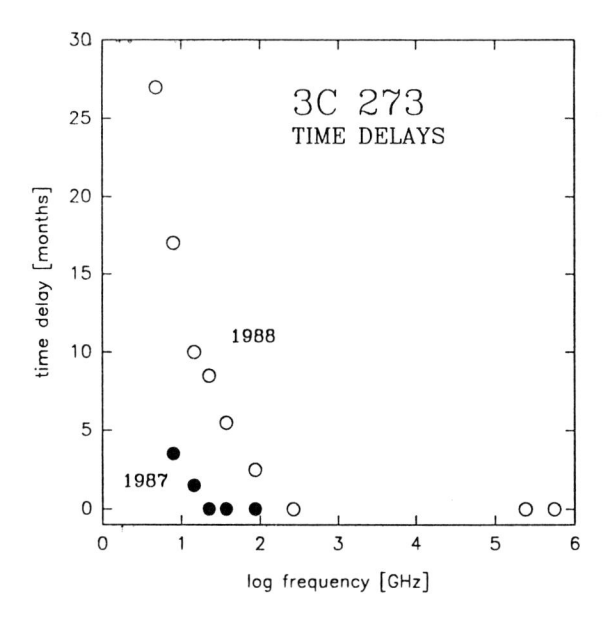

Figure 6. The observed time delays between burst maxima in 3C 273.

the distance is short, then the shock is strong and the optical depth is high. If the distance is longer, then the shock is weaker and the optical depth is lower. If this is the case, then the stronger shock should give larger maximum burst amplitude. The two outbursts in 3C 273 fit well into this picture: The first outburst is the weaker one and the second is the stronger one.

5 CONCLUSIONS

We have argued that the generalized shock model can provide an unifying framework for understand the outbursts in AGN. We have suggested that outbursts in AGN evolve along similar tracks, defined by the motion of the shock spectrum's turnover peak (S_m, v_m) in time, and that most of the observed differences result from variations of a single parameter, the frequency v_r at which the outburst reaches its maximum development. The model can explain both multifrequency lightcurves closely tracking each other, and lower frequency variations trailing after higher frequencies. Time delays or the lack of them can be explained by variations in v_r. In comparing the model with two outbursts in 3C 273, a good agreement between the theory and the data is found.

REFERENCES

Hughes, P.A., Aller, H.D., Aller, M.F.:1989a, Astrophys. J. **341**, 54
Hughes, P.A., Aller, H.D., Aller, M.F.:1989b, Astrophys. J. **341**, 68
Hughes, P.A., Aller, H.D., Aller, M.F.:1991 (preprint)
Marscher, A.P., Gear, W.K.: 1985, Astrophys. J. **298**, 114
Pauliny-Toth, I.I.K., Kellerman, K.I.:1966, Astrophys. J. **146**, 634
Shklovsky, I.S.: 1965, Nature **206**, 176
Teräsranta, H. et al.:1991 (in preparation)
Valtaoja, E. et al.: 1988, Astron. Astrophys. **203**, 1
Valtaoja, E. et al.: 1991 (preprint)
van der Laan, H.: 1966, Nature **211**, 1131

Multifrequency Observations of the Flaring Characteristics of Blazars.

Ian Robson

Dept of Physics and Astronomy
Lancashire Polytechnic
Preston PR1 2TQ
UK

Abstract

This paper reviews the flaring behaviour of Blazars, specifically concentrating on the synchrotron component, manifest mainly in the optical-IR-radio regime. The high frequency and non-synchrotron aspects of flaring behaviour will be covered by other speakers. The requirements from observations will be highlighted, and the great value of monitoring databases stressed. Results to date will be reviewed, overall conclusions presented and areas in which there are still uncertainties will be noted. Finally requirements for the future to provide sufficiently detailed observational frameworks for detailed theoretical modelling will be outlined.

1. INTRODUCTION

The fundamental observational property I intend to discuss is the variability of the radio-UV continuum output from 'Blazars', assumed to be electron synchrotron emission. For the purpose of this review, it is not crucial whether we are considering a classical 'lineless' BL Lac, a lined BL Lac, or an OVV quasar; to a first approximation the 'Blazar' phenomena seems to be the same as far as the synchrotron emission is concerned. Fundamental to all these objects is a flat radio loud spectrum, which turns over around the mm-submm regime to the optically thin spectrum which extends through the infrared (IR) into the optical and in some cases (mostly BL Lacs) is seen to extend smoothly into the UV and X-ray regime. For some objects, different components begin to dominate over the synchrotron at some high frequency. Figure 1 shows two continuum spectra in terms of observer units, ie. flux density.

I am a great believer in declaring personal prejudices, and in what follows I shall assume that BL Lac objects are luminous elliptical galaxies with a radio jet which happens to be

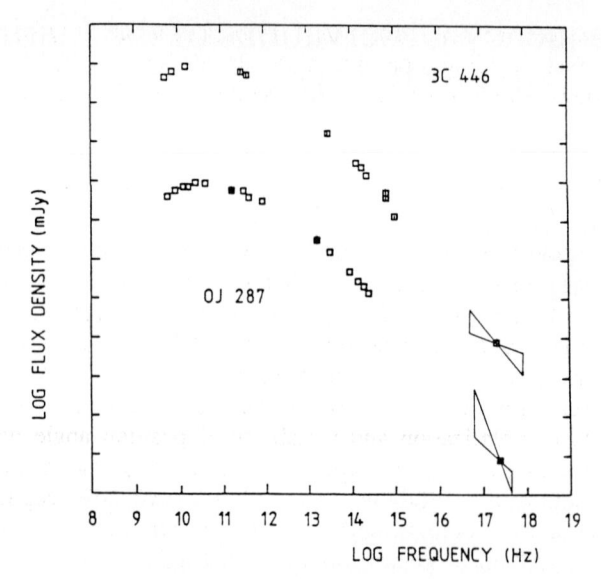

Figure 1 Spectrum of an OVV quasar and a BL Lac adapted from Brown *et al.* (1989a).

pointing towards the line of sight to the observer (see the review by Marie-Helene Ulrich 1990). Likewise OVV quasars are higher luminosity objects, with a clearly defined broad line region, and possessing a radio jet which is again aligned towards the observer. What type of galaxy the quasar resides in I shall not speculate on, or, even whether the entire quasar population resides in the same type of galaxy. We are currently conducting a large IR imaging campaign (Hughes et al 1991) to determine the galaxy environment of radio quiet and radio loud quasars. I will assume that orientation and magnetic field effects are linked to the phenomenon we describe as a 'Blazar'.

Central to the above is the fact that this Blazar phenomenon is contained in a galaxy and the redshift of the emission lines, where seen, is a true indication of the cosmological distance of the galaxy. In the output continuum I take the radiation extending from the cm radio to the mid IR (and often beyond) to be synchrotron emission from relativistic electrons of a jet which may possesses bulk relativistic motions and which is aligned towards the line of sight. The degree of alignment probably affects some of the variability properties of the observed phenomenon and therefore is object dependent.

Because we believe that the Blazar phenomenon is related to a galaxy, then at some wavelength, (and epoch), we may expect to see evidence of the underlying galaxy. At least in these continuum-dominated objects we do not have the horrendous task of massive spectral energy deconvolution in order to unravel such components as an infrared bump, the big blue bump and so on. About the only object for which these complicating factors come into this study is the quasar 3C273, which from other evidence (Impey, Courvoisier and refs therein - both this volume) shows Blazar characteristics.

2. REQUIREMENTS

In order to be able to study the behaviour of the flaring component it is vital that the 'steady-state', or 'quiescent state', (if either in fact exist) is understood. If we restrict ourselves to the large amplitude flaring outbursts then this statement is more meaningful. To make progress in determining the precise details of the physical processes of the flaring component of the emission mechanism requires three observational parameters: (i) the quiescent continuum spectrum, (ii) the spectral and temporal flare development, (iii) the degree of polarization and variability of position angle with time. This latter point must be stressed because it provides another very critical tool for following the detailed behaviour of the variability (see H Aller this volume). Significantly, insufficient coherent polarization data exist on the optical and infrared variability, while currently it is nonexistent in the submillimetre and millimetre regime.

The quasar 3C273 shows many aspects of the Blazar phenomenon. The 1983 outburst had sufficient observations (Robson et al 1983) to allow a new theoretical emission model to emerge, that of Marscher and Gear (1985). I refer to Alan Marscher and Phil Hughes (this volume) for discussions of the jet models. As in many areas, the breakthrough for emission models came with the opening up of a new wavelength regime, the 1mm and submm wavelengths, with the UKIRT telescope.

The appearance of generic Blazar spectra is that they have a 'flat' radio spectrum which becomes optically thin at around 1mm wavelength and then falls to the infrared where there may be steepening due to radiation losses. To pin down the models of emission, the temporal development of the synchrotron turnover frequency, which occurs at about 1mm or in the submm during flaring, takes on a highly important role.

Let us now look at the first part of the required information on the continuum emission from Blazars; the snapshot quiescent emission. Even here we are really asking a number of questions. For a single object we would like to know (i) through what range of spectral characteristics does it track in its variability? (ii) how long does it stay in any particular state - this questions whether there is a most probable state of emission? (iii) what is the minimum emission state? These can be extended to the entire population by: (iv) can we define a series of populations showing the same characteristics? Here we would like to know how, observationally, we might be able to distinguish BL Lacs from OVV/HPQ's as far as the synchrotron variability is concerned.

Now it turns out that the minimum emission state is very important for a number of reasons. First, it is the best chance of observing other components in the emission; components which are otherwise drowned out by the strength of the synchrotron radiation. Obvious examples include: (a) emission lines in BL Lac objects which allow equivalent widths to be deduced and a distance redshift to be determined (eg Sitko & Junkkarinen 1985); (b) thermal emission from heated dust in the bulk of a galaxy (eg Gear et al 1985); (c) the presence of a blue bump (eg Brown et al 1989a) or other aspects of the underlying galaxy (Courvoisier et al 1988).

To make progress with these studies clearly requires monitoring observations at a number of wavelengths. With sufficient data one could, in principle answer whether there is a state of emission which predominates for most of the time, or whether objects transit from lowest to highest phases with some statistical distribution function which is telling us that the fuelling is chaotic. Can we define a characteristic variability time for any wavelength?

3. SNAPSHOT SPECTRA AND MONITORING STUDIES

Dedicated multifrequency monitoring campaigns are the key to success, and without doubt, the key word is *dedicated*. I shall now review the optical through radio wavelength regimes and give a critical appraisal of where we are in relation to dedicated monitoring of Blazars. In this I exclude those multifrequency campaigns organised around specific objects and/or satellite missions (such as the IUE, GINGA, ROSAT etc). Although these are absolutely vital in providing a snapshot spectrum, they are, unfortunately by the nature of things, very much an ad-hoc method of observing. Instead I refer to dedicated monitoring of a sample of Blazars at a selection of wavelengths.

It is clear, that with the drive to large 4-m class telescopes and above in the optical/IR the ability to perform such monitoring is practically non-existent. On the other hand, smaller, (usually 'private') and less time pressured facilities are ideal for monitoring studies provided that they are (a) on an excellent photometric site, (b) sufficiently large to obtain the desired signal-to-noise for all ranges of variability - ie when very faint, (c) adequately staffed so that observations are quasi continuous, (d) provide rapid data reduction for notification of flaring events. These are very stiff criteria and there are no facilities that appear to fulfil them all. I will return to this aspect in more detail under future work. However, the Rosemary Hill Observatory in Florida has built up an impressive track record for optical observations (see Webb this volume for examples).

I have regular infrared monitoring on UKIRT, but pressure for time continues to squeeze the number of sources observed. However, the observatory is always willing to make observations during flaring behaviour and this is potentially a key for future modes of operation. Some quasi-regular monitoring is carried out at a number of observatories, Palomar being the best example, also ESO and the CST at Izana. These provide data, mostly at J, H and K (1.25, 1.65 and 2.2 µm). Observations at 5, 10 and 20 µm are extremely scarce.

There are very few mm-submm telescopes. SEST has a regular programme of 100 and 230 GHz [3.0 mm and 1.3mm] observations of Blazars, IRAM provides a comprehensive list of observations of radio loud objects at the same frequencies from pointing and calibration runs (eg Steppe et al 1988). The James Clerk Maxwell Telescope * (JCMT) has a monthly programme of monitoring a sample of 15 Blazars at 2.0, 1.3, 1.1, 0.8 and 0.45 mm wavelength. This has lasted for the past two years and has one more year to run. It is hoped that this will be sufficient to characterise the emission from the sample and to determine flaring behaviour better for future studies. The sample is shown in Table 1.

Source	z		Source	z	
0235+164	0.85	BL	0420-014		
0735+178	0.42	BL	0736+017	0.19	
0851+202 (OJ287)	0.31	BL	1156+295	0.73	
1226+023 (3C273)	0.16		1253-055 (3C279)	0.54	
1308+326	1.00	BL	1514-241 (AP LIB)	0.05	BL
1641+399 (3C345)	0.60		1921-293 (OV236)	0.35	BL
2200+420 (BL LAC)	0.07	BL	2223-052 (3C446)	1.40	
2251+158 (3C454.3)	0.86				

Table 1. JCMT Blazar monthly monitoring sample (plus 3C84, 3C120, Mkn421 & 521).

Without doubt the radio regime is best served, primarily because there are a range of facilities which have been dedicated to such important observations. Here we have famous examples as the University of Michigan monitoring programmes at 4.8, 8.0 and 14.5 GHz [6.25, 3.75 and 2.07 cm] and the Metsahovi Telescope of our conference hosts, making observations mainly at 22 and 37 GHz [1.36 cm and 8.1 mm]. The results from these two facilities will figure prominently in this volume and a selection of the excellent Michegan data can be seen in the papers by Hugo and Margo Aller and Philip Hughes. Metsahovi results are shown by Harri Terasranta and others.

One thing is clear to those attempting to model the flaring behaviour, apart from the cm radio emission, there never seems to be sufficient data! It is extremely hard to find correlations between the various wavelength regimes, to search for emission lags as a function of time and to determine the stochastic nature of the variability. The plea is always for more and better data. However in spite of the difficulties, much progress has been made and a brief overview of some of the outcomes is instructive.

Firstly let us look at the snapshot spectra of Blazars, the generic smooth curves shown in Figure 1. Are they totally smooth from the cm radio to the IR/optical etc, or are there subtle differences in various regimes. This brings us to one of the controversies in the literature, the shape of the synchrotron spectrum. Some authors attribute the optically thin mm to IR-optical to a power law (eg Brown et al 1989a) whereas others (eg Landau et al 1986) describe the entire cm to optical as a single parabola. It is clear that the infrared and optical suffer curvature for a number of sources at various phases in their variability and so the difference between broken power laws or a smooth parabola is small in the observational sense, but does reflect a difference in the underlying electron energy distribution function.

The key difference is in the mm regime. Are the cm and mm components one and the same and very smoothly connected, if so the parabolic description is valid. On the other hand, if the mm-IR is a different component, although obviously connected, then the parabola description and underlying assumptions fail.

Figure 2, adapted from Brown et al (1989a), provides appealing evidence that the components are separate. In an attempt to settle this question, we have made two campaigns covering the 6 cm through to 0.45 mm regime, using the University of Michegan, Metsahovi, NRAO 12-m and the JCMT telescopes. The second campaign took place in February 1990. The crucial 12-m 3.3mm data are still outstanding but some examples of the data are shown in Figure 3, so far with somewhat inconclusive results. We shall proceed by assuming that the mm-submm-IR-optical emission is a single synchrotron component which presumably comes from the innermost part of the jet. This emission becomes optically thick at about 1mm (the turnover) and falls to longer wavelengths. Here it meets another component which peaks in the near cm regime. The latter of the two components might be identified as the underlying jet seen in the radio, the former being the inner core, which is the more flaring component and dominates at higher frequencies. A prediction is that this component (mm-IR-optical) would respond together during flaring behaviour, whilst opacity effects would put delays into the longer wavelength cm emission.

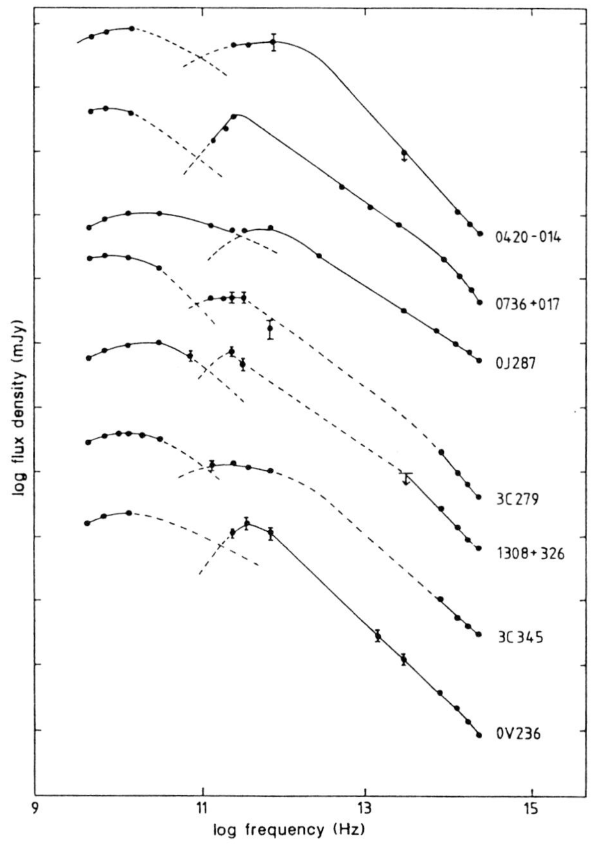

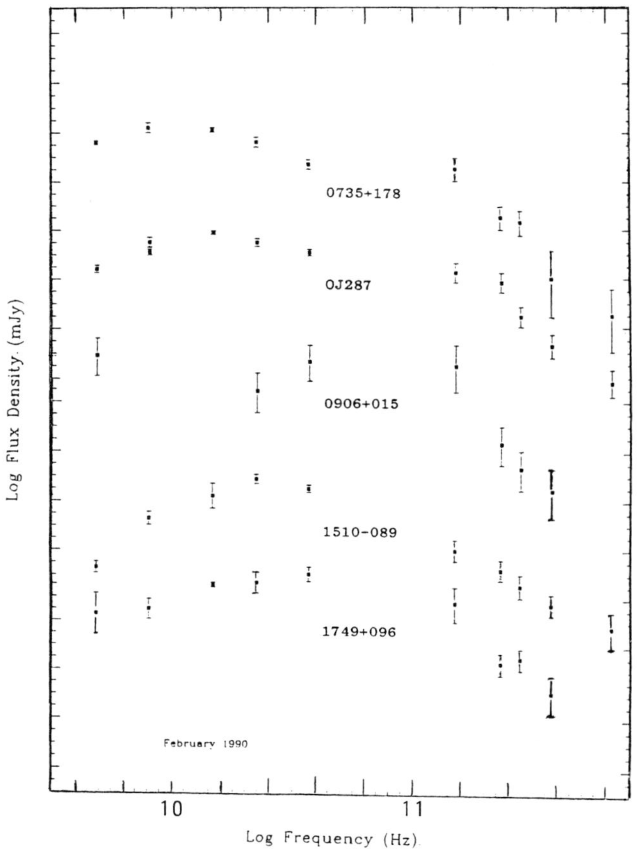

Figure 3 Sample of snapshot mm-cm spectra without the 90GHz data. (Robson et al 1991).

Figure 2 The mm-cm emission from a sample of Blazars showing the lack of continuity from mm to cm wavelengths.

4 VARIABILITY CHARACTERISTICS

4.1 Time Dependence

A number of extensive papers have now appeared in the literature, where much of the available data at most frequencies have been used to produce an analysis of the temporal variability and of the relation between frequencies. Here the key is to take a large temporal dataset at many frequencies and to subject them to various statistical analyses, (power spectrum, structure and fractal) to determine the form of the noise spectrum of the source and then to perform correlation analysis to search for links and lags between the various frequencies. To do this well requires a large time sequence of data and preferably the same density of sampling at each frequency. Also it is quite critical that fast flaring events are not drastically undersampled, this is most important for optical and infrared observations. The failure of these latter two parts is one of the most critical aspects of this entire area and leads to loss of statistical significance of results.

The latest in the series of work along this track is that from Bregman et al. (1990) using 20 years of data for BL Lac at a number of frequencies, mostly using the databases discussed above. This clearly demonstrates the essential value of these databases. For BL Lac, the authors find that there is no time delay between the optical and infrared (within the data sampling times of about 1 day) but the ensuing high frequency radio emission is delayed by a few years, and the lower radio frequency variability by a further few weeks. From our studies of flaring behaviour of individual sources we tend to find the same, although we could not differentiate timescales of less than 12 hours for any optical-IR delay.

One anticipates three potential forms of noise spectra corresponding to: white, flicker and shot noise. A good description of these is to be found in the Como conference paper by Bregman (1989). For BL Lac, using structure function analysis, they find the optical variability characteristically behaves as flicker noise, while the radio behaviour is better characterised by shot noise, at least for timescales less than 600 days. This general pattern would appear to be the same for the small number of sources for which the above analysis has been done.

This demonstrates again that there is a clear difference between the two wavelength regimes, the radio cannot simply be the optical emission expanding spatially until a point is reached where it is transparent to radio photons (otherwise they would both show the same type of noise spectra). Therefore, either some reprocessing of the electrons takes place or the two regions are distinctly different, although connected. Clearly the second possibility fits in well with the optical-IR-submm being one distinct component from the cm radio. Arguably shocks are the most likely medium for the transference of the energy between the two. One way by which the two component model is testable is to investigate the noise spectrum of the 1mm emission. Unfortunately an adequate database does not exist but we hope our mm-submm work will soon enable this to be tested.

Looking at another recent piece of work on the noise spectra of Blazars, Barbieri et al. (1990) have re-addressed the optical emission of 3C446, in which various groups had come to different conclusions, based on different statistical techniques, about how the emission could be classified. Bregman et al (1988) using structure function analysis concluded that the emission was characterised by flickering behaviour on a time scale of days to weeks, with major outbursts which lasted for a month to a year. Barbieri et al. (1985) and others, using power spectra analysis, suggested that there were two periodicities at about 1540 and 2130 days. Periodicities of course, are rather exciting as it could be related to instabilities of the accretion disk surrounding what most of us believe to be the ultimate power source, a massive black hole. However, periodicities are notorious for being easily seen by the eye, homed on by power spectrum analysis, but in fact have little statistical significance and are rarely confirmed. No doubt we shall hear more of such studies later in this conference (see Lehto, this volume).

The upshot was that Barbieri et al.(1990) believed there was no fundamental discrepancy between the results, the observers frame periodicity of 1540 days was strengthened and they predicted the next burst for early 1992. This is excellent science, a testable prediction. However, one of the most interesting and telling comments by the authors is that 'we think to gain further insight on the 3C446 physics it is still necessary to obtain more and better observations.' They go on to say that the observations need to be of higher quality and more regular and note that little is known about the very short timescale variability of minutes to days. This is very important and should not be forgotten, our very observational methodology normally mitigates against finding very short timescale variability.

If we look at one of the most recent compilations of the minimum variability timescale of AGN's (eg figure 7 from Bregman 1990), we see that the BL Lacs cluster towards the lower end of the scale, showing fastest variability. But how complete a picture is this, both for the Blazar and the AGN's themselves? Clearly more work must be done on this particular problem and hopefully we will learn more of this during this conference. One must however also bear in mind the question as to what these short timescales mean physically, particularly if there are significant relativistic effects.

So what about the flaring variability? The radio wavelengths are best sampled but as we have seen above, it is not clear what relation these have to the higher frequency components. Do some of the flares in the optical-mm propagate to the radio faster than others, even in the same object? If so, and I believe that this is true for some of the data on 3C273, then can we really correlate the flares, or do some of the fast optical-mm flares become washed out by the radio? This must pose a grave problem for correlation analysis. (see Hughes this volume).

What is clear from the Michigan database and the Metsahovi work is that in the radio, sources show differing characteristics, as might be expected. Many sources have flaring behaviour occurring on timescales of a few years, and these flares last for a few months (to within the observable resolution). In between flares the source can be relatively quiescent, showing at most a low amplitude, long period variability. On the other hand, one or two sources (eg OJ287) show variability at virtually all epochs and little evidence for any long-term quiescent phase.

The mm database is not yet sufficiently large to make really detailed comparisons, especially if there is a delay of years between the mm and radio. One of the more worrying aspects is whether the monthly mm sampling is sufficient, even worse if we happen to get weathered out for one or two occasions. We will see later that even on these timescales major flaring events could be missed.

In the optical/IR, large flares would seem to occur also on timescales of a few years, with the general conclusion being that much smaller scale flickering is a phenomenon which is occurring most of the time (as deduced by the noise spectra analysis and also by extended observations of a small number of sources, with OJ287 being probably the best example see eg Kikuchi 1989). Later papers in this conference will address the optical flickering behaviour and potential optical-radio linkage.

Once a flare has occurred, then one thing which we feel relatively comfortable about is being able to make some assertions about the cooling time based on synchrotron theory. Let us look at the critical wavelength of the turnover and assume for the moment that it occurs at v_m=1mm. It is easy to calculate the synchrotron cooling time for a reasonable field strength ($B \sim 1G$) and electron energies corresponding to the Compton brightness limit of 10^{12} K; the answer is over two weeks. What do we find in the observations? Normally this timescale is not violated, although observations are patchy, however for 3C273, there are clearly problems.

The overall behaviour of 3C273 from 1982 to the present day at various wavelengths has been reported in a number of papers which also include flare events (Clegg et al 1983, Robson et al 1983, Robson et al 1986, Courvoisier et al 1987, 1990). In 1988 there was a series of 3 flares between February and April, two of which were only 2 days apart (Courvoisier et al 1988). The rapidity of flaring and decline at optical and IR wavelengths indicated an amazing high rate of change of luminosity ($\sim 10^7$ L_O s^{-1}) for isotropic emission. Furthermore, with this speed of event there may have been more flares which were missed altogether. From the available evidence, the optical and IR were simultaneous to within 12 hours.

If the mm-submm is the same emission component as the optical/IR, one would expect a nearly simultaneous event to appear. Inspection of the 800 and 1100 μm data, show that the flares were followed to within at least 1 day as far as the rise times are concerned. 3C273 was observed on two occasions separated by 3.5 hours during one night and both the 1.1 and 0.8 mm flux were found to have increased by 50% between the observations. When 3C273 was again observed 3 nights later, surprisingly the flux had fallen sharply to its quiescent level. Therefore these results indicate a flaring and fading at 1.1 mm of timescales hours to days. This is clearly a problem for synchrotron cooling and the timescale of flaring gives a brightness temperature exceeding the limit of 10^{12} K, indicating that beaming was occurring.

Unfortunately 3C273 was only observed on a small number of occasions at mm wavelengths from then until May, but data from the JCMT and IRAM show that the flux remained at a low level. Considering the IR lightcurve of fig 4, the data end as 3C273 goes into the Sun in June, but the mm-submm observations continued and showed an increase. So we have evidence for a further large flare which was completely missed in the IR and visible, and this will clearly cause problems for correlation analysis when a radio flare is apparently devoid of any optical/IR precursor.

The June flare was relatively short lived, appearing at first sight to fade to its lower level in the space of 1-3 months. However, there appears to be evidence of more rapid fluctuations superimposed on this decline, including one extremely rapid fall of 30% taking place in the space of a few hours. Looking at the declines at mm-submm wavelengths, synchrotron cooling mechanisms clearly hit trouble, requiring extremely large magnetic fields combined with very high values of relativistic beaming to achieve the timescales involved. We believe there is a another explanation.

The optical/IR/mm February results show rapid swings in optical polarization. This, together with the very fast mm variations, suggests that we might not be dealing with synchrotron cooling but rather with geometrical effects. One possible explanation for the extremely rapid variability (for the wavelengths concerned) is that there are wiggles in the relativistic jet which occasionally move the flaring component of the beam out of the observers line of sight. Wiggles in jets are now commonly seen in VLBI maps (see Zensus this volume and are also discussed by Hughes).

The above work on 3C273 (Robson et al 1991) has demonstrated two facets: (1) that unless observations, even at 1mm wavelength, are undertaken on timescales of days to a week, flaring events can be missed; (b) some of the more rapid events (days) are probably best described by geometrical effects of the flaring relativistic beam rather than cooling timescales.

What about the longer wavelength radio emission of 3C273 during this time. We now have data from 0.45mm up to 90GHz and the longer wavelength monitoring from Metsahovi and UMAS. Inspection of the 90GHz data gives us a clue to what is probably happening. Although only a factor of 3 lower in frequency than the 1.1 mm data, the lightcurve fails to show the clear tracking of the optical/IR flare as demonstrated by the 0.8, 1.1, 1.3 and 2.0 mm wavelengths. Furthermore, the 90 GHz maximum of the flare appears to have been delayed by about 70 days.

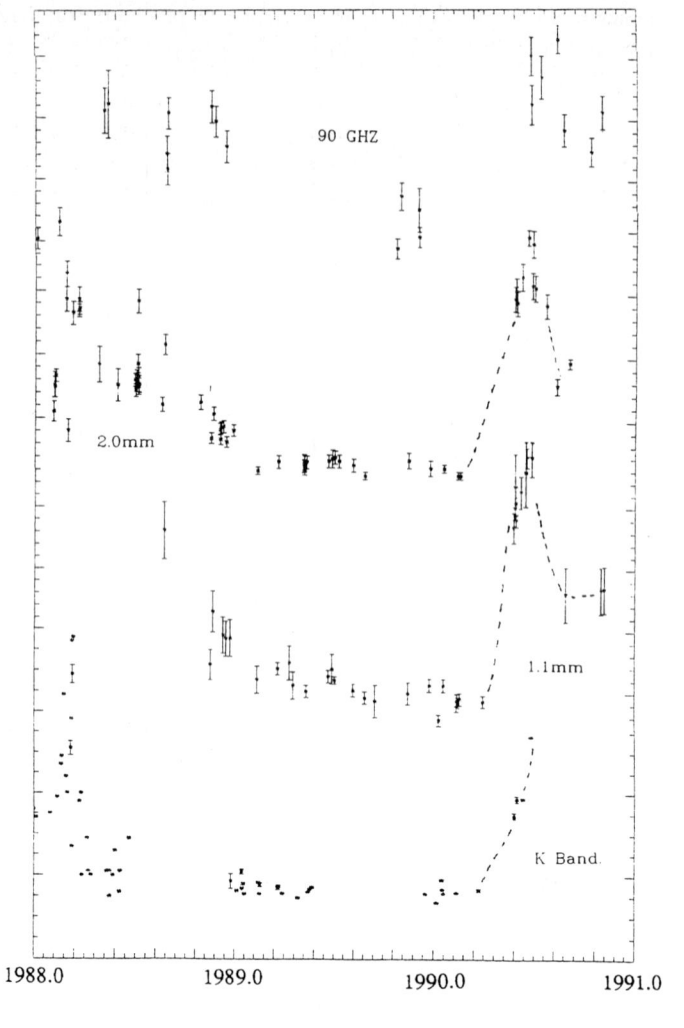

Figure 4

Composite light curve of 3C273 from 1988.0 until 1991 at 2.2µm, 1.1, 2.0 and 3.3 mm, showing the fast flaring of early 1988 and the slower flare in mid 1990.

The 22 & 37 GHz data are particularly interesting, both clearly show a large flare developing in 1988. Assuming this was the same flare as the March IR/optical then the onset was between 4 and 17 days, but the subsequent rise was much more gradual than at the higher frequencies, the maximum being recorded sometime in August, a delay of around 5 months, following what appears to be a linear rise in flux with time. The 22 GHz (1.36cm) flare was yet more delayed, with the maximum being recorded less than 1 month later. At longer wavelengths still, the UMAS 2cm data clearly show the onset of the flare in late May, a delay of around 80 days, with a slow rise to maximum which occurred in late December. The 3.75cm (8 GHz) and 6.3cm (4.8 GHz) light curves show even slower and less pronounced response.

The overall pattern seems to be that the optical/IR/near mm region responds the same; synchrotron cooling times lengthening the flux declines at longer wavelengths and geometrical effects probably playing a significant role at all the wavelengths. From 3mm longwards, the propagation of the energy is different, with a much slower rise time and subsequent delay of flare maximum, although nothing like the years which have been quoted for other beaming objects. The amplitude of variability followed a frequency dependence, with higher frequencies undergoing a much larger amplitude outburst.

4.2 SPECTRAL EVOLUTION

What has been learned of spectral development during flaring behaviour from all the data available? Gear *et al.* (1986) showed that taking the IR flux values for OJ287 the spectrum flattened as the source brightened and steepened as it faded. The interpretation was an injection or re-acceleration of electrons with a harder spectrum. This behaviour was also observed by Brown *et al.* (1989 a,b) and by Tanzi *et al.* (1989) although for a very limited sample. On the other hand, McHardy (1989) showed that for 1156+295, there was almost no change in spectral shape with variability, perhaps a slight steepening with fading. For OJ287, Kikuchi (1989) showed that the Gear *et al.* behaviour was only observed in the optical during periods of flaring behaviour; if a long term view of the source was taken, the correlation vanished. So clearly the behaviour is consistent with the large flaring components. However, Bregman *et al.* (1990) in their study of the long term behaviour of BL Lac claim that the infrared spectrum *softened* during a major outburst in 1980. Clearly the definitive position is very uncertain, mostly because of the relative paucity of data, particularly following a flare on a daily, or even hourly basis. This is one of the areas which merit more extensive work, to determine the energy spectrum of the electron population responsible for the flaring behaviour.

For the mm-submm regime, the situation is much more complex as this is the turnover part of the synchrotron spectrum and the behaviour is much more difficult to analyse with slopes changing from positive to negative as the wavelength changes from optically thin to thick as the flare develops. With 3C273, the flaring behaviour was so rapid that attempting to match the various wavelengths measured has proved very difficult. However, the general development of the flare through the mm region, by 3C273 and other blazars, suggests that the Marscher and Gear model may require some further refinement as the flare is flatter than the assumptions allowed in their model. The mm variability will be published by Robson *et al.* 1991b.

5 THE FUTURE

So where are we now? We have made big advances over the past ten years but we still do not understand the precise relationships between the various wavelength regimes for the Blazar phenomenon overall. On the other hand, clear relationships have been determined for a small number of individual objects, and even here we see a variety of result. We clearly lack monitoring in the X-ray and are oblivious to the shape of the spectrum in the EUV, which for OVV quasars could be a major luminosity regime. In modelling flaring behaviour, advances have been made using shocks in relativistic jets but we still lack sufficiently detailed observational evidence to pin down sufficiently the theories of emission mechanisms. This latter point can however be tackled.

What is required is a continuous monitoring programme in the optical-IR-mm domain. In principle the IR is the best wavelength for overall coverage as the moon does not pose so much of a problem as in the optical and there are a range of telescopes available. However the ease of instrumentation and many more telescopes for the latter makes it the better choice. But this is only part of the story, the monitoring data must be continuously reduced and a coordinated switch-on of telescopes at IR, far-IR, submm, mm and cm must follow rapidly the discovery of a flare. This is something we should definitely strive for, perhaps through the IAU as the natural multinational organisation using EMAIL (when it works) as the medium. I do not wish in any way to belittle the efforts of those who staff the smaller telescopes to provide the hard won monitoring capability which has led to many of the results and advances of today. However, for the future I firmly believe we must go one stage better.

To this aim, some of us in the UK are proposing a robotic telescope of 1-m aperture, to be sited in the Canary Islands, to have a CCD photometer, with artificial intelligence for pattern recognition and source acquisition, and semi-real time data analysis with flags for flaring behaviour above some predetermined flux level. The aim of such a project is to provide an early warning facility so that a world-wide network of major facilities at other wavelengths can respond rapidly. This would allow flaring objects to be observed in over-ride time on major facilities such as UKIRT, the JCMT, IRAM, SEST etc. Flaring behaviour could then be followed in precise spectral and temporal detail giving precise data to which theories can be matched.

As far as the radio through optical spectrum is concerned, the problem with the observational gap in the far infrared will be removed with the launch of the ISO mission sometime in 1993. This should be a tremendous boost to these studies and will provide definitive snapshot spectra and possibly polarization as well. There are already plans within the CORE programme to make detailed photometry of a sample of Blazars. These plans will be published in October 1991, and then guest observations can be applied for to cover any gaps or to perform other observations. Critical to all these will be ground-based pre-mission, simultaneous and follow-up observations and that is something that we as Blazar pundits should ensure is organised well before the event.

REFERENCES
Barbieri,C. *et al.* 1985, Astr.Ap., **142**, 316
Barbieri,C. *et al.* 1990, ApJ, **359**, 63
Bregman,J.N. *et al.* 1988, ApJ, **331**, 746
Bregman,J.N. & Hufnagel, 1989, in BL Lac Objects, eds Maraschi, Maccacaro & Ulrich, Springer-Verlag, p159
Bregman,J.N, *et al.* 1990, ApJ., **352**, 574
Bregman,J.N. 1990, Astron.Astrophys Rev. 2:125
Brown,L,M,J. *et al.*,1989(a), ApJ., **340**, 129
Brown,L.M.J., Robson,E.I.,Gear,W.K. & Smith,M.J. 1989(b), ApJ., **340**, 150
Clegg,P.E. *et al.* 1983, ApJ., **273**, 58
Courvoisier,T.J-L, *et al.* 1987, Astron.Astrophys., **176**, 197.
Courvoisier,T.J-L, *et al.* 1988, Nature, **335**, 330.
Courvoisier,T.J-L., *et al.* 1990, Astron.Astrophys., in press
Gear,W.K.,Gee,G,Robson,E.I. & Nolt,I.G. 1985, Mon.Not.R.astr.Soc., **217**, 281
Gear,W.K,Robson,E.I & Brown,L.M.J. 1986, Nature, **324**, 546
Landau,R. *et al.*1986, ApJ., **308**, 78.
Robson,E.I. *et al.*1983, Nature, **305**, 194.
Robson,E.I. *et al.* 1986, Nature, **323**, 134.
Robson,E.I. *et al.* 1991a, in preparation
Robson,E.I. *et al.* 1991b, in preparation
Steppe,H. *et al.* 1988 Astron.Astrophys.Suppl.Ser., **75**, 317.

The Cm-Wavelength Flux Behavior of AGNs

M. F. ALLER, H. D. ALLER, P. A. HUGHES, and G. E. LATIMER

Department of Astronomy, University of Michigan

ABSTRACT
Results from the long-term monitoring program carried out with the University of Michigan 26-meter telescope are presented. Characteristics of the flux variability are discussed including time scales, the effect of opacity on the observed outbursts, a search for periodicities, and class differences between BL Lac objects and QSOs.

OVERVIEW OF PROGRAM
The University of Michigan 26-meter telescope has been devoted to the study of variable sources for over 25 years. Observations at 8.0 GHz commenced in 1965, and coverage at 14.5 and at 4.8 GHz was added in 1974 and 1978 respectively. In 1977 we automated our telescope, allowing us to operate around the clock; this procedure has dramatically increased the number of sources we observe and significantly improved the homogeneity of the database. A core group of about 60 sources, selected on the basis of current activity, is observed on a weekly basis; the objects are primarily a mix of BL Lacs and QSOs. Additionally, multifrequency observations of special classes of objects are obtained monthly to tri-monthly; included are a sample of low frequency variables as part of a long-term continuing program with Bologna (Padrielli *et al.* 1987), a flux-limited sample of BL Lac objects, and Pearson-Readhead survey sources (Pearson and Readhead 1984). Because of the overlap of the objects in these various programs, quite good coverage is available for over 200 objects and sporadic data are available for many others. The primary goals of the program are: to search for isolated well-defined events in individual objects which may be suitable for modeling and hence for studying the physical conditions in active objects; to identify and quantify common properties in the long-term database (time scales and amplitude of variability, repeatability in a given source, periodicities, spectral evolution during outbursts, differences in behavior as a function of class); to identify active objects for future VLBI studies; and to participate in broadband cooperative studies.

PROPERTIES OF THE VARIABILITY
In Figure 1 we show two-week averages of our data for BL Lacertae. The outbursts modeled by Hughes *et al.* (1989) illustrate the kind of events we are seeking to find in the database, and these particular events are unique for a number of

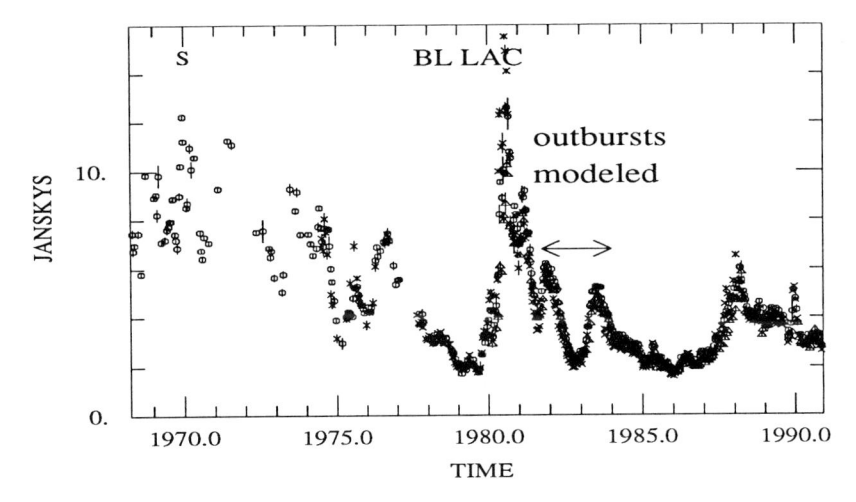

Figure 1. Two-week averages of the total flux density of BL Lacertae versus time. The data at 14.5 GHz, 8.0 GHz and 4.8 GHz are represented by crosses, open circles and triangles respectively; this symbol convention is adopted throughout.

reasons: they are relatively well-resolved in time, the spectral behavior indicates that the emitting region is relatively transparent during the outbursts, and, most importantly, they each have an associated well-defined outburst in polarized flux which provides additional important constraints on the models. The rapid declines seen in the outbursts in this source are characteristic of the behavior seen in BL Lac objects, and it is this feature which allows us to distinguish individual events more readily in BL Lacs than in other AGNs. The activity in OJ 287, shown in Figure 2, can be characterized as continuous, and this source exhibits the most rapid variability we have seen. Here the outbursts exhibit severe overlapping, and self-absorption effects are clearly important. While typically a source is observed only once per week at each frequency, in early 1983 day-to-day observations of a number of BL Lac objects were obtained at 14.5 GHz as part of a special polarization study; at this frequency the variations are generally most rapid and greatest in amplitude. Daily averages for OJ 287 shown in Figure 3 exhibit rapid variability with large percentage changes on times scales of a day to a week, a characteristic we have seen in other objects. Such behavior represents a *flicker* superimposed on the overall activity, and it is a feature of the variability which must be explained by any adequate interpretation of the variability phenomenon.

Several investigators have attempted to ascertain whether or not periodic behavior is present in active extragalactic objects, and, as part of this effort, Philip Hughes has computed Scargle periodograms (Scargle 1982) for the best observed sources in our database. The result for OJ 287 is shown in Figure 4. The vertical lines draw

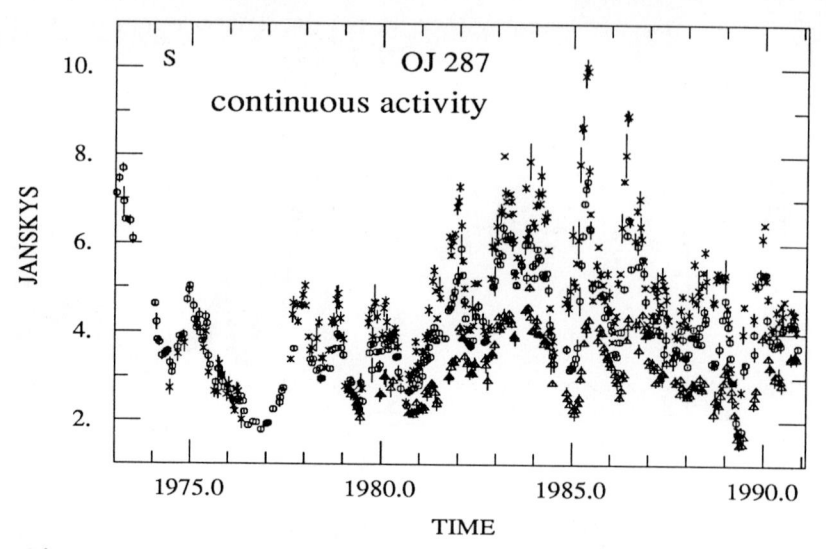

Figure 2. Two-week averages of the total flux density of OJ 287 versus time.

attention to peaks in the periodograms. The rightmost and the leftmost lines mark peaks which we believe are spurious: the first results from the finite length of the data while the last corresponds to a peak in the window function and also appears to be aliasing. The peaks marked by the centermost line are not associated with peaks in the window function, so they are more difficult to discount. Nevertheless, we would have expected the peak in the 8.0 GHz periodogram to be more prominent than that at 14.5 GHz since the data span is longer for this frequency, and that behavior is not seen. We conclude that there is *weak* evidence for periodicity at 1.67 years in this object and note that this is the only source in which we see *any* evidence, albeit weak, for periodicity.

While many BL Lac objects show the rapid behavior seen in the previous figures, this is not true of all BL Lacs. For example, Mkn 501 exhibits only low amplitude variability. The object shown in Figure 5, ON 231, exhibits a different type of behavior, one more commonly seen in QSOs, which can be characterized as fluctuations superimposed on a long-term decay. Thus, within the BL Lac class alone, we find a wide range of behavior exhibited in the light curves.

Another question of interest is whether or not there are changes in the state of activity in a given source. While the data span available at radio wavelengths is substantially shorter than in the optical, such changes clearly exist, and the best example is 3C 120 (Figure 6). Until the mid-1970s this object was one of the most active, exhibiting large, self-absorbed events; subsequently the source has been

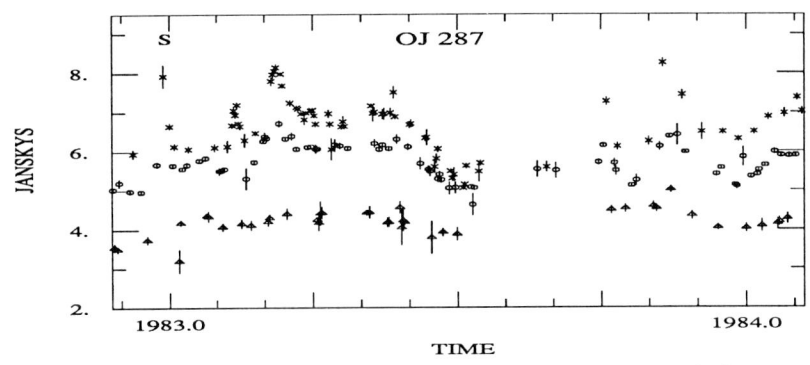

Figure 3. Daily averages of the total flux density of OJ 287 versus time during 1983.

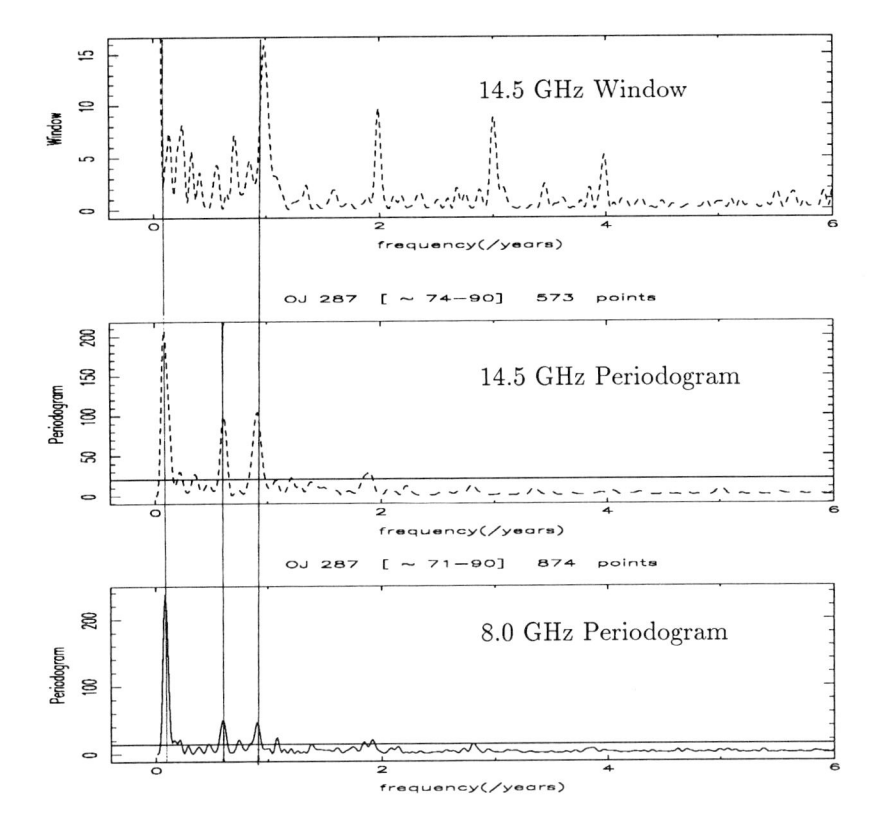

Figure 4. Periodograms for OJ 287. The top panel shows the 14.5 GHz window function, the middle and lower panels show the periodograms for 14.5 and 8.0 GHz. The numbers in square brackets give the date ranges of the data.

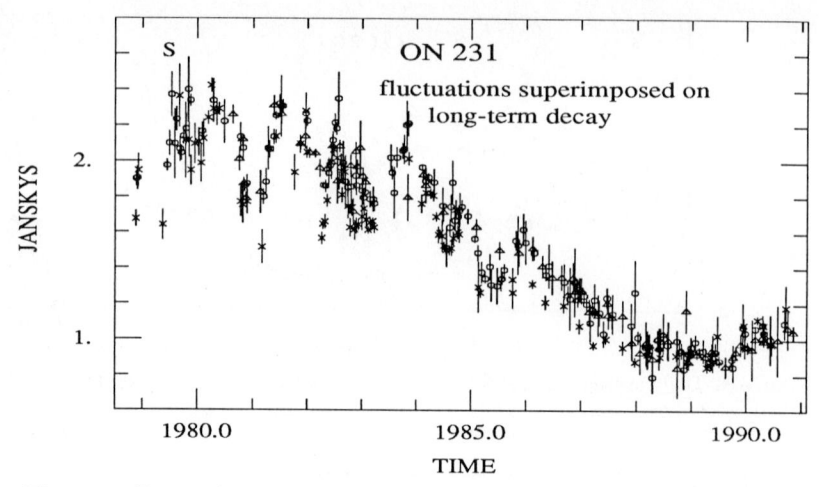

Figure 5. Two-week averages of the total flux density of ON 231 versus time.

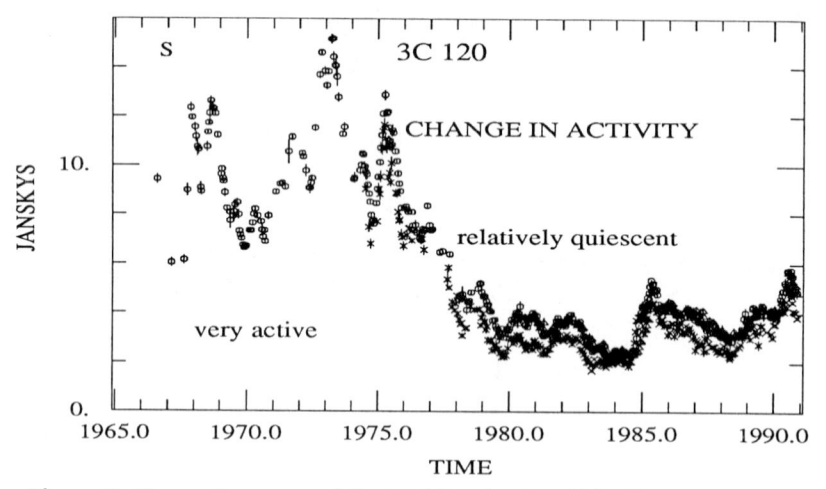

Figure 6. Two-week averages of the total flux density of 3C 120 versus time.

relatively quiescent and the spectrum more nearly flat. Another source which has exhibited a large change is 1308+326 (Figure 7). Through 1986 the flux appeared to fluctuate above an apparent preferred level of 1.7 Jy; thereafter it decreased sharply. We have seen this type of behavior in many sources, where a preferred flux level appears to exist for a substantial period of time followed by a drop, and it illustrates why it is dangerous to remove "baseline fluxes", presumably produced by extended features, based on light curves only; additional information from VLBI on the contribution from extended structure is clearly needed in order to make

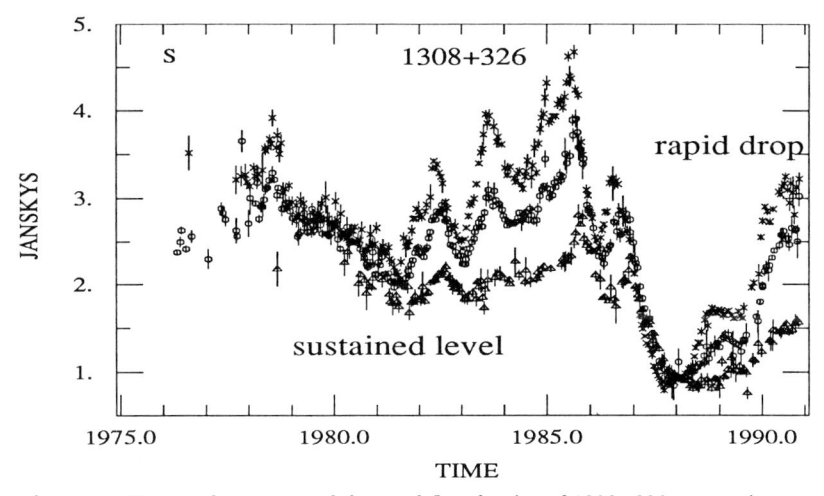

Figure 7. Two-week averages of the total flux density of 1308+326 versus time.

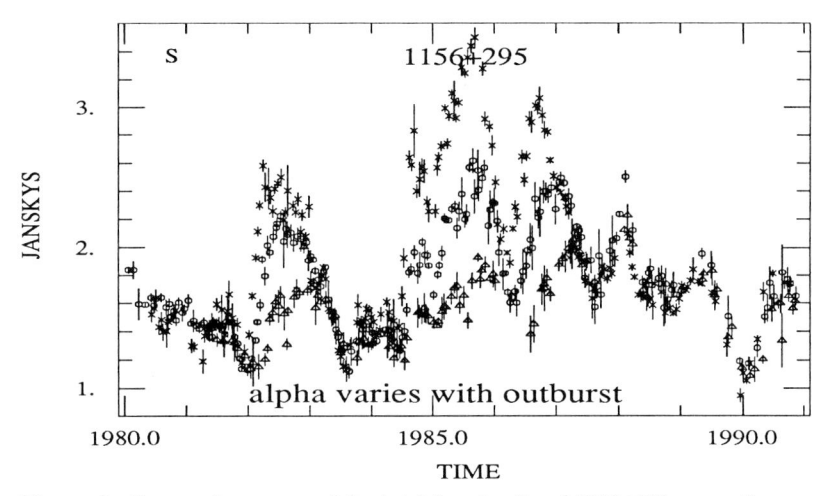

Figure 8. Two-week averages of the total flux density of 1156+295 versus time.

such corrections. Figure 8, which shows the data for 1156+295, illustrates that the spectral behavior in a given source and during a relatively short period, 11 years, can vary significantly. The outbursts in the early and late 1980s are flat while those during 1982-1987 are moderately steep with indices on the order of 0.6. Another feature of the variability is illustrated in Figure 9 which shows both the flux and polarization data for the QSO 2134+004. The light curve shows a slow decrease but no resolved events. In contrast, significant changes are apparent in the polarized flux and polarization position angle. The VLBI observations of Pauliny-Toth *et*

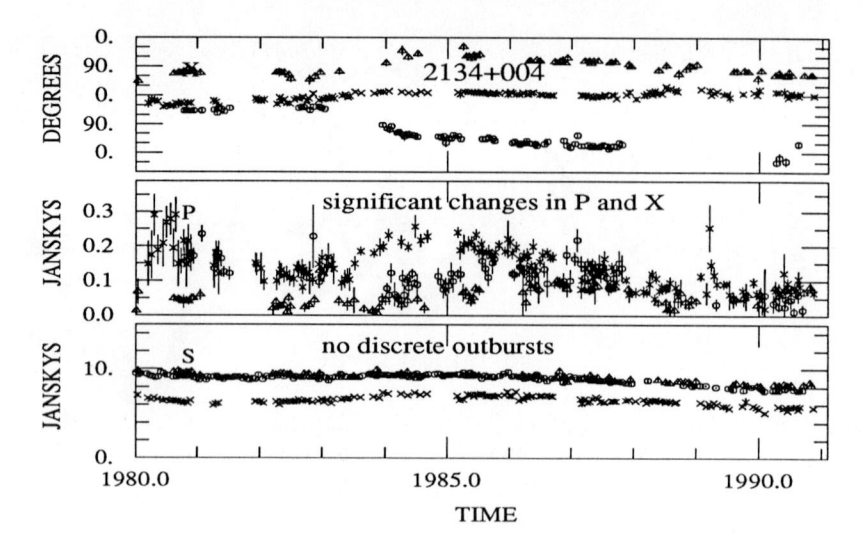

Figure 9. Two-week averages of the total flux density and polarization of 2134+004 versus time.

al. (1990) show that very significant changes occurred in this source subsequent to 1984 in number of components, source extent and orientation. The combined data clearly illustrate that large changes in source structure can be hidden in the light curves either because of the way in which the contributions from individual components combine and/or because of opacity; polarization data are often a more sensitive tool for probing such changes. A similar behavior can be noted in other sources. The QSO 4C 39.25 has exhibited large changes in polarization concurrent with little structure in the light curves; Marscher et al. (1991) have interpreted the behavior in that source in terms of a twisted jet.

Because of the wide range of behavior found, we have sought to characterize the time variability of our long-term data using structure functions. The procedures and results of this analysis are described in Hughes et al. (1991). The slope of the rise portion of a plot of log (structure function) versus log (timelag) gives information about the process responsible for the variability, while the log (timelag) turnover gives a characteristic time scale. The analysis shows that the slopes are independent of both frequency and source classification which suggests that a common process is responsible for all variability; the slope values match those expected for shot noise. The time scales given by the turnovers quantify the result that BL Lacs vary more rapidly than QSOs.

CONCLUSIONS

On the basis of the light curves there are class differences between BL Lac objects and QSOs. The BL Lac objects typically exhibit more rapid, clearly resolved events of higher fractional amplitude compared with QSOs; the structure function analysis quantifies the result that the maximum correlation times are shorter for BL Lacs. We would like to find evidence of a universal outburst shape in our data, but, unfortunately, blending and opacity mask individual events in the light curves; nevertheless, the structure function analysis suggests that a common origin is responsible for all activity. Finally, we find no strong evidence for any periodicities on the order of weeks to a decade.

This research was supported in part by NSF grant AST 88-15678.

REFERENCES

Hughes, P. A., Aller, H. D., and Aller, M. F. 1989, *Astrophys. J.*, **341**, 68.
Hughes, P. A., Aller, H. D., and Aller, M. F. 1991, these proceedings.
Marscher, A. P., Zhang, Y. F., Shaffer, D. B., Aller, H. D., and Aller, M. F. 1991, *Astrophys. J.*, submitted.
Pearson T . J. and Readhead, A. C. S. 1984, in *VLBI and Compact Radio Sources*, ed. R. Fanti, K. Kellermann and G. Setti (D. Reidel: Dordrecht, Holland), p. 15.
Padrielli, L., Aller, M. F., Aller, H. D., Fanti, C., Fanti, R., Ficarra, A., Gregorini, L., Mantovani, F., and Nicolson G. 1987, *Astron. Astrophys. Suppl. Ser.*, **67**, 63.
Pauliny-Toth, I. I. K., Zensus, J. A., Cohen, M. H., Alberdi, A., and Schaal, R. 1990, in *Parsec-Scale Radio Jets*, ed. J. A. Zensus and T. J. Pearson (Cambridge: Cambridge University Press), p. 55.
Scargle, J. D. 1982, *Astrophys. J.*, **263**, 835.

Centimeter-Wavelength Linear Polarization Observations of Active Galactic Nuclei

H.D. ALLER, M.F. ALLER, and P.A. HUGHES

Department of Astronomy, University of Michigan

ABSTRACT
Characteristics of the time variability in the linear polarization of extragalactic sources at centimeter wavelengths are discussed. The polarization position angles are often related to the orientation of compact jets (observed by VLBI) in cores of active extra-galactic objects; the competing effects of shocks, propagating in the relativistic flow, and synchrotron self-absorption complicate the observed polarization behavior. The polarization can serve as a probe of the nature of the flow, and of the characteristics of the shocks, in the radio emitting regions.

SOURCES WHERE A "SINGLE" EFFECT DOMINATES
As discussed in the previous paper (Aller *et al.* 1991), extragalactic active sources exhibit a wide range of variability behavior; the typical source shows repeated outbursts which may or may not overlap in time. These outbursts are also often evident in linear polarization, although the temporal variations of the polarized flux do not mimic the total flux. A common characteristic is that the polarization position angle will remain near the same value from outburst to outburst. This "stability" is illustrated in Figure 1, which shows the University of Michigan data for 3C 120 since 1980. Also typical of most variables, the degree of polarization rarely exceeds five percent, indicating that the magnetic fields are not highly ordered. The polarized flux density, while not mimicking the total flux density, has changed (fractionally) by an even larger ratio than the total flux, but the polarization position angle has generally remained within 10 degrees of the direction perpendicular to the extended structure observed by VLBI (Walker *et al.* 1987). We believe that this is an example where the polarized flux density is dominated by (largely transparent) turbulent cells in the jet flow which have magnetic field directions that are (almost) random: the polarization fluctuations result from different cells "lighting up" as a disturbance propagates down the flow, and the stable position angle results from a slight preference for the magnetic field to be aligned along the flow (Jones *et al.* 1985).

A similar picture holds for BL Lac, (Figure 2), except that during the bursts (*e.g.* in late 1981 and mid 1983), the polarization was dominated by a single source component where an axial compression caused the magnetic field to be strongly

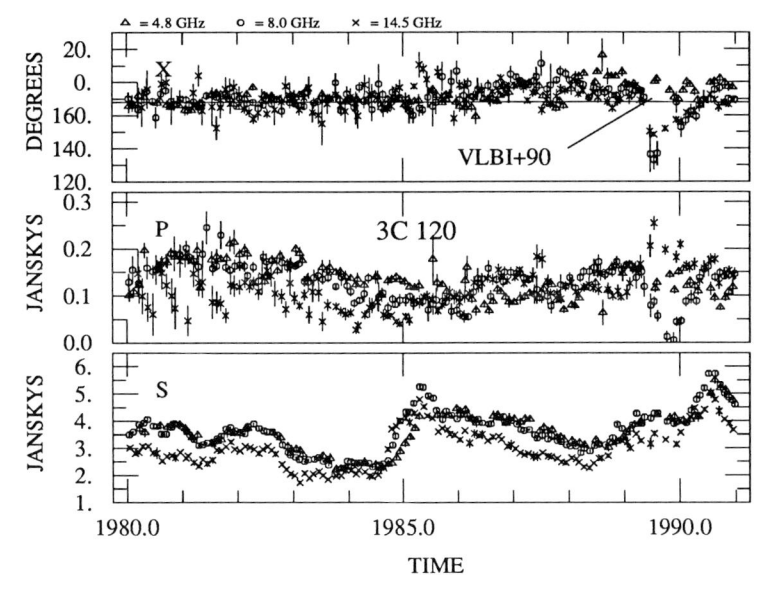

Figure 1. Monthly averages of the total flux density, polarized flux density and polarization position angle of 3C 120 versus time. The horizontal line through the position angle data is 90 degrees (perpendicular) to the axis of the inner jet structure observed by VLBI (Walker 1987).

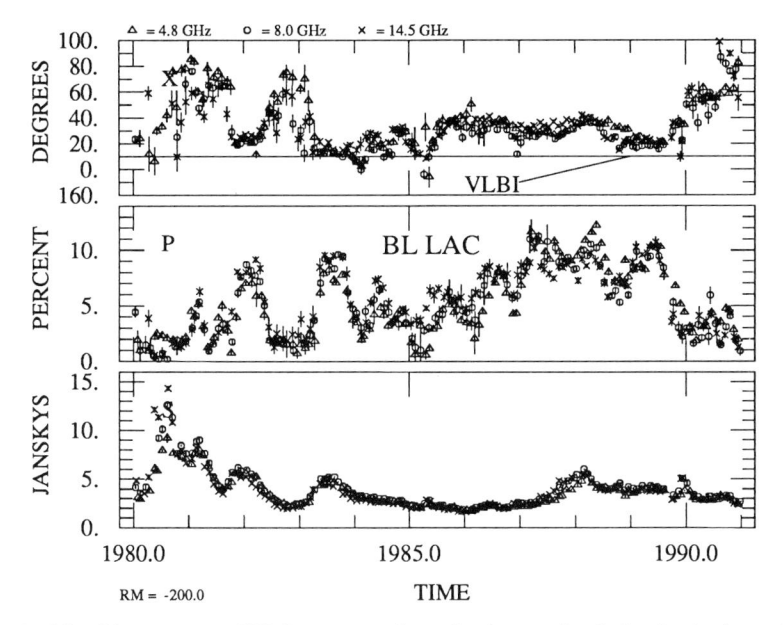

Figure 2. Monthly averages of BL Lac versus time; the degree of polarization is shown in percent. The polarization position angles have been corrected for a rotation measure of -200 rad m^{-2}. The horizontal line shows the orientation of the VLBI jet (Mutel et al. 1990).

aligned perpendicular to the flow direction. This axial compression (which we have successfully modeled as a propagating shock, Hughes, Aller and Aller 1989) causes the electric vector of the polarized emission to become aligned parallel to the relativistic flow. As the shocked region decays by adiabatic losses, the associated polarization rapidly decreases and the polarized emission is again dominated by the conditions in the "quiescent" jet (see Figure 3). In many sources, the polarized emission is controlled by emission from the quiescent jet, and in a few cases we have found individual bursts which are sufficiently isolated in time and strong enough to produce a clear signature in integrated polarization (BL Lac and 3C 279 are the best examples). The majority of sources show a complex behavior which results from the simultaneous effects of: bursts which overlap in time (*i.e.* multiple source components are simultaneously present in the emitting region), opacity in the emitting region, and changes in the direction of flow or in the direction of shock propagation in the source.

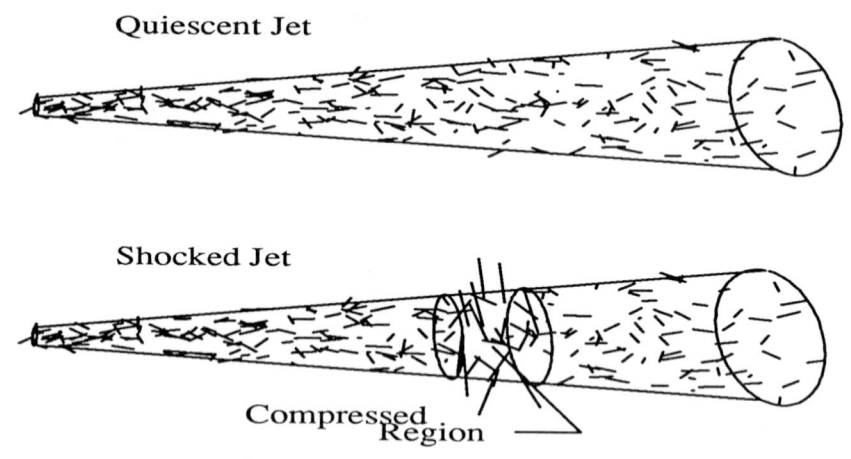

Figure 3. Schematic model of a quiescent and a shocked jet. The short lines represent the random orientations of the magnetic fields in small cells within the jet flow. The effect of an axial compression is to strengthen the component of the magnetic field perpendicular to the axis of the jet. An observer sees the maximum apparent alignment when situated perpendicular to the jet (in the non-relativistic case) or at a small angle to the jet if relativistic aberration is present. An observer along the jet axis sees no field alignment by this process

3C 273: A TYPICAL COMPLEX SOURCE

Figure 4 shows the long-term Michigan data for 3C 273; like 3C 120, the polarization position angle has tended to remain perpendicular to the jet axis observed by VLBI (indicating that the axial magnetic field component is dominating), but fluctuations in position angle are larger and they are frequency dependent. There are multiple source components, and internal opacity is important. Note that the 4.8 GHz polarization (where the opacity is higher) has deviated from that at the two higher

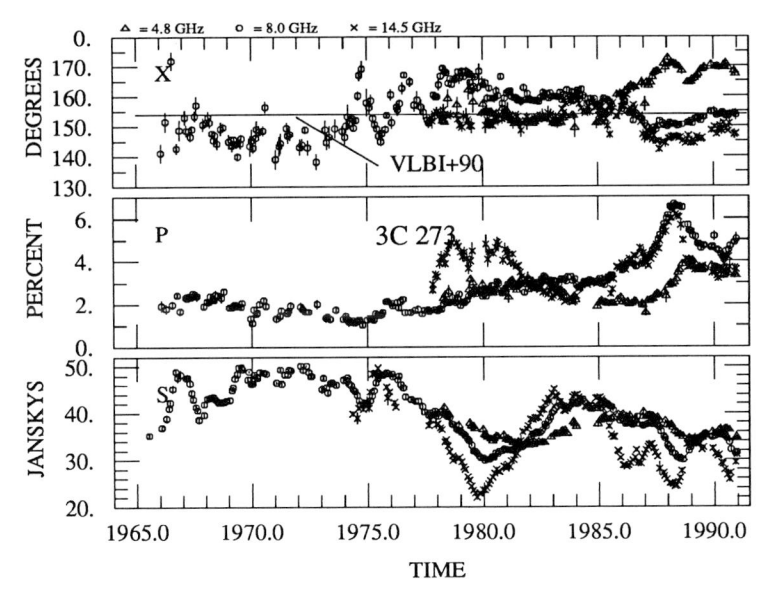

Figure 4. Monthly averages of the total flux density, polarized flux density and polarization position angle of 3C 273 versus time. The horizontal line through the position angle data is 90 degrees (perpendicular) to the axis of the inner jet structure observed by VLBI (Unwin *et al.* 1985).

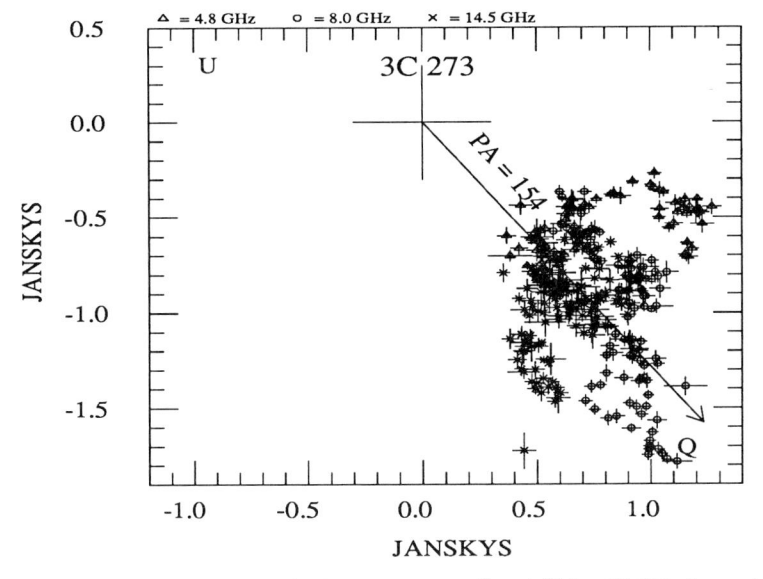

Figure 5. Monthly averages of the Stokes parameters Q and U for 3C 273 observed since 1980.0 shown as a polar plot. The arrow from the origin indicates the direction perpendicular to the VLBI jet.

frequencies since 1985. The nature of the polarization is shown in another way in Figure 5, which is a polar plot (in terms of the Stokes parameters) of the data since 1980. Although the time order is not shown, it is evident that the ranges of the polarization (in terms of direction and amplitude) are quite different at the three frequencies. To account for this complex picture, the source emitting region must contain multiple variable components which have significant differences in the orientation of their magnetic fields; at least some of these components have significant synchrotron self absorption at centimeter wavelengths.

In sources with isolated bursts, such as BL Lac and 3C 279, outbursts have been well described by a simple axial shock model, so that the electric vector of the polarized emission is parallel to the apparent jet axis. However, even in BL Lac, we have observed bursts where the axis of the compression apparently makes a large angle to the inferred (from VLBI) flow direction (Aller, Hughes, and Aller 1991). To investigate the relative amplitude of polarization variations in relation to the jet axis in 3C 273, Figure 6 shows the data in Figure 5 plotted versus time, but with the coordinate system rotated so that the jet axis direction corresponds to positive Q, zero U. One can see that while the variations in Q (which can be accounted for by axial compressions alone) are slightly larger in amplitude, the variations in U are quite significant; the magnetic fields are not confined to being either perpendicular or parallel to the apparent jet axis.

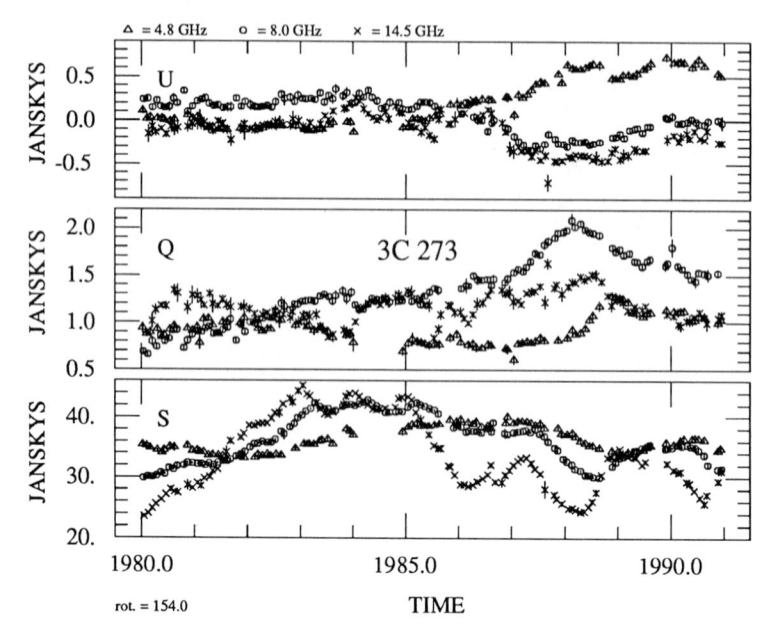

Figure 6. Monthly averages of the total flux density, and the Q and U Stokes parameters for 3C 273 in a coordinate system that is rotated so that the direction perpendicular to the jet axis is along positive Q.

TEMPORAL CHANGES IN SHOCK DIRECTION

As mentioned above, the series of bursts in BL Lac in the early 1980s exhibited axial compressions, but a later burst (which peaked in early 1988) appears to have resulted from a compression along a direction that is almost 40 degrees different (see figures in Aller, Hughes & Aller 1991). The very active BL Lac object OJ 287 also has a relatively stable polarization position angle over many bursts (Figure 7); and like BL Lac the polarization position angle lies near the VLBI orientation. However, during individual bursts, the polarization can be quite different, as shown (Figure 8) by polar plots of the Stokes parameters during two bursts (marked A and B in Figure 7). In each burst the highest polarized flux is seen at 14.5 GHz (where the internal source opacity is least) but the orientation of the polarization is quite different. If propagating shocks (with their associated compression of the quiescent, turbulent magnetic field in a plane perpendicular to the shock axis) are the cause of all bursts in these sources, then the polarization behavior we see in these two sources suggests that the shocks appear with different orientations, or that the orientation of the shock axis may change as a disturbance propagates down the flow.

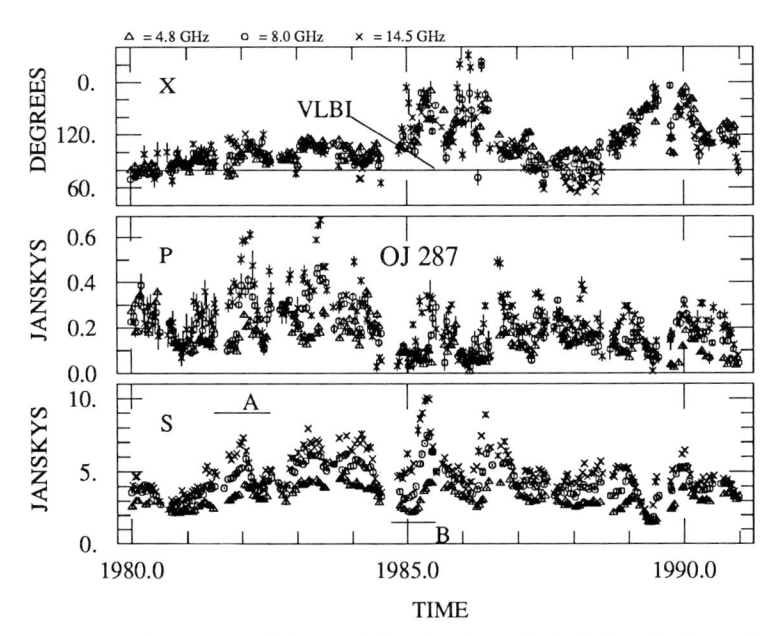

Figure 7. Two-week averages of the total flux density, polarized flux density and polarization position angle of OJ 287 since 1980. The horizontal line through the position angle data is along the axis of the jet structure observed by VLBI (Gabuzda 1991)

The question of the relation between the axis of the propagating shock and the direction of flow is an interesting one which is still unsolved. As illustrated in Figure 9, there are three basic regimes which may exist in these sources. There is frequent "circumstantial" evidence for curved jets (Figure 9a) from the variations in position angle observed by VLBI experiments using different angular resolutions

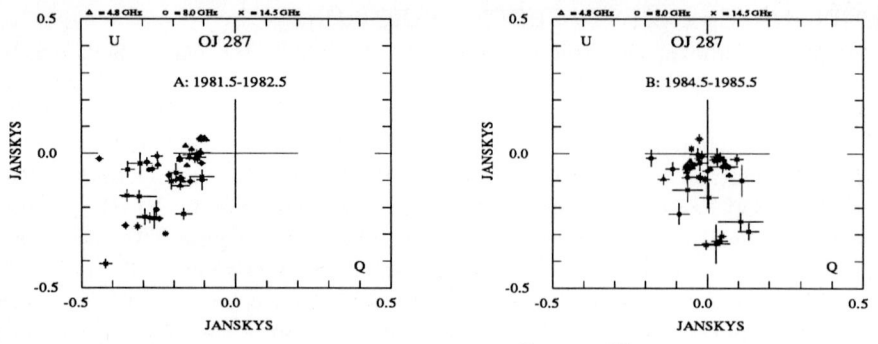

Figure 8. Two week averages of the Stokes parameters Q versus U for OJ 287 in the two time periods marked A and B in Figure 7.

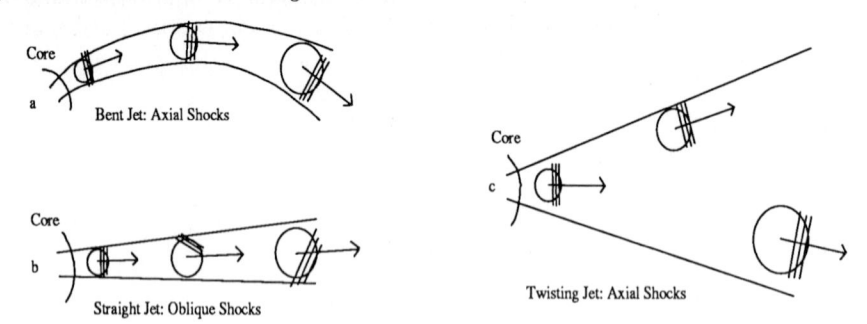

Figure 9. Cartoon of possible jet/shock configurations. The solid lines represent the boundary of a relatively homogeneous region of steady out flow; the local direction of flow is indicated by the arrows. Circles represent shocked regions (which would appear as VLBI components); and the parallel lines indicate the plane perpendicular to the shock axes.

(*e.g.* the source 3C 345, Biretta, Moore, & Cohen 1986); and axial shocks propagating down such a curved flow could reproduce the observed polarization position angle variations. However, dynamical arguments (it appears difficult to change the direction of a relativistic flow) suggest that a more likely scenario is that illustrated in Figure 9c, where the flow is rectilinear (over a relatively wide opening angle), but the "visible" jet follows the instantaneous path of the shock front, which need not propagate in the same direction as the flow. Finally, shocks may form which are oblique to the flow direction (Figure 9b), and this orientation may change as the disturbance propagates down the flow. Although polarization maps of VLA scale jets have not resolved this question, there are reasons to believe that VLB polarization maps at centimeter wavelengths would be able to settle this question for parsec scale jets.

CONCLUSIONS

The range of behavior we have observed in centimeter wavelength polarizations is consistent with the idea that shocks propagating in a relativistic flow are responsible for radio outbursts and moving VLBI components. The orientation of the polarization is either perpendicular to the jet axis if an axial component of the magnetic field is relatively strong, or is parallel to the jet if the emission from axial shocks dominates. We have found objects with long-lived source components and well-defined jets often to be members of the first group, and active BL Lac objects are often in the second category. Some objects, such as 3C 279, exhibit characteristics of both groups, with the polarization position angle flipping back and forth between being parallel and perpendicular to the jet axis. As described here, in many sources we find evidence that either oblique shocks form in the flow or, if the shocks are always along the flow, that given sources exhibit a range of flow directions, either due to actual deflections (difficult to understand) or because the apparent jet only lights up part of a wide flow channel.

This work was supported in part by NSF Grant AST-8815678.

REFERENCES

Aller, H.D., Hughes, P.A., & Aller, M.F. 1991, *In Variability of Active Galactic Nuclei*, ed. H.R. Miller & P.J. Wiita (Cambridge: Cambridge University Press), p. 172.

Aller, M.F., Aller, H.D., Hughes, P.A., & Latimer, G.E. 1991, these proceedings.

Biretta, J.A., Moore, R.L., & Cohen, M.H. 1986, *Astrophys. J.*, **308**, 93.

Gabuzda, D.C. 1991, these proceedings.

Hughes, P. A., Aller, H. D., & Aller M. F. 1989, *Astrophys. J.*, **341**, 68.

Jones, T. W., Rudnick, L., Aller, H. D., Aller, M. F., Hodge, P. E., & Fiedler, R.L. 1985, *Astrophys. J.*, **290**, 627.

Mutel, R. L., Phillips, R. B., Su, B. & Bucciferro, R. R., 1990. *Astrophys. J.*, **352**, 81.

Unwin, S.C., Cohen, M.H., Biretta, J.A., Pearson, T.J., Seielstad, G.A., Walker, R.C., Simon, R.S., & Linfield, R.P. 1985, *Astrophys. J.*, **289**, 109.

Walker, R. C., Benson, J. M., Unwin, S. C. 1987, *Astrophys. J.*, **316**, 546.

Modeling The UMRAO Database: Status and Application to Other Sources

P. A. HUGHES, H. D. ALLER and M. F. ALLER

Department of Astronomy, University of Michigan

ABSTRACT

We describe recent modeling of cm-waveband total and polarized flux variability of compact extragalactic radio sources, and discuss our general conclusions about source orientations and internal conditions. We demonstrate that our results are insensitive to poorly known parameters such as shock direction and filling factor. A statistical analysis of the UMRAO database using structure functions provides evidence that most, or all, of the observed cm-waveband variability is a consequence of shocks in relativistic jets.

RELATIVISTIC JET MODELS

We consider simple flows of synchrotron plasma, characterized by parameters specifying the flow shape and orientation, the magnetic field geometry, the flow velocity, and the details of the radiating particle spectrum. In shocked regions the conditions are determined in terms of the quiescent flow parameters using the relativistic jump conditions of Königl (1980). The modeling involves radiation transfer calculations through many lines of sight, enabling us to predict the behaviour of the total and polarized flux as a function of both frequency and time. Details may be found in Hughes, Aller & Aller (1989a,b). Fitting to the UMRAO data (see Aller et al. 1991) then enables us to determine the model parameters for each source. The total flux rises as a shocked region propagates into the optically thin portion of the flow, and then declines due to adiabatic losses. The high percentage polarization during an outburst arises because the shock compresses a turbulent magnetic field in a plane perpendicular to the flow axis, leading to a significant percentage polarization for radiation seen in this plane in the source frame. Modeling is particularly sensitive to the source orientation, because this determines the profile of the outburst *via* the time delay between the 'near' and 'far' parts of the flow, and the observed percentage polarization *via* relativistic aberration.

We have applied this model to the sources BL Lac, OT 081 and 3C 279. The model light curves are shown in Figures 1, 2 and 3. The best-defined outbursts are seen in BL Lac, for which we are thus able to get the best model fit. The other sources display more complex activity – a sequence of shocks, the total fluxes from which merge to give the appearance of a single 'event'. The modeling has shown trends in Ψ, the angle between flow and observer, α, the optically thin frequency spectral index, γ_i, the low energy spectral cutoff, and ϵ_B^2, the fractional energy in mean field; values of these parameters are presented in Table 1.

Table 1. Model Parameters

Source	Ψ	α	γ_i	ϵ_B^2
BL Lac	38°	0.25	20	0.04
3C 279	29°	0.15	35	0.05
OT 081	40°	0.40	37	0.15

Angle of View

We find an angle of view between flow axis and observer which is much larger than the commonly assumed value of $\lesssim 10°$. We note that our modeling provides very strong evidence for this, at least in the case of BL Lac, because the angle of view that allows the best model fit to the outburst *profile* is the same angle that allows us to correctly fit the *percentage polarization*; these fits are independent. Furthermore, Mutel et al. (1990) find independent evidence for a viewing angle $\gtrsim 30°$ in the case of BL Lac. We do not believe that these results are in conflict with the idea that many variable sources are seen close to the flow axis, and we suggest that there is a strong selection effect that causes us to choose sources for modeling that have large viewing angle: for even modest flow Lorentz factor (*e.g.*, $\gamma \sim 5$) viewing close to the flow axis would produce a Doppler factor significantly greater than one, causing time scales to be compressed (reducing our ability to resolve events), and lowering the source-frame emission frequency (reducing, through opacity, the polarized flux to insignificant levels). However, we stress that the fact that we seem to be viewing sources such as BL Lac at a significant angle means that *BL Lac-like properties cannot be explained as due to viewing close to the flow axis*.

Spectral Slope

We find an optically thin spectral index $\alpha < 0.5$; *i.e.*, the electron spectrum has a slope $\delta < 2.0$. The result is supported by Valtaoja et al. (1988). Studies at higher frequency (optical, X-ray) imply steeper spectra than we find. However, Robson (1991) notes a discontinuity in the spectrum of variable sources between radio and IR/optical wavebands, confirming the lack of a *simple* causal connection between the low and high frequency emissions. Studies of optically thin kiloparsec scale sources imply that $\delta > 2.0$. However, Heavens & Drury (1988) have shown that the first order Fermi acceleration process in a relativistic environment can be very efficient, with the production of flat electron spectra. Perhaps such a mechanism is responsible for the observed parsec scale emission, while a less efficient process re-energizes the electrons on the kiloparsec scale.

Low Energy Cutoff and Magnetic Field

We find that the electron spectra must be cut off at quite high Lorentz factor; otherwise the models would exhibit a frequency-dependent Faraday depolarization

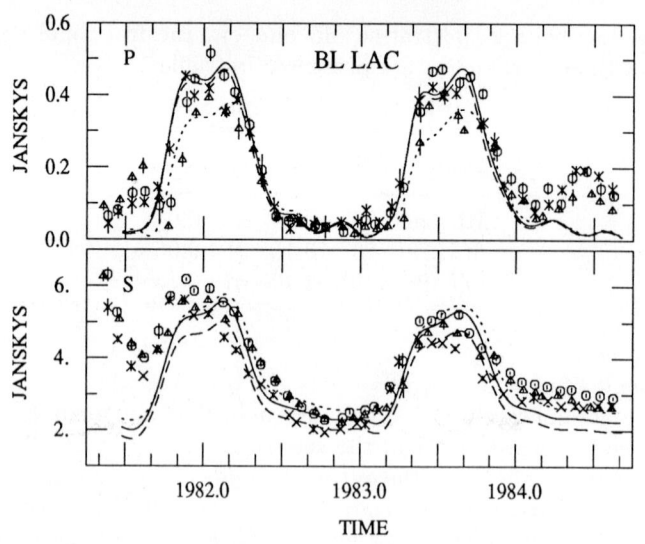

Figure 1. The BL Lac model superposed on monthly averaged data; short dashed lines and triangles: 4.8 GHz; solid lines and circles: 8.0 GHz; long dashed lines and crosses: 14.5 GHz.

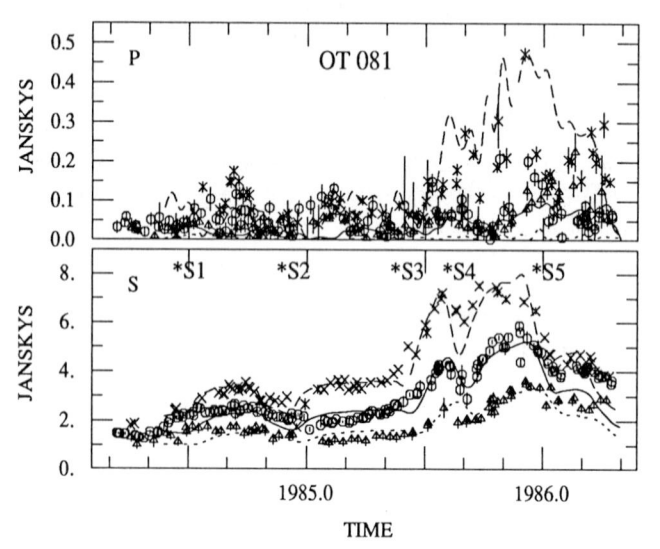

Figure 2. The OT 081 model superposed on daily averaged data; symbols are as in Figure 1; S1, S2 etc. indicate the onset of shock events.

that is not evident in the observed polarized flux. The obvious inference from this is that only very energetic particles are ejected from the environment of the central engine, and that little or no entrainment occurs on the subparsec scale. However, the models require a turbulent magnetic field structure, and if it cannot be gen-

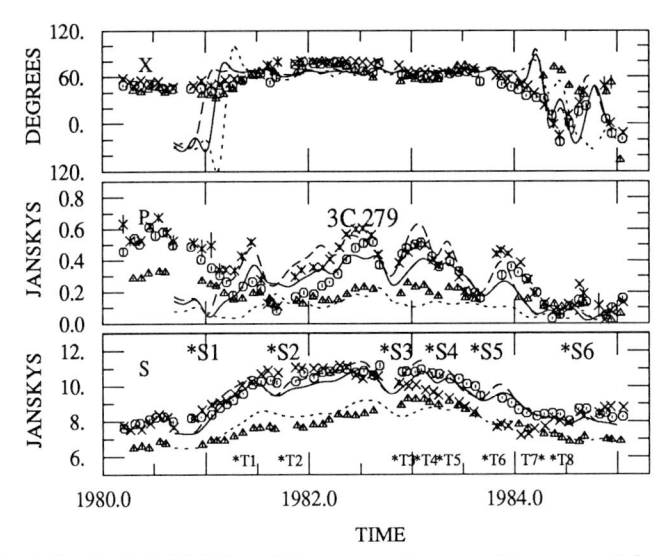

Figure 3a. The fully-shocked 3C 279 model superposed on monthly averaged data; symbols are as in Figure 1; S1, S2 etc. indicate the onset of shock events, and T1, T2 etc. indicate epochs at which VLBI maps exist.

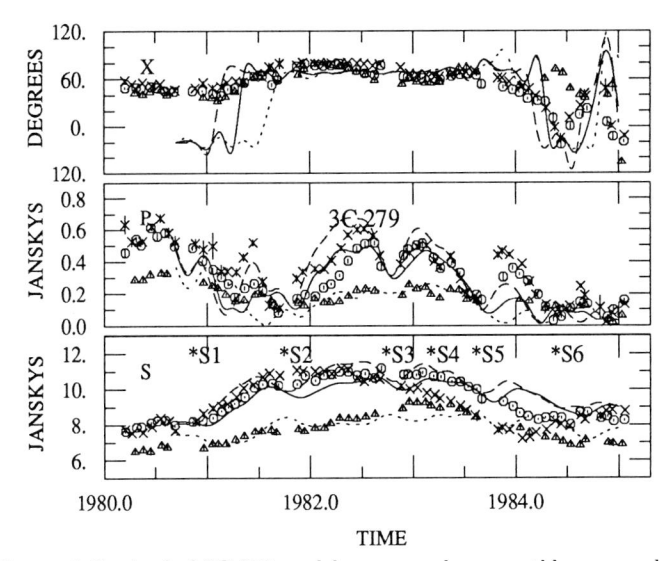

Figure 3b. The partially-shocked 3C 279 model superposed on monthly averaged data; symbols are as in Figure 1; S1, S2 etc. indicate the onset of shock events.

erated locally by, for example, turbulent motions (perhaps induced by streaming instability, and which would necessitate some entrainment), the implication is that the field is tangled all the way back to the central engine. This might have implications for central engine models – for example, providing evidence against those that require an ordered field (*e.g.*, Wiita 1991).

PROBING THE FLOW

The above modeling adopted a very simple description for the plasma flow: simple streamlines within an envelope $r \propto d^p$, with 'slugs' of shocked material propagating along the flow axis. It is important to ask whether deviations from this simple picture will effect our conclusions.

Direction and Volume of the Shock

The shock may be either 'reverse' or 'forward', according to whether the shock plane propagates towards the 'core' in the flow frame, but is advected by the flow away from the core in the observer's frame; or whether the shock plane propagates across the flow so that the shocked material lies on the 'core' side of the shock. The shock plane moves at speed γ_s^o, and the up- and down-stream speeds in the shock frame are $\gamma_{u,d}$, which transform to $\gamma_{u,d}^o$ in the observer's frame. The percentage polarization is determined by shock compression and angle of view, and the compression of an ultra-relativistic shock is a function only of the magnitude of the upstream flow speed in the shock frame. By choosing the same *magnitude* for γ_u for the reverse and forward shocks, and values of γ_s^o that lead to identical values of γ_d^o for the two cases, we can arrange that both the compression and the aberration are the same for the down-stream flows in both pictures. It is thus possible to accommodate identical polarization properties within distinct flow patterns.

Numerical hydrodynamical simulations of jets show that internal shocks do not necessarily occupy the entire width of the flow (*e.g.*, Figure 3a of Norman *et al.* 1982). The detailed flow pattern associated with our propagating shocks is unknown, and we shall parameterize this uncertainty by assuming that only a fraction, ϵ_s, of the flow radius is shocked. As we are concerned primarily with modeling *outbursts*, it is the *relative* Doppler factors for quiescent and disturbed flow that are important. Frequency is transformed by a factor $\mathcal{D}_s \mathcal{D}_{u,d}$; the Doppler factor associated with the shock is common to the two flows, and $\mathcal{D}_{u,d} \sim 1$ because the shocks are not extremely strong, and the observer is viewing, as noted, at a large angle to the flow axis. This means that, particularly as the spectrum is rather flat, there is very little change in the spectral properties of the model in going from reverse to forward shocks. The intensity of either flow transforms as $\mathcal{D}_s^3 \mathcal{D}_{u,d}^2$, if we regard both up- and down-stream flows as bounded by a plane moving at speed γ_s^o. Again, the Doppler factor associated with the shock is common to both flows; the modest change in the relative values of $\mathcal{D}_{u,d}^2$ between reverse and forward shock scenarios can be offset by a modest change in the shock volume.

In summary, if we parameterize our uncertainty about the flow dynamics in terms of 'adjustable' shock direction and volume, the above arguments suggest that we should be able to construct a set of models in which these parameters differ, but which preserve the fit to the data. Figure 3a shows a model for 3C 279 using a reverse shock which occupies the entire radius of the flow ($\epsilon_s = 1$). Figure 3b shows a model, also with a reverse shock, but with $\epsilon_s = 1/\sqrt{2}$; this is achieved primarily by varying the shock strength. We can do this without dramatically increasing the percentage polarization, because after the initial compression, further enhancing the perpendicular field components has only a limited effect. The increased shock strength requires a larger fractional mean field energy to produce the 'nulls' in polarized flux, and we optimized the model fit by slightly increasing both the angle of view and the low energy spectral cutoff. Nevertheless, we have a model of similar quality of fit to that in Figure 3a, without radically changing model parameters. We have also computed a model with a forward shock, and $\epsilon_s = 1$; this is almost identical to the model shown in Figure 3b, because, as discussed above, the change in Doppler boosting is nearly offset by the change in shock volume from the $\epsilon_s = 1/\sqrt{2}$ case. A model involving a forward shock and $\epsilon_s = 1/\sqrt{2}$ yields a significantly worse fit than those shown.

We conclude that we have only limited sensitivity to details of the flow dynamics, which cannot therefore be effectively probed by our modeling, although a detailed comparison of model maps of polarized flux (Hughes, Aller & Aller 1991) and the results of VLB polarimetry (see Gabuzda 1991) might help. However, this does mean that the general picture of a shocked, collimated flow, and the range of values derived from such a model, are robust, and are unlikely to change as our picture of the fluid flow is improved.

STRUCTURE FUNCTION ANALYSIS
The first order structure function is defined as $D(\tau) = \langle (S(t) - S(t+\tau))^2 \rangle$ ($= 2\sigma^2(1 - \rho(\tau))$, for a stationary random process of variance σ^2 and autocorrelation ρ). A typical structure function is shown in the upper left hand panel of Figure 4. For time lags greater than the maximum correlation time scale ($\tau > \tau_c$), $D \sim 2\sigma^2$. This plateauing may be seen to the right of the figure. At short time lags $D \to 0$, but the structure function plateaus to some nonzero level, which is just the value of $2\sigma^2$ associated with the Gaussian noise of the measurements (for which $\tau_c = 0$). Between these limits the structure function has an approximately power law form $D \propto \tau^a$, where the index a is a 'measure' of the noise process (*e.g.*, shot noise, flicker noise etc.).

Total Flux
We have calculated structure functions for 53 well-observed sources from the UM-RAO database, and show six examples for the total flux in Figure 4. The sources in the three left hand panels are BL Lacs; those in the three right hand panels are QSOs. About 80% of the sources are like those shown in Figure 4, and as

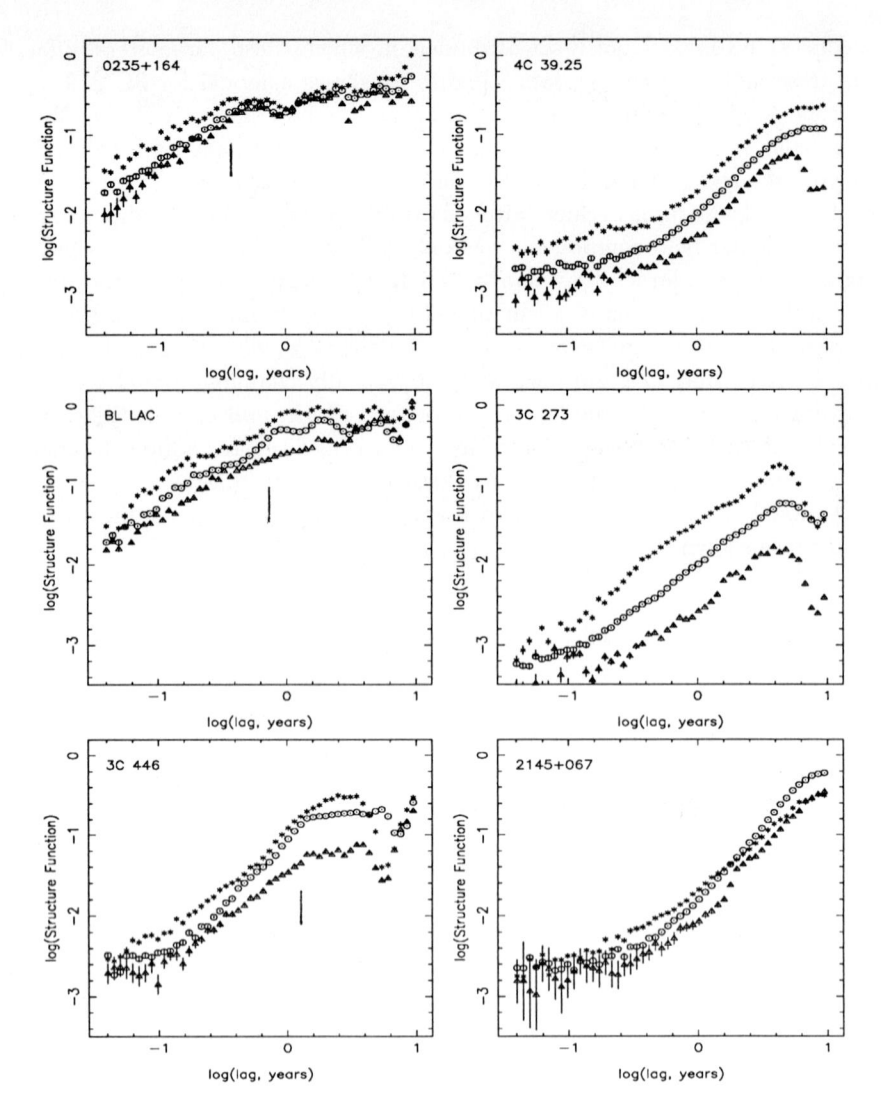

Figure 4. Total flux structure functions. Sources on the left are BL Lacs, those on the right are QSOs. Symbols used are the same as those in Figures 1 – 3. Arrows mark the time lag at which the structure function 'plateaus'.

a general rule the BL Lac structure functions show a turnover at long time lag, whereas the QSOs do not. This enables us to quantify the well-known result that BL Lacs have shorter characteristic time scales compared to QSOs: $\tau_{c\,BL\,Lac} \sim 1$ year $\ll \tau_{c\,QSO} \gtrsim 10$ years. The most important point to come from this analysis, however, is that the slope, $a \sim 1$, is very similar for both BL Lacs and QSOs: it appears that a common process is responsible for activity in these sources. In that

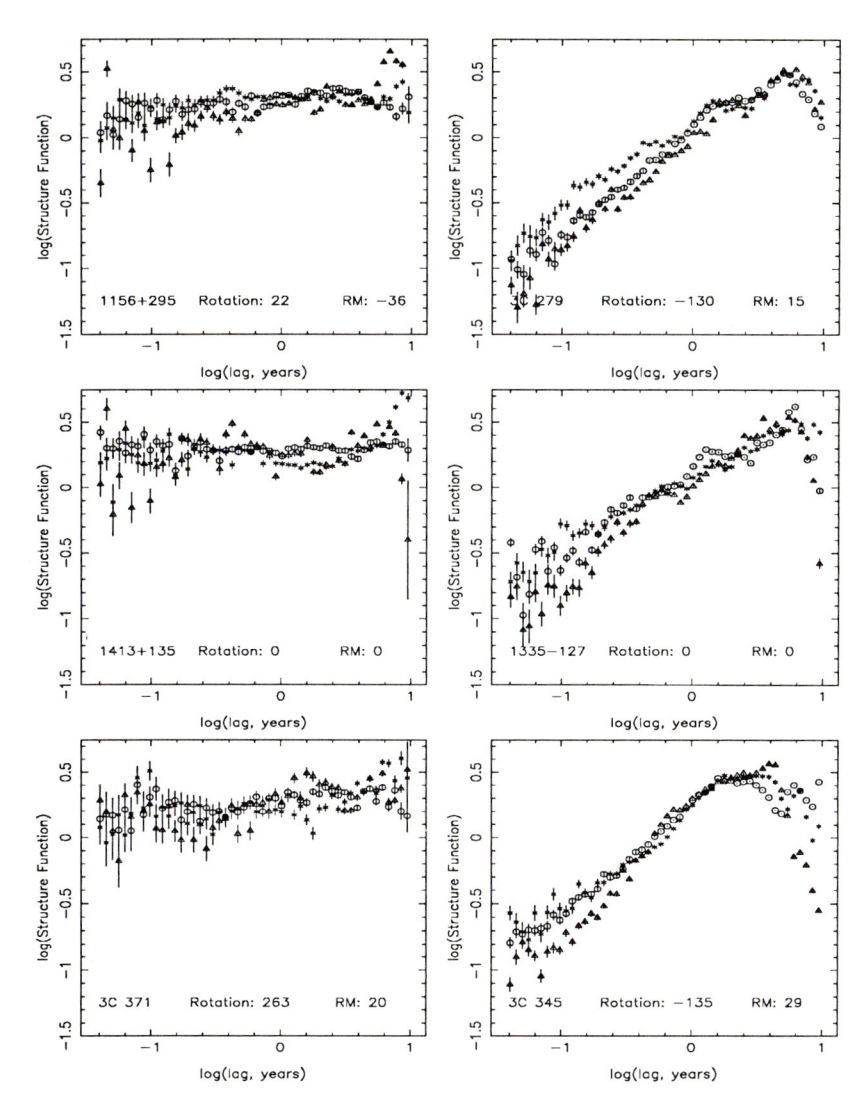

Figure 5. Q structure functions. The annotation on each frame indicates the adopted rotation and RM (a zero implies the value is unknown). Symbols used are the same as those in Figures 1 – 3.

shocks provide a good explanation of activity in BL Lac, 3C 279 and OT 081, we infer that structure functions provide circumstantial evidence for the ubiquity of shocked jets in compact radio sources.

Polarized Flux

We have performed a similar analysis on the polarized flux, in this case calculating structure functions for the Stokes parameters Q and U. This enables us to avoid the spurious fluctuations of P associated with cancellation of orthogonal polarization components, and to rotate the Q-U plane so that the Q axis is parallel to the direction of the VLBI jet (when known); we also correct the observations for external Faraday rotation, when this is known. Figure 5 shows some typical Q structure functions. They divide into two types, apparently uncorrelated with source classification, or with whether or not the VLBI-PA and/or RM are known. The three panels on the left show typical 'flat' structure functions. For these sources $D \sim 2\sigma_{Q,U}^2$, implying an uncorrelated noise process. It might be thought that in these sources the characteristic time scale is short, and information at shorter time lags would show the anticipated power law form. However, we note that these sources have $\sigma_Q^2/\sigma_U^2 \sim 1$, which is different from the type illustrated by the right hand plots. In the latter, there is a well-defined power law portion, and we find that σ_Q^2/σ_U^2 is preferentially $\gtrsim 1$ or $\lesssim 1$. In these sources the correlated fluctuations are related to the direction of the underlying jet. For the latter sources at least, this is further circumstantial evidence for the domination of the source polarization by a shock phenomenon, as this behaviour is just what one would expect from axial compression of a turbulent magnetic field.

This work was supported in part by NSF Grant AST-8815678.

REFERENCES

Aller, M. F., Aller, H. D., Hughes, P. A. & Latimer, G. E. 1991, these proceedings.
Gabuzda, D. C. 1991, these proceedings.
Heavens, A. H. & Drury, L. O'C. 1988, *M. N. R. A. S.*, **235**, 997.
Hughes, P. A., Aller, H. D. & Aller, M. F. 1989a, *Ap. J.*, **341**, 54.
Hughes, P. A., Aller, H. D. & Aller, M. F. 1989b, *Ap. J.*, **341**, 68.
Hughes, P. A., Aller, H. D. & Aller, M. F. 1991, *Ap. J.*, in press.
Königl, A. 1980, *Phys. Fluids*, **23**, 1083.
Mutel, R. L., Phillips, R. B., Su, B. & Bucciferro, R. R. 1990, *Ap. J.*, **352**, 81.
Norman, M. L, Smarr, L., Winkler, K.-H. A. & Smith, M. D. 1982, *Astr. Ap.*, **113**, 285.
Robson, I. 1991, these proceedings.
Valtaoja, E. *et al.* 1988, *Astr. Ap.*, **203**, 1.
Wiita, P. J. 1991, In *Beams and Jets in Astrophysics*, ed. Hughes, P. A. (Cambridge: Cambridge University Press), p. 379.

Observations of Southern Blazars
D BRAMWELL and G D NICOLSON

Hartebeesthoek Radio Astronomy Observatory

Abstract

We report on two studies of variability in flat spectrum radio sources south of +20 degrees declination. One study was of a complete sample strong (S ≥ 1.5 Jy) flat spectrum radio sources over a twenty year time scale at 13 and 6 cm. The other study investigated the variability of blazars at 13, 6, and 3.6 cm over 5 years.

The variability of the sources in both samples is broadly in accord with core, shocked jet models. There is evidence of refractive interstellar scattering in the flux changes of some sources.

1 INTRODUCTION

In 1967, one of us (GDN) began a programme to monitor variations in the flux density of a sample of extragalactic radio sources at 13 cm wavelength. The initial sample, which was intended to be complete, was drawn from the original Parkes 408 MHz Catalogue and contained 57 sources chosen on the following criteria:

1. $-60 < \delta < +20$, b $> 10°$

2. Spectral index $\alpha > -0.5$ $(S \sim \nu^\alpha)$

3. Compact structure indicated by peaked spectrum (e.g. 1934–63)

4. 13 cm flux density $S \geq 1.8$ Jy.

These criteria were sufficient to include all six (!) sources known to be variable at that time. The sample was supplemented with newly discovered sources which met the above criteria and which were found in increasing numbers in high frequency surveys such as the Dwingeloo and Ohio surveys. Many more candidate sources were catalogued in the Parkes 2700 MHz surveys, and in 1975 the sample was extended to include 120 objects stronger than about 1.2 Jy at 13 cm. This will be referred to as the "strong source" sample.

In 1982, DB began a program to investigate the variability of a sample of rapidly varying southern sources ($\delta \leq +20°$). Initially, the sample was drawn from the more active sources in GDN's study and sources found to be variable or possibly variable by Altschuler (1982 and 1983). The list was supplemented with further objects defined as BL Lac objects by various authors (see e.g. Hewitt and Burbidge, 1987, and references therein) and by sources in Wilkes et al. (1983) which those authors found to have lineless optical spectra. The only criterion applied was that the sources be south of declination +20°. While not being complete, this "rapid variable" sample is certainly representative of objects showing blazar characteristics.

The objectives of the programmes were to study the statistical properties of the variable source population, and to investigate the viability of models of source structure and variable components by studying the spectral evolution of bursts.

2 OBSERVATIONS

The observations were made with the Hartebeesthoek 26m radio telescope (originally NASA Deep Space Station 51, but operated by the South African Council for Scientific and Industrial Research as a national research facility for radio astronomy since 1975).

Sources in the strong source sample were observed at intervals of 3–4 weeks. In 1978 the observations were extended to 6 cm with known and established variable sources observed every 2 weeks and others at intervals of four weeks.

By the end of 1987 there were 120 sources in the rapid variable sample, being observed at timescales from daily (in the case of 1144–379, a very active object) through weekly to monthly and 3 monthly in the case of sources shown by earlier, more frequent measurements to be weak or inactive.

The sources were observed at 3.6, 6, and 13 cm (8.4, 4.8 and 2.3 GHz) on most observing sessions, with no 13 cm measurements taken during the day to avoid interference from the Sun. An observation lasted about 12 minutes at 3.6 and 6 cm and about 7 minutes at 13 cm, so the measurements were essentially simultaneous. No source was excluded on the basis of its flux being too low, but the measurement technique and receiver stability led to a lower flux limit of about 0.4 Jy at each frequency.

3 SUMMARY OF RESULTS

A detailed analysis of the data (e.g. structure function and Fourier analysis) has yet to be completed. However, some trends are evident in the data and this paper is essentially descriptive. After the final calibration and error analysis, the flux scale adopted here will change by not more than a few percent.

The results for the strong source sample will be summarised before discussing the observations of the southern blazars obtained by DB from 1982 to 1988.

3.1 Strong Source Sample

In order to study the statistical properties of the variable source population, a complete sample of flat spectrum sources (criterion ii) which have varied above a minimum baselevel of 1.5 Jy at 13 cm over the interval 1967 – 1987 was derived. Because the Parkes 2700 MHz finding survey was complete to less than 0.5 Jy (generally 0.2 Jy), this sample was necessarily complete and was also independent of source variability. The sample contains 67 sources and there were a further 25 sources whose flux density was found to exceed 1.5 Jy for only a part of the interval 1967 – 1987, and which were otherwise weaker than 1.5 Jy. This constitutes a lower limit to the number of weaker sources which may vary above 1.5 Jy. This second group of 25 sources will be referred to as the "floating population".

A variety of statistical tests were applied to determine which of the 67 sources were definitely variable, with the following results:

Nine of the 67 sources (13%) were stable over the period of the observations. Five of these lie in blank fields, two are identified with faint (20 m) quasars and one with a faint galaxy (19.5 m). The remaining source is the well known high redshift quasar OQ172 ($z = 3.53$).

Sixteen sources (24%) show only weak, (10 – 20%) slow, secular changes, with a partial burst, or at most one or two bursts, evident over a 20 year interval. Little more can be said of them other than that they vary.

A further twelve sources (18%) exhibit low level variability characterised by multiple outbursts at a low level (10 – 20%). These tend to follow the behaviour at shorter wavelengths but with reduced amplitude and with a time delay. At shorter wavelengths the outbursts are more structured.

There are 22 sources (33%) which exhibit strong classical outbursts. Many of these are well known sources such as 3C120, 3C279, 1510-08 and 3C454.3, although there are less well known objects such as 0438–43, a high redshift quasar with $z = 2.852$. The bursts are generally optically thick during the initial stages (usually until maximum flux at 13 cm), e.g. 3C454.3. DW1335–12 has consistently had strong outbursts at shorter wavelengths and these are clearly optically thick, suggesting very compact structure.

There are, however, a number of exceptions. Most notable is 2345-16, which had strong optically thick outbursts during 1970 – 1973, but while the level of variablity at 13 cm has remained much the same, the bursts at 13 and 6 cm have closely followed each other since 1978, maintaining a constant spectral index indicating an optically thin variable component.

Finally there is a group of eight sources (12%) which exhibit rapid variability at 13 cm superimposed on slower outbursts. This behaviour disappears when the outbursts fade away and occurs mostly at 13 cm. The outburst at 6 cm is far more smooth, and the bursts are initially optically thick. This behaviour appears to be different to the rapid variations which Aller et al. (1985) have observed at shorter wavelengths near the maximum of outbursts in several sources. The most notable examples are 0528+153 and 0537–441 (see fig. 1), which both show large fluctuations as the outburst develops at 13cm while the 6 cm flux varies smoothly. This variability may not be intrinsic to the source and it is noted that 0537–441 has a superimposed foreground galaxy along its line of sight (Stickel et al.,1988).

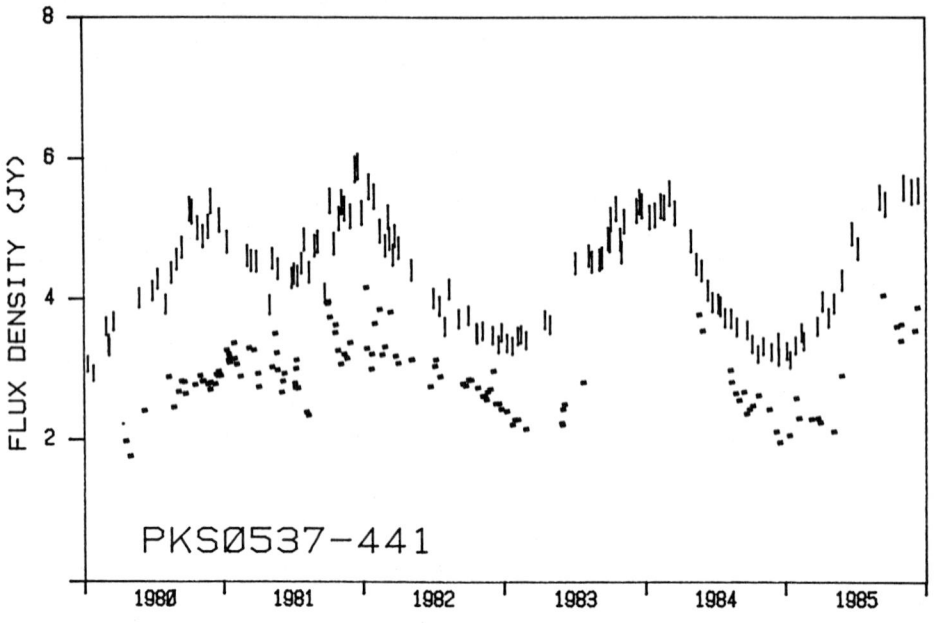

Fig. 1. Flux density variations in PKS 0537-441 at 6 cm (with error bars) and 13 cm (points). The 13 cm points have been displaced by -1.0 Jy for clarity. Error bars at 13 cm are similar to those at 6 cm.

The most extreme behaviour is observed in sources which are members of the floating population. Examples are BL Lac objects such as 0235+164 and OJ287. The most notable source in this group is the southern BL Lac object 1144–379 (fig. 2) which has varied over the range 1 – 6 Jy, with outbursts as short as several weeks, requiring observations at least every day to delineate their form.

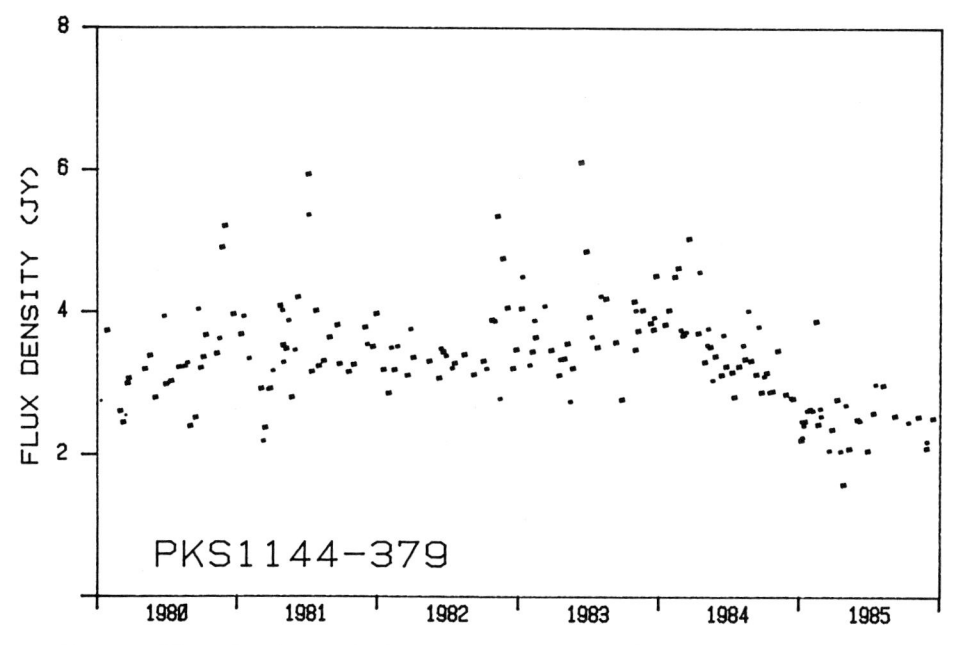

Fig. 2. Flux density variations at 6 cm. Error bars similar to those in fig. 1, but omitted for clarity.

3.2 Rapid Variable Sample

Over the time span of the observations of the rapid variable sample, 5 years for most sources down to 2 years for some, 40% showed some form of variability.

No differences are evident, in terms of the variability timescale, spectral index of outbursts or the extent of the variability as shown by various variability statistics (e.g. χ^2, variability index of Altschuler, 1983, or Kesteven et al. 1976), between sources classified as BL Lacs, HPQ or OVV. This accords with the results of Wiren (these proceedings) that radio measurements alone are unable to distinguish between the various types of blazar.

Five different types of variability are evident in the data, two of which have the signature of being caused by some form of refractive scintillation in the interstellar medium. In general the behaviour of the sources can be explained by the models of a core-jet structure with variability due to shocks propagating in the jet. See e.g., Hughes et al. (1989), Marscher and Gear (1985), Valtaoja et al. (1988), and the same authors at these proceedings.

1. Slow variations with flux changing by the order of 10% per year in smooth increase or decrease over a few years. The source is typically optically thin. There is some evidence for small scale changes of a few percent on the time scale of a few weeks superimposed on the long term trend in some of these sources.

2. "Classical" flares with large scale flux changes taking place over days, weeks or months as shown by M. Aller and others at these proceedings. The outbursts have spectral indices, α, $(S \sim \nu^\alpha)$ in the range $0 < \alpha < 1.4$. The spectral index may change during an outburst, approaching zero indicating that the source is becoming progressively optically thinner. The most rapid changes are shown by the source 1144–379 where the flux varies by up to 20% over a few days. The spectral indices of subsequent flares in any particular source may differ, indicating some reprocessing of the jet by the passage of the shock front.

   In some sources, flicker is seen at the peak of the outburst, particularly at the higher frequencies. This has been explained by Marscher (these proceedings) as due to small scale hydrodynamically unstable knots in the jet ahead of the shock front.

3. Sources in a quiescent state do show some variability of a few percent. If the quiescent state reflects only the emission from the core and jet (Valtaoja et al., 1988), these small scale optically thin variations could represent the passage of the shock front further out in the jet, or alternatively small scale variability in the core.

4. The variability which is probably extrinsic to the source shows 2 forms. One is a flicker that is most prominent at 13 cm with reduced amplitude at the shorter wavelengths. The peak to peak flux flicker in the source 1519 – 27 at 13 cm and 6 cm relate by $\Delta S \sim \lambda^{1.2}$ with the time scale of variability of the order of weeks (fig. 3). This does not conflict with the models and calculations of Rickett (1986). At 3.6 cm, the flicker is below the noise level, in accordance with the model.

5. The second example of refractive interstellar scintillation has the same form as the occultation events reported by Fiedler et al. (1987). The source 1424–41 shows a low plateau followed by a peak at 13 cm, with the flux at 6 cm undergoing rapid flickering (fig. 4). The 3.6 cm measurements are too noisy to contibute to the analysis.

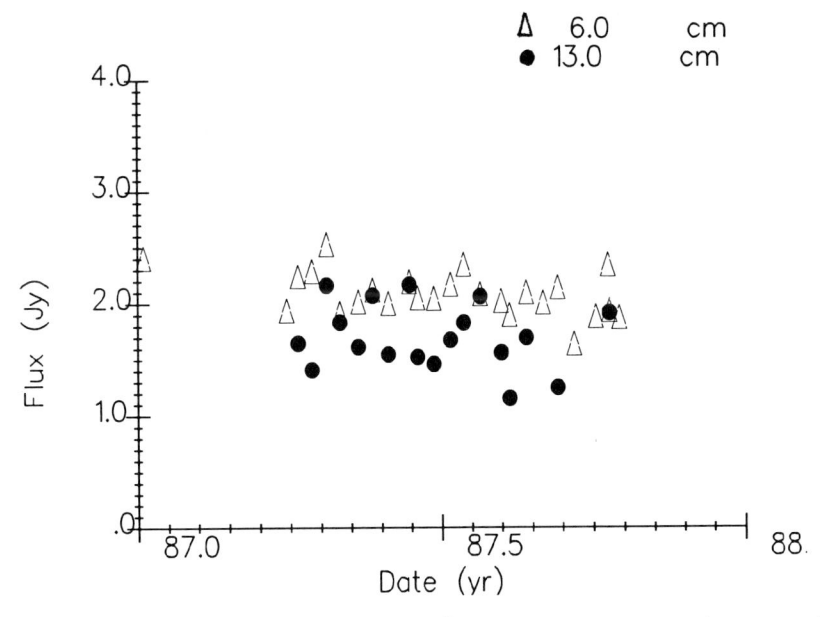

Fig. 3. Flux density variations in PKS 1519–27 at 6 cm and 13 cm. Error bars are less than the symbol size.

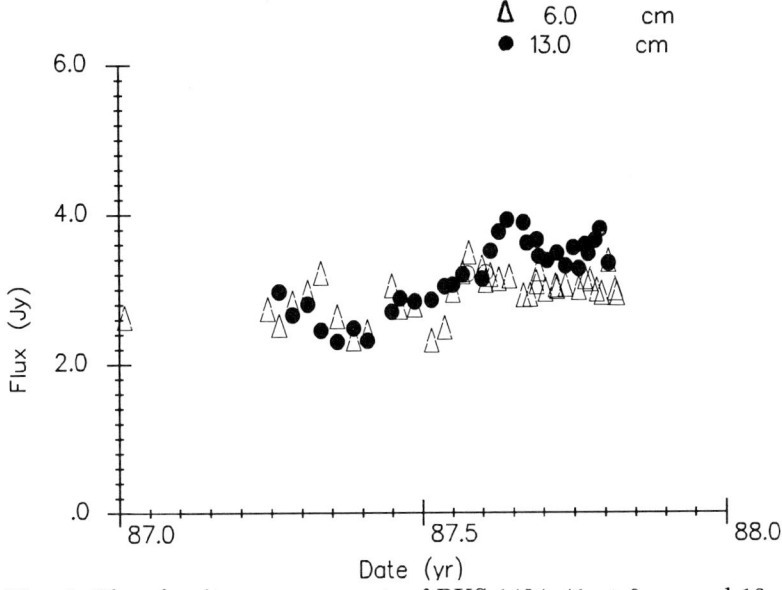

Fig. 4. Flux density measurements of PKS 1424–41 at 6 cm and 13 cm. Error bars are less than the symbol size.

4 CONCLUSIONS

In terms of radio variability, it is concluded that at a wavelength of 13 cm, 87% of flat spectrum sources are variable over a period of 20 years, and that variablity correlates with optical identification. None of the blank field objects was found to vary. At 13 cm, only about 45% of the sources show large outbursts characteristic of the so-called blazars.

The behaviour of the sources broadly confirms the core, shocked jet models that are currently in vogue. Variability that does not fit that description can be explained in terms of refraction of the emission in the interstellar medium.

The extreme variables, 1144–379 and 0851+20, were observed during some of their more active phases for intraday flux variability but this was not detectable to within measurement errors.

The source 0426–380, chosen from the catalogue of Wilkes et al. (1983) purely on the basis of its lack of emission lines, has shown a large scale flux change of the classical type.

REFERENCES

Aller, H. D., Aller, M. F., Latimer, G. E., Hodge, P. E. 1985, *Ap. J. Suppl.*, **59**, 513.

Altschuler, D. R. 1982, *A. J.*, **87**, 387.

Altschuler, D. R. 1983, *A. J.*, **88**, 16.

Fiedler, R. L., Dennison, B., Johnston, K., J., Hewish, A. 1987, *Nature*, **326**, 675.

Hewitt, A. and Burbidge, G. 1987, *Ap. J. Suppl. Ser.*, **63**, 1.

Hughes, P., Aller, H. D., Aller, M. F. 1989, *Ap. J.*, **341**, 54.

Kesteven, M. J. L., Bridle, A. H., Brandie, G. W. 1976 *A. J.*, **81**, 919.

Marscher, A. P. and Gear, W. K. 1985, *Ap. J.*, **298**, 114.

Rickett, B. J. 1986, *Ap. J.*, **307**, 564.

Stickel, M., Fried, J. W., Kuhr, H. 1988, *Astron. Astrophys.*, **206**, L30

Valtaoja, E. et al. 1988, *Astron. Astrophys.*, **203**, 1.

Wilkes, B. et al. 1983, *Proc. Astron. Soc. Aust.*, **5**, 2.

Ten Years Monitoring of Blazars at Metsähovi

H. TERÄSRANTA[1], M. TORNIKOSKI[1], K. KARLAMAA[1],
E. VALTAOJA[1,2], S. URPO[1], M. LAINELA[2], J. KOTILAINEN[2],
S. WIREN[2], S. LAINE[2], K. NILSSON[2], A. LÄHTEENMÄKI[2],
R. KORPI[2], M. VALTONEN[2].

1) Metsähovi Radio Research Station, Helsinki University of Technology, Otakaari 5A, SF-02150 Espoo, Finland
2) Tuorla Observatory, University of Turku, SF-21500 Piikkiö, Finland

ABSTRACT

We give an overall presentation of the monitoring programme of extragalactic sources at Metsähovi Radio Research Station. The monitoring has now lasted for 10 years and most of the 12000 observations are at 22 and 37 GHz, thus giving the largest existing database at these frequencies.

1 INTRODUCTION

The Blazar monitoring programme of Metsähovi Radio Research Station and Turku University Observatory started in 1980. The purpose of the programme was to monitor flux densities of a sample of extragalactic sources at 22 and 37 GHz. Similar monitoring has been done by the Michigan group at lower frequencies (4.8, 8.0 and 14.5 GHz) since 1965, Aller et al.(1985).

The Metsähovi Radio Research station is located at 60N 24E and thus sources with declination higher than -10 degrees can be observed. The station is located at an elevation of 50 metres, some 20 km from the coast. Typical winter temperatures are from 0 to -15°C and during the summer 15-25°C. Yearly rainfall is 600-700 mm and it comes on all seasons. Observing conditions are such, that at 37 GHz typically 50% of the observing time can be used, at 22 GHz about 40% and at 87 GHz less than 15%. About half of the antenna´s observing time has lately been used for this monitoring programme.

2 OBSERVATIONS

The diameter of the radome inclosed antenna is 13.7 metres. The surface accuracy of the antenna is about 0.3 mm rms and the antenna can be operated up to 115 GHz. In Table 1 are presented the receivers used in the observations and their performance. All observations are done with dual beam ON-ON mode. Stronger sources (typically S>5 Jy) are observed with a five point pattern to correct for small pointing errors. Typically the pointing is within 12" of the source. All observations are relative and calibrated against DR 21 (our primary calibrator), 3C 274, Jupiter and Mars. The flux densities of DR 21 were given by Baars et al. (1977) and Ulich (1980). Calibration sources are observed 4-5 times a day. Attenuation of the atmosphere is estimated from sky-dip measurements or from local weather parameters. A noise tube calibration is performed every second hour to check the receiver gain stability.

Receiver	T_n	rms	HPBW
12 GHz (1982-)	500 K	0.20 Jy	7.5'
22 GHz (1980-)	800 K	0.15 Jy	4.0'
37 GHz (1979-)	500 K	0.09 Jy	2.4'
87 GHz (1984-)	200 K	0.25 Jy	1.1'

Table 1. Receivers used with Metsähovi radio telescope ,with noise temperatures, typical rms values for 1000 second integrations and antenna HPBW for the frequency.

The error estimates are obtained by adding the rms value of the observation, the uncertainty of the atmospheric correction and the error of the calibration quadratically together. At 37 GHz typical errors are 0.1 Jy for weak sources, which mainly originates from the rms error, and 2-3% of the flux for stronger sources, which comes mostly from the uncertainty of the atmospheric correction. With 22 GHz and 87 GHz receivers typical errors are about 0.2 and 0.3 Jy, respectively. The first 5 years´ observations were presented by Salonen et. al. (1987). The number of yearly observations has been increasing and reached 2500 in 1990 (Figure1).

3 THE MONITORING PROGRAMME
Currently we have 63 extragalactic sources, which should be observed at least once a month at 22 and 37 GHz. In Table 2 is the source list with the number of observations until the end of 1990 and the lowest and highest flux densities observed at 37 GHz. Our source list includes basically all Northern extragalactic flat spectrum sources which have at least sometimes had flux > 2 Jy at 22 GHz. As every even month is completely our own time, also shorter time scale variability can be studied. Typically, if the weather permits, all sources can be observed at both frequencies (22 and 37 GHz) during 4 days. As one third of the time is used for 22 GHz observations, this gives us the possibility of 3 day sampling at 37 GHz and 6 day sampling at 22 GHz, if the weather remains good.

Yearly statistics at Metsähovi
31.12.1990

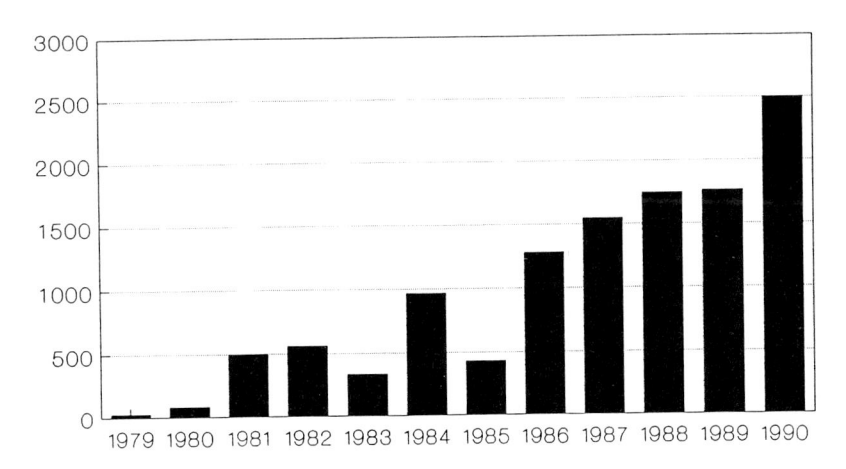

Figure 1. Yearly QSO observations at Metsähovi

The odd months are used for 87 GHz observations and typically one week of this time is devoted to quasar observations. The 87 GHz observations are more affected by the weather conditions and therefore systematic monitoring is rather difficult in this observatory.

From our ten years' observations it can be noted that typically the outbursts last from some months to 2 years. In Figure 2 is a example of the 1990 outburst of 3C 273. Observed at 22, 37, 90 and 230 GHz, the last two being observed with the SEST telescope. The outburst started at higher frequencies and also the duration is much shorter there. At 22 and 37 GHz the rise time is in this case about six months, thus indicating that our monthly sampling would easily detect all similar events. The sampling at higher frequencies (90 and 230 GHz) should be denser and weekly observations would be needed in this case. In Figure 3 is another example of an outburst, this time BL Lac during 1988. The top of the picture shows 22 GHz observations and the bottom 37 GHz observations. The ouburst seen at 37 GHz during September is missed in the 22 GHz data due to poor sampling. If the same trend is valid as in the case of 3C 273, the observations at 230 GHz should be nearly daily to catch also this kind of shorter outbursts.
In Figure 4 is our 11 year record of the source 3C 279 at 22 and 37 GHz. This source is now having the biggest outburst recorded by us.

Source	N	min Jy	max Jy	Source	N	min Jy	max Jy
III ZW 2	93	0.27	3.08	ON 231	127	0.44	1.93
OC 012	98	0.61	3.64	3C 273	612	15.65	51.43
0109+224	23	0.42	1.16	3C 279	323	6.56	19.04
DA 55	144	1.33	3.09	1308+32	196	0.45	4.16
0149+218	8	0.48	1.03	1413+135	28	0.95	4.55
0202+149	35	1.67	4.15	OQ 530	107	0.49	2.48
0224+671	5	1.01	2.39	1510-089	132	1.12	5.83
0234+285	15	2.90	3.62	4C 14.60	113	0.57	2.11
0235+164	192	1.23	4.48	1611+343	27	1.68	2.23
0248+430	58	0.34	1.30	4C 38.41	223	1.03	3.79
3C 84	569	23.98	60.53	3C 345	399	3.61	17.03
NRAO 140	62	0.62	2.30	OS 562	32	0.67	2.31
NRAO 150	199	1.87	8.07	1727+502	14	0.00	0.48
OA 129	129	3.69	7.06	1739+522	19	1.60	2.69
0422+004	42	0.27	1.67	1741-03	41	1.38	3.30
3C 120	181	1.56	4.42	1749+096	142	1.37	4.47
0458-02	40	1.89	4.03	1803+784	25	0.84	2.86
DA 193	21	4.68	5.69	3C 371	70	0.60	1.84
OH 471	173	1.62	3.60	1845+797	10	0.17	0.46
0716+714	12	0.43	1.53	1928+738	17	1.75	2.83
0735+178	217	0.91	5.26	2005+40	265	1.66	3.80
0736+017	70	0.66	2.53	2007+776	14	0.73	2.34
OI 090.4	79	0.33	2.27	OW 637	41	0.40	1.62
0804+499	19	0.36	1.61	OX 057	142	2.58	4.92
0814+425	13	0.51	1.26	OX 169	53	0.58	2.24
0846+513	29	0.18	1.18	2145+067	145	5.86	9.18
OJ 287	683	1.66	9.18	BL Lac	632	1.17	13.75
4C 39.25	172	1.80	5.98	2201+315	121	2.00	6.38
0953+25	14	1.05	2.13	3C 446	132	3.24	10.86
0954+658	13	1.01	1.62	CTA 102	66	1.62	3.86
OL 093	144	1.57	4.60	3C 454.3	332	4.33	13.69
4C 29.45	123	0.86	3.86				

Table 2. Sources included in the Metsähovi monitoring programme with the number of observations and lowest and highest flux values measured at 37 GHz.

Our monitoring has shown that growing and decaying shocks in relativistic jets seem to be able to account for most radio variations (Valtaoja et al. 1988, 1991a,b). However we do not yet possess realistic physical models for the structure of the evolution of the shocks. Perhaps the main obstacle is that there are so few observations of the initial growth stages, typically occurring above 50-100 GHz. High-frequency observations at SEST enable us to monitor radio variations also in these initial stages of shock evolution. Further development and observational testing of shock models will remain the main goal of our monitoring program.

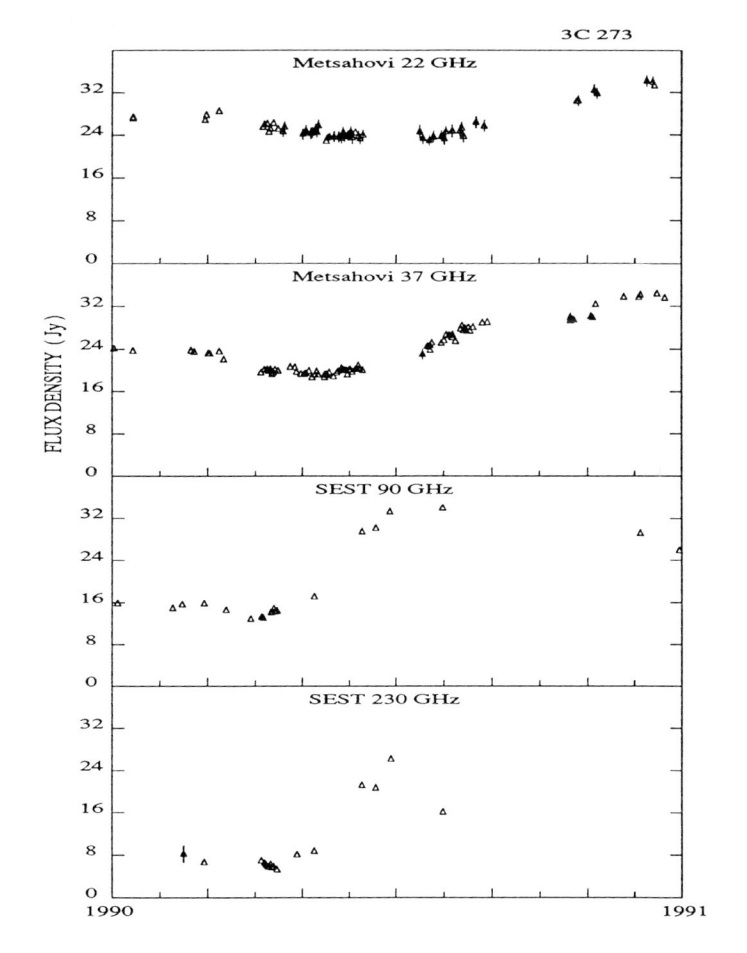

Figure 2. The outburst of 3C 273 during 1990 observed at Metsähovi (22 and 37 GHZ) and SEST (90 and 230 GHz).

4 OTHER PROGRAMMES

Besides the long term monitoring, we also have many other projects which can utilize basically the same data.

Multifrequency monitoring of 3C 273 is coordinated by T. Couvoisier. Our observations cover the frequencies 22 and 37 GHz at Metsähovi and 90 and 230 GHz at SEST, e.g., Courvoisier et al. (1990).

Multifrequency studies with the I. Robson team, who cover the submillimeter band (2.1 to 0.7 mm), e.g., Brown et al. (1989).

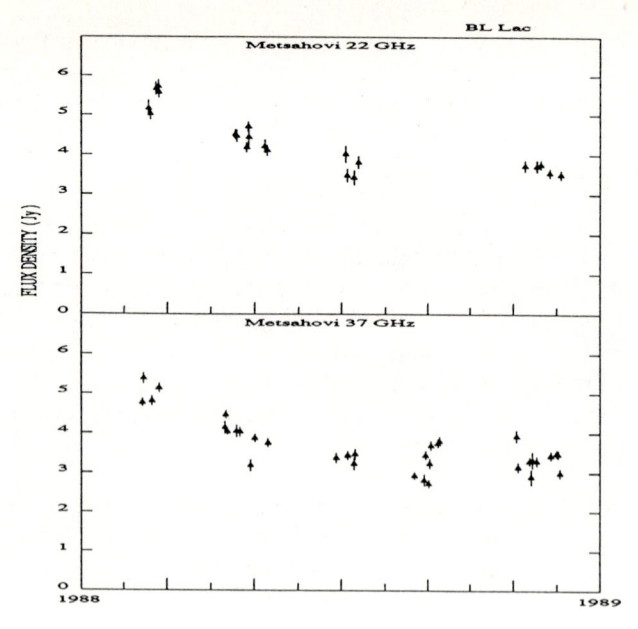

Figure 3. Observations of BL Lac during 1988 at Metsähovi.

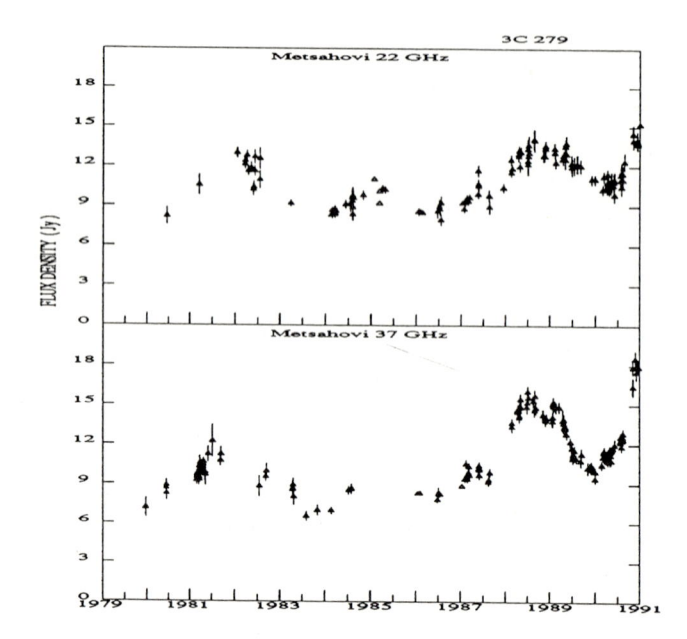

Figure 4. 11 years monitoring of 3C 279 at Metsähovi.

Surveys of selected samples, e.g., Wiren et al. (1991) (these proceedings).

Simultaneous observations and target optimizing with the Ginga X-ray satellite, as in the case of 3C 279, where a good correlation was found between our 37 GHz and X-ray fluxes, (Makino et al.1990).

Continuum observations of the spectra and variability can also be used to select the most compact sources for VLBI observations, to interpret VLBI data and to relate VLBI maps to the evolutionary history of the sources. High frequency continuum observations will also form an essential part of the RADIOASTRON and VSOP space VLBI satellites´ ground support. Our 2 Jy monitoring sample includes all the Northern hemisphere primary targets of the RADIOASTRON satellite, and probably also all the other sources suitable for high frequency ground and space VLBI observations, given the sensitivity limits of the present-day instrumentation.

There is increasing evidence that at least in some cases the optical and radio variations are synchronous : the spectrum of the radio flare reaches all the way to the optical regime. Coordinating the optical and radio observations we will be able to study the radio-optical connection more closely (Valtaoja et al. 1991).

Cooperation with the Crimean group, who observe basically the same sources at 22 and 37 GHz, giving a better sampling when our data are combined.

We also take part in other multifrequency campaigns, whenever possible for us.

5 FUTURE PLANS

To achieve an even better sampling of the sources selected, we would need improvements in our receivers´ sensitivity. The plan is to get new front ends during 1992 and get the noise temperatures down to 250 K at 22 GHz and to 300 K at 37 GHz. This should allow us to shorten the integration times so that all the sources could be observed within one day.

As the number of yearly observations keeps growing but the manpower and money involved does not, automation of the observing procedure is the only solution. The software needed is under test and will be utilized in full this same spring.

Instrumentation for polarization observations will also become available in the near future, giving us a better chance to explain burst structures.

REFERENCES

Aller H.D., Aller M.F., Latimer G.E.,Hodge P.E.: Astrophys. J. Suppl. Ser. 59, 513-768, 1985.

Baars J.W.M., Genzel R., Pauliny-Toth I.I.K., Witzel A.: Astron. Astrophys. 61, 99, 1977.

Brown L.M.J., Robson E.I., Gear W.K., Hughes D.H., Griffin M.J., Geldzahler B.J., Schwartz P.R., Smith M.G., Shepherd D.W., Webb J.R., Valtaoja E., Teräsranta H., Salonen E.: Astrophys. J. 340, 129, 1989.

Courvoisier T.J.-L., Robson E.I., Blecha A., Bouchet P., Falomo R., Maisack M., Staubert R., Teräsranta H., Turner M.J.L., Valtaoja E., Walter R., Wamsteker W.: Astron. Astrophys. 234, 73-83, 1990.

Makino F., Kii T., Hayashida K., Ohashi T., Turner M.J.L., Sadun A.C., Urru C.M., Neugebaur G., Matthews K., Teräsranta H., Aller M.F.: ISAS research note 449, 1990.

Salonen E., Teräsranta H., Urpo S., Tiuri M., Moiseev I.G., Nesterov N.S., Valtaoja E., Haarala S., Lehto H., Valtaoja L., Teerikorpi P., Valtonen M.: Astron. Astrophys. Suppl. Ser. 70, 409-435, 1987.

Ulich B.L.: Astron. J. 86, 1619, 1981.

Valtaoja E., Haarala S., Lehto H., Valtaoja L., Valtonen M., Moseev I.G., Nesterov N.S., Salonen E., Teräsranta H., Urpo S., Tiuri M.: Astron. Astrophys. 201, 1, 1988.

Valtaoja E. et al.: 1991a (preprint).

Valtaoja E. et al.: 1991b (preprint).

Valtaoja L., Valtaoja E., Shakhovskoy N.M., Efimov Yu. S., Sillanpää A.: Astron. J. 101, 78, 1991.

A study of long term variability of blazars in multifrequency spectral monitoring program

A.B. BERLIN, YU.A.KOVALEV, YU.YU. KOVALEV,
G.M. LARIONOV, N.A. NIDGELSKI AND V.A. SOGLASNOV

Astro Space Center, Lebedev Physical Institute
Profsoyuznaya 84/32, Moscow 117810, USSR

ABSTRACT

Some results of a ten years study are discussed. Multifrequency spectral variability of BL Lac and 3C 345 during 7 years is in agreement with the "Hedgehog" model. It is possible that these sources may be used for the testing of the cosmological parameters in the future.

1 MONITORING OBSERVATIONS

This program was started in 1979 by Institute for Space Research in collaboration with Special Astrophysical Observatory at RATAN-600. The program produces data for (i) a study of the nature of spectral evolution, (ii) a choice of interesting sources for the RADIOASTRON mission, (iii) a choice of sources for determining the extragalactic distance and cosmological parameters.

15 sources were observed in 1979 – 1983 and about 50 sources in 1983 – 1985. The source list consists of 116 variable and calibration sources after 1985. Each year we have 4 periods of observations at 5 – 7 wavelengths: 1.4 cm, 2.1 cm, 3.9 cm, 7.6 cm, 8.2 cm, 13 cm, 31 cm. The horn localization is linear. As a result when a source moves in the sky by Earth's rotation we have a quasi simultaneous multifrequency response (during 1.5 – 2.5 minutes). All sources are observed during each period for one week continuously (2 – 6 times for each source of the list). 55 – 60 sources of the list are observed each day of the period.

2 RESULTS

Data processing is not finished yet. Main preliminary results of the spectral variability study during the 10 years can be summarized as follows.

1. There may exist 3 types of spectral variability, at least:
   a) flux densities vary quasi synchronously at approximately all frequencies or at high frequency regions of the spectrum ("synchronous variability"), b) spectral variability is like "a wave", the HF region of which is moved to the low frequency ("wave variability"), c) spectral variability occurs in random at the different frequencies ("random variability"). It should be pointed out that these results may depend on time and frequency selection effects for the present.

2. At least some blazars with "synchronous" and "wave" variabilities may be in agreement with the so called 'Hedgehog" model for variable sources (a cloud or a jet of relativistic electrons is dissipating in quasi radial magnetic fields and radiates synchrotron emission). This model was suggested by Kardashev (1969) and was developed by many authors (references in Kovalev and Mikhailutsa, 1980).

3. We have strong evidence of this agreement for 2 sources at 7 – 8 years interval. They are BL Lac and 3C 345 (and maybe 3C345.3 and OJ 287). The evidence for BL Lac at 3 years is discussed in Kovalev, 1984.

3 COSMOLOGICAL CONSEQUENCES

If the fit to the data is the result of the nature of sources (but not chance!) and the number of the fitted sources is more than 5...10, then it may be possible to have a) the radio scale for the distances R to these sources, independently from the optical observations; b) the estimate of cosmological parameters if we will use the sources with the known redshifts z. Why?

It can be shown that the model has a distance limit – the distance, in which a variable component can move from the nucleus of the source to be visible to an observer. At larger distances the component does not radiate towards an observer because of the evolution of pitch-angles of electrons (see in details Kovalev, 1980). The absolute value of this distance can be estimated from the observations by the model fitting. As a result we will calculate the distance R to the source if the angular evolution data is known from the VLBI observations. Then we can estimate the Hubble constant H and the deceleration parameter q from the well known equation for the Universe $R = f(z, q, H)$. Thus, formally, two objects are sufficient for calculating H and q, if R and z have been measured accurately enough. In fact, many more objects may be required because of the large expected dispersion of R.

Thus, a way for determining the distance and the cosmological parameters is as follows: 1) to choose the source, which is in agreement with the model (with the spectral evolution); 2) to fit the angular evolution to the model (from the VLBI data); 3) to fit H, and q (from the equation mentioned above).

References

Kardashev, N.S. 1969, Editors note to Burbidge G., Burbidge M. 1969, *Quasars* (in Russian), Moscow: Nauka Publishers.
Kovalev, Yu.A. and Mikhailutsa, V.P. 1980, *Sov. Astronomy* **57**, 696.
Kovalev, Yu.A. 1980, *Sov. Astronomy* **57**, 22
Kovalev, Yu.A. 1984, Preprint of Space Research Institute, No. 879.

A study of Quasars, BL Lacertae objects and Active Galactic Nuclei at 22 and 43 GHz

L.C.L. BOTTI

Centro de Radio-Astronomia e Aplicações Espaciais, CRAAE
Escola Politécnica
Universidade de São Paulo, SP, Brazil

Z. ABRAHAM

Instituto Astronõmico e Geofísico
Universidade de São Paulo, Brazil

ABSTRACT

The radiosources 0735+178, 3C273, Centaurus A and 3C454.3 were observed during a period of 10 years (1980 – 1990) in the frequencies of 22 and 43 GHz with the Itapetinga radiotelescope (Brazil). The aim of this work has been to detect variability in the intensity, and its possible association with VLBI components. The shock model in a relativistic jet (Marscher and Gear, 1985) was applied to 3C454.3, with good accordance.

1 OBSERVATIONS AND RESULTS

The observations were made in the frequencies of 22 and 43 GHz with the 13.7 m radome enclosed Itapetinga radiotelescope, using drift scans through the source. See Abraham et al. (1986) and Botti and Abraham (1988) for further details about Itapetinga system and observational method.

1.1 – 3C273

The quasar 3C273 has been studied since July 1980 (Abraham et al. 1982, Botti and Abraham 1987, Botti and Abraham 1988, Abraham and Botti 1990) at 22 and 43 GHz. We can see in Figure 1 the behaviour of 3C273 at these frequencies. The observed outbursts seen at 22 and 43 GHz between 1982 and 1984 can be associated with C7 and C8 VLBI components seen at 22 GHz (Zensus et al., 1990). We suggest an association

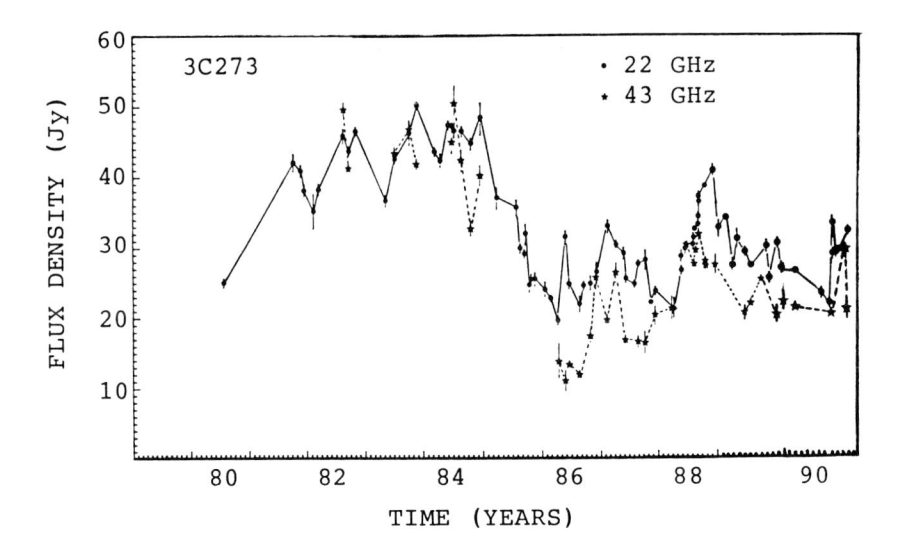

Figure 1. Time behaviour of 3C273 at 22 and 43 GHz.

between the E4 VLBI components at 100 GHz (Bååth, 1990) and the 1988 radio flare observed at 22 GHz and 43 GHz.

1.2 – 3C454.3

We began the observational program in November 1983, before the second activity phase (figure 2). The first big outburst propagated at several frequencies between 1981 and 1983. At the end of 1985 3C454.3 had an increase by a factor of 3 in its flux density relative to the October 1983 intensity.

We applied the Marscher and Gear (1985) model to the principal flare with good accordance. We estimated the radio extent of the shock as 3.5 pc at 22 GHz, 5 pc at 14.5 GHz, 6.6 pc at 10.7 GHz, 8.4 pc at 8 GHz, and 12.8 pc at 4.8 GHz. The value of 6.6 pc at 10.7 GHz is compatible with the scales observed in the 10.7 GHz VLBI maps (Pauliny-Toth, 1986). Morphological VLBI changes seen in 1983.8 might be associated with the increase at 22 GHz, 5 months between 14.5 and 4.8 GHz, 4 months between 10.7 and 4.8 GHz, all compatible with the observational delays observed by Botti and Abraham 1988, Salonen et al. 1987, Aller et al. 1985 and Haddock 1984.

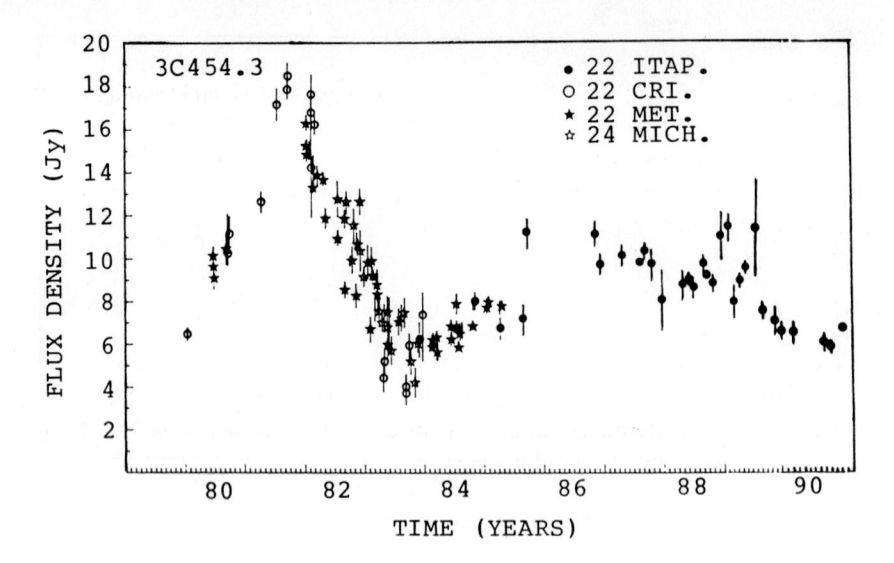

Figure 2. Time behaviour of 3C454.3 at 22 and 24 GHz (Itapetinga; Salonen et al. 1987, Haddock 1984).

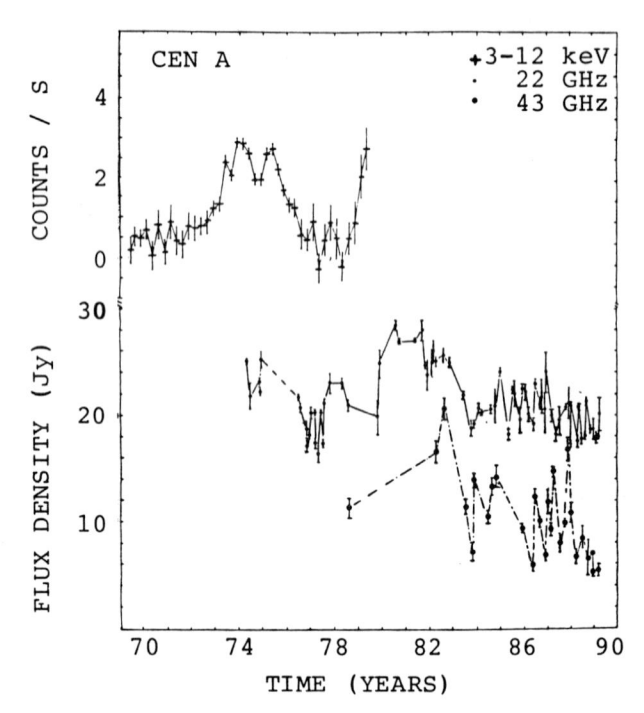

Figure 3. Time behaviour of Centaurus A at 22, 43 GHz (Itapetinga) and X-rays (Terrell, 1986).

1.3 – Centaurus A

This object has been studied since 1974 (Kaufmann et al. 1977, Kaufmann and Raffaelli 1979, Kaufmann et al. 1981, Abraham et al. 1982, Botti and Abraham 1987). From Figure 3 we can see similar behaviors between 22 GHz, 43 GHz and X-rays. The X-rays and radio behaviors show a good correlation, allowing the use of synchrotron self-Compton model as in Jones et al. (1974).

1.4 – 0735+178

We have observed 0735+178 at 22 GHz since 1980, obtaining a variability with time scales of about 1 year and amplitude of about 2 Jy (Figure 4). In 1989 we detected a large flare, when the flux density increased by a factor of 3. This was the largest outburst seen in this source. A new outburst started in 1990.

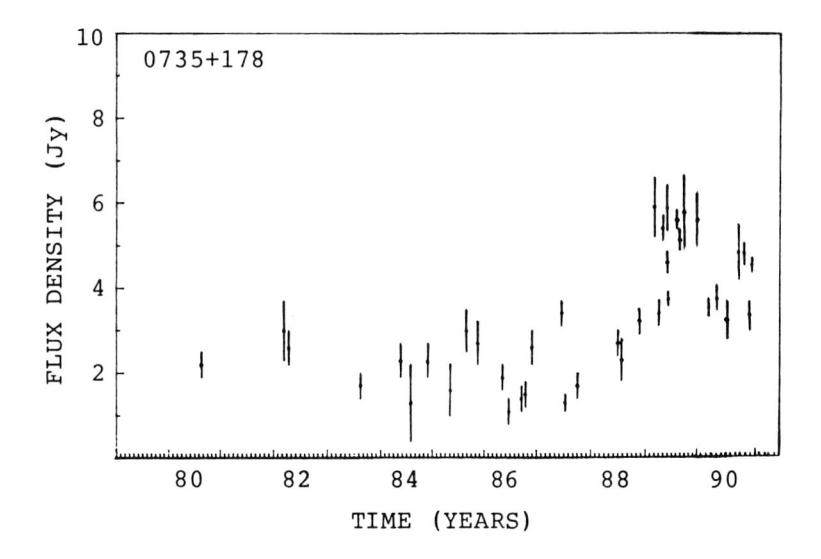

Figure 4. Time behaviour of 0735+178 at 22 GHz.

3 ACKNOWLEDGMENTS

This work was partially supported by the Brazilian agencies, FAPESP and CNPq. Special thanks to Dr. Esko Valtaoja and Turku University for financial support.

REFERENCES

Abraham Z., Botti L.C.L., Del Ciampo L.F. 1986, *Revista Mexicana de Astronomia y Astrofisica* **12**, 414.
Abraham Z., Kaufmann P., Botti L.C.L. 1982, *A.J.* **87**, 532.
Abraham Z., Botti L.C.L. 1990, In Symposium held in Socorro, USA, ed. Zensus J.A., Pearson T.J. (Cambridge University Press).
Aller H.D., Aller M.F., Latimer G.E., Hodge P.E. 1985, *Ap.J.Suppl.* **59**, 513.
Botti L.C.L., Abraham Z. 1987, *Revista Mexicana de Astronomia y Astrofisica* **14**, 100.
Botti L.C.L. 1988, *Astron. J.* **96**, 465.
Bååth, L.B. 1990, In Symposium held in Socorro, USA, ed. Zensus J.A., Pearson T.J. (Cambridge University Press).
Haddock T.F. 1984, *Ph.D. thesis*, Michigan University
Jones T.W., O'Dell S.L., Stein W.A. 1974, *Astrophys.J.* **188**, 353.
Kaufmann P., Santos P.M., Raffaelli J.C., Scalise Jr. E. 1977, *Nature* **269**, 311.
Kaufmann P., Raffaelli J.C. 1979, *M.N.R.A.S.* **187**, 23.
Kaufmann P., Strauss F.M., Coe M.J., Carpenter G.F. 1981, *Astr. Ap.* **100**, 189.
Marscher A.P., Gear W.K. 1985, *Ap.J.* **298**, 114.
Salonen E., Teräsranta H., Urpo S., Tiuri M., Moiseev I.G., Nesterov N.S., Valtaoja E., Haarala S., Lehto H., Valtaoja L., Teerikorpi P., Valtonen M. 1987, *Astr. Ap. Suppl.* **70**, 409.
Zensus J.A., Unwin S.C., Cohen M.H., Biretta J.A. 1990 *A. J.* **96**, 120.

SEST Observations of Southern AGN

M.TORNIKOSKI[1], H.TERÄSRANTA[1], E.VALTAOJA[1,2],
J.KOTILAINEN[2], M.LAINELA[2], L.C.L.BOTTI[3]

1) Metsähovi Radio Research Station, Helsinki University of Technology, Otakaari 5 A, SF-02150 Espoo, Finland
2) Tuorla Observatory, University of Turku, SF-21500 Piikkiö, Finland
3) Centro de Radioastronomia e Aplicações Espaciais, Universidade de Saõ Paulo, Brazil

ABSTRACT

We describe our first continuum observing projects of active galactic nuclei (AGN) with the Swedish-ESO submillimetre telescope. We discuss the main questions for which we seek solutions with our projects, and in more detail present the results of one of our projects, the survey of Southern compact sources.

1 THE SWEDISH-ESO SUBMILLIMETRE TELESCOPE

The Swedish-ESO submillimetre telescope, acronym SEST, is situated on the European Southern Observatory (ESO) site of La Silla, Chile, at an altitude of 2300 m. SEST is an IRAM (Institut de Radio Astronomie Millimetrique) designed, 15 m diameter Cassegrain antenna and it has been operational since 1987. Scheduled observations at SEST began in April 1988. The telescope is funded and operated on a 50/50 basis by the Swedish Natural Science Research Council and ESO, and a separate Nordic agreement entitles Finland to 10% of the Swedish time.

At present there are two receivers, centered on 100 and 230 GHz, which were built at Onsala Radio Observatory, Sweden, and the backend can be chosen from one of the two acousto-optical spectrometers (AOS) built at the University of Cologne. One of them is a high resolution AOS with a bandwidth of 100 MHz, and the other is a wide band AOS with 1 GHz bandwidth. So far also the continuum observations are done using the wide band AOS. The future plans include a new 350 GHz receiver and a 230 GHz bolometer, which, according to a preliminary schedule, should be operational in April 1991 and August 1991, respectively. A more detailed description of SEST and its equipment can be found in Booth et al. (1989).

2 SEST CONTINUUM OBSERVATIONS OF AGN

2.1 General

Continuum observations occupy only a small fraction of total observing time used at SEST. Our group has used SEST to extend our continuum monitoring programme of AGN to higher frequencies and to Southern sources. Most of the sources are selected so that they can also be observed from other, supporting observatories at lower frequencies, enabling us to get multifrequency lightcurves of the sources.

Since very little was known of the high-frequency behaviour of Southern sources, we first started out gathering basic information of those sources: spectra, variability and compactness. These data can then be used to select the most interesting sources for further investigation and monitoring.

The high-frequency observations at SEST enable us to monitor radio variations at the initial stages of shock evolution and thus help us in developing physical models of the shocks. We are also looking for evidence for the orientation-dependent scheme of different classes of sources: can the differences between BL Lacs, highly polarized quasars and "ordinary" quasars be explained with the unified theories of AGN as resulting from geometrical effects.

Here we describe three of our projects run at SEST. At the end of this article we present the preliminary results of the first completed project, the Southern survey.

2.2 Survey of a Complete Sample of Southern Compact Extragalactic Sources

This project was started in 1988. The aim was to obtain at least two-epoch spectra at 22, 43, 90 and 230 GHz of a complete sample of flat-spectrum ($\alpha > -0.5$) bright ($S(22) > 1$ Jy) sources in the Kühr 1 Jy catalogue within the declination range $0° - -25°$. The 90 and 230 GHz observations were to be done at SEST and the 22 and 43 GHz observations at Itapetinga in Brazil. In reality we obtained at least two-epoch spectra of all the sources at 90 GHz, but only 20% of the sources at 230 GHz. The small number of observations at 230 GHz is mainly due to the sensitivity requirements and pointing problems. Observations at 230 GHz require long integration times even under good conditions, and for the weakest sources no reliable results could be obtained at all. Instead of observing all the sources at 230 GHz we concentrated in getting a good coverage of the sources at 90 GHz. In this paper we concentrate on the results from the SEST observations as the processing of the Itapetinga observations has not been finished yet.

The aims of this project were to increase our knowledge of these relatively rarely observed sources and of the high-frequency spectra of radio sources in general, and to use the sample for statistical studies (variability vs. classification, spectral shape vs. classification and variability). In addition to these, the observations also enable us to identify the most compact bright cores in these sources. This information can be used to choose suitable sources for VLBI and space VLBI observations.

2.3 Monitoring of Highly Active Blazars

Our SEST monitoring of highly active blazars has been continuing since 1987. It is closely tied to international projects (Courvoisier's radio-to-X-ray 3C273 collaboration and Robson's IR observations at JCMT (Courvoisier et al. 1990, Brown et al. 1989), Botti's observations in Itapetinga, Brazil, plus monitoring at Crimea) and to our own regular monitoring in Metsähovi. Its aim is to study the basic properties of AGN cores with a variety of telescopes and wavelengths.

The sourcelist consists of 11 highly active blazars which are almost equatorial, so that they can be observed from both the Southern (SEST, Itapetinga) and Northern (Metsähovi, JCMT, Crimea) observatories.

SEST-observations of the sources extend our knowledge of the spectra and variability to higher frequencies. This is essential in trying to understand the physics of AGN cores, because the early phases of the outbursts can be followed only at mm-wavelengths.

Radio monitoring also supports other kinds of observations: since most theoretical models suppose a close connection between high frequency radio and X-ray fluxes, radio monitoring can give warnings to X-ray observers. It can also be used to select suitable targets for VLBI observations, because we have real-time information of the flux levels and activity of the sources.

2.4 Survey of Southern BL Lacs and Highly Polarized Quasars

The complete sample of this survey consists of all the known or suspected Southern blazars: 34 BL Lacs and 29 highly polarized quasars. Our aim is to obtain at least two-epoch spectra of these sources at 90 GHz and 230 GHz in order to determine their high frequency spectral shape and variability.

Once this work is completed we will have three complete sets of radio sources for comparisons: all Southern BL Lacs and HPQs from this survey, and all bright flat-spectrum quasars from the $0°$ — $-25°$ survey. In addition to these we also have similar classes of sources from our Northern survey (Wiren et al, these proceedings). Thus we will be able to compare the properties of variously classified radio sources.

3 THE 0° — -25° SURVEY

3.1 Sources

The sample originally consisted of 50 sources. Here we present the results obtained from 7 BL Lacs, 13 highly polarized quasars (HPQs) and 26 low polarized quasars (LPQs) or "ordinary" quasars. The remaining 4 sources were either galaxies or without identification. The sourcelist is presented in table 1.

BL Lac	HPQ	LPQ	
0048-097	0238-084	0003-06	1148-00
0138-097	0336-019	0112-017	1243-072
0743-006	0403-132	0122-00	1302-102
1514-24	0420-014	0135-247	1354-152
2131-021	0454-234	0202-17	1406-076
2155-152	0458-020	0414-189	1741-038
2223-052	0605-085	0440-00	2008-159
	1253-055	0511-220	2126-15
	1334-127	0834-20	2128-123
	1504-167	1032-199	2203-18
	1510-089	1045-18	2216-03
	2243-123	1127-14	2227-08
	2345-167	1145-071	2354-11

Table 1. Sourcelist.

3.2 Data Analysis

The analysis was carried out to find out if there are significant differences in the average radio properties of different classes of sources, detectable by statistical analysis. We were looking for correlations between different radio properties to find out how this all fits the orientation dependent scenarios.

For the analysis we computed spectral indexes, fractional variablility and timescales of variability. We used spectral indexes ($S \propto \nu^{\alpha}$) between 10 and 90 GHz and between 90 and 230 GHz, 10 GHz fluxes being from the Kühr catalogue (Kühr et al. 1981) and 90 and 230 GHz fluxes from the SEST observations. These indexes measure the overall shape of the spectrum. The fractional variability index $\Delta S = (S_{max} - S_{min}) / S_{min}$ indicates how much the flux changes between observations and it is the estimate of the relative strength of the variable and nonvariable components in the source. We had more than two observations for 84% of the sources at 90 GHz, in which case we calculated the average variability. In order to determine typical timescales of variability, we took a look at the variability indexes in different timescales ($t < 100$ d, 100 d $\leq t < 250$ d, $t \geq 250$ d).

The statistical tests that we used were the Kruskall-Wallis significance test to evaluate degree of association between any groups of samples, and the Spearman rank correlation to see if there are any correlations between the variability and the spectral index of any classes of sources.

3.3 Spectrum vs. Classification

The 10—90 GHz spectral index distributions of different classes of sources are shown in figure 1.

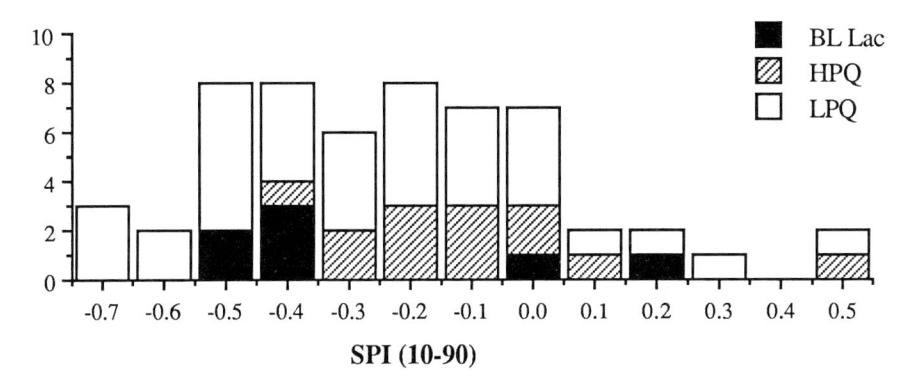

Figure 1. Spectral index 10 GHz — 90 GHz vs. classification.

It can be seen that there is increasing steepness in the spectra when proceeding from HPQs and BL Lacs towards LPQs. When testing the differences between different classes of sources, it turned out that differences between HPQs and LPQs are statistically significant, but it is more difficult to distinguish BL Lacs from either LPQs or HPQs.

These results were obtained using the average spectral index of several observations. When comparing these with the results obtained by using the spectral index calculated from the lowest flux of all observations, the assumed quiet spectral index, the results were basically the same.

Figure 2 shows the 90—230 GHz spectral index distribution. With only very little data at 230 GHz it is difficult to draw any conclusions of differences between classes, but it seems that all classes have steep spectra at this frequency. The lack of flat high frequency spectra indicates that there are no strong components peaking at mm/submm regions. Even the most compact components have turned over at these frequencies.

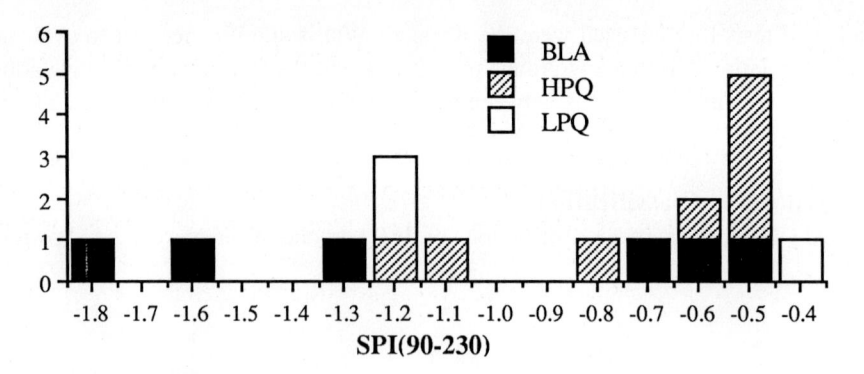

Figure 2. Spectral index 90 GHz — 230 GHz vs. classification.

3.4 Variability vs. Classification

Figure 3 shows the average variability of different classes of sources. We observed a tendency for the LPQs and HPQs to be more variable than BL Lacs. It seems that the variability of BL Lacs is not very strong at high frequencies.

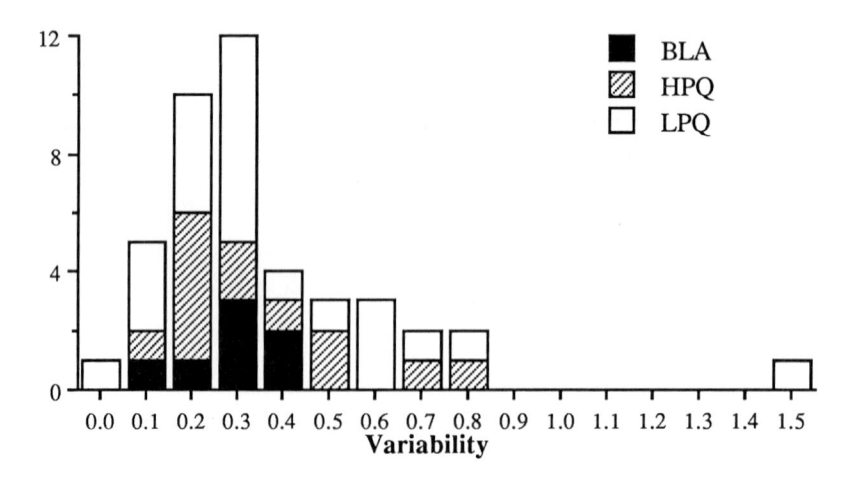

Figure 3. Average fractional variability.

3.5 Variability vs. Spectrum

The variability vs. spectrum plot is shown in figure 4. Here we obtained an interesting result: with LPQs there is a clear correlation (P=98%) between the spectral index and the variability, so that the variability surprisingly increases with steeper spectrum. Even more interesting is to notice that the same correlation was found in the Northern survey (Wiren et al. 1991). This is against both the simplest predictions of the orientation-dependent models (cf. Eckart et al. 1989, Valtaoja et al. 1991) and conventional wisdom, according to which the flattest spectrum sources are more variable. However, it must be noted that the true steep-spectrum sources ($\alpha < -0.5$) were excluded from our sample.

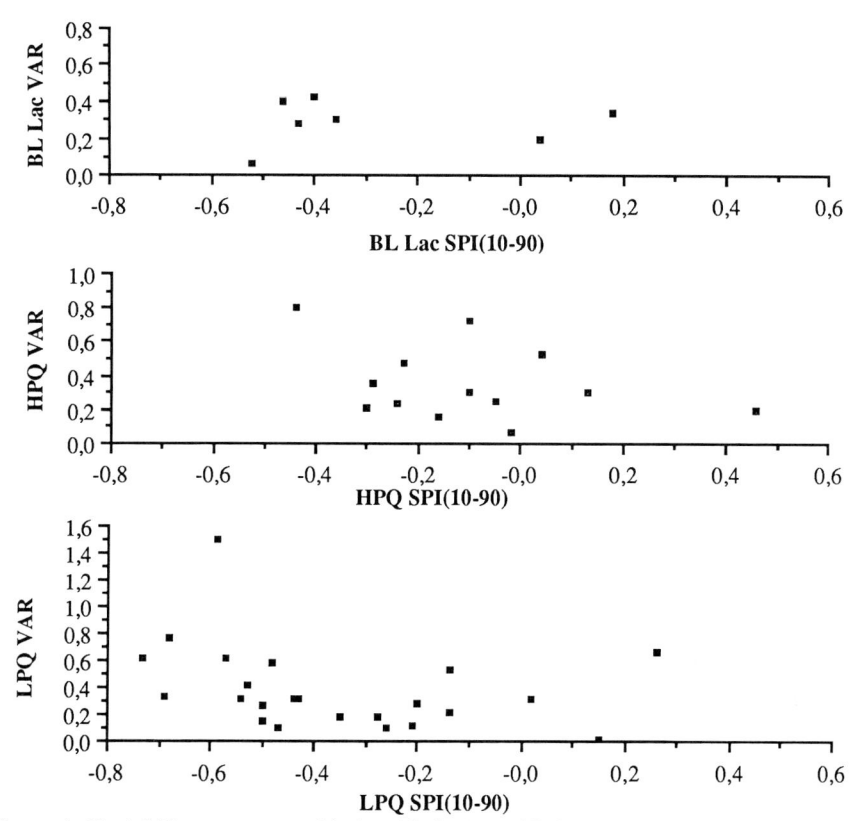

Figure 4. Variability vs. spectral index 10 GHz — 90 GHz.

3.6 Variability Timescales

The variability during three different ranges of time between observations is shown in figure 5.

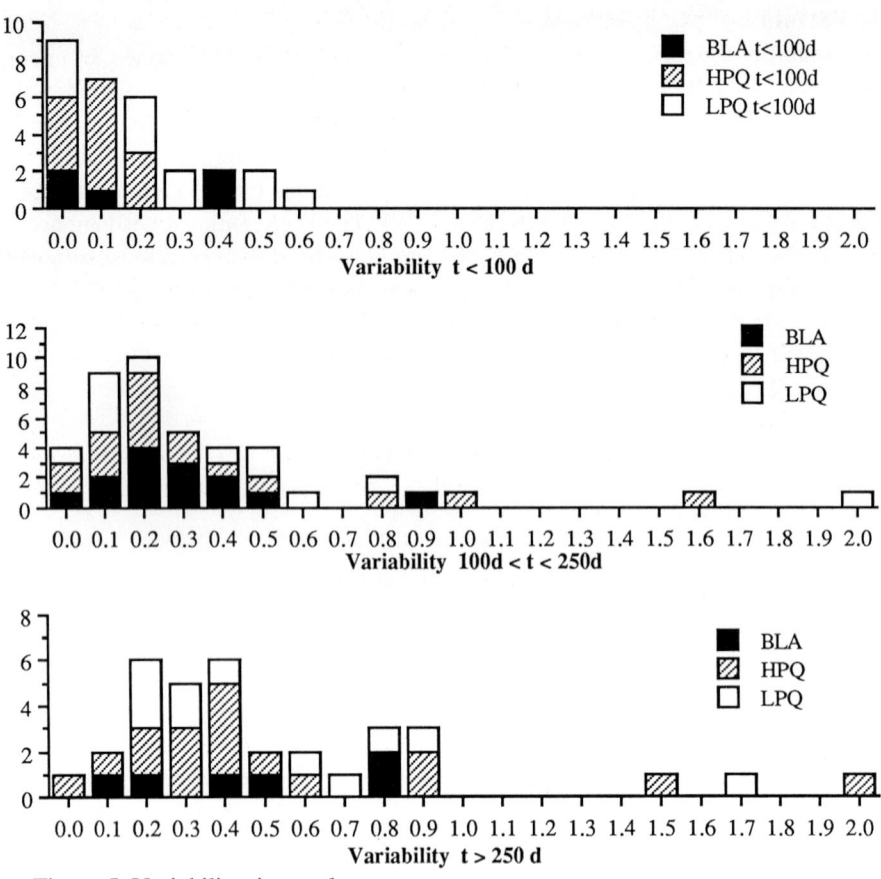

Figure 5. Variability timescales.

The variability increases with time only up to some 200 days for all classes of sources. The explanation to this may be that the high frequency variations are dominated by repeated outbursts of duration of some months. Secular changes in the flux levels are small, indicating the long-time stability of the underlying core component.

4 SUMMARY

We have completed one of the three projects we have been running at SEST, the survey of Southern AGN between the declination range $0° - -25°$. The results from the 90 GHz and 230 GHz SEST observations show increasing steepness in the spectra between 10 and 90 GHz when proceeding from HPQs and BL Lacs towards LPQs and overall steep spectra between 90 and 230 GHz. The variability of ordinary, low polarization quasars tends to increase with steeper spectrum. With all classes the variability increases with time only up to some months, showing the importance of frequent monitoring to guarantee reliable results. Of the three classes of sources, HPQs and LPQs are easier to distinguish from each other than from BL Lacs. This is especially true when measured by the spectral index.

We wish to emphasize the fact that these results are so far based on a relatively small sample. We will need more observations on a great number of sources with sufficiently sensitive equipment to obtain good data points at all frequencies in the desired frequency range 22 GHz — 230 GHz to verify these results. Especially the possibility for bolometre observations at SEST will provide us with reliable data using only short integration times.

REFERENCES

Booth, R.S., Delgado, G., Hagström, M., Johanson, L.E.B., Murphy, D.C., Olberg, M., Whyborn, N.D., Greve, A., Hansson, B., Lindström, C.O. and Rydberg, A. (1989) *Astron. Astrophys.* **216**, 315-324.

Brown, L.M.J., Robson, E.I., Gear, W.K., Hughes, D.H., Griffin, M.J., Geldzahler, B.J., Schwartz, P.R., Smith, M.G., Smith, A.G., Shepherd, D.W., Webb, J.R., Valtaoja, E., Teräsranta, H., Salonen, E. (1989) *Astrophys. J.* **340**, 129-149.

Courvoisier, T.J.-L., Robson, E.I., Blecha, A., Bouchet, P., Falomo, R., Maisack, M., Staubert, R., Teräsranta, H., Turner, M.J.L., Valtaoja, E., Walter, R., and Wamsteker, W. (1990) *Astron. Astrophys.* **234**, 73-83.

Eckart A., Hummel, C.C., Witzel, A. (1989) *MNRAS* **239**, 381.

Kühr, H., Witzel, A., Pauliny-Toth, I.I.K. and Nauber, U. (1981) *Astron. Astrophys. Suppl.* **45**, 367-430.

Valtaoja, E., Teräsranta, H., Urpo, S., Nesterov, N.S., Lainela, M., Valtonen, M. (1991) *Astron. Astrophys.* submitted.

Wiren, S., Teräsranta, H., Valtaoja, E. and Kotilainen, J. (1991), *These proceedings.*

Northern Hemisphere Survey of Quasar Variability

S. Wirén[1], H. Teräsranta[2], E. Valtaoja[1],[2], J. Kotilainen[1]

(1) Tuorla Observatory, University of Turku, SF-21500 Piikkiö, Finland
(2) Metsähovi Radio Research Station, Helsinki University of Technology, Otakaari 5A, SF-02150 Espoo, Finland

Background

Many of the observed differences between the properties of radio galaxies and various types of active galactic nuclei may be explainable with orientation effects. The smaller the viewing angle, the bigger the activity, because beaming magnifies the relativistic effects in the jet. Barthel (1989) proposed a specific model, according to which FR II radio galaxies appear as ordinary quasars and finally as blazars as the viewing angle decreases. Such relativistic beaming models can be tested with respect to spectral flatness and variability, since both should be greatest in the most beamed objects. Using the Kühr 1 Jy catalogue data (Kühr et al. 1981), Eckart et al. (1989) found that the variability seemed indeed to be greatest in flat-spectrum blazars and smaller in ordinary quasars with steeper spectra. However, similar studies using high radio frequency data of complete samples have been lacking. We present here a two-epoch survey of a complete sample of flat-spectrum quasars and BL Lacs observed at 22 and 37 GHz.

Aims of the survey

The sources of our survey were selected from the Kühr 1 Jy catalogue with the criteria that all the objects were in the northern hemisphere, the objects had a measured flux $S \geq 1$ Jy at any frequency above 10 GHz, and a spectral index $\alpha > -0.5$ ($S \propto \nu^\alpha$). In addition, we included three prominent flat-spectrum sources not in the Kühr catalogue: NRAO 150, OH 471 and 2005+40. The sample is a complete sample, with a total of 113 sources. All the sources were observed twice at both 22 and 37 GHz with the 14-m radio telescope of the Metsähovi Radio Research Station between 12.4.1988 and 15.5.1989. Here we consider only the sources for which $S/\sigma > 5$.

We have investigated if there are any statistically significant differences between the radio properties of ordinary low polarization quasars (LPQs), highly polarized blazar quasars (HPQs) and BL Lac objects. We also searched for possible differences between LPQs and blazars (HPQs and BL Lacs) and between quasars (LPQs+HPQs) and BL Lac objects. Finally, we searched the different groups for correlations between spectral properties and flux changes. There were between 12 and 62 objects in each group.

Results

For each object we calculated the relative flux changes $\Delta S/S_{min}$ at both frequencies from the two-epoch measurements. We also calculated three spectral indices; $\alpha(22,37)$, $\alpha_{Kühr}(2.7,5.0)$ and $\alpha(5.0,37)$ from measurements in Metsähovi (22 and 37 GHz) and from the Kühr catalogue data (2.7 and 5.0 GHz). Next, we used the Kruskal and Wallis significance test to see if there are differences in the distribution of the relative flux changes or the spectral indices between the groups mentioned above. We also tested, using the Spearman rank correlation, if there is any correlation between the relative flux change and the spectral index.

We did not find any statistically significant or even probably significant differences in variability or in spectra between LPQs, HPQs, BL Lac objects, or between any combinations of these groups. This means that we have no way of distinguishing between flat-spectrum quasars, blazars, or BL Lac objects using our high radio frequency measurements. When we use the Spearman correlation test, the only statistically significant correlation with the spectral index $\alpha_{Kühr}(2.7,5.0)$ seems to exist only in BL Lacs: the probability of getting by chance alone the observed correlation between $\alpha_{Kühr}$ and the relative flux change at 37 GHz is 2.2 %.(Figure 1.) The same probability between $\alpha(5,37)$ and the relative flux change is 0.6 % in LPQs at 22 GHz and 33.9 % at 37 GHz. No correlations can be found for the other groups, nor using the spectral index $\alpha(22,37)$.

For the low polarization quasars the correlation seems to be statistically proven between $\alpha(5,37)$ and $\Delta S_{22}/S_{22min}$. Surprisingly, LPQs with *steeper* overall spectra seem to be more variable, contrary to what is usually assumed to be the case from lower frequency studies and what would also be expected from simple orientation models (Eckart et al. 1989; Valtaoja et al. 1991). The correlation can be seen in figure 2.

In order to be able to confirm this unexpected result to higher statistical accuracy we need more than just two-epoch measurements of the sources in our sample. The lack of a similar correlation in the 37 GHz data rises the possibility that the 22 GHz result may be a statistical fluke. However, in our recent SEST survey of a sample of Southern sources at 100 GHz (Tornikoski et al., these proceedings) we also found that LPQs with steepest spectra were the most variable. We are looking forward to making larger surveys and getting new data in the near future.

References

Barthel, P.D.: 1989, *Astrophys. J.* **336**, 606
Kühr, H. *et al.*: 1981, *Astr. Astrophys. Suppl.* **45**, 367
Eckart, A., Hummel, C.A., Witzel, A.: 1989, *M.N.R.A.S.* **239**, 381
Valtaoja, E. *et al.*: 1991 (preprint)

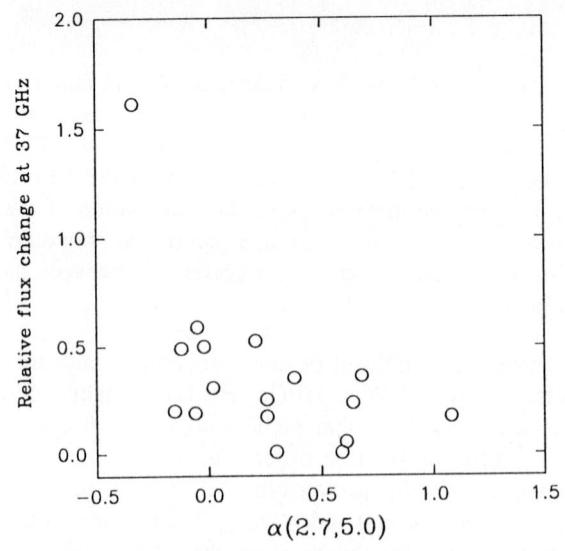

Figure 1. The spectral index $\alpha(2.7,5.0)$ from the Kühr 1 Jy catalogue versus the relative flux change at 37 GHz for BL Lacs.

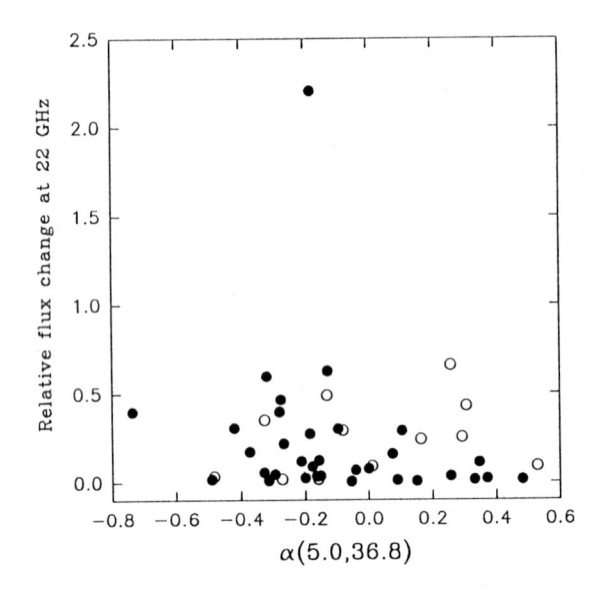

Figure 2. The overall spectral index $\alpha(5.0, 36.8)$ versus the relative flux change at 22 GHz for BL Lacs (open circles) and low polarization quasars (filled circles).

VLBI Observations of Blazars

J. A. Zensus
National Radio Astronomy Observatory
P.O. Box O, Socorro, New Mexico 87801, U.S.A.

Abstract

Results from VLBI observations of compact radio jets in sources with blazar properties are reviewed.

1 Introduction

Many compact radio sources with blazar properties exhibit parsec-scale radio jets that can be imaged with VLBI. From such observations, evidence for relativistic motion can be inferred from the apparent superluminal expansions that are frequently seen (provided one accepts the redshift as a cosmological distance indicator). Whereas the relationship between the blazar properties of a source and the appearance and evolution of its compact radio structure is not yet clearly understood, there are nevertheless indications of a close connection.

In this paper, I will highlight some of the properties of the parsec-scale structure in blazars (i.e., BL Lac objects and OVVs). An excellent overview of the radio structure of BL Lac objects is given by Mutel (1990). For overviews of the subjects of superluminal motion and parsec-scale radio jets see Zensus and Pearson (ed., 1987 and 1990). At this meeting, several papers discussing the compact structure of blazars were presented (e.g., by Akujor, Bååth, Gabuzda, Krichbaum, and Schalinski).

2 VLBI Studies of Compact Radio Sources

So far, only few blazars have been extensively monitored with VLBI (see Table 1). However, systematic VLBI studies of samples of compact sources are underway that include BL Lac objects and OVVs: Witzel and colleagues (Witzel et al. 1988) are investigating a sample of 13 flat-spectrum sources from the MPIfR 5-GHz survey with $\delta > 70°$, $b_{\mathrm{II}} > 10°$, and $S_{5\,\mathrm{GHz}} > 1$ Jy. Pearson and colleagues (Pearson and Readhead 1988, and references therein) are studying a sample of 65 objects with $\delta > 35°$, $b_{\mathrm{II}} >$

10°, and $S_{5\,\text{GHz}} > 1.3$ Jy. Cohen, Wehrle, and colleagues are monitoring a sample of strong sources selected at 5 GHz (Wehrle, Cohen, and Unwin 1990). Gabuzda, Wardle, Roberts, and colleagues have been monitoring the polarization structure of a large sample of compact sources (Gabuzda, this Volume; Gabuzda et al. 1989b; Roberts et al. 1990). Bååth and colleagues have been investigating a sample of BL Lac objects (Bååth, this Volume).

The parsec-scale structures found in compact sources can be sorted in a small number of morphological classes (e.g., Pearson and Readhead 1988): (a) the "core-jet" sources, that are characterized by a compact flat-spectrum core, and a typically one-sided steep-spectrum jet; (b) the "compact steep-spectrum" sources that have a steep spectrum below 5 GHz and a (relatively) small overall size $\lesssim 15$ pc; (c) the "compact double" sources containing two separated regions of emission of comparable flux density and spectrum. In addition to this double morphology, this group is characterized by gigahertz-peak component spectra and faint galaxy identification.

Typically, blazars have compact structure of core-jet character. BL Lacs tend to be strongly core-dominated (lobe-dominated exceptions are 0716+714, 1400+162). In contrast, OVV objects are usually lobe-dominated. In BL Lac objects, jet structures are typically considerably misaligned from the overall source structure, whereas OVVs tend to be better aligned. Pc-scale jets in BL Lac objects do typically not appear to be strongly curved, but this finding may be affected by the limited quality of many VLBI images.

3 Superluminal Motion

The systematic studies show that apparent superluminal motion is a common phenomenon in quasars and BL Lac objects (e.g., Zensus 1989; Schalinski, this Volume; Witzel et al. 1988; Pearson and Readhead 1988). Among the morphological groups, it occurs most frequently in core-jet sources. It is also seen in compact steep-spectrum sources, but never in compact-double objects. The motions always appear as expansion, i.e., outward, never as contraction. They typically occur aligned with or even along an underlying smooth jet. Different velocities have been measured for individual components in some sources, as well as accelerations and decelerations. The observed tracks can be curved, and in some cases appear to differ within a source. The observed speed may of course reflect a "phase velocity" of the brightness peak within a jet, not a physical speed.

3.1 VLBI Monitoring of Blazars

Table 1 lists the blazars for which I am aware of multiple VLBI observations existing at the time of this meeting (cf. Mutel 1990). Of the eight BL Lacs with known redshift, seven are superluminal and one, 1803+78, is subluminal (throughout, $H_0 = 100h$ km s^{-1} Mpc^{-1}, $q_0 = 0.5$ is assumed). In comparison, four out of the five well-studied OVV sources are superluminal.

There appears to exist a trend towards slower motion in BL Lac objects as compared to quasars and OVVs, but this needs still to be confirmed (see Mutel 1990). On average, BL Lacs and OVVs have velocities of $\beta_{\text{app}}h = v/c \sim 3\text{--}5$, whereas for quasars,

Table 1: Superluminal Motion in Blazars

Source	Type	z	N^1	$\beta_{\rm app}h^2$	Refs.
0212+735	OVV	2.37	3	2.2	5,11
0235+164	BL	0.94	4	37–45:[3]	1,2,3
0454+844	BL	–	2	≥ 1.6 :[4]	4,5,6
0716+714	BL	–	2	≥ 2.3 :[4]	5
0735+178	BL	> 0.42	4	> 6.2	1,4,7
OJ 287	BL	0.31	3	2.4,3.2	8
Mk 421	BL	0.03	4	1.36,1.9	9,10
3C 279	OVV	0.54	5	2.9,9.2	13,14
1749+701	BL	0.77	3	5.7	5,6,9
1803+784	BL	0.68	3	3.9	4,5,6
3C 371	OVV	0.05	2	5.9:	3,6,15
2007+77	BL	0.34	4	3.6	5,11
BL Lac	BL	0.07	16	3.3–3.7	3,6,12
3C 446	OVV	1.4	3	3.4:	16
CTA 102	OVV	1.04	3	< 10	17,18

References: 1. Bååth, Zhang, and Nicolson 1990. 2. Jones et al. 1984. 3. Gabuzda et al. 1989a. 4. Gabuzda et al., in preparation. 5. Witzel et al. 1988; Schalinski et al., this Volume. 6. Pearson and Readhead 1988. 7. Gabuzda, Wardle, and Roberts 1989b. 8. Gabuzda, Wardle, and Roberts 1989a. 9. Bååth 1984. 10. Zhang and Bååth 1990. 11. Eckart et al. 1987. 12. Mutel et al. 1990. 13. Unwin et al. 1989. 14. Cotton et al. 1979. 15. Lind 1987. 16. Cohen et al., in preparation. 17. Bååth 1987. 18. Wehrle and Cohen 1989.
Notes: [1] Number of VLBI observations. [2] for $H_0 = 100h$ km s^{-1} Mpc^{-1}, $q_0 = 0.5$. [4] may be gravitationally lensed. [3] based on upper limit $z > 0.3$.

the average is $\beta_{\rm app}h \sim 6$ (e.g., Zensus 1989). Especially the work of Gabuzda and colleagues (this Volume) will be important in this respect. Such a difference in speeds has been explained by an especially close orientation along the line-of-sight (e.g., Impey 1989; cf. Mutel 1990). The evidence for this hypothesis is still inconclusive. However, note that (*a*) the redshift distribution of BL Lacs appears to differ from that of other superluminal sources (but see the results of Impey on HPQs/LPQs); (*b*) the angle to the line of sight for BL Lac itself is large; (*c*) less core-dominated sources should be faster; but 2007+77 is least core-dominated BL Lac object and lies below the average with $\beta_{\rm app}h = 2.3$.

4 Three Examples

I will briefly highlight results from VLBI observations of three superluminal sources: BL Lac, 3C 273 (a low-polarization quasar, with some degree of "blazar" activity), and 3C 345 (a high-polarization quasar).

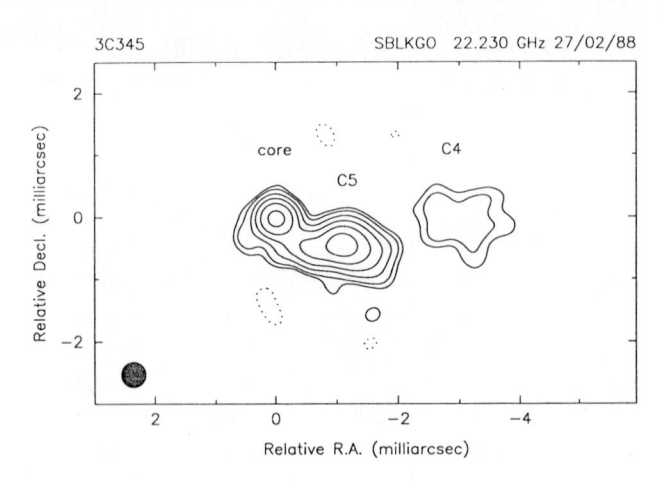

Figure 1: VLBI image at 22 GHz of the superluminal quasar 3C 345, in February 1988 (Zensus, Unwin, and Cohen 1991, in preparation). The image shows the compact (stationary) core, and the two superluminal features C4 and C5. Contours are −2.5, 2.5, 4, 8, 16, 32, and 64% of the peak (1.65 Jy per beam). The size of the CLEAN beam is 0.4 mas (FWHM).

4.1 BL Lac

BL Lac itself is still the best studied example of a BL Lac object (Mutel *et al.* 1990). At least four structure components have been followed with VLBI so far, all of which show apparent superluminal motion at $\beta_{\text{app}}h = 3.3$–$3.7c$ away from an always unresolved core component. Mutel and his coworkers were able to show that the occurrence of each of these components can be associated with a distinct outburst of the source's total flux density. There is no compelling evidence for the existence of continuous jet emission, and only marginal indication of jet bending. As in other superluminal sources, the components expand as they move away from the core.

Hughes, Aller, and Aller (1989*a*, *b*; see also their contributions to this Volume) have shown that the total flux evolution of BL Lac can be adequately explained by a weak-shock model with Lorentz-factor $\gamma \sim 4$ and viewing angle of about 30° to the line-of-sight (see below).

4.2 3C 345

The quasar 3C 345 ($m_b = 16$, $z = 0.595$) is a prototype of core-dominated superluminal radio sources (Zensus, Unwin, and Cohen, in preparation). The milliarcsecond structure of this source contains a core that is unresolved at the highest frequencies and stationary (Figure 1). Multiple features are moving (at apparently superluminal speeds of $\beta_{\text{app}}h \sim 1.4$–9.5, for $H_0 = 100h$ km s^{-1} Mpc^{-1}, $q_0 = 0.5$) from the core in the direction of the outer structure and the best images have confirmed the existence of an underlying complex jet (Unwin and Wehrle, in preparation). Extensive monitoring at cm wavelengths has provided a detailed record of the structural evolution. Observations at mm-wavelengths (discussed by Krichbaum and Bååth at this meeting) provide the relatively new possibility to study the evolution of the superluminal

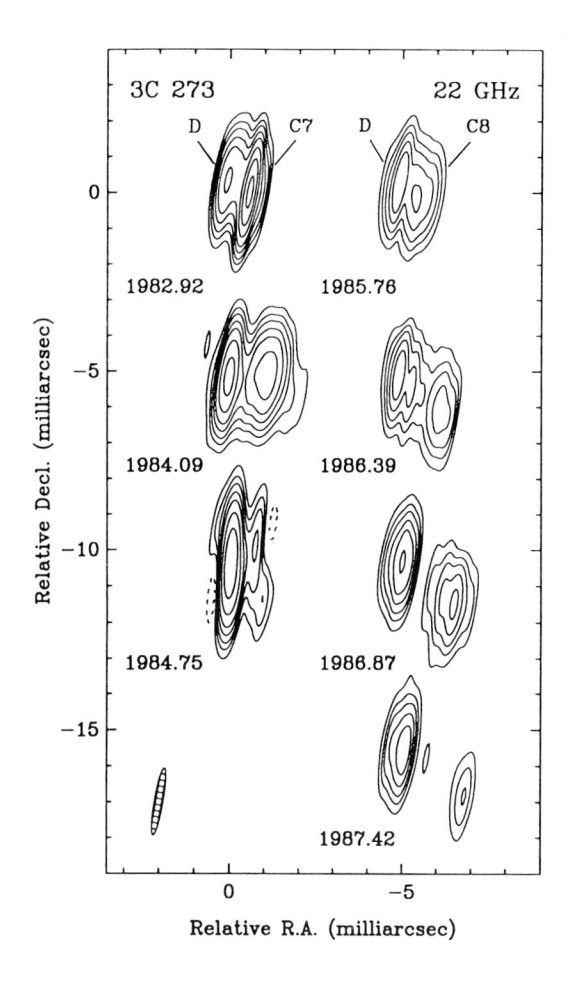

Figure 2: VLBI images at 22 GHz of the nucleus of 3C 273 at seven epochs from 1982–1987 (Figure 2; Zensus et al. 1990). The sequence illustrates the superluminal motion of the jet components C7 and C8 away from the core D. The convolving beam for all images (shown as a shaded ellipse) is a Gaussian with FWHM 0.25×1.9 mas, in P.A. $-10°$. Contours are drawn at -0.25 (dashed), 0.25, 0.5, 0.8, 1.3, 2.0, 3.5, 6.0, and 10.0 Jy/(beam area). The peak brightness (epoch 1982.92) is 13.6 Jy/(beam area).

components in immediate vicinity of the core.

At least 4 superluminal components have been found in 3C 345 so far, and the mm-observations indicate that new components continue to emerge from the compact core. Component sizes and core-separations are frequency dependent. Accelerations are seen and the component sizes increase with core distance. Whereas the outer components appear to roughly follow the same path towards the outer jet, distinctly different trajectories are found for the inner components C4 and C5. These trajectories appear to be non-monotonically curved.

Qian *et al.* (in preparation) have developed a kinematic model of 3C 345 that is based on component motion along several helical magnetic field structures, arranged around the (curved) main jet axis.

4.3 3C 273

The quasar 3C 273 ($z = 0.158$) is another classic example of sources showing apparent superluminal motion (Figure 2; Zensus *et al.* 1990). VLBI observations have shown components in the narrow jet to separate from the core region with apparent speeds of $v/c = 5-7h^{-1}$ (for $H_0 = 100h$ km s^{-1} Mpc^{-1}, $q_0 = 0.5$). This superluminal motion can be traced to at least $46h^{-1}$ pc from the core. The jet is non-monotonically curved, extends continuously to more than 150 milliarcsec ($275h^{-1}$ pc) from the core (Unwin 1990), and is closely aligned with the position angle of the 23″ jet.

Once again, the superluminal components appear to emerge from the core region coincident with major flux density increases at mm-wavelengths indicating a causal connection. The trajectories of components appear to be determined by the ridge line of the jet which undergoes repeated bending, probably amplified by projection and beaming effects. There is no significant evidence against one single path of the motion.

The compact structures seen in the VLBI images—especially the superluminal features—are clearly related to the regions of the jet which cause the variability in total flux density. Shock models, which have been successful in explaining total flux variability in other sources (e.g., BL Lac) offer also an attractive interpretation of the structure of 3C 273. These models would interpret the moving components as disturbances in an underlying relativistic flow, rather than, e.g., relativistically moving clouds. The observations so far are consistent with some of the predictions of such a scenario, although alternative models have not been ruled out.

5 Interpretation of the Source Structure

It is now widely accepted that the apparent superluminal motion and other properties of compact radio sources are a consequence of the presence of bulk relativistic motion in a plasma jet originating in the nucleus of the source (e.g., Blandford and Königl 1979; Blandford 1987). In this picture, the "core" in VLBI images represents the optically thick base of the jet (located near the apex of the jet cone), and the superluminal "components" are regions of enhanced emission moving down the jet. The latter can be modeled as clouds of plasma in the jet which are by some mechanism accelerated to relativistic speeds. For example, this approach has been applied to 3C 345 and reasonable physical parameters consistent with observed properties can be derived

(Zensus, Cohen, and Unwin, in preparation). In particular, the expected frequency dependence of the radial positions of components measured relative to the "core" (due to varying opacity of that region; e.g., Königl 1981) is indeed seen in 3C 345.

When the axisymmetry is broken and the jet is curved, as seen in many core-jet sources, frequency-dependence of positions transverse to the jet can also occur. The sense of the effect, for optically-thin moving components, is that higher frequency images show the component peaks on the *outside* of the bend, relative to lower frequencies. In general, for jets which have greatly increased curvature at small radii (like 3C 273 and 3C 345) opacity effects will cause the jet to appear more curved at higher frequencies. This is separate from any effects of varying resolution.

Recently, the "shock-in-jet" models (which are based on the model of Blandford and Königl) have been remarkably successful in reproducing the radio variability behavior of several sources (cf. Marscher, this Volume; Daly and Marscher 1988; Marscher 1990; Hughes, Aller, and Aller 1989a,b). These include 3C 273: Marscher and Gear (1985) modeled the flux outburst that occurred in 1983. In these models, the underlying jet flow is assumed to be relativistic and confined to a cone of constant opening angle, and the physical conditions are functions solely of distance from the jet's apex. Variability of total flux and polarization, and moving or stationary features in the jet (but likely not the core itself) correspond to shocks which are caused by disturbances of the underlying steady jet flow. The magnetic field is thought to be random or to some degree parallel to the jet flow originally, but perpendicular in the regions of the shocks, owing to compression of the field structure, which has been confirmed with polarization VLBI in some sources (e.g., Roberts *et al.* 1990).

A consequence of the shock models is that the Lorentz factor of the shocks can be different from that of the underlying flow, and it can vary between different shocks according to the strength of the disturbances (e.g., Marscher 1990). From the measured apparent speeds β_{app}, we can derive only limits for the Lorentz factor γ and the component motion's angle to the line of sight θ (Pearson and Zensus 1987): $\gamma_{min} = (1 + \beta_{app}^2)^{1/2}$, and $\cot(\theta_{max}/2) = \beta_{app}$. According to Marscher and Gear (1985), the lowest Lorentz factors which would be observed for the shocks correspond to those of the undisturbed flow. The actual values are expected to be higher in the early shock phases, although only by about a factor of two for the most extreme disturbances.

The shock model predicts that the superluminal components become resolved at centimeter wavelengths in the adiabatic phase of the shock's evolution (cf. Marscher and Gear). They are then expected to expand isotropically, and their Lorentz factors should decrease with time until they equal that of the flow, i.e., the components should appear to decelerate as they separate from the core. The expansion of components is consistent with our images of superluminal sources, but deceleration is generally not observed; however, depending on the shock properties this may be explained by the fact that they can typically be followed only for 2–3 yr.

The "wiggling" observed in several compact radio sources could be due to precession of a decelerating jet as might be the case for the kpc-scale structure of 3C 273 (Davis *et al.* 1985). For large inclination angles, suitable magnetic field configuration (e.g., Königl and Chouduri 1985) could cause the observed bends as well. Finally, Kelvin-Helmholtz instabilities from interaction with the environment and oblique reflection shocks may be relevant in this context (Hardee and Norman 1989).

I am grateful to the hosts of the conference for support and to Esko Valtaoja for his patience. The National Radio Astronomy Observatory is operated by Associated Universities, Inc. under cooperative agreement with the U. S. National Science Foundation.

References

Bååth, L. B. 1984, in *IAU Symposium 110, VLBI and Compact Radio Sources*, ed. R. Fanti, K. Kellermann, and G. Setti (Dordrecht: Reidel), p. 127.
Bååth, L. B. 1987, in *Superluminal Radio Sources*, ed. J. A. Zensus and T. J. Pearson (Cambridge: Cambridge University Press), p. 206.
Bååth, L. B., Zhang, F. Z., and Nicolson, G. 1990, *Astron. Astrophys.*, submitted.
Blandford, R. D. 1987, in *Superluminal Radio Sources*, ed. J. A. Zensus and T. J. Pearson (Cambridge: Cambridge University Press), p. 310.
Blandford, R. D., and Königl, A. 1979, *Astrophys. J.*, **232**, 34.
Cotton, W. D., Counselman, C. C., III, Geller, R. B., Shapiro, I. I., Wittels, J. J., Hinteregger, H. F., Knight, C. A., Rogers, A. E. E., Whitney, A. R., and Clark, T. A. 1979, *Astrophys. J. (Letters)*, **229**, L115.
Daly, R. A., and Marscher, A. P. 1988, *Astrophys. J.*, **334**, 539.
Davis, R. J., Muxlow, T. W. B., and Conway, R. G. 1985, *Nature*, **318**, 343.
Eckart, A., Witzel, A., Biermann, P., Johnston, K. J., Simon, R., Schalinski, C., and Kühr, H. 1987, *Astron. Astrophys. Suppl.*, **67**, 121.
Gabuzda, D. C., Cawthorne, T. V., Roberts, D. H., and Wardle, J. F. C. 1989a, *Astrophys. J.*, **347**, 701.
Gabuzda, D. C., Cawthorne, T. V., Roberts, D. H., and Wardle, J. F. C. 1989b, in *BL Lac Objects*, ed. L. Maraschi, T. Maccacaro, and M.-H. Ulrich (Berlin: Springer), p. 22.
Gabuzda, D. C., Wardle, J. F. C., and Roberts, D. H. 1989a, *Astrophys. J. (Letters)*, **336**, L59.
Gabuzda, D. C., Wardle, J. F. C., and Roberts, D. H. 1989b, *Astrophys. J.*, **338**, 743.
Hardee, P. E., and Norman, M. L. 1989, *Astrophys. J.*, **342**, 680.
Hughes, P. A., Aller, H. D., and Aller, M. F. 1989a, *Astrophys. J.*, **341**, 54.
Hughes, P. A., Aller, H. D., and Aller, M. F. 1989b, *Astrophys. J.*, **341**, 68.
Impey, C. 1989, in *BL Lac Objects*, ed. L. Maraschi, T. Maccacaro, and M.-H. Ulrich (Berlin: Springer), p. 149.
Jones, D. L., Bååth, L. B., Davis, M. M., and Unwin, S. C. 1984, *Astrophys. J.*, **284**, 60.
Königl, A. 1981, *Astrophys. J.*, **243**, 700.
Königl, A., and Chouduri 1985, *Astrophys. J.*, **289**, 188.
Lind, K. R. 1987, in *Superluminal Radio Sources*, ed. J. A. Zensus and T. J. Pearson (Cambridge: Cambridge University Press), p. 180.
Marscher, A. P. 1990, in *Parsec-Scale Radio Jets*, ed. J. A. Zensus and T. J. Pearson (Cambridge: Cambridge University Press), p. 236.
Marscher, A. P., and Gear, W. K. 1985, *Astrophys. J.*, **298**, 114.
Mutel, R. L. 1990, in *Parsec-Scale Radio Jets*, ed. J. A. Zensus and T. J. Pearson (Cambridge: Cambridge University Press), p. 98.

Mutel, R. L., Phillips, R. B., Su, B., and Bucciferro, R. R. 1990, *Astrophys. J.*, **352**, 81.
Pearson, T. J., and Readhead, A. C. S. 1988, *Astrophys. J.*, **328**, 114.
Pearson, T. J., and Zensus, J. A. 1987, in *Superluminal Radio Sources*, ed. J. A. Zensus and T. J. Pearson (Cambridge: Cambridge University Press), p. 1.
Roberts, D. H., Wardle, J. F. C., Brown, L. F., Gabuzda, D. C., and Cawthorne, T. V. 1990, in *Parsec-Scale Radio Jets*, ed. J. A. Zensus and T. J. Pearson (Cambridge: Cambridge University Press), p. 110.
Unwin, S. C. 1990, in *Parsec-Scale Radio Jets*, ed. J. A. Zensus and T. J. Pearson (Cambridge: Cambridge University Press), p. 13.
Unwin, S. C., Cohen, M. H., Biretta, J. A., Hodges, M. W., and Zensus, J. A. 1989, *Astrophys. J.*, **340**, 117.
Wehrle, A. E., and Cohen, M. 1989, *Astrophys. J. (Letters)*, **346**, L69.
Wehrle, A. E., Unwin, S. C., and Cohen, M. H. 1990, in *Parsec-Scale Radio Jets*, ed. J. A. Zensus and T. J. Pearson (Cambridge: Cambridge University Press), p. 49.
Witzel, A., Schalinski, C. J., Johnston, K. J., Biermann, P. L., Krichbaum, T. P., Hummel, C. A., and Eckart, A. 1988, *Astron. Astrophys.*, **206**, 245.
Zensus, J. A. 1989, in *BL Lac Objects*, ed. L. Maraschi, T. Maccacaro, and M.-H. Ulrich (Berlin: Springer), p. 1.
Zensus, J. A., and Pearson, T. J. (ed.) 1987, *Superluminal Radio Sources* (Cambridge: Cambridge University Press).
Zensus, J. A., and Pearson, T. J. (ed.) 1990, *Parsec-Scale Radio Jets* (Cambridge: Cambridge University Press).
Zensus, J. A., Unwin, S. C., Cohen, M. H., and Biretta, J. A. 1990, *Astron. J.*, **100**, 1777.
Zhang, F. J., and Bååth, L. B. 1990, *Astron. Astrophys.*, **236**, 47.

High Resolution Images of Blazar Cores

L.B.BÅÅTH
Onsala Space Observatory, S-439 00 Onsala, Sweden

1. INTRODUCTION

Observations with Very Long Baseline Interferometry (VLBI) at mm wavelengths is currently our only possibility to make images of the cores of active galactic nuclei, quasars, blazars, or whatever name we choose to use. The resolution we can now obtain with such instruments is 50 µas (microarcsecs.) or better, resulting in linear resolution, diameter, of 10^{15} - 10^{17} cm. Such observations therefore gives us direct insight into the core areas on size scales of the same order as the accretion disc itself. These images can then be combined with the observations of the changes in the continuum spectra to make better models how radio jets are created and sustained and of how components, or shocks propagate through these. The resolution we achieve is equivalent to that of a 2km diameter optical telescope which is diffraction limited. The angle of 50 µas is what a golf-ball on the moon would fill as observed from earth.

VLBI observations at 100GHz (λ3mm) started around 1982, but 1988 was the very first epoch where we could produce hybrid maps. This could be achieved because of new scheduling strategy, where we for the first time aimed for a full mapping session, but mostly because of the development of new data reduction technique (Bååth 1991, Bååth et al. 1991a). Since 1988 we have observed three epochs, in March 1988, March 1989, and April 1990. These observations involved telescopes in Sweden, Japan, and USA. The epoch of 1990 also for the first time in VLBI included the SEST telescope in Chile. The 1990 epoch therefore includes a telescope on the southern hemisphere, and the resolution in the north-south direction is drastically improved. This paper will, for the first time, show some of the images obtained in 1990.

The two epochs 1988 and 1989 have been published and discussed in two recent papers (Bååth *et al.* 1991a and Bååth *et al.* 1991b). I refer the reader to these papers for a more detailed discussion on the individual sources. All following linear sizes will be calculated assuming cosmological redshift, H_0 = 100 km sec^{-1} Mpc^{-1} and q_o = 0.05.

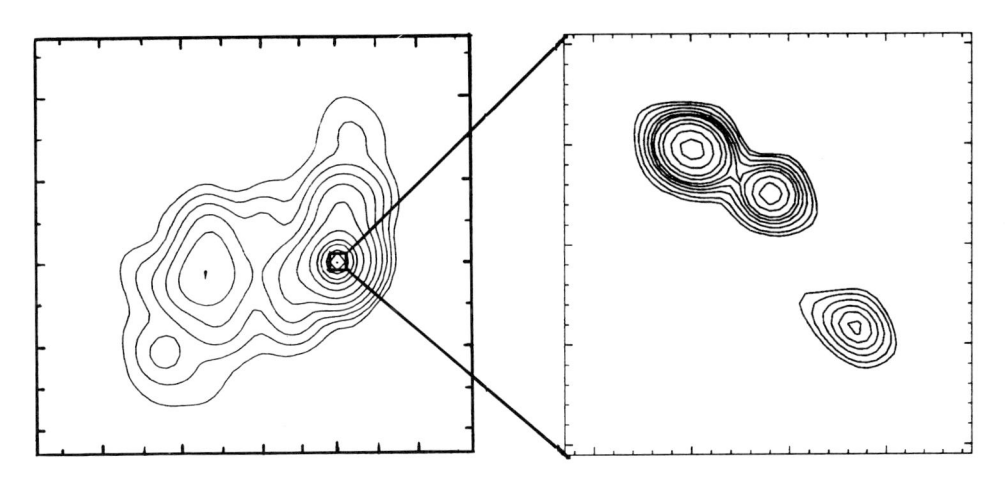

Figure 1. The "blazar" 3C446 observed with global VLBI arrays at 5GHz (left: from Wehrle, Cohen & Unwin 1990) and 100GHz (right). Tick marks are at 1mas (left) and 20µas (right).

2. RESULTS

3C446 is a good example of how our resolving power compares to that of classical VLBI instruments. Fig. 1 shows a 5GHz VLBI map (from Wehrle, Cohen & Unwin 1990) made with the highest available resolution. Our image from 100GHz data shows only the area very close to the core. 3C446 was not in our original list of objects and was therefore not observed by us until the 1990 epoch when the inclusion of the SEST telescope made low declination sources more attractive. 3C446 is truly a blazar with very rapid temporal variations in continuum as well as line emission (CIII]). Our 100GHz map shows a clear jet, not unlike those seen in other sources with similar behaviour (e.g. 3C273 and 3C279). This jet has to bend around by 120° in order to continue as suggested by the VLBI image at 5GHz. This source was in fact one the most luminous in our sample. Only 3C273 was more powerful on the longest baselines. We will continue to observe 3C446 in the future as one of our main target sources. Unfortunately I believe that we cannot draw any more conclusions from our only present epoch than that the radio jet exists and supposedly will curve towards the east.

OJ287 is of special interest for our hosts. This source is the most compact of all sources we observe at 100GHz and is unresolved by us in 1988 and 1989. The 1990 epoch data have not yet been reduced, but will of course be of particular interest since our observation was close to an outburst. The linear size we obtain for this source is $5 \cdot 10^{16}$ cm at the position angle of the major axis of the binary system suggested by Sillanpää *et al.* (1988). This size is also of the same order as the diameter of the orbit of the two proposed black holes. It is unlikely that two diverging beams from this kind of system would not have been observed by us at any of the two epoches 1988 or 1989.

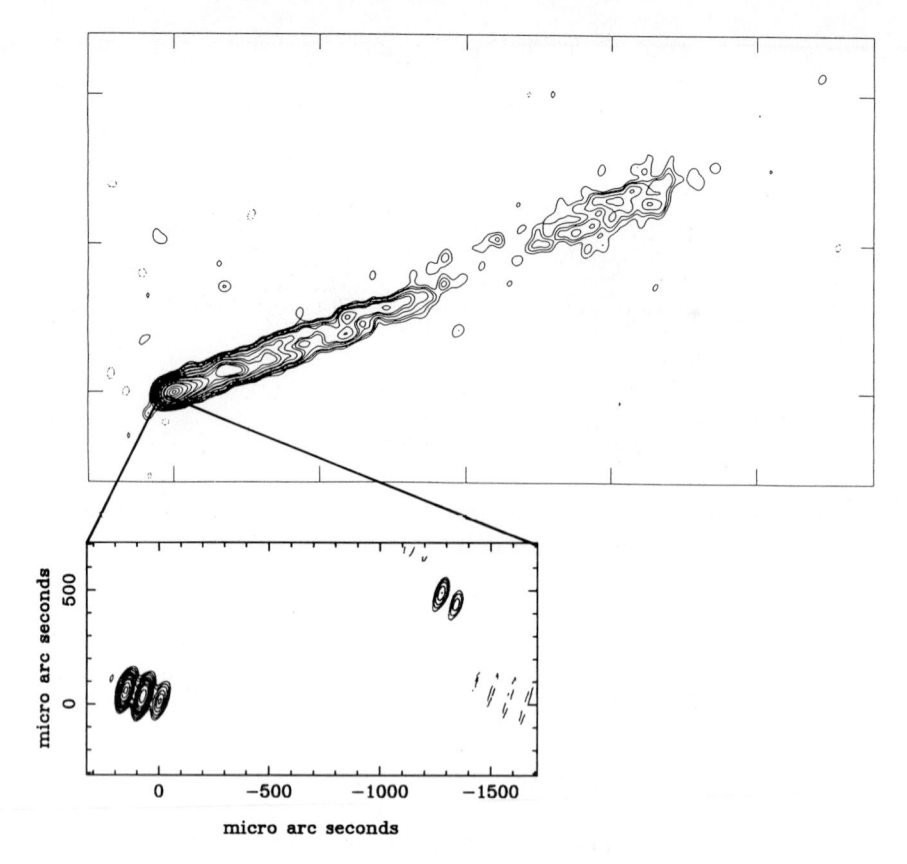

Figure 2. The radio jet of the active galaxy 3C274 (M87) observed with global VLBI arrays at 1.6GHz (upper: from Reid *et al*. 1989; tick marks are at 50mas) and 100GHz (lower).

Therefore the binary black hole model seems unlikely, and instead OJ287 appears to be a more standard radio source except that its jets may be observed more end-on than is usual.

3C274 (≡ M87)4 is not a blazar, but the most nearby of all AGN's and can therefore give us more insights and even more details of the core region of such objects. 3C274 is indeed a very powerful radio source with a very long and straight radio jet. Figure 2 shows this jet observed with VLBI at 1.6GHz (from Reid *et al*. 1989) compared to our 100GHz image with 60 times higher resolution. The image by Reid *et al*. (1989) shows that the jet is limb brightened and that the substructure consists of a helical type pattern.

Our map clearly indicates that the helical pattern continues all the way into the core and keeping the same phase of winding. The overall position angle is also kept the same at about -70°. The 100GHz map in Fig. 2 should not be overinterpreted though. The individual subcomponents are not in themselves reliable, but two main complexes exist and have sub-structure along position angle about -110°. The very fact that such an helical pattern can exist makes modelling of blazars, quasars, and other AGN's much more complicated than many previous models have assumed.

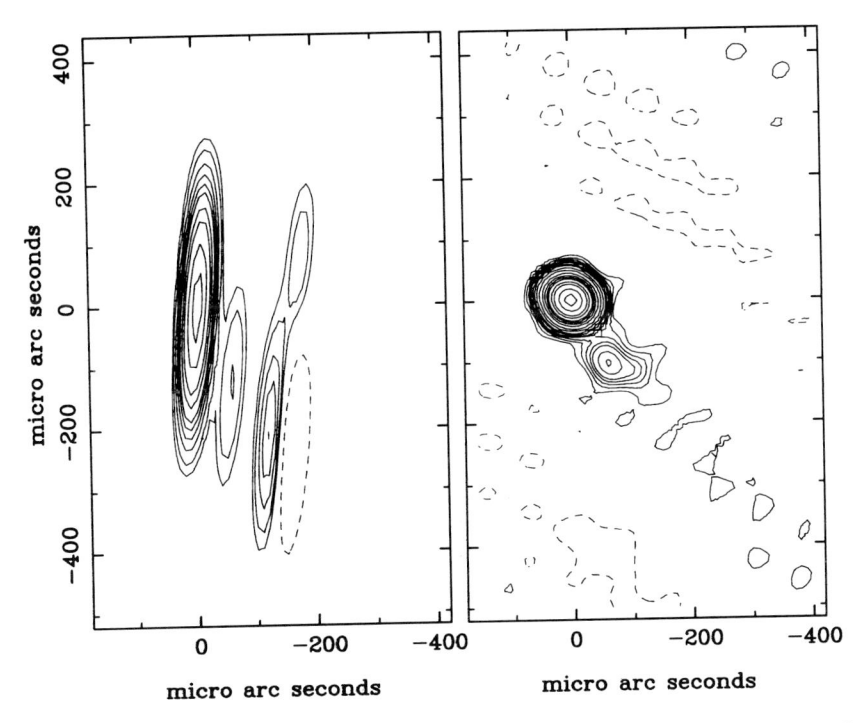

Figure 3. The radio jet of the blazar 3C279 observed with global VLBI arrays at 100GHz in March 1989 (right) and April 1990 (lower). Restoring beams were 280x50 μas in 1989 and 50x50 μas in 1990.

The position angle of the axis of the helix, the tightness of the windings, and the phase of the helix at the start of the jet will all conspire to create apparent classes of objects.

3C279 is one of our main target sources. Figure 3 shows images of this source from 1989 and 1990. The difference in the north-south resolution by adding the SEST telescope in 1990 is truly significant. This radio jet is in sharp contrast to the ones we have discussed earlier in that it is remarkably straight. There is no significant offset from the position angle of -145° at any scale, going from ours at 100 μas to 10" (dePater & Perley 1983). There are some indications of wiggles in our map from 1990, but these are on a much smaller scale then in e.g. 3C273 (Bååth *et al.* 1991a). At the same time we did observe 3C279 in 1989 less than one year after a major outburst in flux density, but no indication of any sharp and intense new component is visible in our image.
The integrated flux density was 17Jy at this epoch, as compared to about 3Jy (e.g. Gear *et al.* 1985) at more quiet periods. We would have expected to see a compact component at about 10Jy, but the brightest component is only about 2.8Jy. This is at the level we expect for the quiet core from an extrapolation of the optically thin part of the spectra of the D-component at lower frequencies (Unwin *et al.* 1989). The outbursts in this source have been modelled by Hughes, Aller & Aller (1989) as initially thin shocks which propagate out through the plasma in the jet and eventually expand adiabatically. Such a shock would not have any sharp edge if it is observed coming directly towards the observer. However, this is not probable since :

1. The deprojected radio jet would be extremely long, longer than 40Mpc for a viewing angle <1°.

2. Any small deflection of the jet from a straight line would be grossly exaggerated by relativistic effect.

3. No, or only small, superluminal expansion would be observed.

The third point is directly contradicted by observations. 3C279 was in fact the very first source in which superluminal motion was detected (Whitney *et al.* 1971). In fact the angular expansion speed in this source is measured to be quite large, 0.5mas/year (Cotton *et al.* 1979). The other two points would make 3C279 a truly outstanding object in the universe. This is possible, but hardly probable. The model by Hughes, Aller & Aller (1989) suggest an alternative possibility. New components seem to be associated with changes in polarization angle and polarized flux rather than outbursts in integrated flux density. A number of such components can be masked by a single outburst in flux density and therefore we may not be correct in assuming that a new component should be seen emerging from the core in conjunction with the outburst.

Outbursts may instead be originating further out in the jet where the path curves more closely towards the observer. Such an event has been observed in 0735+178 (Bååth, Zhang & Chu 1991). Very small changes of the viewing angle can create such an effect. Very small deviations from a straight line may not be visible further out in the jet since the angular resolution at the lower frequencies need to observe the shocks at this late stage of development, is not sufficiently good to resolve this pattern.

BL Lac is another of our main target sources. This is one of the few sources where we have been able to observe a component which could be identified on lower resolution maps. Figure 4 shows our images from all three epochs, but also a larger area convolved down to the 10 times lower resolution as appropriate to compare with VLBI observations at 10GHz. Aller (these proceedings) has discussed the outbursts of this source and I have labelled the individual components accordingly. The component S5 observed at 100GHz is somewhat closer to the core and at slightly different position angle than what is observed at 10GHz. S5 in our map of 1988 is at 1486±5 µas at position angle 201.6±0.5° as compared to 1950±160 µas at position angle 195.9±3.2° at 10GHz (Mutel *et al.* 1990). The position of our S5 is very close to the position of the previous component S4 at the time when this component was stable for a few years. This place may therefore well be where the deflection of the jet, as suggested by Aller (these proceedings), is situated.

Our observations at 100GHz shows two new components extending from the core. These fit well with the appearances of S6 and S7 as suggested from polarization properties by Aller (these proceedings). The velocities of these components are rather low though. S6 is moving with about 90±5 µas/year which is subluminal. If these components are later seen to be moving superluminally then they either have to speed up significantly, or apparently speed up by a change in viewing angle. The latter seems more probable in view of the fact that the jet is bent in the plane perpendicular to the line of sight: the components closer to the core are extending to the west while the milliarcsecond jet is more to the south.

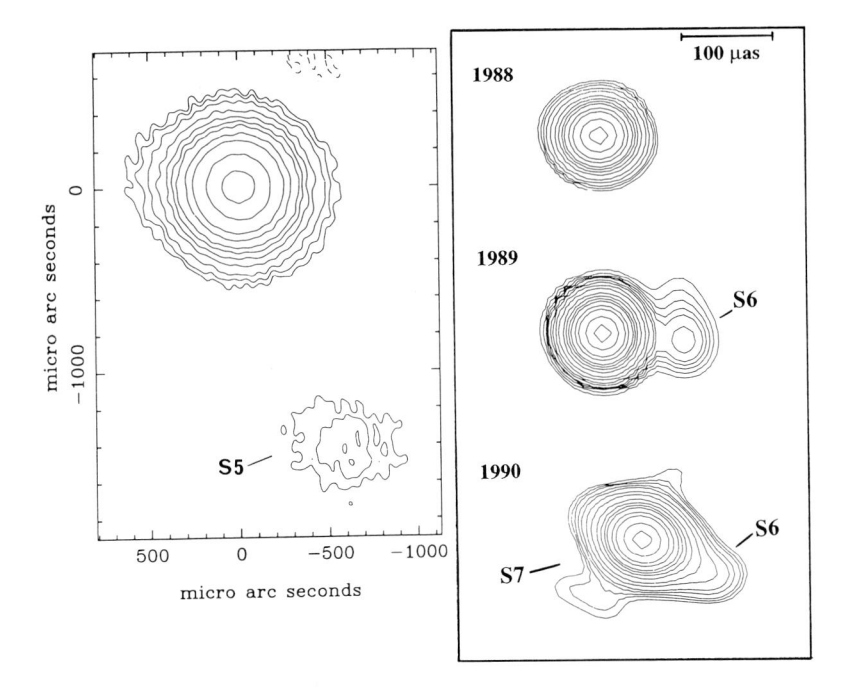

Figure 4. BL Lac observed with global VLBI arrays at 100GHz. Left: March 1988 (upper), March 1989 (middle) and April 1990 (lower). Restoring beam was 50x50 µas. Left: a larger area from 1988, convolved with a beam of 500µas.

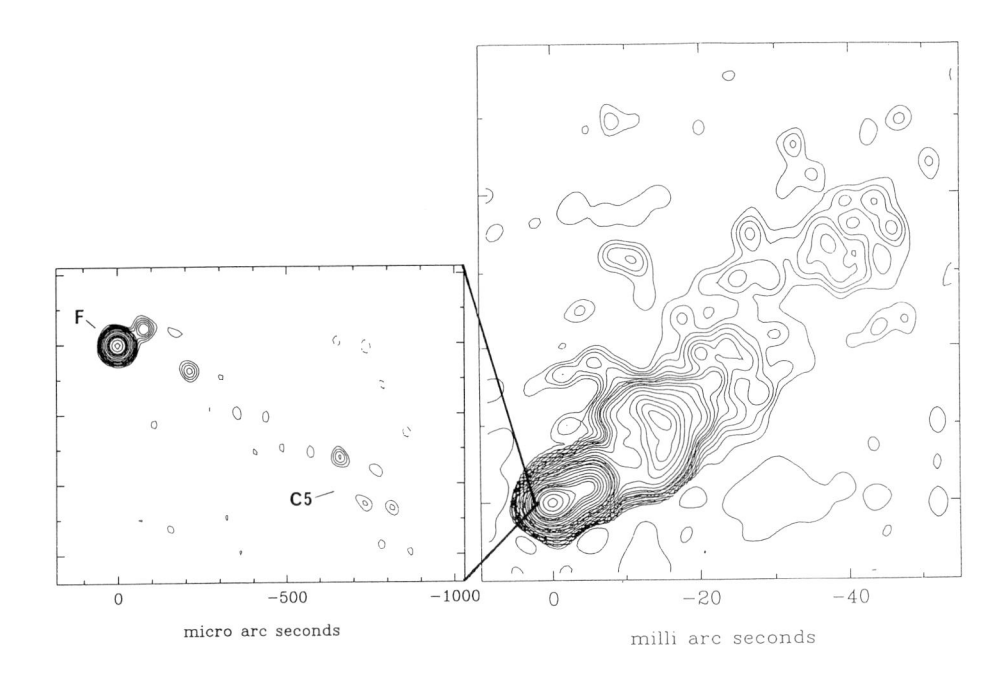

Figure 5. 3C345 observed with global VLBI arrays at 100GHz (left) and 1.6GHz (right). The 1.6GHz image is from Rantakyrö et al. (1991).

3C345 is one of the most studied of all superluminals. The radio jet is curved, and a component has been shown to move along this curved trajectory (Moore *et al.* 1983). Figure 5 shows maps obtained with VLBI at 1.6 and 100GHz. The wiggles in the jet are obvious from these images and fits well with the model suggested by Hardee (1990). The component C5 is partially resolved at 100GHz, but its position is closer to the core than what has been reported at 22 and 43GHz (Zensus these proceedings). This can have several explanations:

- A component created from a shock front moving down the jet may well have a spectral index gradient along the jet. Thus the peak brightness of the component may be dependent on the observing frequency.
- Likewise the position of the core itself may be frequency dependent.
- The VLBI observations at 22 and 43GHz have much poorer resolution than ours at 100GHz and the peak may therefore be shifted because more of the lower surface brightness emission is included in the image.

There are no indications of any other path such as have been suggested from data at 22GHz (Zensus these proceedings). This can be explained if the position of the core is shifting with time. The core at 8GHz has been proved to be stationary relative another quasar, but this is at another frequency, and *the formal standard errors are larger than our resolution*! If in fact the core is merely new shock fronts moving out from the central engine, and have not yet reached a point where they can be distinguished from each other, then the position of the core is likely to change with time. Such a scenario would phase-shift the helical pattern and a new component would apparently have another trajectory.

Our VLBI observations at 1.6 GHz (Rantakyrö *et al.* 1991) shows that the jet must be curved in three dimensions and probably is curling away from the line of sight outside about 5mas from the core. Therefore a shift in the position angle of the path of a new component as suggested by Zensus (these proceedings) also would include a shift in viewing angle. Any such shift in viewing angle would have dramatic consequences in the velocity and Doppler boosting of that component. Since these effects seems to be roughly unchanged between components there cannot be any large changes in viewing angle and the change of path is probably merely apparent.

3. DISCUSSION

There are some major points I want to stress from these observations:
1. The core of these objects is very small, on the order of $10^{16} - 10^{17}$ cm.
2. The curvature observed with lower resolution and at larger distances from the core, continues and is in many cases further enhanced closer to the core.
3. We observe very little structural changes with time at this close distance from the core.

The radius of the core region is only some 50-500 times the Schwarzchild radius of the $4 \cdot 10^8$ M_o black hole which has been suggested for the central engine (Band & Malkan 1989). This is of the same size scale as the proposed accretion disc and the nonthermal source observed by Band & Malkan (1989). Therefore radiation pressure from the central region is still a viable source of energy for the relativistic plasma within the radio jet. The components we observe fit well with the scenario that shock waves move down stream through this plasma and creates peaks of emission in their wakes. The velocity of such a shock can be several times larger than the velocity of the plasma itself. Therefore if we do mostly resolve the shocks at 100GHz, then we would expect to observe structural changes more appropriate to those of the plasma flow rather than the shocks. Fast expanding shocks would then be apparently moving slowly with high resolution and at high frequency, while at lower resolution and lower frequency they would appear to move faster once they are at a distance where they can be separated from the core. We would expect to observe more heavy shocks as a thin edge, thin along the direction of the jet and long perpendicular to this. We have indeed observed such a component in 3C273 (Bååth *et al.* 1991; also these proceedings).

It is important here to note that the components we observe are components of *emission*. These do not have to follow the same track as the shock, but are the sum of the acceleration of particles behind the shock and the structure of the magnetic field. It is quite conceivable that the shock moves down the plasma flow, but we observe only a pattern along the rim of this flow. Thus a helical pattern can be formed, and even change in phase, without any change in viewing angle relative the motion of the shock.

The existence of significant wiggles in the jet makes it probable that such wiggles exist in all three dimensions. The result is that the Doppler boosting factor may change significantly along the jet. The appearance of a source dominated by such a radio jet would be very much dependent on the geometry. If the jet is turned towards the observer at the beginning, than outbursts would appear very highly Doppler boosted at high frequencies. If, on the other hand, the jet is pointing away from the observer at the start and then turns towards him, than the high frequency outburst would not be seen, but a low frequency outburst would be observed. The nature of these two types of outburst would appear to be quite different. In the second scenario the shocks would have had time to expand and the component would be much larger. If the jet is more wiggling closer to the core, than we would expect that at some point it would be directed towards the observer and there would be an excess of the first type of sources described above as compared to the second type.

A helix seems at the moment to be the best working hypothesis for a jet model. A true helix is not necessary and any small deviation from this would not significantly alter the discussion. Models of the outbursts observed in "blazars" now have to take such a helical path of components into account. It is quite possible that differences in this helix between sources may explain most of the different groups of sources we have discussed. How tight the helix is wound, how large the opening angle is, how much the axis itself curves in space, and the phase at the beginning are all new and important parameters which any model have to deal with in the future. It is true that "a horse is not some kind of a cow" (Impey these proceedings), but maybe these additional parameters will make it possible to breed a horse that looks very much like a cow from a distance.

It is unlikely that we observe the central engine itself at 100GHz. Therefore we cannot tell the true distance between the "engine", with its black hole and accretion disc, and the start of the radio jet which we usually label the "core" for convenience. The radio structure at 100GHz would fit inside the broad line emission region if in fact the high frequency radio core is close to the true nucleus. It is quite possible though, that the radio core is instead formed in the BLR. Only VLBI images at even higher frequency, where the accretion disc also become visible to our instruments, will tell the true geometry.

The images presented here are clear evidence that VLBI observations at 100GHz now are in full operation and can produce results of very good quality. It is my hope that the next few years will see such images produced at 230 and 350GHz as well.

REFERENCES

Bååth,L.B., Padin,S., Woody,D., Rogers,A.E.E., Wright,M.C.H., Zensus,A., Kus,A.J., Backer,D.C., Booth,R.S., Carlstrom,J.E., Dickman,R.L., Emerson,D.T., Hirayabashi,H., Hodges,M.W., Inoue,M., Moran,J.M., Morimoto,M., Payne,L., Plambeck,R.L., Predmore,C.R. & Rönnäng,B.: 1991a, *Astron.Astrophys.*, **241**, L1

Bååth,L.B., Rogers,A.E.E., Inoue,M., Padin,S., Wright,M.C.H., Zensus,A., Kus,A.J., Backer,D.C., Booth,R.S., Carlstrom,J.E., Dickman,R.L., Emerson,D.T., Hirayabashi,H., Hodges,M.W., Moran,J.M., Morimoto,M., Payne,L., Plambeck,R.L., Predmore,C.R., Rönnäng,B. & Woody,D.: 1991b, *Astron.Astrophys.*, submitted

Bååth,L.B. Zhang,F.J. & Chu,H.S.: 1991, *Astron.Astrophys.*, submitted

Band,D.L. & Malkan,M.A.: 1989, *Astrophys.J.*, **345**, 122

Cotton,W.D.: 1979, *Astrophys.J.(Letters)*, **229**, L115

dePater,I. & Perley,R.A.: 1983, *Astrophys.J.*, **273**, 64

Hardee,P.E.: 1990, *Parsec scale radio jets,* Zensus and Pearson (eds.), Cambridge University Press, p266

Hughes,P.A., Aller,H.D. & Aller,M.F.: 1989, *Astrophys.J.*, **341**, 54

Moore,R.L., Readhead,A.C.S. & Bååth,L.B.: 1983, *Nature*, **306**, 44

Rantakyrö, F.T., Bååth,L.B., Pauliny-Toth,I.I.K., Matveyenko,L.I., Unwin,S.C. & Cohen,M.H.: 1991, *Astron.Astrophys.*, submitted

Reid,M.J., Biretta,J.A., Junor,W., Muxlow,T.W.B. & Spencer,R.E.: 1989, *Astrophys.J.*, **336**, 112

Sillanpää,A., Haarala,S., Valtonen,M.J., Sundelius,B. & Byrd,G.G.: 1988, *Astrophys.J.*, **325**, 628

Unwin,S.C., Cohen,M.H., Biretta,J.A., Hodges,M.W. & Zensus,J.A.: 1989, *Astrophys.J.*, **340**, 117

Wehrle,A.E., Cohen,M.H. & Unwin,S.C.: 1990, *Parsec scale radio jets,* Zensus and Pearson (eds.), Cambridge University Press, p49

Whitney,A.R., Shapiro,I.I., Rogers,A.E.E., Robertson,D.S., Knight,C.A., Clark,T.A., Goldstein,R.M., Marandino,G.E. & Vandenberg,N.R.: 1971, *Science*, **173**, 225

Structural Variability of Active Galactic Nuclei at 43 GHz

T. P. Krichbaum and A. Witzel

Max-Planck-Institut für Radioastronomie, Bonn, Germany (F.R.G.)

Abstract

We present and discuss results from recent global VLBI-observations at 43 GHz. Here we show maps and models of the sources 3C 84, 3C 273 and 3C 345 with an angular resolution of down to 0.1 mas. We observed changes of the source structures during the three observing epochs. The 43 GHz VLBI-observations indicate – in consistency with data obtained at adjacent frequencies – non-linear structural variations and an increased jet-bending in the proximity of the cores.

1. Introduction

The expression

$$\theta \propto \lambda \cdot S^{1/2} \cdot T_B^{-1/2}$$

relates the size θ of the emitting region, the observing wavelength λ, the flux density S, and the brightness temperature T_B. This dependence of the source size on the observing wavelength was the prime reason for the astronomers interested in the small-scale structures of AGN to promote mm-VLBI measurements, since at longer wavelengths optical-depth effects do not allow any direct imaging of the self-absorbed core regions.
At present radio-telescopes in Europe, the USA, and Japan form an aperture-synthesis VLB-interferometer at 43 GHz frequency, which allows imaging with an angular resolution of down to 90 μas. This corresponds to a linear scale of 0.38 pc for a source with a redshift of $z = 1.0$ ($H_0 = 100$ km s^{-1} Mpc^{-1}, $q_0 = 0.5$). Figure 1 shows the location of the telescopes, which participated in the experiments described here.
At present 7 mm-VLBI is only possible as a really joint effort of the scientists and technicians at the participating institutes. In Table 1 we listed the colleagues, without whose cooperative efforts the results discussed below would not have been possible.

Technical difficulties at mm-wavelengths, with receivers and telescopes, and the limited number of antennas still restrict the number of observable sources. Table 2

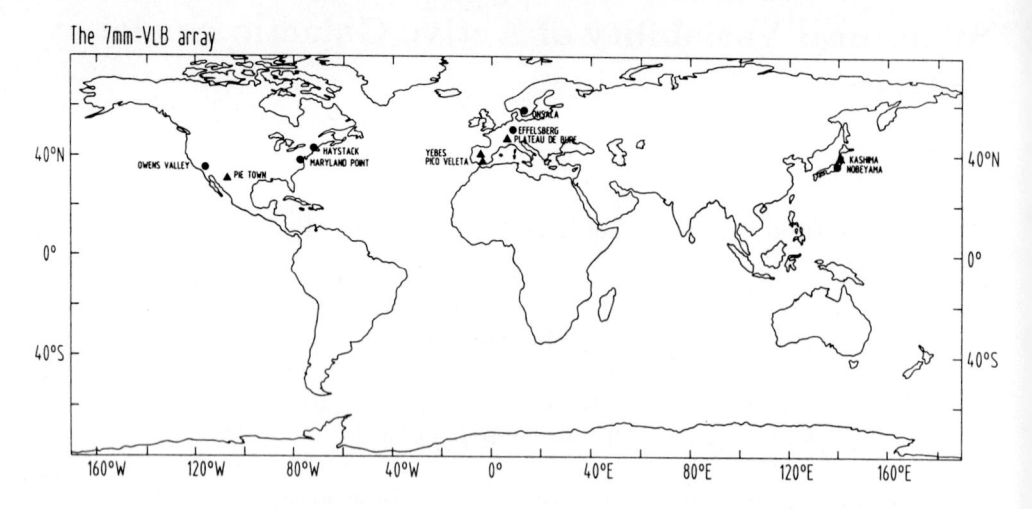

Fig. 1: The location of antennas presently participating in 7 mm-VLBI. Some future mm-VLBI observatories are marked by triangles.

summarizes the antennas, their abbreviation, their diameters D [m], present (1990) values for the system-temperatures T_{sys} [K], antenna aperture-efficiencies η_A [%] and types of receivers used. In Table 3 the 7σ-detection thresholds in [Jy] (Rogers, 1991) for each interferometer baseline is given, adopting a typical coherent integration time of 25 sec and an observing bandwidth of 56 MHz. It is obvious that target sources for present 7 mm-VLBI observations should exhibit a correlated flux density of at least 1 Jy (within the beam of the interferometer) to be mapped succesfully.

Table 4 lists the 20 sources detected so far during the 7 mm-VLBI campaigns since 1985 (cf. Dhawan, 1987, for a summary of earlier observations). Seven out of these sources have been adequately observed to allow imaging, while for the remaining 13 sources only snapshot observations (short observing intervals on some sensitive interferometer baselines) have been made yet. Columns 2-7 of Table 4 summarize the results for any individual source at the various observing epochs.

The main subject of this paper are those objects for which structural variability on the sub-parsec scale can be studied on the basis of repeated observations, and for which maps and/or modelfits were obtained: 3C 84, 3C 273, 3C 345 and 1803+78.

2. Results

In the following chapter we will first describe the results of our 43 GHz VLBI-observations for the individual sources and then briefly discuss their relevance in the context of more recent theoretical work.

Table 1: The participants of 7 mm-VLBI

Institution	Collaborators
California Institute of Technology, Pasadena, C.A., USA	A. C. S. Readhead, C. R. Lawrence, R. Vermeulen
Harvard-Smithsonian Center for Astrophysics, Cambridge, M.A., USA	I.I. Shapiro, N. Bartel
Instituto de Astrofisica de Andalucia, Granada, Spain	J. M. Marcaide, A. Alberdi
Max-Planck-Institut für Radioastronomie, Bonn, FRG	A. Witzel, T. P. Krichbaum, D. A. Graham, I.I.K. Pauliny-Toth, C.A. Hummel, A. Quirrenbach[1]
National Radio Astronomy Observatory, Socorro, N.M., USA	J. A. Zensus
Naval Research Laboratory, Washington, D.C., USA	K. J. Johnston, J. H. Spencer
NEROC Haystack Observatory, Westford, M.A., USA	A. E. E. Rogers
M.I.T., Cambridge, M.A., USA	B. F. Burke, V. Dhawan [2]
Nobeyama Radio Observatory, Japan	M. Morimoto, M. Inoue, H. Hirabayashi [3]
Onsala Space Observatory, Sweden	R. S. Booth, B. O. Rönnäng, A. Kus [4], F. Colomer

[1]: now at U.S. Naval Research Laboratory, Washinghton, D.C., USA.
[2]: now at Raman Research Institute, Bangalore, India.
[3]: now at Institute of Space and Astronautical Science, Tokyo, Japan.
[4]: now at Torun Radio Astronomy Observatory, Poland.

Table 2: The Antennas

Station	Code	D [m]	T_{sys} [K]	η_A [%]	Receiver
Effelsberg	B	60	400	29	CS
Nobeyama	X	45	300	56	SIS,HEMT
Owens Val.	O	40	140	7	M
Haystack	K	36	280	16	M
Mar.Point	N	26	740	5	US
Onsala	T	20	450	48	SIS

CS= Cooled Schottky-Diode mixer, US= Uncooled Schottky-Diode mixer, M= Maser amplifier, SIS = SIS-tunnel-junction mixer, HEMT = HEMT-amplifier receiver. For the 100m-telescope in Effelsberg the illuminated diameter is given. Owing to the linear polarization of the Effelsberg receiver, the "effective" system-temperature is $2 \cdot T_{sys}$.

Table 3: The 7σ-detection thresholds

$7\sigma[Jy]$	X	O	K	N	T
B	0.5	0.9	0.9	3.8	1.2
X	-	0.5	0.6	2.3	0.7
O	-	-	1.2	4.9	1.6
K	-	-	-	5.1	1.7
N	-	-	-	-	6.7

Table 4: The sources detected so far

Source	1985.16 [1]	1986.35 [2]	1987.44 [3]	1988.49 [4]	1989.19 [5]	1990.48 [6]
3C 84	S	Map/Mod.	Map/Mod.	Map/Mod.		Map/Mod.
4C 39.25						Map/Mod.
3C 273	S	S		Map/Mod.	Map/Mod.	
3C 345	S	S	Mod.	Map/Mod.		Map/Mod.
1803+78	S	S	Mod.	Map/Mod.	Map/Mod.	Mod.
BL Lac						Mod.
3C 454.3		S	Mod.			
0133+47	S	S				
0212+73						S
0234+28	S					
NRAO 150	S	S				
0716+71			S	Mod.		
0804+49						S
0917+62						S
OJ 287	S	S				
3C 279			S			
1308+32	S	S				
NRAO 512			S			
1928+73			S			
2007+77			S			S

[1]: see also Marcaide et al., 1985
[2]: see also Dhawan, 1987; Bartel et al., 1988
[3]: see also Dhawan et al., 1990; Krichbaum, 1990a
[4]: see also Krichbaum and Witzel, 1991; Zensus et al., 1991, Krichbaum et al., 1991
[5]: see also Krichbaum, 1990b; Krichbaum et al., 1990
[6]: data analysis not yet complete

Map/Mod. : map and Gaussian modelfit obtained
Mod. : amount of data allows only modelfits
S : snapshot observations yielding detections

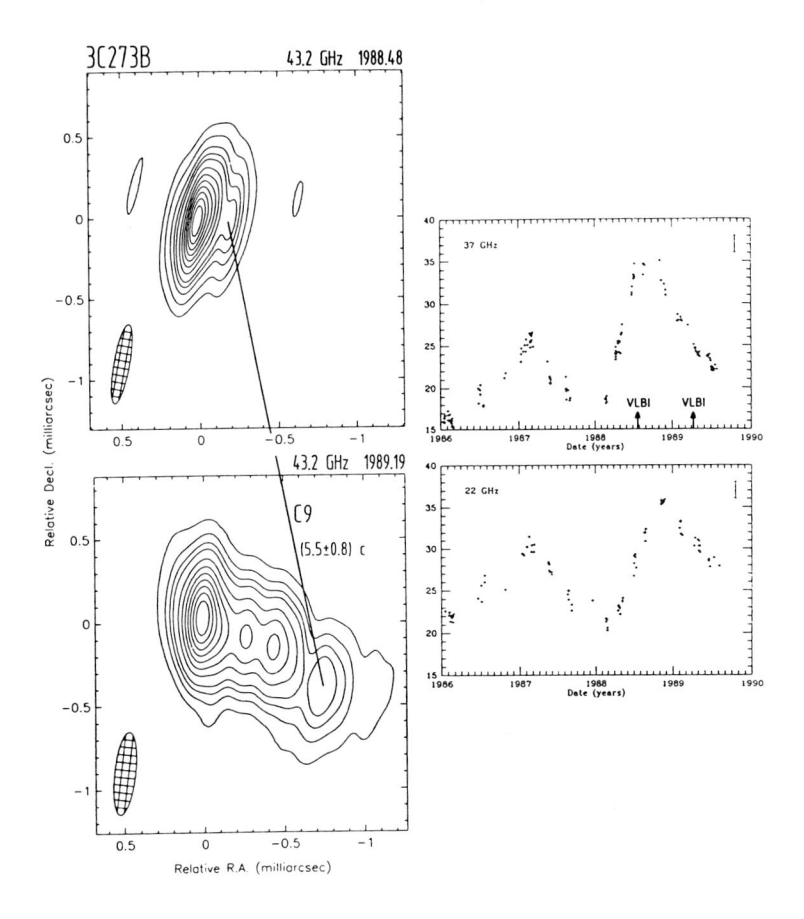

Fig. 2: 43 GHz-maps of 3C 273 at 1988.48 and 1989.9. The most plausible identification of the jet component C9 is indicated by a straight line. Lightcurves at 22 GHz and 37 GHz (Courvoisier *et al.*, 1990) show an flux density outburst related to an optical outburst starting in early 1988).

2.1 Observations of 3C 273

The quasar 3C 273 ($z = 0.158$) shows a pronounced and well collimated jet observable in the radio and optical regime (e.g. Davis *et al.*, 1985; Röser and Meisenheimer, 1986). The jet has been investigated in detail from arcsecond- to mas-scale (e.g. Fraix-Burnet and Nieto, 1988; Zensus, 1987 and references therein). From repeated VLBI-observations at cm-wavelengths apparent superluminal motion of the jet components is deduced (Unwin *et al.*, 1985, Cohen *et al.*, 1987, Unwin, 1990 and references therein). Recent VLBI-observations at 5 GHz show that the ridgeline of the jet-axis of 3C 273 is (quasi-sinusoidally) curved and that superluminal components move on curved trajectories. Between about 1 mas and 30 mas one common path seems to be applicable for the motion of components at 5 GHz (Zensus et al., 1988) and 10.7 GHz (Unwin et al., 1985; Cohen et al., 1987).

At 43 GHz two maps have been obtained for 3C 273 in 1988.48 and 1989.19. Krichbaum et al., (1990) showed that between the two observing epochs the jet evolved and that a new component C9 – we follow the commonly accepted nomenclature for this source – was ejected with $\beta_{app} = (5.5 \pm 0.8)$ (Figure 2). Back-extrapolation of the motion of C9 yields 1988.2 $\pm$ 0.2 as starting point of the expansion. This value coincides very well with the time of a major flux density outburst first observed in the optical domain (Courvoisier, *this workshop*; Courvoisier et al., 1988) and later propagating in the radio band (see Figure 2; and Courvoisier et al., 1990; Valtaoja, 1990; see also Robson, *this workshop*). In the documented history of the structural evolution in 3C 273 the ejection of superluminal jet components can be attributed to flux density outbursts (Krichbaum et al., 1990; Zensus et al., 1990; Botti and Abraham, 1988), however apparently not every outburst leads to the ejection of a new component (e.g. no VLBI-component of 3C273 could be found yet which could be related to the well documented outburst in 1983 (Robson et al., 1983; Marscher and Gear, 1985)). Frequent VLBI-observations more closely spaced in time are required to clarify, whether there are flux density outbursts which do not lead to the ejection of new superluminal features.

From a comparison of the apparent velocities of the jet components C2 - C9 it is obvious that successive components move with different velocities (see Figure 3 in Krichbaum et al., 1990, Zensus et al., 1990 and references therein). In Figure 3 we display β_{app} of the different components as a function of ejection time t_0, which we extrapolate under the assumption of constant component velocity. Although the errors are still large a systematic variation in the velocity pattern of the jet of 3C273 seems to be present in the data. The peak-to-peak variations of the velocity pattern $\beta_{app}(t_0)$ range between $4.3 \pm 0.6\,c$ (for C7) and $8.0 \pm 0.2\,c$ (for C5).

Between the sub-mas and arc-second region the mean ridgeline of the jet bends from $P.A. \simeq 270°$ at $r \simeq 0.24\,\text{mas}$ (Krichbaum et al., 1990) to $P.A. \simeq 220°$ at $r = 20''$ (Davis et al., 1985). Superimposed on this general curvature of the jet are 'quasi-sinusoidal' oscillations of the jet axis, clearly visible in VLBI-images obtained at 5 GHz (Zensus et al., 1988) and 10.7 GHz (Cohen et al., 1987). In Figure 4 we display position angle $P.A.$ and separation r from the core of the inner jet components C4 - C9 for the various observations at cm-wavelengths. We further include the positions of new components visible in our recent 43 GHz-maps (for details and a complete reference list see Krichbaum et al., 1990). As indicated by straight lines in the plot, the curvature of the jet axis increases with decreasing core separation (see also the map obtained at 3 mm wavelength, L. Bååth, *this workshop*) and the jet-components seem to move along curved trajectories with alternating sign of the curvature $d(P.A.)/dr$. It is not yet clear, whether component C9 will join the path defined by C7 and C8 and whether the path of the components C7 and C8 will be the same as that of C4 and C5. Some datapoints at $r = 2...5\,\text{mas}$ suggest that the

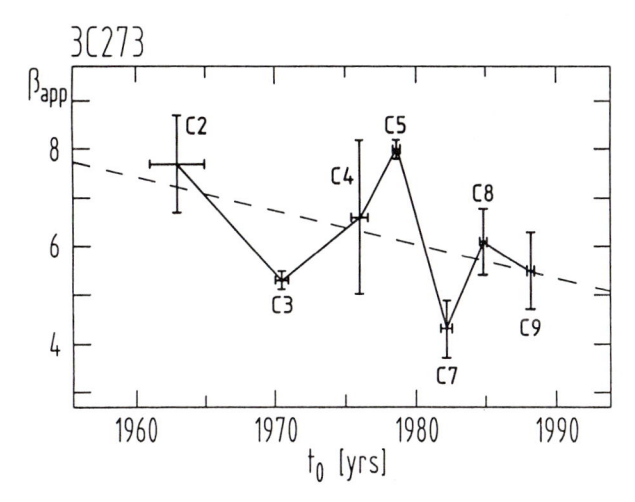

Fig. 3: The apparent velocity β_{app} of the jet components C2 - C9 of 3C 273 plotted versus the (linearly extrapolated) time of their ejection t_0. The dashed straight line represents a linear fit to the data with slope $d\beta_{app}/dt = -(6.9 \pm 6.1) \cdot 10^{-2} yr^{-1}$.

path of C7/C8 may be different from the trajectory of the earlier components C4/C5.

2.2 Observations of 3C 345

The quasar 3C 345 (z = 0.595) is the archetype of a superluminal source, in which jet components do neither move with constant speed (Biretta *et al.*, 1986) nor move along straight trajectories (A. Zensus, *this workshop*). VLBI-monitoring at cm-wavelengths for more than two decades makes it one of the best observed sources (e.g. Unwin *et al.*, 1983; Zensus, 1990).

At core separations $r \leq 3$ mas the source consists of a unresolved core D at 22 GHz and two superluminally moving components C4 and C5. Both components move with non-constant apparent velocities on curved paths, which are clearly different from each other (A. Zensus, *this workshop*; Krichbaum, 1990a).

Due to its declination, intensity, and also due to its comparatively simple source structure (cf. the more complex structure of 3C 84) 3C 345 is at present one of the most promising objects for 7 mm VLBI-observations. 43 GHz VLBI-observations in 1988.49 showed that a new component C6 appeared close to the core ($r = (0.15 \pm 0.03)$ mas, P.A. = $(245 \pm 7)°$) (see Figure 5 a, and Krichbaum, 1990b; Zensus *et al.*, 1991 *in preparation*). The map also shows that the components C4 and C5 continue to move along their curved paths derived from the 22 GHz data (A. Zensus, *this workshop*). A new map of 3C 345 is now obtained from observations in 1990.48 (Figure 5 b). The still preliminary map indicates that C4 and C5 have moved further out, and that at $r \leq 1$ mas the source structure clearly changed since 1988. With respect

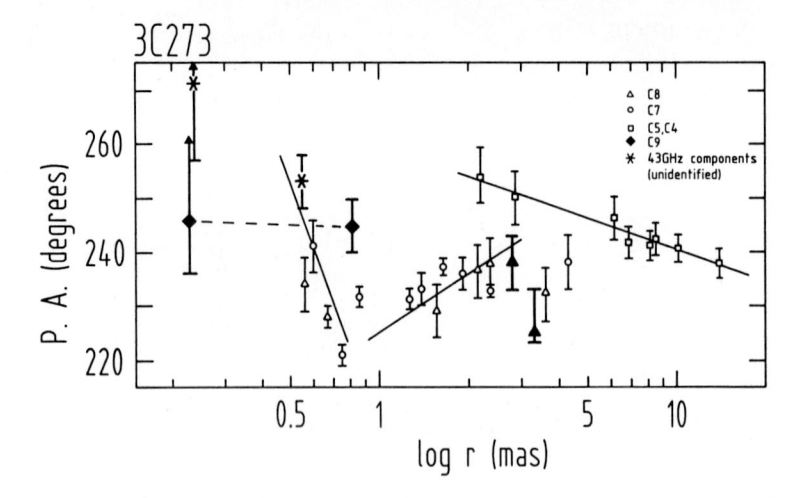

Fig. 4: The location of the jet components of 3C 273 in a plot of position angle versus core distance. Symbols denote different jet components: open squares are for C4 and C5, open circles for C7, triangles for C8 (filled triangles from the 43 GHz experiments), filled diamonds for C9, asterisks for two new, not yet identified 43 GHz components. 43 GHz data-points are marked by enlarged symbols.

to the brightest component, which we identify as the core D, two components at $r \simeq 0.22$ mas, $P.A. \simeq 302°$ and $r \simeq 0.64$ mas, $P.A. \simeq 284°$ are visible. It seems to be most reasonable to identify the component at $r \simeq 0.6$ mas with the component C6 first observed in 1988. Its apparent velocity derived from the two observing epochs is $\beta_{app} = (4.3 \pm 0.3)$, which is in good agreement with the apparent velocities of other jet components. In this picture the component at $r \simeq 0.2$ mas is a new component (C7), most recently ejected from the core. Monitoring of the total flux density of 3C 345 at 37 GHz (E. Valtaoja, *private communication*) indicates after years of more or less steadily decreasing source brightness (since 1982), a recent increase in total intensity (starting near 1989.5). The ejection of the component C7 from the core then could be related to the beginning of a new period of source activity, similar to the recently observed event in 3C 273 (Krichbaum *et al.*, 1990).

In Figure 6 we combined the data from the 22 GHz VLBI-observations (Zensus *et al.*, 1991 *in preparation*; and A. Zensus, *this workshop*) with the new data obtained at 43 GHz. We plotted the position angles of the components C4 - C7 at the various observing epochs as a function of their relative separation with respect to the core D. We added lines to indicate the direction of motion. It appears that the *different* trajectories of C4 and C5 are non-monotonically curved ($d(P.A.)/dr$ not constant) and that the curvature changes its sign for component C4. The two observations of C6 suggest a motion to the north-west, parallel, but not identical with the path

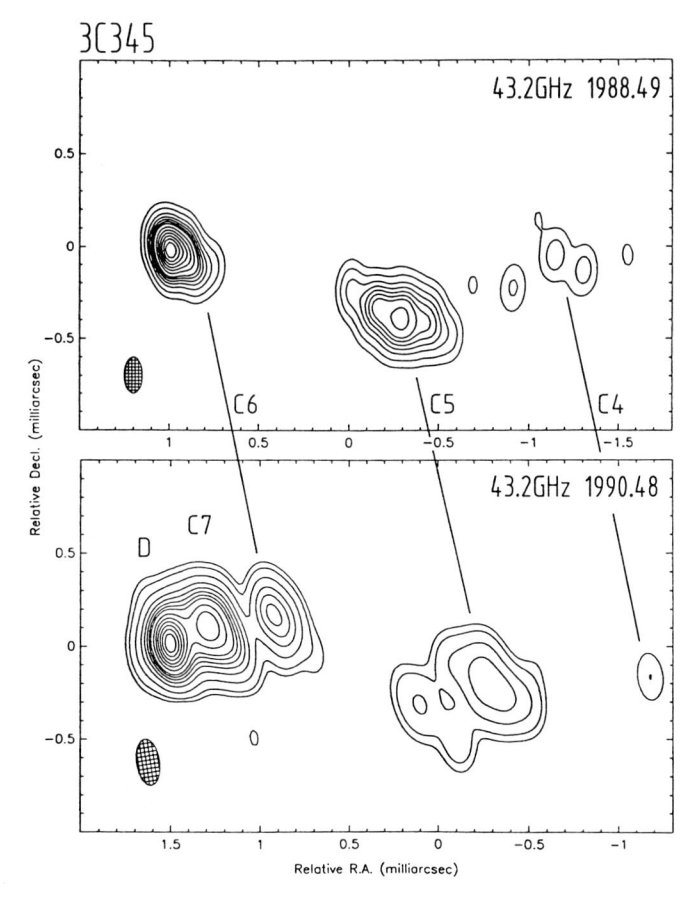

Fig. 5: Maps of 3C 345 at 43 GHz at observing epochs 1988.49 (Fig. 5 a, top) and 1990.49 (Fig. 5 b, bottom). Labels and lines identify the components D (the core), C7, C6, C5 and C4 (for 22 GHz maps including C4 and C5 see also A. Zensus, *this workshop*). Contour levels are 1, 2, 5, 10, 15, 20, 25, 30, 40, 50, 60, 70, 80, 90 % of the peak flux density (1988.49: 0.91 Jy/beam; 1990.48: 1.11 Jy/beam). The 1% level has been omitted in Fig. 5 a. The size of the restoring beam is (0.20 x 0.10) mas (Fig. 5 a), and (0.25 x 0.13) mas (Fig. 5 b).

of C4 at comparable r. Since there is only one measurement for C7 yet, no firm conclusion can be drawn for this component. However it should be noted that with respect to the other components C4, C5 and C6 its position is located significantly further north (at similar r ($r \simeq 0.2$ mas) the position angles are: C7: $P.A. \simeq 300°$, C6: $P.A. \simeq 250°$, C5: $P.A. \simeq 270°$, and C4: $P.A. \simeq 230°$). Further continous monitoring with good time-coverage is needed to accurately determine this complex velocity pattern in the inner jet of 3C 345.

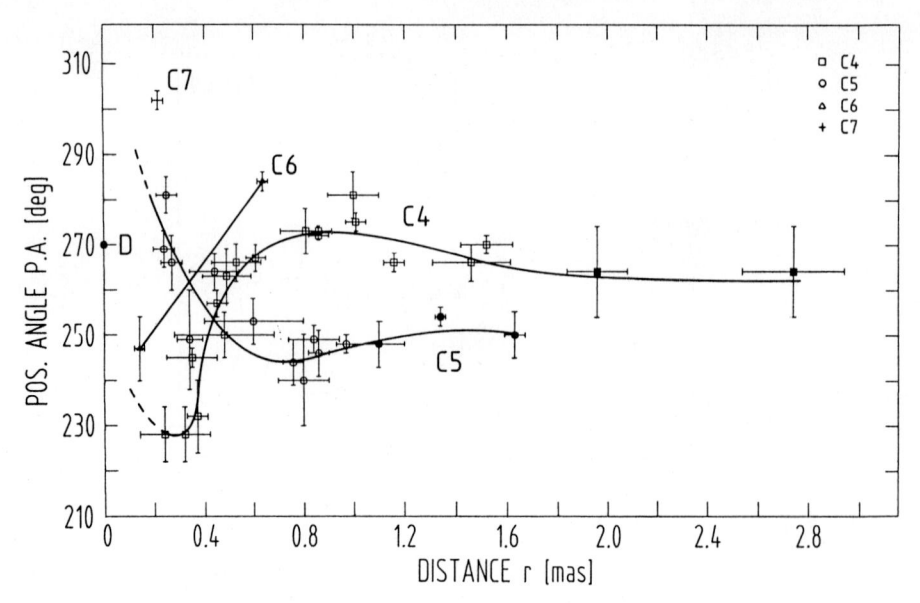

Fig. 6: A plot of position angle versus core separation with respect to the core D of the components C4, C5, C6 and C7 in 3C 345. Open symbols denote for data obtained at cm-wavelengths (Zensus et al., *this workshop* and references therein), filled symbols denote for 43 GHz measurements. We added lines to sketch a possible mean path for the components C4 - C6.

2.3 Observations of 3C 84

In contrast to the quasars 3C 273 and 3C 345 the peculiar radio-galaxy 3C 84 (NGC-1275) at its moderate distance (z = 0.018) shows subluminal motion in its mas-structure (Romney et al., 1984). VLBI-observations at cm-wavelengths showed the source to consist of a compact core elongated along $P.A. = 225°$ and a southern extended region of complex emission (Unwin et al., 1982), which we will call the southern lobe. Two separate velocity patterns coexist in the southern lobe. Bright knots of emission, observed with VLBI at 22 GHz, move with $\beta_{app} \simeq 0.5$ (Marr et al., 1989). They are embedded in the more diffuse lobe, which itself separates with $\beta_{app} \simeq 0.2$ from the core (Romney et al., 1984; Backer, 1987). Back-extrapolation of this expansion of the lobe-structure in time, yields $t_0 \simeq 1959$ as zero separation time of the motion, which is consistent with the beginning of a very pronounced increase of flux density (amplitude $\Delta S \simeq 40 - 50$ Jy at cm-wavelengths) (O'Dea et al., 1984). 7 mm VLBI-campaigns in 1986, 1987, and 1988 resulted in three maps and models of the source structure, which are presented and discussed in detail in Krichbaum et al., (1991) (see also Bartel et al., 1988; Dhawan et al., 1990; Krichbaum and Witzel, 1991). Due to the very complex source structure of 3C 84, we feel that the models shown in Figure 7 give a more adequate and clear representation of the mas- to

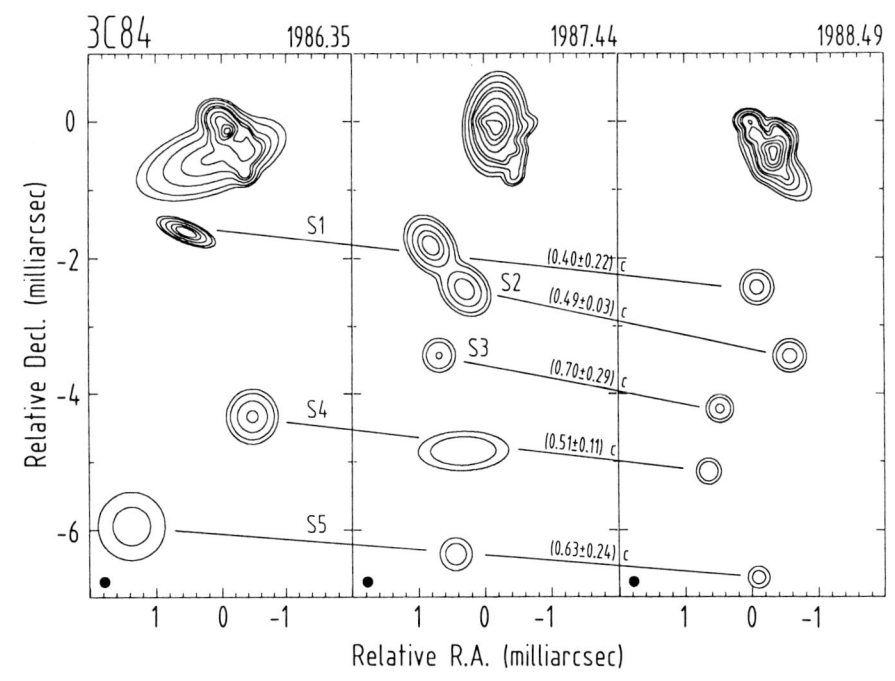

Fig. 7: Models of Gaussian components of the brightness distribution of 3C 84 obtained from three observing epochs 1986.35, 1987.44 and 1988.49 at 43 GHz. Contour levels are 1, 2, 5, 10, 15, 30, 50, 70, and 90 % of the peak flux density (1986.35: 1.2 Jy/beam, 1987.44: 2.1 Jy/beam, and 1988.49: 2.7 Jy/beam). A circular restoring beam of size 0.15 mas was used for the representation of the models. Straight lines indicate a plausible identification and motion of components embedded in the southern lobe, numbers give apparent velocities.

sub-mas structure than CLEAN-maps could provide. These models are in good agreement with corresponding models and maps obtained from recent VLBI-observations at 22 GHz (Readhead et al., 1990; Marr et al., 1990).

At 43 GHz the core-region of 3C 84 consists of a jet-like structure oriented at $P.A. \simeq 220°$, which is embedded in a much brighter and more extended region of diffuse emission. Describing this halo-like structure by an elliptical Gaussian modelfit component, we find – consistent with data obtained at 3 mm wavelength (Wright et al., 1988) – that its mean size $(FWHM)_{mean}$ decreased between 1986.35 and 1988.49 with a rate of $d(FWHM)_{mean}/dt = -(0.14 \pm 0.03)$ mas/yr ($\beta_{app} = -(0.12 \pm 0.03)$). South of the northern core region our data reveal emission in the region $(1 < r < 7)$ mas, which can best be described by up to 5 compact structural components. The still limited uv-coverage of our observations at present does not allow to map more extended regions of emission, probably also present in this area (see the 22 GHz maps of Readhead et al., 1990 and Marr et al., 1989). The most plausible identification of the individual components at the 3 observing epochs is sketched in Figure 7. In this scheme the components separate from the northern core A (see below) with apparent velocities ranging between $\beta_{app} = (0.4 \pm 0.2)...(0.7 \pm 0.3)$. These numbers compare very well with the velocity of the 'fast moving knots' observed at 22 GHz (Marr et al., 1989). However, since at 10.7 GHz intensity peaks at the southern boundary of the lobe change their position with $\beta_{app} \simeq 0.2$, it seems plausible that either two velocity systems coexist in the southern lobe or a north-south directed velocity gradient is present in this region.

Readhead et al., (1990) concluded on the basis of their 22 GHz VLBI-observations (3 observations between 1985 - 1986) that any motion within the northern core region has been smaller than 0.2 mas/yr. In Figure 8 a we show Gaussian models of the northern core region from our observations at 43 GHz between 1986 and 1988. In this representation the diffuse halo-like emission has been omitted to avoid blending effects and the source structure has been rotated counter-clockwise by about 50°. Component A has been identified with the core on the basis of its intensity variability and its inverted spectrum between 22 GHz and 43 GHz. Linear cuts through the jet-like brightness distribution, which measure component separations with respect to A are shown in Figure 8 b. The obvious structural evolution is readily explained by a component identification as indicated by dashed lines (Figure 8 a). In this scenario the components B, C, and D separate from the core A with velocities $\beta_{app} = (0.10 \pm 0.02)$, $\beta_{app} = (0.13 \pm 0.03)$, and $\beta_{app} = (0.16 \pm 0.03)$, respectively. These values are consistent with the upper limit of motion in the northern core region derived by Readhead et al.. In 1988.49 a new component N appeared in the proximity of the core. The component E, visible in 1985.35, cannot be traced at later epochs, suggesting its fading with time.

3. Discussion

As the maps and models shown before demonstrate, all sources observed with VLBI-techniques at 43 GHz show marked deviations from linear structures. Also the maps for 1803+78 (Krichbaum, 1990b) and 4C 39.25 (unpublished data) exhibit bent structures. In general, the components and their kinematics found at 43 GHz are in good agreement with results obtained at neighbouring frequencies (cf. 3C 273: L. Bååth, *this workshop*; 3C345: A. Zensus, *this workshop*; 3C 84: Readhead et al., 1990, Marr et al., 1990, 1803+78: Schalinski et al., 1988), thus forming a consistent picture of the sub-parsec scale structures. In all sources sufficiently well observed (3C 84, 3C 273 3C 345, 1803+78), we find indications that the degree of curvature increases with decreasing core distance. A similar result is also found for the quasar 1928+73 at 22 GHz (Hummel et al., 1991; and C. Schalinski et al., *this workshop*).

Motion along distinct curved trajectories has been clearly demonstrated in the case of 3C 345. Subsequently ejected components (C4, C5, C6) seem to move on different paths (Figure 6). Similar curvature of the paths has also been seen in 3C 273 (Figure 4), but at present the observations do not allow a firm distinction between common

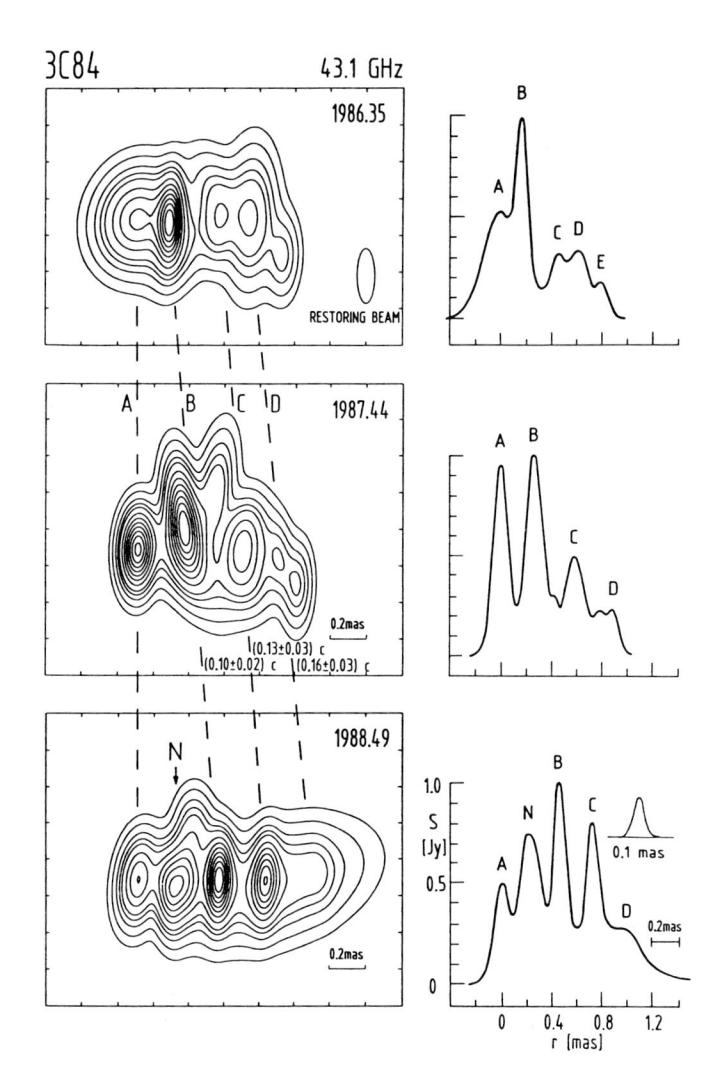

Fig. 8 a, left: Gaussian component models of the nucleus of 3C 84 for 1986.35, 1987.44, and 1988.49. A surrounding halo-component is omitted in the plots to avoid blending effects. Tickmarks are at 0.2 mas intervals, contour levels are at 2, 5, 10, 15, 20, 30, 40, 50, 60, 70, 80, and 90 % of the peak flux density (1986: 1.43 Jy/beam; 1987: 0.72 Jy/beam; 1988: 1.64 Jy/beam). A common restoring beam of (0.1 x 0.3) mas with minor axis parallel to the x-axis (= mean jet axis) was used. Straight lines and labels identify jet-components, numbers are apparent velocities.

Fig. 8 b, right: To peak flux density normalized profiles measured along the jet-axis of the brightness distributions in Figure 8 a. The motion of the brightest component B is obvious.

or different trajectories for the inner 10 mas. For $r > 10$ mas Zensus *et al.*, (1988) find from VLBI-observations at lower frequencies (and thus with less angular resolution) an oscillating jet ridgeline, indistinguishable for successive components. In 3C 84 a major kink of the jet-axis separates two regions of different velocities (at $r < 1.5$ mas: $v \simeq (0.1...0.2)$ c, and at $r \geq 1.5$ mas: $v \simeq (0.2...0.5)$ c). There is some evidence that in the region of the kink components migth move on curved paths (Marr *et al.*, 1990). The observed velocity changes could then be interpreted as an effect of motion along a spatially curved path (Qian *et al..*, 1991).

It is tempting to use recent theoretical work on jets to explain the observed quasi-periodic oscillations in jets, the motion of jet components along curved paths, and the observed changes of apparent velocities. Within the simple relativistic jet-model (Blandford and Königl, 1979) velocity changes could be interpreted in terms of a changing inclination of the velocity vector with respect to the observers line of sight (differential Doppler-boosting, see also A. Marscher *et al.*, *this workshop*). The joint appearance of velocity changes and bends in jets indicate a three-dimensionally curved jet-axis. Independent evidence for helical structures in jets also comes from the observations of 'intraday variability' (Witzel, 1990; Krichbaum, Quirrenbach and Witzel, *this workshop*). In scenarios invoking magnetic fields (magneto-hydrodynamic jet physics, e.g. Königl and Choudhuri, 1985) a helical field structure, possibly anchored in a rotating accretion disk (Camenzind, 1989; Camenzind, 1990) could be used to explain the observed properties. A different approach uses oblique reflection shock modes in Kelvin-Helmholtz instabilities (Hardee and Norman, 1989) in order to explain non-linear velocity patterns and oscillating jet morphologies.

Acknowledgements

We thank all colleagues listed in Table 1 and apologize if we have forgotten any collaborator in there. We also thank the members of the staff of the observatories who helped in the 7mm-VLBI observations.

References

Backer, D.C., 1987, in: *Superluminal Radio Sources*, ed. J. A. Zensus and T. J. Pearson (Cambridge University Press), p. 76.
Bartel *et al.*, 1988, *Nature*, **334**, 131.
Blandford, R.D., and Königl, A., 1979, *Astrophys. J.*, **232**, 34.
Botti, L.C.L., and Abraham, Z., 1988, *Astron. J.*, **96**, 465.
Biretta, J.A., Moore, R.L., and Cohen, M.H., 1986, *Astrophys. J.*, **308**, 93.
Camenzind, M., 1989, in: *Accretion Disks and Magnetic Fields in Astrophysics*, ed. G. Belvedere, Kluwer Academic Publishers, p. 129.
Camenzind, M., 1990, in: *Reviews in Modern Astronomy 3, Accretion and Winds*, ed. G. Klare, Springer (Berlin), p. 3.

Qian, S.J., Krichbaum, T.P., and Witzel, A., 1991, *Acta Astronomica*, submitted.
Cohen, M.H. *et al.*, 1987, *Astrophys. J. (Letters)*, **315**, L89.
Courvoisier, T.J.-L., *et al.*, 1988, *Nature*, **335**, 330.
Courvoisier, T.J.-L., *et al.*, 1990, *Astron. Astrophys.*, **234**, 73.
Davis, R.J., Muxlow, T.W.B., and Conway, R.G., 1985, *Nature*, **318**, 343.
Dhawan, V., 1987, Ph.D. Thesis, Massachusetts Institute of Technology.
Dhawan, V., *et al.*, 1990, *Astrophys. J. (Letters)*, **360**, L43.
Fraix-Burnet, D., and Nieto, J.-L., 1988, *Astron. Astrophys.*, **198**, 87.
Hardee, P.E., and Norman, M.L., 1989, *Astrophys. J.*, **342**, 680.
Hummel, C.A., *et al.*, 1991, *Astron. Astrophys.*, submitted.
Königl, A., and Choudhuri, A.R., 1985, *Astrophys. J.*, **289**, 188.
Krichbaum, T.P., 1990a, Ph.D. Thesis, University of Bonn.
Krichbaum, T.P., 1990b, in: *Parsec-scale radio jets*, ed. J. A. Zensus and T. J. Pearson (Cambridge University Press), p. 83.
Krichbaum, T.P., *et al.*, 1990, *Astron. Astrophys.*, **237**, 3.
Krichbaum, T.P., *et al.*, 1991, *Astron. Astrophys.*, submitted.
Krichbaum, T.P., and Witzel, A., 1991, in: *Frontiers of VLBI*, ed. H. Hirabayashi, M. Inoue and K. Kobayashi (Universal Academy Press, Tokyo), in press.
Marcaide, J.M., *et al.*, 1985, in: *Proc. IRAM-ESO-Onsala Workshop on (Sub-) Millimeter Astronomy*, ed. Shaver,P.A. and Kjär,K., p. 157.
Marr, J.M., *et al.*, 1989, *Astrophys. J.*, **337**, 671.
Marr, J.M., Backer, D.C., and Wright, M.C.H., 1990, in: *Parsec-scale radio jets*, ed. J. A. Zensus and T. J. Pearson (Cambridge University Press), p. 78.
Marscher, A.P., and Gear, W.K., 1985, *Astrophys. J.*, **298**, 114.
O'Dea, C.P., Dent, W.A., and Balonek, T.J., 1984, *Astrophys. J.*, **278**, 89.
Readhead, A.C.S., *et al.*, 1990, in: *Parsec-scale radio jets*, ed. J. A. Zensus and T. J. Pearson (Cambridge University Press), p. 71.
Robson, E.I., *et al.*, 1983, *Nature*, **305**, 195.
Röser, H.J. and Meisenheimer, K., 1986, *Astron. Astrophys.*, **154**, 15.
Rogers, A.E.E., 1991, in: *Frontiers of VLBI*, ed. H. Hirabayashi, M. Inoue and K. Kobayashi (Universal Academy Press, Tokyo), in press.
Romney, J.D., *et al.*, 1984, in: *IAU Symposium 110, VLBI and Compact Radio Sources*, ed. R. Fanti, K. Kellermann, and G. Setti (Dordrecht: Reidel), p. 137.
Schalinski, C.J., *et al.*, 1988, in: *IAU Symposium 129, The Impact of VLBI on Astrophysics and Geophysics*, ed. M. J. Reid and J. M. Moran (Dordrecht: Kluwer), p. 359.
Unwin, S.C., *et al.*, 1982, *Astrophys. J.*, **256**, 83.
Unwin, S.C., 1983, *Astrophys. J.*, **271**, 536.
Unwin, S.C., *et al.*, 1985, *Astrophys. J.*, **289**, 109.
Unwin, S.C., 1990, in: *Parsec-scale radio jets*, ed. J. A. Zensus and T. J. Pearson (Cambridge University Press), p. 13.
Valtaoja, E., 1990, in: *Frontiers of VLBI*, ed. H. Hirabayashi, M. Inoue and K. Kobayashi (Universal Academy Press, Tokyo), in press.

Witzel, A., 1990, in: *Parsec-scale radio jets*, ed. J. A. Zensus and T. J. Pearson (Cambridge University Press), p. 206.

Wright, M.C.H., *et al.*, 1988, *Astrophys. J. (Letters)*, , L329,61.

Zensus, J.A., 1987, in: *Superluminal Radio Sources*, ed. J. A. Zensus and T. J. Pearson (Cambridge University Press), p. 26.

Zensus, J.A., *et al.*, 1988, *Nature*, **334**, 410.

Zensus, J.A., 1990, in: *Parsec-scale radio jets*, ed. J. A. Zensus and T. J. Pearson (Cambridge University Press), p. 28.

Zensus, J.A., *et al.*, 1990, *Astron. J.*, **100**, 1777.

Monitoring of the milliarcsecond-structure of S5-1928+738:
Apparent superluminal motion along a fixed path ?

SCHALINSKI[1,2], C.J., WITZEL[1], A., HUMMEL[1], C.A., KRICHBAUM[1], T.P.,
QUIRRENBACH[1], A., & JOHNSTON[3], K.J.

[1] Max-Planck-Institut für Radioastronomie, Bonn, F.R.G.
[2] Geodätisches Institut der Universität Bonn, F.R.G.
[3] Center for Advanced Space Sensing, Naval Research Laboratory, Washington,D.C.,U.S.A.

Abstract The structure of the 1928+738 jet may be followed out to $\sim$ 250mas on the mas-scales. From multi-epoch VLBI-observations at 23, 5 and 1.6GHz the following velocity profile is found: a region with common velocity (6.5 ± 0.5 c) at $r > 6mas$ of components following a fixed path, an inner region ($< 4.5mas$) with lower velocity (4c), and a transition region characterized by curvature and non-uniform velocities. The latter is interpreted as emission region at larger viewing angle, thus indicating a helical geometry. Structural and kinematical constraints on models, e.g. involving weak reflection mode shocks, are discussed.

1. INTRODUCTION

Since bulk relativistic motion is found to be a common phenomenon in extragalactic radio sources (e.g. Witzel, 1987; Witzel et al., 1988), the major effort is now being focused on a detailed analysis of the jets of active galactic nuclei. As this not only requires high-dynamic-range mapping with milliarcsecond (mas-)resolution, but also frequent observations at different wavelengths sampling the structure at many epochs, only a few sources have been studied extensively over the past ten years, e.g. the superluminals 3C273 (e.g. Unwin, 1990, and refs.) and 3C345 (e.g. Zensus, 1990, and refs.). The quasar 1928+738[1] (z=0.30; 1mas=2.75pc, $H_0 = 100$ km/s/Mpc, $q_0 = 0.5$) is another suitable candidate for these investigations, since it displays jet-like radio emission on scales of a few parsec to kpc (e.g. Johnston et al., 1987), and structural variability has been established on the basis of multiple components in VLBI-maps at 3 wavelengths (λ1.3cm, λ6cm, and λ18cm: Eckart et al., 1985; Hummel, 1990; Schalinski, 1990).

2. MORPHOLOGY OF THE PARSEC-SCALE EMISSION

The arcsecond structure is dominated by a bright, compact component and a double-sided, strongly bent jet, which on the northern side ends in a diffuse lobe (Fig.1). The VLBI-maps at $\lambda\lambda 90cm - 1.3cm$ trace the radio emission of 1928+738 on milliarcsecond scales and can be followed out to $\sim$ 250 mas from the compact "core", with no indication of a counterjet (dynamic range 200:1).
At 0.25mas resolution (23GHz) the structure is defined by an unresolved northern component and wiggles in a knotty jet with an amplitude of about 0.2mas. At 1mas resolution (5GHz) the source displays a curved structure with at least 14 distinguishable features within the inner 18mas (Fig.1). The amplitude of the wiggles rises to $\sim$ 2mas, with at least two significant kinks (change of p.a. from 166^0 to

---
[1]member of the complete S5-VLBI-sample, e.g. Witzel (1987), s. also Schalinski et al., this vol.

159°) at about 5 and 10 mas from the compact "core". Emission in the λ90cm-map points towards 210° at $\sim$ 40mas, but bends back to $\sim$ 180° in a lower resolution (EVN-)image at λ18cm.

Thus 1928+738 meets the criteria of a *jet*: continuous emission on all scales observed, unresolved transversly to the ridgeline, i.e. brightness features ("components", used as reference points for the kinematical analysis), being connected by "bridges" of weak emission.

3. KINEMATICS AND GEOMETRY

We determined the following velocity profile along the jet (Fig.2): on the basis of 5 components identified in all four 5GHz-epochs a common velocity of 6.5 ± 0.5 c (0.56 ± 0.04 mas/a) is derived in the range between $\sim 6 \ldots 18$mas (zone III). This is independently confirmed by the results obtained from the two 1.6GHz-epochs. Thus apparent superluminal motion can be followed out to a projected distance of 50pc, comparable to the value of 46pc derived for the 3C273-jet by Zensus et al. (1990). The inner part of the jet (0...4.5mas) contains evidence for separational changes; from maps with higher resolution at 23GHz (6 epochs) Hummel et al. (1991) determine a mean velocity of 4 ± 0.6 c (0.34 ± 0.05 mas/a) (zone I). A "transition zone" (II), between 4.5...6 mas (the position of the two major kinks in the 5GHz-images), shows significantly lower values (0.19, 0.26 mas/a), which is confirmed by a further epoch at 8.4GHz (1988.89) (0.21mas/a).

The components beyond the kinks ($> 6mas$) appear to move along a fixed channel defined by the curved structure. Further evidence for non-ballistic motion comes from the fact that at several positions along the jet components of different epochs are identified, component positions at adjacent epochs display weak emission (e.g. "B5", Fig.2), and trajectories of single components match the ridgeline of the curved structure (esp. at r=10..12mas).

Within the relativistic beaming concept comparable velocities of components with almost identical trajectories argue strongly in favour of a *uniform* Lorentz factor of the jet ($\gamma_{min} = 6.6$, with $\theta_{LOS} = 9°$). The correlation of (lower) non-uniform velocities and curvature in the "transition zone" then can be attributed to directional changes of the relativistic jet towards larger viewing angles $\theta_{LOS} = 18° \ldots 41°$, adopting moderate Doppler factors ($\delta \leq 2.5$) as derived from SSC-calculations and flux density variability. Furthermore, a common intrinsic velocity ($\beta \geq 0.988$) throughout the jet can also solve the apparent contradiction between the one-sided (boosted) pc-jet and the two-sided kpc-structure, assuming $\theta_{LOS}^{kpc} = 64° \ldots 64.4°$ (taking the jet-counterjet-ratio R=16 for the "inner" kpc-jet into account). The alternative (extreme) solution is *constant* $\theta_{LOS} = \theta_{LOS}^{VLBI} \leq 17°$, implying β is subrelativistic on kpc-scales ($\beta^{VLA} = 0.45 \ldots 0.43$).

There is an increasing number of sources displaying increased curvature of the jet axis in regions close to the "core" (see Krichbaum & Witzel, this vol.). One model of the observed morphology and kinematics to be tested in detail is based on helically magnetic field structures originating in a rotating magnetosphere anchored in an accretion disk (e.g. Camenzind, 1990). The generation of quasi-sinoidal oscillations of the inner 4mas-jet may be alternatively interpreted in terms of a binary black whole system. Further high frequency monitoring is necessary to eventually discriminate between models of ballistic and non-ballistic component trajectories. Constraints may also arise from the magnetic field structure in the jet. From depolarization measurements of the VLA-core (Rusk, priv. com.; Wrobel et al., 1988), a polarization position angle $\chi_0 = -97° \pm 10°$ is derived, consistent with the magnetic field being aligned with the pc-scale jet.

The non-ballistic motions along "stationary", curved trajectories (zone III) can be qualitatively described by 2nd order reflection mode shocks (e.g. Hardee & Norman, 1989). The "decoupling" of the observed ($\beta_{obs} = 6.5\pm0.5$) from the intrinsic velocity β (and δ) allows for larger θ_{LOS}: with $\delta_{SSC} = 1.7$ and $\beta = 0.8$ a *phase velocity* $\beta_{ph} = 1.11$ and $\theta_{LOS} = 36°$. On the other hand phase velocities are not required to explain the high velocity within the standard relativistic beaming model, and represent a contradiction to the non-detection of a counterjet on the mas-scale in case of larger θ_{LOS} and smaller β. (Details s. Schalinski, (1990), and Schalinski et al., in prep.)

References

Camenzind, M., 1990: in *Rev. in Mod. Astron.* **3**, G. Klare (ed.), (Berlin: Springer Verlag), p.3

Eckart, A., Witzel, A., Biermann, P., Pearson, T.J., Readhead, A.C.S., Johnston, K.J., 1985: *Ap.J.* **296**, L23

Hummel, C.A., 1990: PHD-thesis, university of Bonn

Hummel, C.A., Schalinski, C.J., Krichbaum, T.P., Rioja, M.J., Quirrenbach, A., Witzel, A., Muxlow, T.W.B., Johnston, K.J., Matveyenko, L.I., & Shevchenko, A., 1991: *Astron. Astrophys.*, submitted

Hardee, P.E., & Norman M.L., 1989: *Astrophys. J.* **342**, 680

Johnston, K.J., Simon, R.S., Eckart, A., Biermann, P., Schalinski, C.J., & Strom, R.G., 1987: *Astrophys. J.* **313**, L85

Schalinski, C.J., 1990: PHD-thesis, university of Bonn

Unwin, S.C., 1990 : in *Parsec-Scale Radio Jets*, J.A. Zensus & T.J. Pearson (eds.), (Cambridge: Cambridge University Press), p. 13-19

Witzel, A., 1987: in *Superluminal Radio Sources*, J.A. Zensus & T.J. Pearson (eds.), 1987 (Cambridge: Cambridge Univ. Press), p. 83

Witzel, A., Schalinski, C.J., Johnston, K.J., Biermann, P.L., Krichbaum, T.P., Hummel, C.A., & Eckart, A., 1988: *Astron. Astrophys.* **206**, 245

Wrobel, J.M., Pearson, T.J., Cohen, M.H., Readhead, A.C.S., 1988: *IAU* **129**, M.J. Reid & J.M. Moran (eds.), (Dordrecht:Kluwer), p. 165

Zensus , J.A., Unwin, S.C., Cohen, M.H., & Biretta, J.A., 1990: *AJ* **100** (6), p. 1777-1784

Zensus, J.A., 1990 : in *Parsec-Scale Radio Jets*, op. cit., p. 28-31

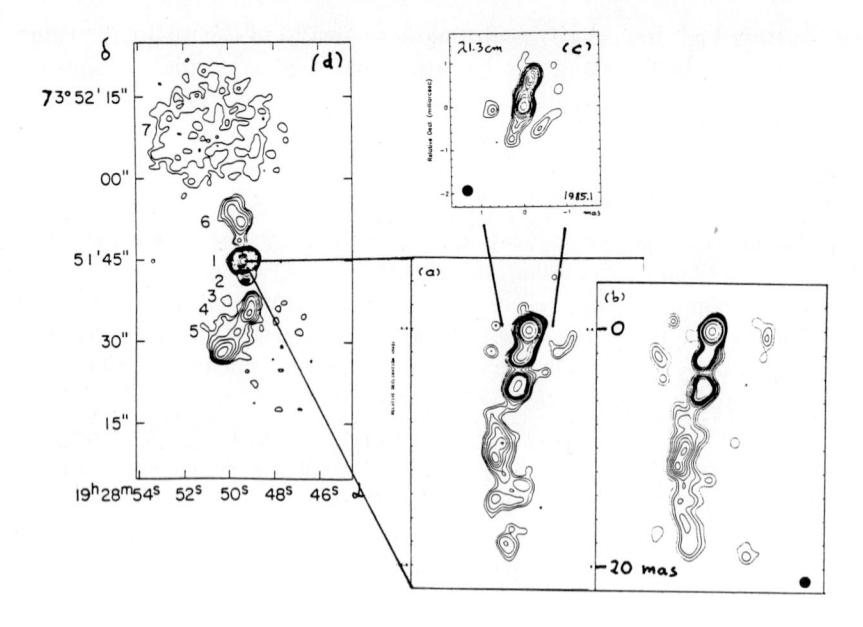

Fig. 1: the structure of 1928+738. (a),(b): 5GHz-VLBI; (c): 23GHz-VLBI (from: Schalinski, 1990); (d): 1.5GHz-VLA (from: Johnston et al., 1987).

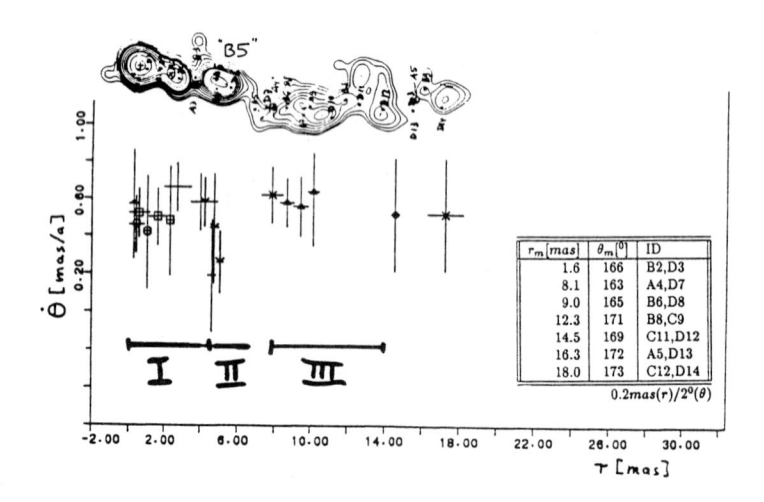

r_m [mas]	θ_m [°]	ID
1.6	166	B2,D3
8.1	163	A4,D7
9.0	165	B6,D8
12.3	171	B8,C9
14.5	169	C11,D12
16.3	172	A5,D13
18.0	173	C12,D14

0.2mas(r)/2°(θ)

Fig. 2: velocity profile of the mas-jet of 1928+738 at 5GHz. The table lists positions of the jet corresponding to component positions at different epochs (labeled A-D). Discussion of the "zones" I-III s. text.

Structural Variability of Blazars from the Complete S5-VLBI-Sample

SCHALINSKI[1,2], C.J., WITZEL[1], A., KRICHBAUM[1], T.P., HUMMEL[1], C.A., QUIRRENBACH[1], A., & JOHNSTON[3], K.J.

[1]: Max-Planck-Institut für Radioastronomie, Bonn, F.R.G.
[2]: Geodätisches Institut der Universität Bonn, F.R.G.
[3]: Center for Advanced Space Sensing, Naval Research Laboratory, Washington, D.C., U.S.A.

Abstract The investigation of a complete, flux-density limited sample of sources from the S5-survey reveals bulk relativistic motion as common phenomenon among flat spectrum radio sources: 12 out of the 13 sources display jet components on the mas-scale separating with apparent superluminal velocities. This is consistent with independent evidence for bulk relativistic motion derived from the deficiency of observed inverse-Compton X-rays and the flux density variability of core components. An increasing number of sources is found to exhibit "stationary" components or different velocities of components, which may be correlated with bending. Selected blazars are briefly discussed.

1. INTRODUCTION

The complete VLBI-sample of northern ($\delta \geq 70^0$), compact flat spectrum ($\alpha_{6cm}^{11cm} \geq -0.5$, $S_\nu \propto \nu^{+\alpha}$) radio sources from the S5-survey (e.g. Kühr et al., 1981) has been investigated from radio to X-rays, and by the radio interferometric technique with spatial resolutions from kpc to $\sim$ pc (e.g. Eckart et al., 1986; Witzel, 1987; Schalinski et al., 1987).
From repeated VLBI-observations at 5 (and partly 43, 23, and 8.4)GHz 12 out of the 13 sources show components separating at apparent velocities $v > c$ (s.Tab.1, and Witzel et al., 1988; Schalinski et al., in prep.). Independent evidence for bulk relativistic motion comes from the lack of inverse-Compton X-rays and the flux density variability (5GHz-core components). Thus is has been shown that bulk relativistic motion is a common phenomenon for flat spectrum radio sources, with $\gamma_{min} = 2h^{-1} \ldots 10h^{-1}$ ($h = [H_0 = 100 km/s/Mpc]$, $q_0 = 0.5$), and angles to the line of sight $\theta = 9^0 h \ldots 44^0 h$. An overview of source parameters is given in Tab.1.

2. DISCUSSION

At least five sources contain evidence for components with constant separations (or at least subrelativistic velocities (upper limits), "S" in Tab.1); the jets of two sources display different velocities of components. Although these sources are either classified as quasars (0016+71, 0153+74, 0836+71, 1928+73), or BL-Lac objects with "quasar spectra" (s. Mutel, 1990) (1749+70, 1803+78), at present it is unclear whether this effect is an artifact caused by the different dynamic range of images or can be attributed to a certain class of objects. Thus there is growing evidence for the co-existence of moving (confirmed for 1749+70 and 0836+71) and stationary components, as is observed in the "prototype"-source 4C39.25 and interpreted by a phenomenological model involving stationary shocks in a curved relativistic jet (Marcaide et al., 1989).

Tab. 1: The *S5-VLBI-Survey*. All sources but 4 () were observed at least three times by means of VLBI at 5GHz.

Source	ID^a	z	m_R^b	S_X^c	$v[c]^{d,e}$	θ^f	θ_{SSC}^g
Blazars							
0212+73	OVV	2.37	19.0	0.49	2.2 ± 1.3:	25^0	25^0
0454+84	BL	....^{i}	20.3	0.095	≥ 1.6	32^0	28^0
(0716+71)	BL	....^{i}	13.2	0.31	(≥ 2.3)	44^0	46^0
1749+70	BL^h	0.77^i	16.5	0.22	5.7 ± 0.6,S	9.3^0	50^0
1803+78	BL^h	0.68	16.4	0.19	3.9 ± 2.6:,S	14^0	13^0
2007+77	BL	0.34^i	16.7	0.18	3.6 ± 1.2	16^0	21^0
Quasars							
(0016+73)	Q	1.76	18.0	0.12	8.3 ± 2.0:,S	6.9^0	7.2^0
0153+74	Q	2.34	16.0	≤ 1.0	≤ 1,S	-	-
(0615+82)	Q	0.71	17.5	≤ 0.2	(≤ 2.2)	24^0:	50^0
0836+71	Q	2.16	16.5	≤ 1.0	5.9-9.5,S	$5.5^0 - 11^0$	14^0
(1039+81)	Q	1.26	16.5	≤ 1.0	≤ 2.2	12^0:	28^0
1150+81	Q	1.25	18.5	≤ 0.2	≤ 4.8	16^0	27^0
1928+73	Q	0.30	15.5	0.59	6.5 ± 0.5^j	8.7^0	36^0

a: optical identification, BL: BL-Lac-Object, Q: Quasar, OVV: optically violently variable
b: R-band-magnitude, measurements with the 90-inch telescope of the Steward-Observatory in 1985 by A. Eckart (priv. com.)
c: flux densities [μJy] at 1 keV (Witzel, 1987; Biermann et al., priv. com.)
d: VLBI at 5GHz (and partly 43,23,8.4GHz); "S": jet contains "stationary" components; note that with β_{obs}: $\gamma_{min} = \sqrt{\beta_{obs}^2 + 1}$
e: $H_0 = 100 km/s/Mpc, q_0 = 0.5$, $1 mas = 2.75 pc (z = 0.3)...4.31 pc (z = 1.26)$
f: angle to line of sight corresponding to γ_{min}
g: maximum θ corresponding to δ_{min}^{SSC}
h: QSO-type spectra (e.g. Kühr, priv. com.; Mutel, 1990)
i: Kühr (priv. com.); 0454+84/0716+71: $z_{min} = 0.3$
j: $r \geq 6$ mas (s. Schalinski et al., this vol.)

Tab.1 shows a good agreement of the angles to the line of sight (θ) corresponding to the minimal Lorentz factors calculated from the observed apparent superluminal motion and the upper limits on θ independently derived from the minimal Doppler factors of SSC-models: $\theta(\gamma_{min}) \leq \theta_{max}^{SSC}$. Within the framework of relativistic beaming models this indicates $\theta \leq \theta(\beta_{obs}^{max})$ (for a given intrinsic β). In the case of $\theta(\gamma_{min}) < \theta_{max}^{SSC}$ (e.g. 1928+73) zones of lower apparent velocities would fit into a model with constant (high) γ and varying (larger) θ throughout the jet, without the requirement of differential Doppler-boosting or varying γ.

The sources with different component velocities display significant curvature of the jet on the pc- (and kpc-) scale. However, the data are not yet sufficient to discriminate between a variation of the intrinsic Lorentz factor (e.g. acceleration) or pure geometrical effects (change of θ_{LOS}). The jet of 2007+77, which has not shown significant differences in p.a., exhibits similar velocities of individual components (on the basis of the analysis of the 5GHz and 23GHz data: Schalinski, 1990, and Schalinski et al., in prep.) (s.Tab.1). The good agreement of θ_{LOS} (s.a.), and the co-existence of "stationary" and moving components as well as the occurrence of different velocities argue in favour of "geometrical" models, assuming constant Lorentz factors, but varying aspect angles (e.g. caused by helical structures) of the higly relativistic (part of the) jet. Further high frequency VLBI-monitoring is required to constrain physical parameters.

3. COMMENTS ON INDIVIDUAL BLAZARS:

1803+78 (z=0.68): at 5GHz and 8.4GHz the mas-structure is dominated by a compact "core" and a (frequency dependent) secondary component at distance 1.2mas (1.4mas: 8.4GHz) and p.a. $\sim 270^0$. On the basis of 4 epochs at 5 GHz and 7 epochs at 8.4 GHz no significant change of separation could be measured ($0.004 \pm 0.028 mas/a : 0.09 \pm 0.61c$ at 8.4GHz; $0.006 \pm 0.046 mas/a : 0.13 \pm 1.0c$ at 5GHz). The non-detection of apparent superluminal motion is in contradiction to Dopplerfactors ~ 4 derived from SSC-calculations and flux density variability, but there is evidence for structural variations close to the core at $\lambda 7mm$ and $\lambda 1.3cm$ (Krichbaum et al., in prep; Schalinski, 1990). This source shows a large misalignment between mas- and arcsecond structure (mapped with WSRT (Strom, priv.com., s.Fig.1), which has an extent of at least 186 kpc (projected).

1749+70 (z=0.77): so far this source was mapped by us twice at 5GHz and once at 1.6GHz. Additional maps at 5GHz were performed by Baath (1984, and priv. com.) and Pearson & Readhead (1988). A component (r= 2.08mas, p.a.= 310^0 at epoch 1981.3 and r= 3.27mas, p.a.= 310^0 (1985.8)) displays a separational change of 0.26 ± 0.03 mas/a ($v = 5.7 \pm 0.6$), and a feature at ~ 1 mas appears to be stationary within the errors. The core-compactness parameter is about 50% at different epochs (5GHz), the rest of the source flux density is predominantly distributed in a halo-like (VLBI-)component. 1749+70 may be a prototype of a "broad ballistic mas-jet" (s. Daly & Marscher, 1988).

REFERENCES

Baath, L., 1984: *IAU* 110 R. Fanti, K. Kellermann, & G. Setti (eds.),(Dordrecht:Reidel),p.127

Daly, R.A., & Marscher, A.P., 1988: *Ap.J.* **334**, 539

Eckart, A., Witzel, A., Biermann, P., Johnston, K.J., Simon, R., Schalinski, C.J., Kühr, H., 1986: *Astron. & Astrophys.*, 168, 17

Kühr, H.,Witzel A.,Pauliny-Toth I.I.K., Nauber U., 1981: *Astron. & Astrophys. Suppl.* **45**, 367

Marcaide, J.M., Alberdi, A., Elósegui, P., Schalinski, C.J., Jackson, N., und Witzel, A., 1989: *Astron. & Astrophys.* **211**, L23

Mutel, R.L., 1990: in *Parsec Scale Radio Jets*, J.A. Zensus & T.J. Pearson (eds.), (Cambridge: Cambridge Univ. Press), p. 98

Pearson, T., and Readhead, A.C.S., 1988: *Ap. J.* **328**, 114

Schalinski, C.J., Biermann, P., Eckart, A., Johnston, K.J., Krichbaum, T.P., & Witzel, A., 1987: *IAU* **121**, E.Ye. Khachikian, J. Melnick, & W. Fricke (eds.), (Dordrecht: Reidel), p. 287

Schalinski, C.J., 1990: PHD-thesis, University of Bonn

Witzel, A., 1987: in *Superluminal Radio Sources*, J.A. Zensus & T.J. Pearson (eds.), 1987 (Cambridge: Cambridge Univ. Press), p. 83

Witzel, A., Schalinski, C.J., Johnston, K.J., Biermann, P.L., Krichbaum, T.P., Hummel, C.A., & Eckart, A., 1988: *Astron. Astrophys.* **206**, 245

228 *Structural Variability of Blazars*

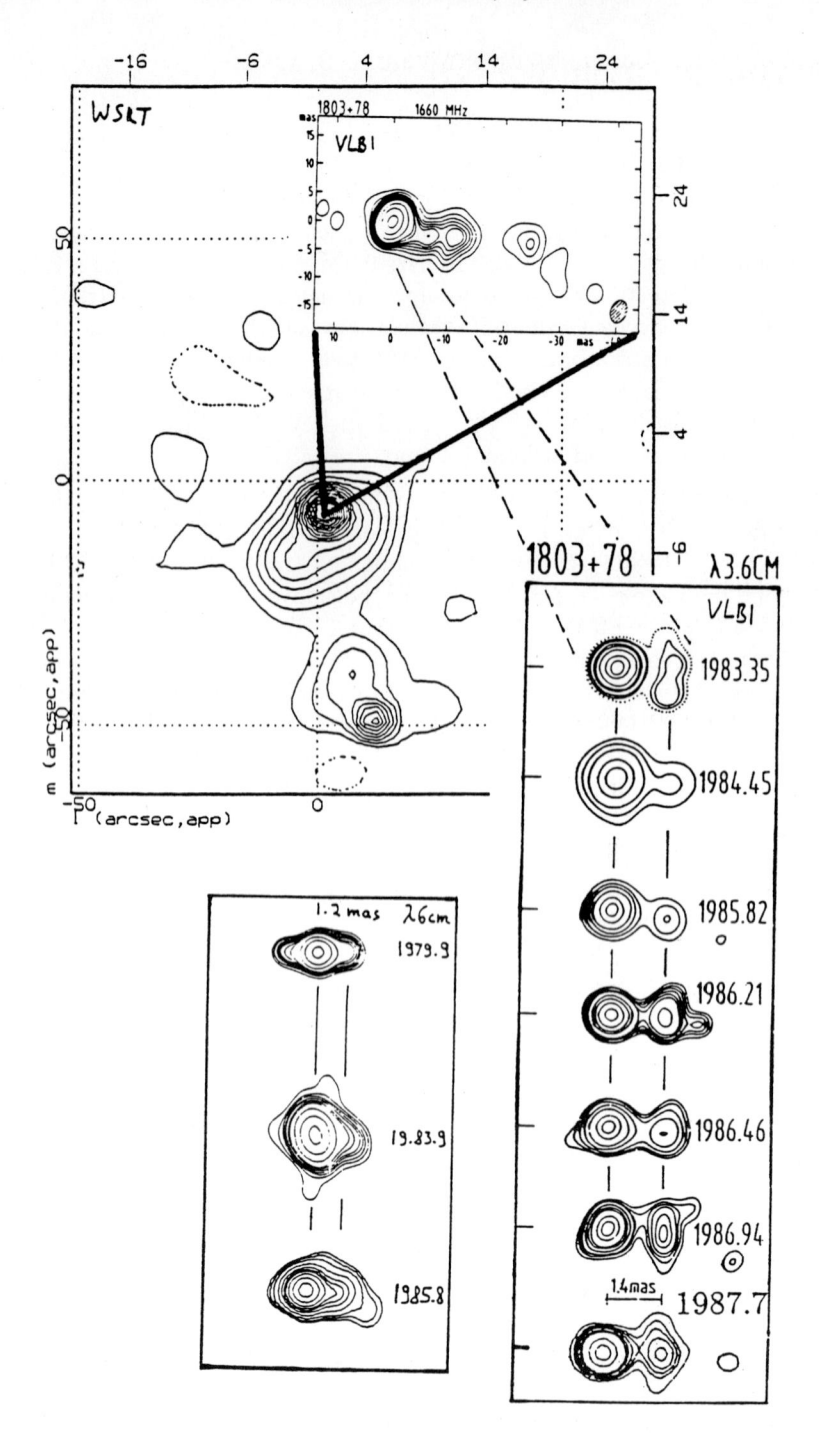

Fig. 1: the radio structure of 1803+78 on kpc- to pc-scales (s. Schalinski, 1990 and refs. therein).

Varying gamma in 3C273

L.B.BÅÅTH
Onsala Space Observatory, S-439 00 Onsala, Sweden

1. INTRODUCTION

One of the primary assumptions in most models of radio jets is that the Lorentz factor (γ) is constant with time. This is for several reasons. Models become more complicated if γ is allowed as a free parameter, and a change of γ with time is also more physically difficult to achieve than a change in viewing angle arising from a bent jet. It is more likely that different outbursts within the same source could arise from shocks of different nature which do not have to travel with the same γ.

3C273 has recently suffered several outbursts of different character. This talk will discuss some of the properties deduced from the observed data on the period of outbursts during 1986 - 1988. This source is one of the best studied of all quasars, and is therefore well suited for such a discussion.

2. THE OUTBURSTS

The history of integrated flux density of 3C273 shows a number of outbursts during the later part of the 1980's. The major outburst was in January 1988 and was studied over a wide range of frequencies (e.g. Courvoisier *et al.* 1988). The change of the spectrum with time fitted well with the model of a travelling thin shock front as suggested by Marscher and Gear (1985). In this model the shock starts from the core as initially very thin and travels downstream the jet. The shock will stay thin for some time, but will eventually expand adiabatically. Since it will stay thin for some length of time we also get a reasonable chance to observe it with very high resolution VLBI.

Our first epoch of 100GHz VLBI for hybrid mapping was planned for, and took place in, March 1988. This was only some 60 days after the outburst, but the timing was coincidental and unfortunately not due to our foreseeings. The resulting hybrid map did indeed show the brightest component to be unresolved along the axis of the jet and elongated perpendicular to this. The size scales were 192x17 light-days (using $H_0 = 100$ km sec^{-1} Mpc^{-1} and $q_0 = 0.05$). These sizes are again compatible with the model by Marscher and Gear (1985). The fact that the shock is indeed observed as thin clearly indicates that we do not see it coming towards us head-on, but at a small angle. Any such small tilt would mean that one of the rims of the jet would be more Doppler boosted than

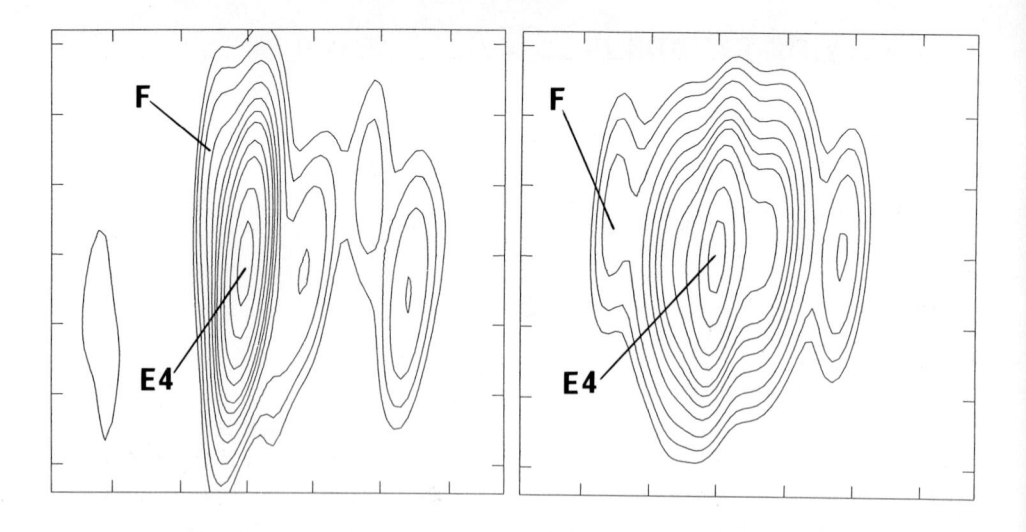

Figure 1. The quasar 3C273 observed with global VLBI arrays at 100GHz in March 1989 (left) and 43GHz in June 1989 (right). Tick marks are at 100μas. Both maps are convolved with a restoring beam of 280x50μas at P.A. -4.4°.

the other and a thin edge would be observed. The component (=E4) was also situated ≈128μas from the north-east end of the jet (the "core component" = F). This distance corresponded well with the scenario that the new component had been born in the core (F) about 60days prior to our observations and had travelled with a mean speed of ≈800 μas/year.

Independent of the VLBI observations at 100GHz (Bååth *et al.* 1991), 3C273 was also observed with VLBI at 43GHz (Krichbaum *et al.* 1991) in June the same year. The map on 3C273 at 43GHz confirmed the existence of most features in our 100GHz map. The brightest component (=E4) was clearly visible and offset from the core (F) by an amount which again corresponded to a speed of ≈800 μas/year. This speed was further confirmed with a new epoch of observation at 43GHz in March 1989 (Krichbaum *et al.* 1991).

The resolution at 43GHz was a factor of 2.5 less than at 100GHz, making comparisons of the two maps difficult. The group working on the 43GHz data kindly borrowed me the data though, and I made a *super-resolved* map at 43GHz by using the 100GHz map as initial guess. The quest was to create the best model at 43GHz which was most closely related to the 100GHz image. The two maps, 100GHz and 43GHz, are shown in Figure 1 here. Both maps clearly shows the "core" component F to the north-east of the brightest component E4.

The major point I want to raise in this talk is the question what the other components (E1 - E3) in the 100GHz map of 1988 were, from where they came and where they went. Most of them are also visible in the 43GHz map from 1988 and they do

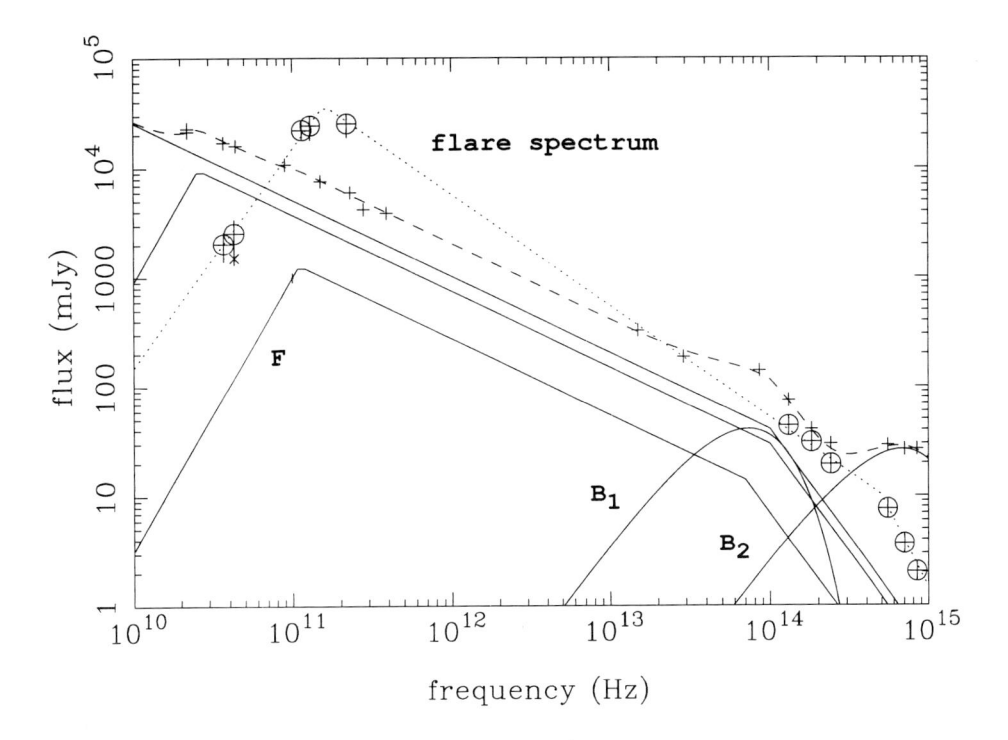

Figure 2. The spectrum of the quasar 3C273 in 1986. Individual components are discussed in the text.

have the general appearance of true components. It is not possible to identify all components between the two maps, but spectral index effects and evolution of components between the two epochs are both likely to play a significant role in this. There are also additional components visible in the later 43GHz map of 1989 (Krichbaum *et al.* 1991), but these are *closer* to the core than the brightest component at that time. The difference in brightness between the components in the 1989 map is large, with the brightest at 5.3Jy and the others at about 1-2 Jy (Krichbaum *et al.* 1991). Comparing with the 1988 map therefore suggests that either one of the other components has become much brighter while at the same time the component E4 has become dimmer, or E4 has *passed* the other components on its way out. The two maps in Figure 1 indicates that the latter probably is the case. The 43GHz map from the later epoch of June 1988 shows the brightest component (E4) to be more in the centre of the complex structure than in the 100GHz map of March 1988. The positions also fit with a model where E4 moves along the wiggling pattern which is clear from the 100GHz map.

If the component E4 is indeed moving with a larger speed than the other components (E1-E3) then we would have to look for the birth of these components about a year prior to the outburst associated with the birth of E4. There were indeed some minor outbursts during 1986-1987 followed by a larger one in 1987 as precursors to the major event of 1988. The 1987 outburst was significantly different from the outburst in

1988 (Valtaoja these proceedings) in that it peaked below 22GHz while the 1988 outburst peaked at above 100GHz. Therefore it is likely that the shocks responsible for these outbursts also have different nature. A heavier and thinner shock would create a high frequency outburst, would be more compact, and also survive for a longer time period.

The existence of shocks with different Lorentz factors have severe consequences for our modelling of outbursts. An additional parameter will of course result in that models more easily can fit the observed data, but also in that each outburst may have to be treated in its own right as a separate event. It is only by studying a large number of such outbursts and associated components that we can hope to learn anything about the central power-house from where they all emerge.

The high frequency (above 10GHz) end of the spectrum of 3C273 can be broken down into some major components as is shown in Figure 2. The crosses represent data from mid 1986 when the source was in a reasonable steady state (from Courvoisier *et al.* 1988). The broken curve is a fit to the observed data and is the sum of the components discussed below. The solid lines show components spectra as:

B_1 is a black body spectrum having a temperature of 1250°K and a radius of $9 \cdot 10^{17}$ cm.

B_2 is also a black body spectrum having a temperature of 11500°K and a radius of $2.6 \cdot 10^{16}$ cm.

F represents the high frequency core observed at 100GHz.

The other two solid lines represent components extrapolated from lower frequency VLBI. The core component F observed with 43GHz VLBI is represented with a star in Fig. 2. The flux of this component is too high to fit the synchrotron spectra F as in Fig. 2. However, 3C273 had an outburst also later in 1988 - beginning of 1989 and it is likely that the new component will contaminate the flux of the core already in the June the same year so it would be brighter than in a quiet state.

The "quiet" spectrum can be subtracted from the flux densities observed in March 1988, during the flare period. The circles represent such a subtraction of data at radio (Matsuo *et al.* 1989) and optical frequencies taken at close epochs (Courvoisier *et al.* 1988). This is clearly a synchrotron spectrum which becomes optically thick due to self-absorption at around 130GHz and have a high frequency cut-off at about 500THz. It is interesting to note that the size of the new component across its thin part is about $3 \cdot 10^{16}$ cm in radius, which is on exactly the same order as the black body component B_2, which is generally referred to as the UV-bump. This is intriguing, but may be purely coincidental.

Since the major outbursts are relatively short lived, they seem to develop significantly during one month, it will in the future be extremely important that flux measurements over the full range of the spectrum are coordinated with high resolution VLBI observations. This is especially true if each individual outbursts has its own characteristics.

REFERENCES

Bååth,L.B., Padin,S., Woody,D., Rogers,A.E.E., Wright,M.C.H., Zensus,A., Kus,A.J., Backer,D.C., Booth,R.S., Carlstrom,J.E., Dickman,R.L., Emerson,D.T., Hirayabashi,H., Hodges,M.W., Inoue,M., Moran,J.M., Morimoto,M., Payne,L., Plambeck,R.L., Predmore,C.R. & Rönnäng,B.: 1991, *Astron.Astrophys.*, **241**, L1

Courvoisier,T., Robson,E.I., Blecha,A., Bouchet,P., Hughes,D., Krisciunas,K. & Schwarz,H.: 1988, *Nature*, **335**, 330

Krichbaum,T.P., Booth,R.S., Kus,A.J., Rönnäng,B.O., Witzel,A., Graham,D.A., Pauliny-Toth,I.I.K., Quirrenbach,A., Hummel,C.A., Alberdi,A., Zensus,J.A., Johnston,K.J., Spencer,J.H., Rogers,A.E.E., Lawrence,C.R., Readhead,A.C.S., Hirayabashi,H., Inoue,M., Morimoto,M., Dhawan,V., Bartel,N., Shapiro,I.I., Burke,B.F. & Marcaide,J.M.: 1991, *Astron.Astrophys.*, **237**, 3

Marscher,A. & Gear,W.: 1985, *Astrophys.J.*, **298**, 114

Matsuo,H., Matsumoto,T., Murakami,H., Inoue,M., Kawabe,M., Tanaka,M. & Ukita,N.: 1989, *P.A.S.J.*, **41**, 4

What is responsible for the new outburst of 0735+178

F.J. Zhang[1,*], L.B. Bååth[2], H.S. Chu[3]

January 1991

[1]Nuffield Radio Astronomy Laboratories, Jodrell Bank, Cheshire, U.K.
(* on leave from Shanghai Observatory, Academia Sinica, Shanghai, China.)
[2]Onsala Space Observatory, Onsala, S-43900, Sweden
[3]Purple Mountain Observatory, Academia Sinica, Nanking, China.

Abstract

A radio outburst of the BL Lac-object 0735+178 began from early 1988 and is still in progress. It was observed with a global VLBI array at 22 GHz last year. The resulted hybrid map consists of two parts: a core-jet like structure at the centre and a weak, diffuse emission region located to the east of the core. Two new components have been produced during this outburst. They could be moving out with superluminal motion. The core could have embraced another new component. The resolution used can not be able to resolve out this component from the core at that moment. The weak emission region could be caused by an old component reached the point at which the Doppler Boosting reached its maximum.

0735+178 has been optically identified with a stellar object (Blake 1970). The measured redshift z=0.424 is based on a single absorption doublet of MgII (Burbidge & Hewitt 1987). In addition to its characteristics common to BL Lac-objects, this source has a remarkably flat spectrum in the radio range may be a result of a "cosmic conspiracy", that is a superposition of several VLBI radio components. 0735+178 has been observed with VLA, MERLIN and VLBI (VLA: Ulvestad & Johnston 1984; Antonucci & Ulvestad 1985. MERLIN: Bååth & Zhang 1990. VLBI: Shaffer 1978; Bååth 1984; Gabuzda, Wardle & Roberts 1989, 1990; Bååth & Zhang 1990a, hereafter BZ90a; Zhang & Bååth 1990, hereafter ZB90).

The structure of 0735+178 has been partially resolved in most observations to date. The 13-cm VLBI observations of earlier period (Shaffer 1978) indicated that either the source expanded with v/c more than 2 or the structure became more complex. Both the effects occured in the results of 6-cm VLBI observations (BZ90a). The outer component (designated B) of the jet was stationary on the 6-cm maps and an inner component (designated C0) moved out with superluminal motion. The superluminal motion of the components of 0735+178 are confirmed by 5-GHz VLBI polarization observations (Gabuzda, Wardle & Roberts 1990) and 1.35-cm VLBI observations (ZB90). The situation of 0735+178 is similar to that of 4C39.25 (Shaffer and Marscher 1987) where one component is moving between two more stationary components. A model, in which the trajectory of the components of 0735+178 probably is bent, and the jet could curl to turn towards the observers and reaches a viewing angle of nearly 0^0 at the location where component B is initially seen, has been suggested (BZ90a; Bååth & Zhang 1990b). The 1.35-cm morphologies of 0735+178 at epochs 1982.9 and 1984.0 are simpler than that we expected and only include its core and a component (ZB90). The component B, which is located at ~4.2 mas to the east of the core on the 5-cm maps, has fully been resolved on the 1.35-cm maps.

0735+178 suffered two radio outbursts in the early 1970's, and was quite stable from 1980 to early 1988. A new radio outburst began from 1987/1988 has gone on for a number of years and is still in progress (*Fig*.1). What causes this outburst? The behavior of the total flux and linear polarization for BL Lac-objects is highly correlated with the emergence of new component (Mutel et al. 1989). On the other hand, we have predicated that the extrapolation of the motion of C0 suggests that C0 would reach B around 1989 at that point the viewing angle would be smallest and at the same time the Doppler boosting would be at its peak (BZ90a) and therefore expect that the latest outburst of 0735+178 could partially be caused by this event. For looking for new components emerging from the core during the outburst and checking the model described above, a 1.35-cm VLBI observation was carried out in June, 1990 (1990.42).

The resulted hybrid maps are shown in $Fig.2$ in which $Fig.2(a)$ was obtained with the normal way, and for $Fig.2(b)$, we put more weights on the short baselines than on the long baselines for searching the effect of C0. The details of data deduction and mapping are described in another paper (Zhang, Bååth & Chu 1991). Comparing with the previous 1.35-cm results, we can find that two new components, labelled C1 and C2, were borne during this outburst. Supposing these components are a sequence of shock waves and each one was produced at the time of the corresponding minimun of the polarization, we could estimate the birth-epoch for each component, and then infer their angular velocities and apparent transverse speeds. The corresponding speeds are 4.7±0.4c and 6.1±0.2 for C1 and C2 respectively. We used model fitting for getting parameters for each component. The flux density of the 'core' is 1.89 Jy. During the 'quiet' period of 0735+178, the average

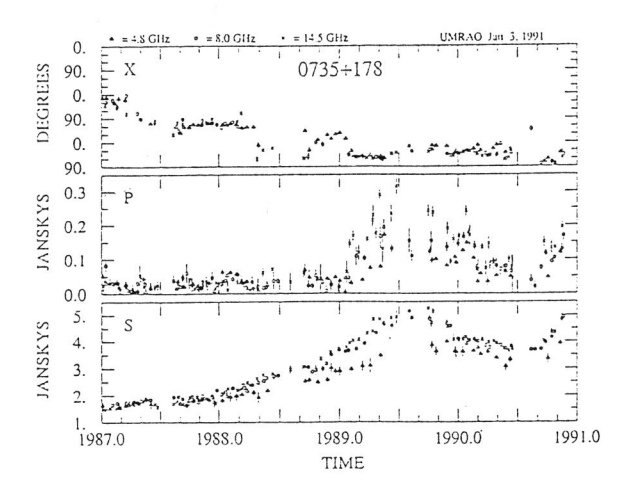

Figure 1. Monthly averages of the total flux density, degree of polarization, and polarization position angle of 0735+178 (courtesy of H.D. Aller and M.F. Aller)

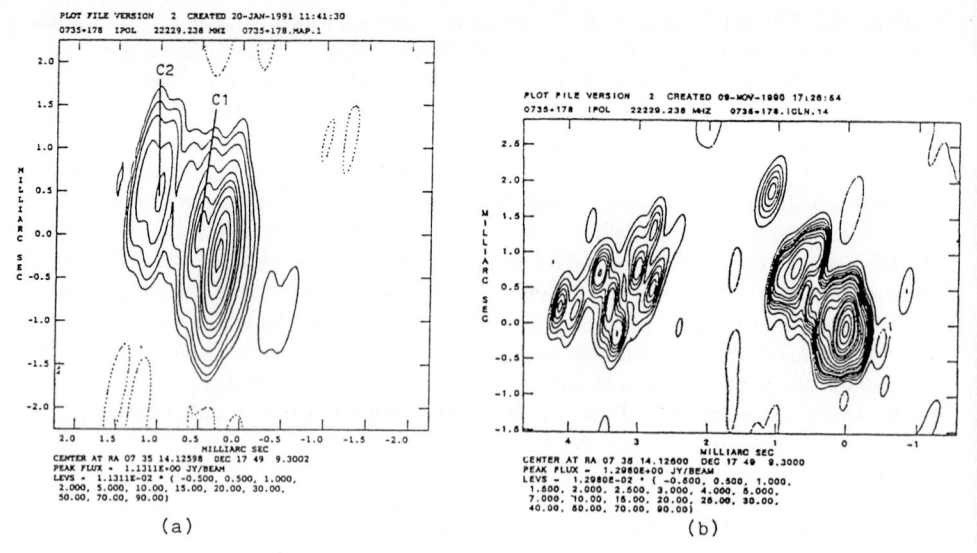

Figure 2. The hybrid maps of 0735+178 at epoch 1990.42

flux density of the core is about 0.75 Jy. Thus the 'core' contains an extra flux density of ¿1 Jy at epoch 1990.4. On the other hand, the data of total flux density in $Fig.1$ shows that the total flux density has passed a peak and droped out about 1.5 Jy from the peak value of ~ 5 Jy while we observed it last June. There is another outburst started and is undergoing now. Thus we can infer that the core could contain another new component. It is too close to the core in last June, and the resolution used was still too low and was not able to resolve it out from the core.

On $Fig.2(a)$, there is a relatively weak emission region located to the east of the core. It is super-resolved and does not appear on the previous 1.35-cm maps. The indication reveals that something has happened at this location. It is possible that this part could be caused by the component C0 reached the point B at where the Doppler boosting would be at its peak. If 0735+178 was observed at 6 cm in last June or earlier, the sign would be more clear. Unfortunately, there was not any 6-cm VLBI observation carried out last year for checking the event which is expected to be happened near the location of B.

Acknowledgements: FJZ is financially supported by the Royal Society K.C. Wong Fellowship, U.K., for which FJZ is very grateful. He thanks prof. R. Davies for making the visit to NRAL possible. FJZ also thanks financial support by the organizers of the meeting for making the attend to the meeting possible.

References

Antonucci, R.R.J. and Ulvestad, J.J.S. 1985. *Astrophys. J.,* **294**, 158.

Blake, G.M. 1970, *Astrophys. J.*, **6**, 201.
Burbidge, G. and Hewitt, A. 1987. *Astron. J.*, **92**, 1.
Bååth, L.B. 1984. in *IAU Symposium 110, VLBI and Compact Radio Source*, ed.
    R. Fanti, K. Kellermann, and G. Setti (Dordrecht: Reidel), p.127.
Bååth, L.B. and Zhang, F.J. 1990a. *Astron. Astrophys.*, in press.
Bååth, L.B. & Zhang, F.J. 1990b. in *Compact Steep-spectrum & GHz-peaked Spectrum Radio Sources,* eds. C. Fanti, R. Fanti, C.P. O'Dea & R.T. Schilizzi (Consiglio Nazionale delle Ricerche: Istituto di radioastronomia - Bologna), p.147.
Gabuzda, D.C., Wardle, J.F.C., and Roberts, D.H. 1989. *Astrophys. J.*, **338**, 743.
Gabuzda, D.C., Wardle, J.F.C., and Roberts, D.H. 1990. *Astrophys. J.*, in press.
Mutel, R.L., Phillips, R.B., Su, B.-M., and Bucciferro, R.R. 1990. *Astrophys. J.*, **352**, 81.
Shaffer, D.B. 1978. in *Pittsburg Conference on BL Lac Objects,* ed. A.M. Wolfe,(Pittsburg: University of Pittsburg), p.68
Shaffer, D.S. and Marscher, A.P. in *Superluminal Radio Sources*, eds. J.A. Zensus and T.J. Pearson (Cambridge: Cambridge University Press), p.67.
Ulvestad, J.S. and Johnston, K.J. 1984. *Astron. J.*, **89**, 189.
Zhang, F.J. and Bååth, L.B. 1990. *Mon. Not. R. astr. Soc.*, in press.
Zhang, F.J., Bååth, L.B. & Chu, H.S. 1991. in preparation.

Milliarcsecond Polarization Structure and the Classification of "Blazars"

D. C. Gabuzda

Jet Propulsion Laboratory, California Institute of Technology

T. V. Cawthorne

Harvard-Smithsonian Center for Astrophysics

1. INTRODUCTION

The linear polarization of the milliarcsecond-scale radio emission in active galactic nuclei (AGN) yields information about the magnetic field structure, densities of both relativistic and thermal electrons, and kinematics in these sources. VLBI polarization images for some 30 quasars, radio galaxies, and BL Lacertae objects have now been made (Cotton *et al.* 1984; Wardle *et al.* 1986; Roberts, Gabuzda, and Warde 1987; Gabuzda, Wardle, and Roberts 1989; Gabuzda *et al.* 1989; Gabuzda *et al.* 1991; Cawthorne *et al.* 1991); see Table 1. One of the most striking results to come out of this work is the discovery that these three different classes of object each have surprisingly characteristic and distinctive VLBI polarization structure. Such systematic differences in polarization structure suggest that these classes contain intrinsically different objects.

2. DESCRIPTION OF THE OBSERVATIONS

The calibration and mapping of VLBI polarization data have been pioneered by the astrophysics group at Brandeis University (Roberts *et al.* 1984; Wardle *et al.* 1986). The antennas used in the polarization-sensitive VLBI observations yielding the results presented below were the Medicina (26 m, L), Effelsberg (100 m, B), Haystack (37 m, K), Green Bank (43 m, G), Fort Davis (26 m, F) and Owens Valley (40 m, O) antennas; the phased Westerbork Synthesis Radio Telescope (14 X 25 m, W); and the phased Very Large Array (27 X 25 m, Y). The observations were made in December 1981 and December 1982 (KGYO), March 1984 and October 1984 (BKGFYO), and May 1987 (LBWKGFYO). In all cases, the observing frequency was 5 GHz. Left and right circular polarization (LCP

Table 1: Sources Currently Imaged With Polarization VLBI

Source	Alternate Name	ID	Redshift	Reference
0016+731		Q	1.78	8
0108+388		G	----	8
0133+476		Q	0.86	8
0153+744		Q	2.34	8
0212+735		Q	2.37	8
0235+164		B	0.94	5
0300+470		B	----	7
0306+102		B	----	7
0454+844		B	----	5
0711+356		Q	1.62	8
0735+178		B	0.42	4
0754+100		B	----	7
0836+710	4C71.07	Q	2.17	8
0851+202	OJ287	B	0.31	3
0954+658		B	0.39	7
1219+285		B	0.10	7
1226+023	3C273	Q	0.16	6
1308+326		B	1.00	7
1538+149		B	----	7
1624+414		Q	2.55	8
1633+382		Q	1.81	8
1641+399	3C345	Q	0.60	2
1652+398		B	0.03	7
1749+096	OT081	B	----	5
1749+701		B	0.77?	7
1803+784		B	0.68	7
1807+698	3C371	B	0.05	5
1823+568		B	----	5
1828+487	3C380	Q	0.69	8
1928+738		Q	0.30	8
2200+420	BL Lac	B	0.07	5
2251+158	3C454.3	Q	0.86	1
2351+456		Q	----	8
2352+495		G	0.24	8

Refs: (1) Cotton et al. 1984; (2) Wardle et al. 1986; (3) Roberts, Gabuzda, andWardle 1987; (4) Gabuzda, Wardle, and Roberts 1989b; (5) Gabuzda et al. 1989; (6) Roberts et al. 1990; (7) Gabuzda et al. 1991; (8) Cawthorne et al. 1991 and RCP) were detected at all stations, except for the 1981 and 1982 experiments, in which all four stations detected LCP but only G and Y detected RCP. The data were calibrated as described in Roberts *et al.* (1984) and Wardle *et al.* (1986). Hybrid maps of the distribution of total intensity I were made using standard self-calibration techniques (Cornwell and Wilkinson 1981). Maps of the linear polarization P were made by referencing the cross-hand visibilities to the parallel-hand visibilities using the antenna gains determined in the hybrid mapping process, Fourier transforming the cross-hand visibilities, and performing a complex CLEAN. One by-product of this procedure is to register the total intensity and polarization maps to within a small fraction of a fringe spacing, so that corresponding maps may be meaningfully superimposed.

3. SUMMARY OF RESULTS

There are now some 18 sources which have been described by at least some authors as BL Lacertae Objects for which VLBI polarization images have been made (Roberts, Gabuzda, and Wardle 1987; Gabuzda, Wardle, and Roberts 1989; Gabuzda *et al.* 1989; Gabuzda *et al.* 1991); in 15 of these sources, a high fraction of integrated I and P flux is found in the VLBI images. Of these 15 sources, 9 have been found to have resolved polarization structure on VLBI scales. In every one of these BL Lacertae object in which polarization structure has been detected, the polarization position angle in knots in the jet is nearly parallel to the VLBI structural axis, as can clearly be seen in Figure 1a. The magnetic fields inferred by this orientation are in most cases very nearly perpendicular to the direction of the jet. A map of the BL Lacertae object 1219+285, in which this characteristic transverse magnetic field structure is seen, is shown in Figure 2. The degree of polarization of these jet components ranges from ~0-60 %. In contrast, the direction of polarization in the VLBI cores of BL Lacertae Objects appears to be random (see Figure 1b), with degrees of polarization ranging from ~2-5 %.

VLBI polarization images of some 14 quasars (most in the Pearson-Readhead (1981, 1988) sample of bright extragalactic radio sources) have been made to date (Cotton *et al.* 1984; Wardle *et al.* 1986; Cawthorne *et al.* 1991); of these, 12 show appreciable polarized flux on milliarcsecond scales. There is a wide scatter in the offsets between the polarization position angles in the jet components and the VLBI structural axes. The observed offsets range from 30°-90°, corresponding to a projected magnetic field that ranges in orientation from 60° to parallel to the jet axis. The degree of polarization of quasar jet components ranges from ~0-60%. The direction of polarization in the cores of quasars appears to be random (see Figure 1d); the degree of polarization in these components is typically only ~2% or less.

Figure 1. Histograms of the distribution of offset between polarization position angle and VLBI structural axis in a) jet components in BL Lacertae objects, b) core components in BL Lacertae objects, c) jet components in quasars, and d) core components in quasars. In each case, the horizontal axis is divided into bins of ten degrees, and the vertical axis is the number of sources in each bin.

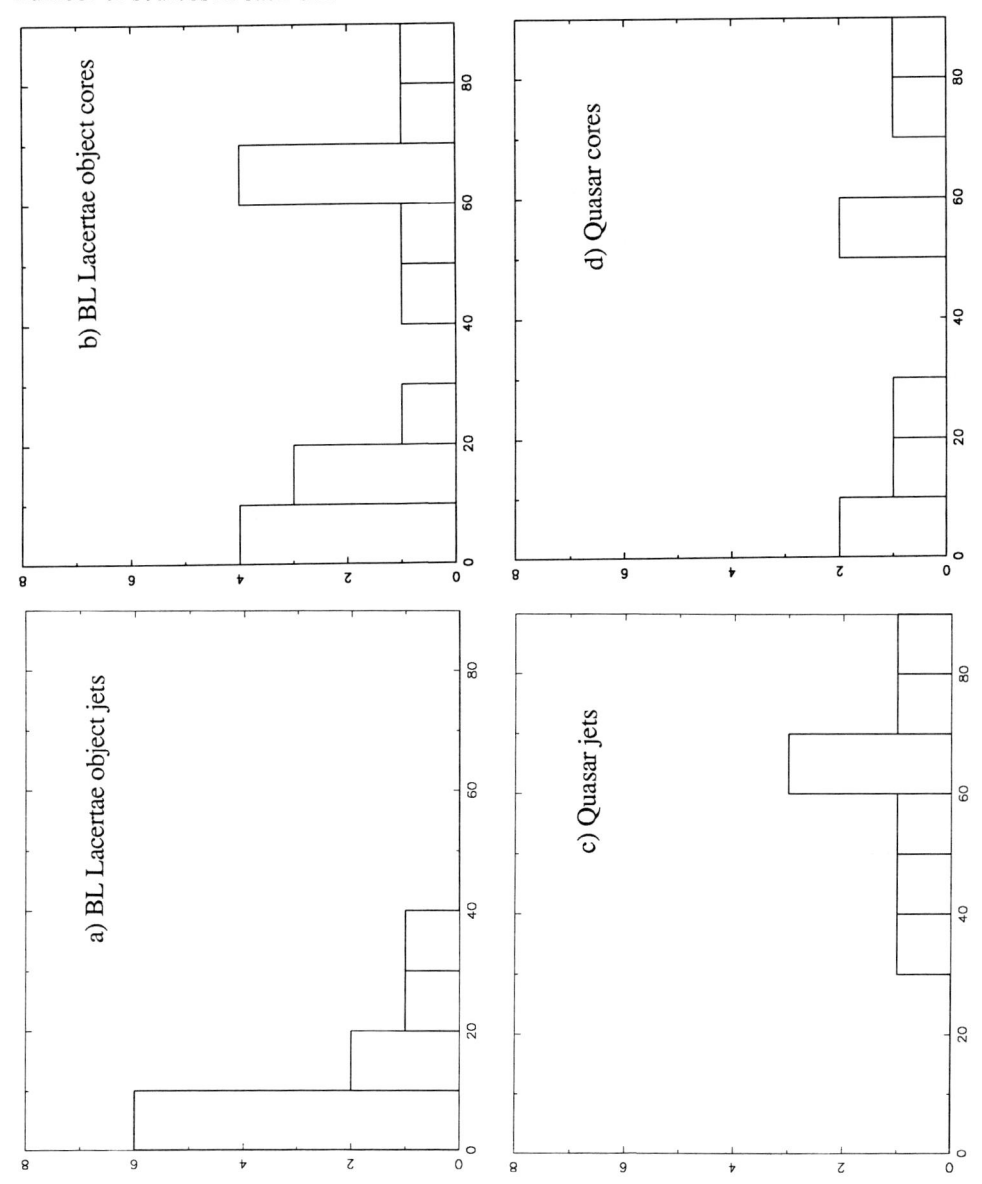

Figure 2. VLBI hybrid maps of the BL Lacertae object 1219+285 at 5GHz, epoch 1987.41. *Top:* Total intensity, with contours at -2.8, 2.8, 4.0, 5.6, 8.0, 11.0, 16.0, 22.0, 32.0, 45.0, 64.0, and 91.0 % of the peak brightness of 250 mJy/beam. *Bottom:* Linear polarization, with contours of polarized intensity at 14, 20, 28, 40, 56, and 80 % of the peak brightness of 26 mJy/beam.

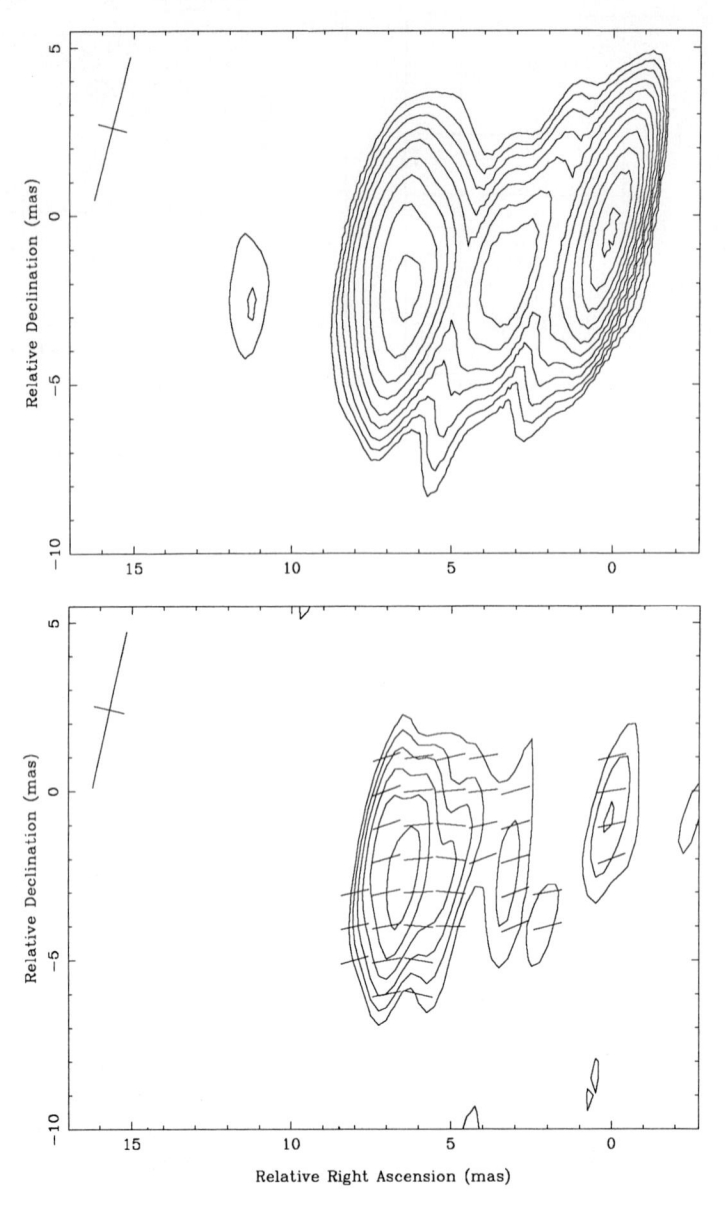

4. DISCUSSION

4.1 The Polarization of Core Components

The distribution in Figure 1c shows that at 5 GHz there is no obvious preferred direction for the polarization in the core components of either quasars or BL Lacertae objects. In the case of the quasars, this is not particularly surprising, since these components are rather weakly polarized. Figure 3 shows the quasar 4C71.07 (0836+710), in which the bright core seen in total intensity is less than 1% polarized. The cores of BL Lacertae objects, however, are appreciably polarized, typically ~2-5 %, and we find it somewhat surprising that there is not some preferred direction for this polarization. The origin of this behaviour is not obvious; one possibility is that there is in fact some preferred direction of polarization relative to the local jet direction, but that the local jet direction is unknown due to bending of the jet on scales much smaller than our resolution. There are commonly large misalignments between the milliarcsecond and arcsecond structural axes in BL Lacertae objects, and it is possible that the jets begin curving on very small scales. Alternatively, it may be that any preferred polarization direction is hidden by the presence of appreciable Faraday rotation in the core region. The strong tendency for polarization in the jet components to be oriented parallel to the VLBI structural axis indicates that the amount of Faraday rotating material in and around the jets is small; if this were not so, a much wider range of orientations would be observed. Therefore, if the random distribution of polarization orientations in the cores is due to appreciable Faraday rotation, the Faraday rotating material must be largely confined to the core region.

The degree of polarization observed in BL Lacertae object VLBI cores is considerably larger than that observed in most quasars, which is usually $\leq 2\%$. One possibility is that what is observed as the VLBI "core" at 5GHz is actually the superposition of a weakly polarized core similar to those observed in quasars and a highly polarized knot (or knots) that is not resolved from it.

Observations of eight BL Lacertae objects made at 8.4GHz in June 1990 should help resolve some of these questions. If the random orientation of the core polarization is associated with Faraday rotation close to the nucleus, then this effect should be substantially reduced at 8.4 GHz, whereas if it is a purely geometrical effect, it should not. In addition, the better resolution attained at the higher frequency should place more stringent constraints on any substructure within the cores of these sources.

Figure 3. VLBI hybrid maps of the quasar 4C71.07 at 5 GHz, epoch 1984.23. *Top:* Total intensity, with contours at -2.0, 2.0, 2.8, 4.0, 5.6, 8.0, 11.0, 16.0, 22.0, 32.0, 45.0, 64.0, and 91.0 % of the peak brightness of 750 mJy/beam. *Bottom:* Linear polarization, with contours of polarized intensity at 20, 28, 40, 56, and 80 % of the peak brightness of 25 mJy/beam.

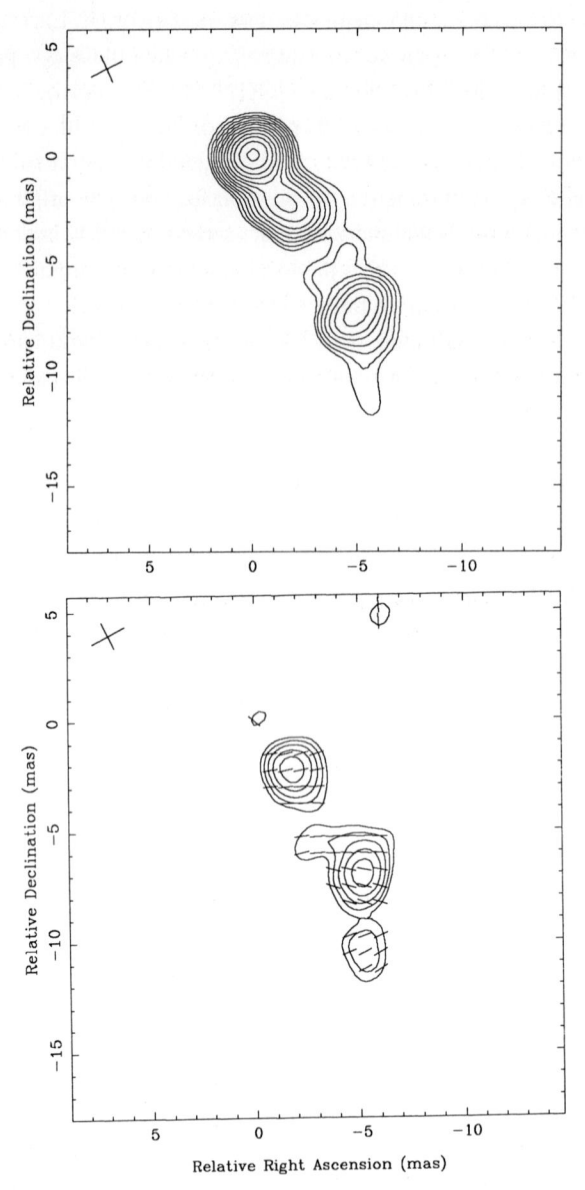

4.2 The Polarization of Jet Components

It is clear that there is a rather strong tendency for the polarization in the jet components of BL Lacertae objects to be aligned with the structural axes. The most plausible reason for this alignment is that these knots are associated with plane relativistic shocks that propagate down the jet. If an initially random magnetic field is compressed by such a shock, the projected magnetic field in the shocked region is perpendicular to the plane of compression, *i.e.*, perpendicular to the jet (Laing 1980). This picture has successfully been used to model integrated cm-wavelength polarized outbursts in BL Lac and 1749+096 (Hughes, Aller, and Aller 1985, 1989, these proceedings), and the VLBI polarization structure and kinematics of OJ287 (Cawthorne and Wardle 1988). The collected VLBI polarization data for BL Lacertae objects now available (discussed in more detail in Gabuzda *et al.* 1991) and the success of these theoretical models strongly suggest that plane relativistic shocks are common in the VLBI jets of BL Lacertae objects. In addition, the narrowness of the jet component polarization position angle distribution argues that rotation measures in these jets are rather small; if this were not the case, the distrubution would be more "smeared out" by Faraday rotation.

Figure 1c illustrates that the distribution of quasar polarization position angles measured from the direction of the VLBI jets is broad, and shows only a weak tendency to peak (at ~60-70° offset). This is very different from the corresponding distribution for the BL Lacertae objects, which is narrow and strongly peaked towards 0°. Another interesting result is that in a number of sources, the polarization position angles, while offset at some seemingly random angle to the structural axis, are fairly uniform throughout the source. Figure 3 shows the quasar 4C71.07 (0836+710), which illustrates this behaviour. The mean polarization position angle in this source is 55° (from the dominant jet axis) and shows a variation ~20° around this mean.

Since the arcsecond-scale jets in luminous radio sources often seem to have polarization position angles indicating longitudinal magnetic fields (Bridle and Perley 1984), it is interesting to ask why the distribution of offsets in Figure 1c is not more strongly peaked towards 90°. (Two quasars in which longitudinal magnetic fields *are* observed on VLBI scales are 3C345 (Wardle *et al.* 1986, 1988) and 3C273 (Roberts *et al.* 1990).) The origin of the polarization position angle offsets seen in quasars is not at present clear; it is possible that there is significant local Faraday rotation in these sources.Without simultaneous multi-frequency VLBI polarization measurements it is not possible to directly determine the distribution of rotation measure on milliarcsecond scales; the best one can do is assume that the integrated rotation measure should be applied to all VLBI components. For the sources in question, integrated measurements of total and polarized flux made with the VLA during the VLBI runs indicate that there is significantly greater polarized flux measured by the VLA than appears in the VLBI images; in this case, rotation measures based on integrated measurements may not be appropriate.

4.3 Weakly Polarized Quasars and BL Lacertae Objects

In the sample of eighteen sources which have been classified as BL Lacertae objects by at least some authors, fifteen have been found to have quite high fractions of their integrated polarized fluxes on milliarcsecond scales while three (1652+398, 1749+701, and 3C371) are essentially unpolarized on milliarcsecond scales. These three sources would not stand out from the other fifteen sources considered here solely on the basis of their total intensity images. Both 1652+398 and 3C 371 are at comparatively low redshift (z~0.03), and if they were at redshifts more typical of the sources considered here, they would be too dim to be included in this sample of objects. In addition, both these sources have emission lines that are rather stronger than is usual for BL Lacertae objects. These facts both suggest that a direct comparison between 1652+398 and 3C 371 and the other sources considered here may not be appropriate, so that their different VLBI polarization properties become less surprising. The case of 1749+701 is less easy to understand, however. It's spectrum is very nearly featureless (possibly indicating a redshift of z=0.77), so that from this point of view it appears more similar to the other BL Lacertae objects in the sample.

The two quasars 0153+744 and 1624+416 have also been found to be essentially unpolarized on milliarcsecond scales. It is interesting that both of these sources also have dramatically curved jets (through ~90° or more) compared to the minor bends typically seen in quasars. Is this a coincidence, or are the lack of polarization and large degree of bending related? It is possible, for instance, that there may be gas present which both causes bending of the jet and Faraday depolarization of the emission.

4.4 Is There More Than One Type of BL Lacertae Object?

Burbidge and Hewitt (1987,1989, these proceedings) suggest that there are actually three populations of BL Lacertae objects: those that lie in the centers of luminous elliptical galaxies (type 1), those that have optical properties of quasars and are at redshifts greater than ~.3 (type 2), and those for which the redshift is unknown (type 3). Of the BL Lacertae objects considered here, there are four of type 1 (1219+285, 1652+398, 1807+398, and BL Lac), nine of type 2 (0235+164, 0735+178, OJ287, 0954+658, 1308+326, 1749+096, 1749+701, 1803+784, and 1823+568), and five of type 3 (0300+470, 0306+102, 0454+844, 0754+100, 1538+149). Sources that display the characteristic polarization structure described above (appreciably polarized core components with no preferred electric vector orientation and jet components polarized with electric vector nearly parallel to the VLBI structural axis) occur in all three categories; of the three sources which are very weakly polarized on VLBI scales, two are of type 1 and the third is of type 2. There is no clear distinction in milliarcsecond polarization properties based on this categorization. Mutel (1990) examined various other radio properties of a slightly larger set of BL Lacertae objects of types 1 and 2 and also found no statistically significant differences between the two groups. It may be that there is some useful subdivision which may be applied to the class of BL Lacertae objects, but these results shed doubt on whether the particular division offered by Burbidge and Hewitt (based primarily on redshift) is appropriate.

4.5 Is It Possible to Unify BL Lacertae Objects and Quasars?

One way in which BL Lacertae objects and quasars could be unified would be if a significant fraction of BL Lacertae objects were actually images of more distant OVV quasars microlensed by stars in intervening galaxies, as has been suggested by Ostriker and Vietri (1985, 1990). This would be a natural way to account for the weakness of the line emission in these sources, since the more compact continuum region would be lensed while the larger line-emitting region would not be. Ostriker and Vietri (1990) have pointed out recent evidence that has suggested that the unusual behaviour of the BL Lacertae objects 0235+164 and 0537-441 may be associated with microlensing (Stickel, Fried, and Kuhr 1988); in addition, they claim that the redshift distribution of BL Lacertae objects may be understood as a consequence of microlensing.

Based on a comparison of the earliest VLBI polarization images of BL Lacertae objects and quasars which were made, Gabuzda *et al.* (1989) argued that the systematic differences in VLBI polarization structure observed in BL Lacertae objects and quasars ruled out this possibility, and pointed out that systematic differences in the arcsecond-scale structure and in the X-ray spectral indices similarly conflict with the picture of BL Lacertae objects as microlensed quasars. VLBI polarization information is now available for a much larger number of objects, and systematic differences between the polarization structures of BL Lacertae objects and quasars are even more apparent (Gabuzda *et al.* 1991; Cawthorne *et al.* 1991). In addition, the characteristic BL Lacertae VLBI polarization structure described above is observed in sources with a range of redshifts. While it seems quite possible that a few BL Lacertae objects are in fact microlensed images of more distant quasars, it is not the case that characteristic properties of BL Lacertae objects as a class may be explained in this way.

Perhaps the strongest candidate for a BL Lacertae object in which it appears lensing is likely is 0235+164. We detect no structure whatsoever in this source, so this is not one of the BL Lacertae objects displaying the VLBI polarization structure we find to be characteristic; therefore, the possibility that this particular object is lensed is not in conflict with our statement that BL Lacertae objects in general are not lensed. In fact, it is possible that lensed BL Lacertae objects may be detected through polarization-sensitive VLBI measurements, since such sources should display characteristics typical of BL Lacertae objects, with the exception that their VLBI polarization structure should be "quasar-like" (see above discussion).

Unification schemes suggesting that BL Lacertae objects are quasars oriented more closely to the line of sight (see E. Valtaoja, these proceedings, for example) suffer from the same problem: it is difficult to imagine such a scheme which is consistent with the largely disjoint nature of the VLBI jet polarization position angle distributions for the two classes. In addition, modelling done by Hughes, Aller, and Aller (1985, 1989, these proceedings) renders evidence that the angle between the jet axis and the line of sight for BL Lac and 1749+096 may be ~30-40°, suggesting that BL Lacertae objects may not in general be particularly close to the line of sight.

5. COMING ATTRACTIONS

Second epoch 6cm observations for ten BL Lacertae objects made in June 1989 are currently being analyzed, and should yield interesting information about the variability of VLBI polarization structure in these sources. In addition, these data should help resolve the question of whether characteristic component speeds in BL Lacertae objects and in quasars are different, as is suggested by the few BL Lacertae object component speeds which have thus far been determined (see Mutel 1990). Observations of some eight BL Lacertae objects at 3.6cm taken in June 1990 are also currently being analyzed; information about the frequency dependence of the characteristic VLBI polarization structure found at 6cm will help considerably in understanding the origin of this structure, particularly the polarization of the core components. Also in June 1990, Roberts *et al.* made nearly simultaneous (within a few days) observations of the quasars 3C345 and 3C273 at 6 and 3.6cm, which in principle should make it possible to make maps of the distribution of rotation measure over these sources. In November 1990, Cawthorne *et al.* made observations of the three quasars 4C71.07, 3C380, and 1928+738 in order to more accurately map their rich I and P structure. It is anticipated that sometime in 1991 the authors will make observations of several sources which have been found to vary in I and P on intraday timescales (Quirrenbach *et al.* 1989; Witzel 1990), with the intention of making independent I and P maps every ~4-6 hours; we hope this will enable us to directly determine the properties of varying components.

6. ACKOWLEDGEMENTS

We thank the US and European VLBI Networks for generous allocations of observing time, and the staff of the Mark III Correlator Facility at Haystack Observatory for their help and good humor. We also wish to thank our colleagues and collaborators Leslie Brown, David Roberts, and John Wardle at Brandeis University, who have been closely involved in the work described here.

7. REFERENCES

Aller, M. F., Aller, H. D., and Hughes, P. A. 1988, in *IAU Symposium 129, The Impact of VLBI on Astrophysics and Geophysics,* ed. M. J. Reid and J. M. Moran (Dordrecht: Reidel), p. 83.
Bridle, A. H. and Perley, R. A. 1984, *Ann. Rev. Astron. Astrophys.*, **22**, 319.
Burbidge, G. and Hewitt, A. 1987, *A.J.*, **92**, 1.
Burbidge, G. and Hewitt, A. 1989, in *BL Lac Objects,* ed. L. Maraschi, T. Maccacaro, and M.-H. Ulrich (Springer-Verlag), p.412.
Cawthorne, T. V., and Wardle, J. F. C. 1988, *Ap. J.*, **332**, 696.
Cawthorne, T. V., Gabuzda, D. C., Brown, L. F., Roberts, D. H., and Wardle, J. F. C. 1991, in preparation.
Cornwell, T. J., and Wilkinson, P. N. 1981, *M.N.R.A.S.*, **196**, 1067.
Cotton, W. D., Geldzahler, B. J., Marcaide, J. M., Shapiro, I. I., Sanroma, M., and Rius, A. 1984, *Ap.J.*, **286**, 503.
Gabuzda, D. C., Wardle, J. F. C., and Roberts, D. H. 1989b, *Ap. J.*, **338**, 743.
Gabuzda, D. C., Cawthorne, T. V., Roberts, D. H., and Wardle, J. F. C. 1989, *Ap.J.*, **347**, 701.
Gabuzda, D. C., Cawthorne, T. V., Roberts, D. H., and Wardle, J. F. C. 1991, in preparation.
Hughes, P. A., Aller, H. D., and Aller, M. F. 1985, *Ap. J.*, **298**, 301.
Hughes, P. A., Aller, H. D., and Aller, M. F. 1989, *Ap.J.*, **341**, 68.
Laing, R., 1980, *M.N.R.A.S.*, **193**, 439.
Mutel, R. L. 1990, in *Parsec-scale Radio Jets*, ed. J. A. Zensus and T. J. Pearson (Cambridge: Cambridge University Press), p. 98.
Ostriker, J. P. and Vietri, M. 1985, *Nature*, **318**, 446.
Ostriker, J. P. and Vietri, M. 1990, *Nature*, **344**, 45.
Pearson, T. J., and Readhead, A. C. S. 1981, *Ap. J.*, **248**, 61.
Pearson, T. J., and Readhead, A. C. S. 1988, *Ap. J.*, **328**, 114.
Quirrenbach, A., Witzel, A., Qian, S. J., Krichbaum, T., Hummel, C. A., and Alberdi, A. 1989, *Astron.Astroph.*, **226**, L1.

Roberts, D. H., Gabuzda, D. C., and Wardle, J. F. C 1987, *Ap. J.*, **323**, 536.
Roberts, D. H., {Potash, R. I., Wardle, J. F. C., Rogers, A. E. E., and Burke, B. F. 1984, in *IAU 110, VLBI and Compact Radio Sources*, ed. R. Fanti, K. Kellermann, and G. Setti (Dordrecht: Reidel), p. 117.
Stickel, M., Fried, J. W., and Kuhr, H. 1988, *Astron.Astroph.*, **198**, L13.
Wardle, J. F. C. and Roberts, D. H. 1988, in *IAU 129, The Impact of VLBI on Astrophysics and Geophysics*, ed. M. J. Reid and J. M. Moran (Dordrecht: Reidel), p. 143.
Wardle, J. F. C., Roberts, D. H., Brown, L. F., and Gabuzda, D. C. 1988, in *IAU 129, the Impact of VLBI on Astrophysics and Geophysics*, ed. M. J. Reid and J. M. Moran (Dordrecht: Reidel), p. 163.
Wardle, J. F. C., Roberts, D. H., Potash, R. I., and Rogers, A. E. E. 1986, *Ap.J. (Letters)*, **304**, L1.
Witzel, A. 1990, in *Parsec-Scale Radio Jets*, ed. J. A. Zensus and T. J. Pearson (Cambridge: Cambridge University Press), p. 206.

Radio Structures of Selected BLAZARS at Sub–arcsec Resolutions

Chidi E. Akujor

Astrophysics Group, Department of Physics and Astronomy, University of Nigeria, Nsukka, Anambra State, Nigeria.
and
University of Manchester, Nuffield Radio Astr. Labs, Jodrell Bank, U.K.

Abstract We present some results from our current radio imaging of blazars from a sample which we are studying based on MERLIN and VLA observations. A wide range of structures of jets and extended emission is found in blazars. Also, blazars variously exhibit properties usually associated with other types of radio sources such as flat spectrum, core–dominated quasars and compact steep–spectrum sources.

1 Introduction

The 'unification' of BL Lacertae type objects and optically violently variable (OVV) quasars under the 'blazar' umbrella based on some common properties – rapid optical and radio variability and strong and variable polarisation (O'Dell 1986) has remained debatable. One reason for the debate is the apparent disregard for the nature of optical emission lines in the unification and the general knowledge that the cores of all quasars including some lobe–dominated objects are probably highly variable and polarised.

However, one way of testing the efficacy of such 'unification' and all other associated 'schemes' is by comparing group properties including those that are thought to be independent of orientation, such as the nature and intensity of the associated extended radio emission. We are currently obtaining multi–frequency images of several powerful blazars in order to compare the extended radio and jet structures of the OVVs, BL Lacs and other flat spectrum quasars. Here we present images of four prototype blazars; 3C216, 3C446, 4C49.22 and 3C371 obtained with MERLIN and the VLA.

2 The Images

a) 3C216 is a complex radio source associated with a blue stellar object with a redshift, z=0.688. It has been classified as a blazar (Angel and Stockman, 1981;

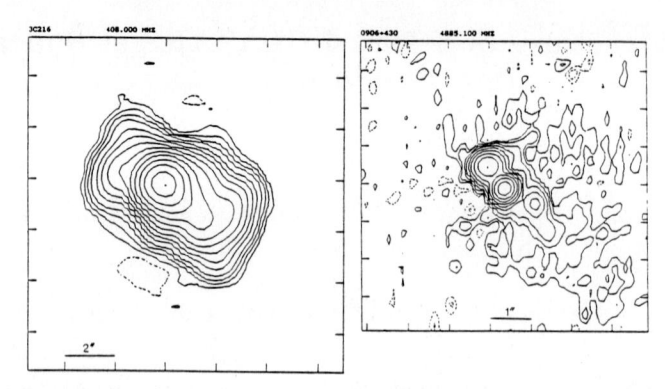

Figure 1: (a) MERLIN map of 3C216 at 408 MHz; res. beam is 1 arcsec; contour levels are 0.01, 0.025, 0.05, 0.1....99 % of peak value, 3.24 Jy/beam (b) VLA map at 5 GHz; res. beam is 0.35 arcsec; cont. levels are 0.25, 0.5, 1...99 % of peak value, 0.85 Jy/beam.

Impey and Neugebauer, 1988) due to its optical variability and strong, variable optical polarisation (Kinman, 1978). Its radio spectrum, α =0.9 (S$\propto \nu^{-\alpha}$), and compact radio structure, $\theta = 2.5$ arcsec, suggest it is a compact steep–spectrum source (CSS), (Fanti et al. 1990). Figure 1a is a VLA map at 5 GHz which shows three dominant components while the MERLIN map at 408 MHz shows an extended halo ~ 10 arcsec across. An EVN map (Akujor & Porcas, in preparation) shows a bent jet that is inclined at $\sim 90°$ to the main source axis.

b) 3C446 is a quasar with redshift, z=1.404 usually described as both an OVV and a BL Lac because it shows rapid variability at cm wavelengths and is erratic in showing emission lines. Our 5 GHz MERLIN image (Fig 2a) shows an extended jet with a sharp bend and another extension northwards from the core (counter–jet ?), while the 1.6 GHz (Fig 2b) image shows a surrounding halo which is more closely aligned with the northern extension. The arcsec radio structure of 3C446 had been mapped by Browne et al. (1982) at 408 MHz and clearly reveals the extent of this halo.

c) 4C49.22 is quasar with redshift, z=0.334 and usually classified as a blazar (Antonucci & Ulvestad, 1985). Fig 3a shows an extended bent and knotty jet in 4C49.22, while a tapered VLA map at 1.4 GHz reveals a very fat halo with a well defined boundary. Many blazars have been found to have core–halo morphology (Antonucci & Ulvestad, 1985) expected from the projection of twin lobes of the extended sources whose radio emission is predominantly directed towards the observer (Blandford & Rees, 1978). The core–halo structure of 4C49.22 therefore fits well with this scenerio, but it needs be mentioned that 4C49.22 barely satisfies the three primary properties of blazars (Impey & Neugebauer, 1988): a) strong compact radio emission, b) optical variability and c) high optical linear polarisation.

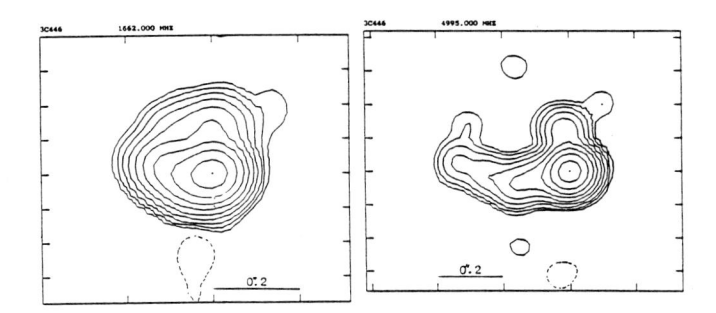

Figure 2: MERLIN maps of 3C446 at (a) 1.6 GHz; res. beam is 0.2 arcsec; contour levels are 0.25, 0.5, 1,...99 % of peak value, 3.16 Jy/beam (b) 5 GHz; res. beam is 80 mas and contour levels are 0.25, 0.5, 1,...99 % of peak value, 2.14 Jy/beam

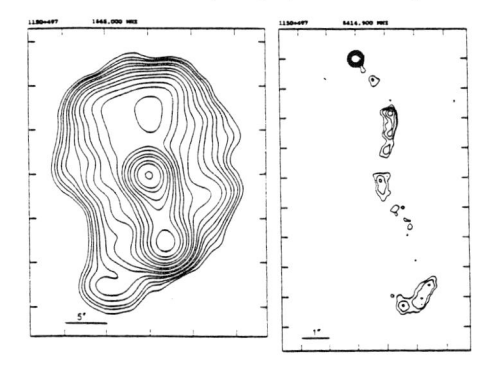

Figure 3: (a) VLA maps of 4C49.22 at: a) 1.4 GHz, res. beam is 2 arcsec; cont. levels are 1, 1.5, 2, 3, 4, 6, 8, 12,...388 mJy/beam and peak value is 407 mJy/beam. (b) 8.4 GHz; res. beam is 180 mas; cont. levels are 0.1, 0.2, 0.4,...25.6 % of peak value, 511 mJy/beam

d) 3C371 is a well known BL Lac (z=0.0508) associated with an elliptical galaxy, which Wrobel & Lind (1990) have recently shown to have a faint twin lobes straddling an extended jet. Our maps (Fig. 4) only detect the jet out to 5 arcsec from the core, after which it fades in brightness but extends out to about 25 arcsec as seen in the Wrobel & Lind VLA maps. The jet in 3C371 is knotty and wiggly like the jet in 4C49.22.

3 A CSS Connection?

There seems to be a lot in common between some blazars and compact steep–spectrum sources typified by the classification of 3C216 (and others like CTA102)

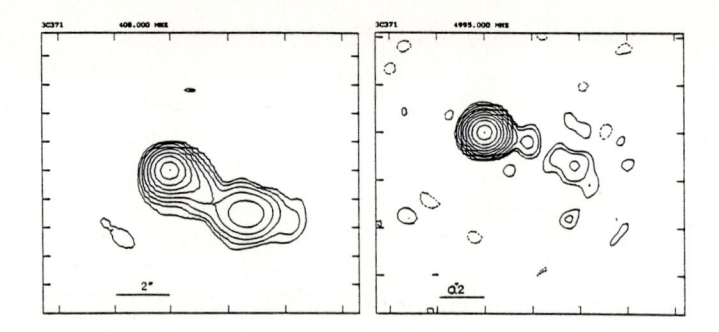

Figure 4: MERLIN maps of 3C371 at a) 408 MHz; res. beam is 0.75 arcsec; cont. levels are 0.5, 1, 2, 4,...99% of peak value, 980 mJy/beam (b) 5 GHz; res. beam is 80 mas and cont levels are 0.1, 0.25, 0.5, 1,...99 % of peak value, 1.427 Jy/beam

as both. Whereas 3C446 is a small source ($\sim 10kpc$), has a steep low frequency spectrum and a structure remarkably similar to 3C216, it is not classified as a CSS. The large bends of the inner jets and the misalignments between the kpc and large–scale structures of both sources are typical of CSS objects. Moreover, there are other blazars and core–dominated sources, e.g. 3C345 and 1642+690 (with extended haloes at lower frequencies and small linear sizes) which could come into CSS samples depending on the selection frequency. Fejes, Porcas & Akujor (1991) argue these classes need not be mutually exclusive as the percieved structural differences may depend strongly on core prominence which determines blazar behaviour.

Also Wehrle *et al.* (1991) find that CSSs are on the average larger in size than the flat–spectrum quasars, which are in turn larger than BL Lacs. This comparison is based on the premise of a homogeneous class of CSS quasars with extended structure. We find that there appears to be two classes of CSSs quasars: (a) those with extended low frequency structure which tend to have prominent cores and show superluminal behaviour, e.g 3C380, 3C147 and 3C309.1, some of which may have blazar cores, and (b) the CSS objects with very weak cores, no large–scale haloes, sometimes with compact lobes, e.g. 3C138 and 3C298. The later group seems to be related more with CSS galaxies.

4 Disrupting Jets?

Akujor & Garrington (1991) explain the peculiar jet structure of 4C49.22–bright knots, wiggles, bends without bright terminal hotspot–as possibly due to either disruption of hydrodynamically unstable jet or the relatistic boosting of the jet over and above that of the hotspot. The earlier possibility is inferred because polarisation data suggests that the surrounding medium of this source is unusually sparse. The jet in 3C371, 3C446 and to some extent 3C216 also show the same characteristics, fading drastically after a few arcsecs and end without a bright hotspot.

But a number of prominent blazars do show such jet wiggles, knots and bends, and terminate with bright hotspots. Since jet instabilty and disruption depends on the relative densities of the jet and its surrounding medium, it is not surprising that a wide range of jet features are observed in blazars, as they may be sitting in different sorts of media, perhaps unrelated to the blazar action. It is, however necessary to image a large number of sources in total intensity and polarisation in order isolate the factors responsible for the different structures.

Acknowledgment I am grateful to Drs I. Browne, R.W. Porcas & S. Garrington for useful discussion and permission to use their data and E. Valtaoja for conference support and sauna.

References

Akujor, C.E & Garrington, S.T. 1991, *Mon. Nots Roy. Astr. Soc.*, **250**, 644.

Antonucci, R.R.J. & Ulvestad, J.S. 1985, *Astrophys. J.*, **294**, 158.

Angel, J.R.P. and Stockman, H. S. 1980, *Ann. Rev. Astr. Astrophys.*, **18**, 321.

Blandford, R.A. & Rees, M.J. 1978, *Phys. Scripta*, **17**, 265.

Browne, I.W.A., Clark, R.R., Moore, P.K., Muxlow, T.W., Wilkinson, P.N., Cohen, M.H., and Porcas, R.W. 1982, *Nature*, **299**, 788.

Fanti, R., Fanti, C., Spencer, R.E., Nan Rendong, Parma, P., Van Breugel, W.J.M., Venturi, T. 1990, *Astron. Astrophys.* **231**, 333.

Fejes, I., Porcas, R.W. & Akujor, C.E. 1991, *Astron. Astrophys.*, in press.

Kinman, T. 1976, *Astrophys. J.*, **205**, 1.

O'Dell, S.L. 1986. *Publ Astr. Soc. Pac.*, **98**, 140.

Wehrle, A.E., Cohen, M.H. & Unwin, S.C., Aller, H.D., Aller, M.F. & Nicolson, G. 1991, preprint.

Wrobel, J.M. & Lind, K.R. 1990, *Astrophys. J.*, **348**, 135.

Jets in a Sample of Superluminal Active Galactic Nuclei.

IRENE CRUZ-GONZALEZ AND RENE CARRILLO

Instituto de Astronomía, UNAM. México.

INTRODUCTION

Relativistic jet models (*i.e.* Blandford, Mc Kee, Rees 1977, Marscher 1978, Blandford and Königl 1979) have been proposed to describe the observed properties of variable extragalactic radio sources associated with the nuclei of galaxies, quasars and blazars. In this work, we assume naïvely that the jet models are correct and decided to use as input parameters, observational data available in the literature for a large sample of jets in AGNs. We compared the observational parameters of several jets and the predictions of the relativistic jet models, *i.e.* using the observational parameters of the sources and their jets: jet velocity β_{ob}, redshift z, luminosity distance to the source D_l, flux density S_{ob}, opening semiangle of the jet ϕ, and jet size R; we deduced the parameters predicted by the models for each source jet: the source luminosity, the maximum brightness temperature, the characteristic frequency of the synchrotron radiation and the magnetic field. We present a sample of 33 extragalactic sources with radio jets and superluminal motions, that includes different types of AGNs: 22 quasars, 6 blazars, 1 Seyfert galaxy, 3 radiogalaxies, and 1 empty field radio source.

THE MODEL

The relativistic jet model was later developed by Blandford and Königl (1979), hereafter BK. In the present paper we assume that the BK model is corrects and pursue a test. With this in mind, we used the available observational data from the literature of several sources to determine the parameters required by the expressions given by BK. The expressions used for the model application are briefly described in Cruz-González and Carrillo (1991), where a full version of the present work is presented.

SUMARY OF RESULTS

1. The observed luminosities at 10 GHz (L_{ob}) and the relativistic jet model luminosities (L_{th}) follow a correlation: $L_{th} = 2.0 \times 10^{17} L_{ob}^{0.65}$ ($r^2 = 0.85$). This result strongly favours the models.

2. The maximum brightness temperatures predicted by the models are in the range $10^{11} - 10^{13}$ °K. All the sources have temperatures close to the inverse Compton catastrophe values, indeed most sources have $T_{max} < 10^{13}$ °K. As other authors have shown from variability studies the temperature derived are as high as 10^{15} °K. It seems that the relativistic jet models predict lower temperatures and may solve a long standing problem on the physics of this sources.

3. The observed velocity, β_{ob}, and the brightness temperature, T_{max}, are correlated, c.f. Fig. 1: $(T_{max}/10^{12}) = 0.19\beta_{ob} + 0.08$ ($r^2 = 0.92$), i.e. the higher the velocity the higher the predicted temperature.

4. The observed velocity and the characteristic frequency are correlated: $v = 1.92 \times 10^{10} \beta_{ob}^{2.23}$ ($r^2 = 0.83$), i.e. the higher the observed velocity the higher the maximum frequency.

5. The Lorentz factor, γ_{jet}, for each source in our sample was calculated using the approximation $\gamma_{jet} \approx \beta_{ob}$, and is compared with the observed luminosities at 10 GHz, c.f. Fig. 2, 60 μm, c.f. Fig. 3, 1 Kev. We notice that most sources seem to follow a trend, i.e. $\gamma_{jet} = 3.76 \times 10^{-12} L_{10\ GHz}^{0.28}$ ($r^2 = 0.51$), $\gamma_{jet} = 5.59 \times 10^{-15} L_{60\ \mu m}^{0.3}$ ($r^2 = 0.47$). Although, this possible trend needs confirmation, we are tempted by this result because it indicates that more luminous sources produce more powerful jets, in such a way that the jets in Seyfert galaxies are much weaker than in other AGNs due to a less powerful central source. Furthermore, it links the emission processes with the kinematics of the jets.

6. The characteristic frequency when compared with the above luminosities also shows an indication of a possible correlation, in the sense that larger characteristic frequencies correspond to larger luminosities (10 GHz, 60μm, and 1 Kev).

7. The radio luminosity at 10 GHz when compared with the other band luminosities (L_{IRAS}, and L_X, shows a correlation.

The last two results indicate a possible relation between the emission mechanisms that are producing the radio properties of the superluminal jet sources and their X-ray and FIR observed luminosities.

REFERENCES

Blandford, R.D., Königl, A.: 1979, *Astrophys.J.*, 232, 34.

Blandford, R.D., McKee, C.F., Rees, M.J.: 1977, *Nature*, 211, 468.

Cruz-González, I., Carrillo, R.: 1991, submitted to *Rev. Mexicana Astron. Astrof.*

Marscher, A.P.:1978, *Astrophys. J.*, 219, 392.

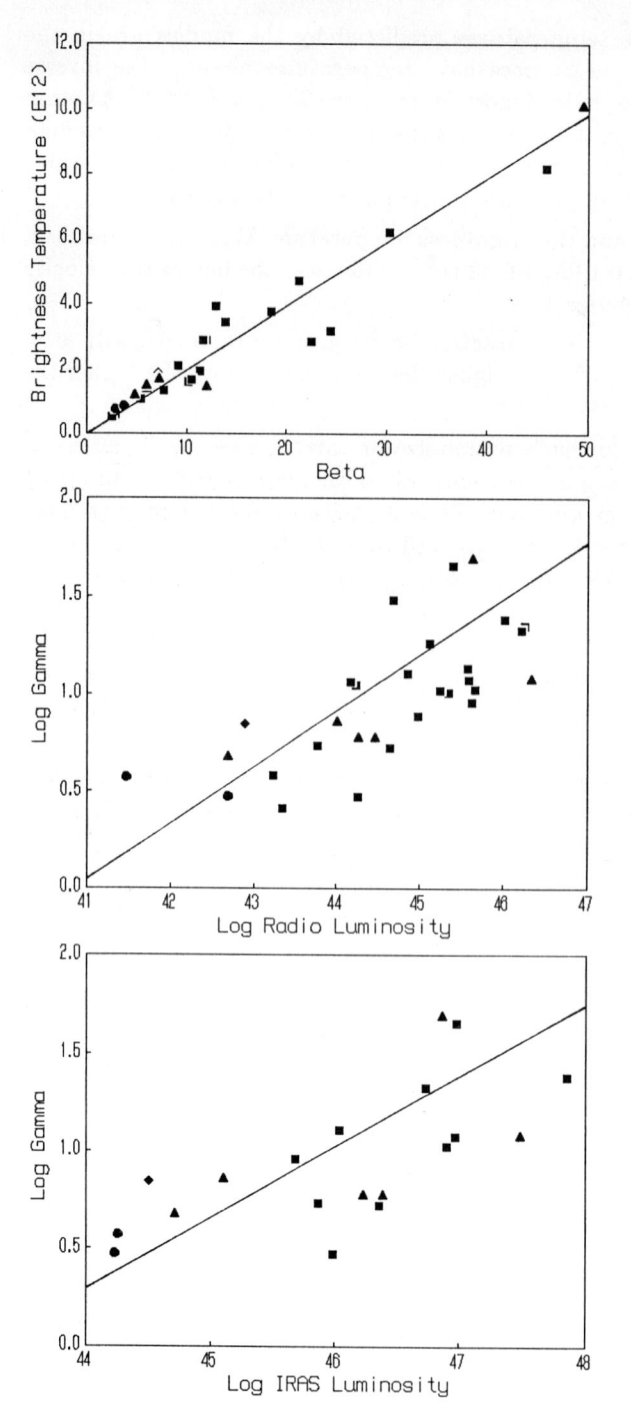

Figure 1. Correlation between the observed velocity β_{ob} and the maximum brightness temperature T_{max}. The line represents the predicted correlation: $T_{max}/10^{12} = 0.19\,\beta_{ob} + 0.08$ ($r^2 = 0.92$) [■ quasars, ▲ blazars, ● radiogalaxies, ◆ Seyfert Galaxy].

Figure 2. Comparison between the Lorentz factor γ_{jet} and the 10 GHz luminosity. The line represents the correlation $\gamma_{jet} = 3.76 \times 10^{-12} L_{10GHz}^{0.28}$ ($r^2 = 0.51$).

Figure 3. Comparison between the Lorentz factor γ_{jet} and the 60 μm luminosity. The line represents the correlation $\gamma_{jet} = 5.59 \times 10^{-15} L_{60\mu m}^{0.3}$ ($r^2 = 0.47$).

Radio Structure of Variable AGN and Cosmological Tests

L. I. GURVITS, N. S. KARDASHEV, A. P. LOBANOV

Astro Space Center of P. N. Lebedev Physical Institute
Leninsky pr., 53, Moscow, 117924, U.S.S.R.

ABSTRACT Some expected features of the radio structure in blazars are described as a target for cosmologically aimed VLBI observations. The effects of relativistic distortion of the apparent shape of the radio structures of blazars (and other types of active galactic nuclei) are proposed to be a combination of standard yardstick and standard candle for measurements of cosmological parameters.

1 INTRODUCTION

Observational cosmology encompasses a very wide field of experimentation that spans all time scales of the age of the Universe since its first seconds until the present time. This contribution is restricted to a narrow understanding of observational cosmology that operates in those parts of the Universe where one may find discrete sources. At this time, the appropriate range of redshifts is from 0 to about 5. (The authors embrace the common cosmological interpretation of redshifts but would be pleasantly surprised to see comprehensive and noncontradictory evidence for other explanations of redshifts.) We will assume that "cosmological" tests mean at least measurement, with the highest accuracy possible, of such important parameters as the Hubble constant H_0 and the deceleration parameter q_0.

The best way to solve the abovementioned problem is to identify a standard yardstick or a standard candle. There have been many attempts to use different classes of extragalactic sources as standard objects – galaxies, their nuclei and clusters. However, the main problem in such investigations concerned confusion between "evolution-free model tests" and "model-free evolution tests" (von Hoerner, 1974). As a result, there are no reasonably reliable objects, distributed over a wide enough range of redshifts, that can be considered as standard candles or yardsticks.

An interesting approach to the problem has been proposed by Pelletier and Roland (1989, hereafter referred to as PR). Their idea is to find some combination of a standard candle and standard yardstick to apply to active galactic nuclei (AGN). The philosophical background of their approach is based on well-known ideas of

Doppler boosting (which could explain a jet's apparent asymmetry) by Shklovsky (1963, 65), variability models by Rees (1966, 67), and the phenomenon of apparent superluminal velocity in some AGNs that was theoretically predicted by Rees (1967) and observationally discovered in 1972. The model proposed by PR is based on the use of the jet/counterjet brightness ratio (a standard candle in some sense) in the VLBI structure of an AGN to choose those objects whose jets lie in a plane perpendicular to the line of sight. Then, if the velocity of matter in the jets is assumed, the distance can be estimated by monitoring the material flow during a fixed time (standard yardstick). The efficacy of proposed method is demonstrated by PR who estimated the Hubble constant as $H_0 = 100 \pm 30$ km s^{-1} Mpc^{-1} (for $q_0 = 1/2$ and published VLBI data).

We devote our attention to the Pelletier and Roland (1989) paper because our approach is logically similar to theirs: to find some reasonable combination of astrophysical objects or phenomena that may serve as standard candles or yardsticks with which to measure H_0, q_0, and, maybe even more sophisticated cosmological parameters such as Λ-term.

The appearance of this paper in this volume is explained by the fact that we propose to use blazars as such objects (or phenomena, as it is agreed by a major part of this conference).

2 PHENOMENOLOGICAL MODEL OF AN EXPANDING RELATIVISTIC SOURCE

To build up the phenomenological model of an expanding relativistic source, we take into account the most common observable features of AGN phenomena to fit them by simple suggestions on emission properties. We assume that these sources eject spherical blobs that expand and move simultaneously along the line of ejection with relativistic velocity. These blobs are optically thick in their radio emission ($\tau \simeq 10^2$–10^9, Begelman et al., 1984) up to certain time scales and dimensions. Since we are interested only in the stages near the beginning of expansion, the large values of τ are appropriate. Considering only the general radio brightness of these blobs obviates the need to account for any polarization effects or the spectral distribution of emitting radiation.

With these constraints, we assume a Gaussian profile for radio brightness distribution in the blob, described by the following formula

$$I(x) = \begin{cases} I_0 \cdot \exp(-x^2/2\sigma^2) \ , & |x| \leq r \\ 0 \ , & |x| > r \end{cases}$$

where I_0 is the peak emission intensity, r is the angular radius of the blob in the picture plane, and x is the angular coordinate in the picture plane.

Let us assume that the blob described above grows and/or moves with a velocity (moving plus expanding ones) that reaches relativistic value. Figures 1 and 2 show examples of brightness distribution profiles obtained for different kinematic and geometrical parameters of the source model. These examples are simulated using the same simple formula (Rees, 1967) that explains apparent superluminal velocities in some sources. Both pictures were drawn with arbitrary units.

One needs considerable imagination to recognize an initial Gaussian distribution in Figure 1, where different curves correspond to the different values of blob's expansion velocity. As one can see, there is a sharp edge in the brightness distribution on the radius that corresponds to the polar angle $\cos^{-1}(v/c)$ with the line of sight in the source frame. (Inner parts of the blob beyond the maximal apparent size are not seen due to the requirement $\tau > 1$.) The obtained distribution can be approximated as a quasi-rectangular distribution for a wide range of kinematic parameters.

Figure 2 shows the dependence of the brightness profile from σ parameter in the Gaussian distribution (in units of co-moving source radius r).

The relativistic distortion on the structural pattern of the source may result in the visible size of a source being significantly larger than the one for the nonrelativistic source. However, the most important feature of the brightness distribution of the relativistically growing blob is the existence of the sharp edge and quasi-rectangular profile.

3 OBSERVATIONAL MANIFESTATIONS

The angular sizes of the blobs discussed above are typical of those measured with VLBI. An interferometric technique applied to source structure investigation in mathematical terms is described by Fourier transformation: the source structure is a core of transformation and observable data are the amplitude and phase of Fourier transformation (we omit here details that are not of principal interest in the context of the paper). The quasi-rectangular distribution shown in Figure 1 should be readily detected because it is so different from the other "usual" cases of sources with smooth edges. And peculiarities in the brightness distribution such as those we considered are so specific that they could be detected in principle even if the observations are performed with only a two-element interferometer.

Now let assume that we do detect such a source with a sharp edge. Its relativistically expanded nature will allow us to observe the growth of the visible angular diameter of the source at the usual time- and size- scales for superluminal sources (milliarcseconds during few years). Thus, the possibility exists to apply a known approach to extract cosmological parameters H_0 and q_0 from measurements of

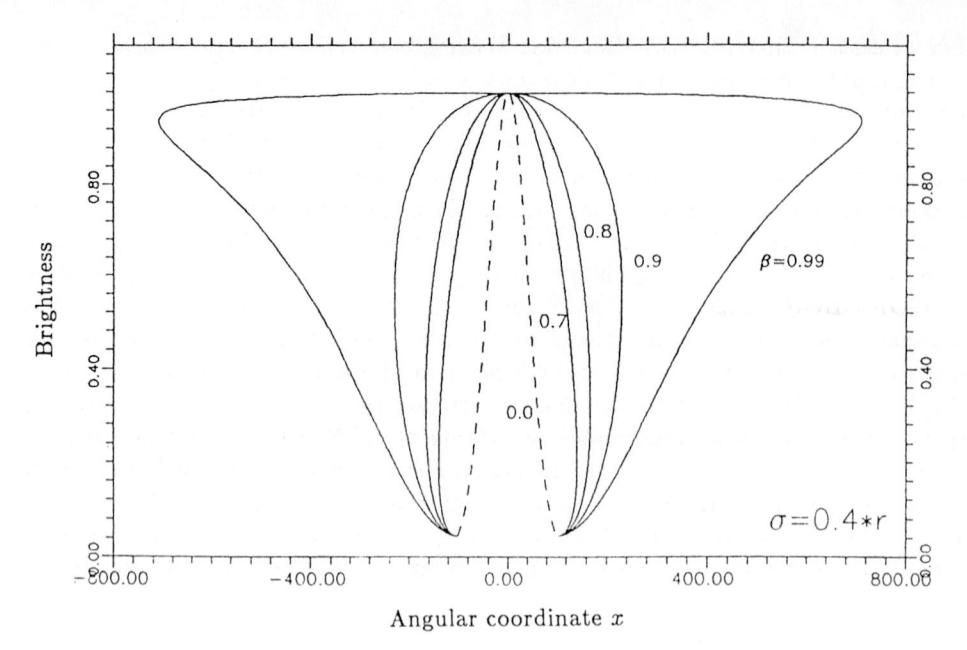

Fig. 1. Profiles of the brightness distribution for an expanding blob. The initial Gaussian nonexpanding profile is shown by a dashed line. Solid lines show profiles for the different expansion velocities $\beta \equiv v/c$. Axis scales are arbitrary.

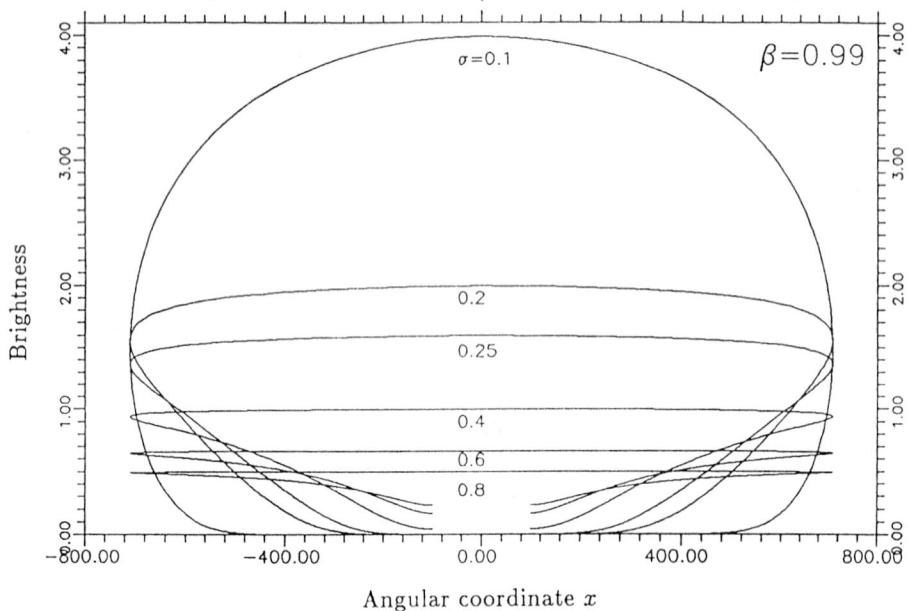

Fig. 2. Profiles of the brightness distribution for a blob expanding with velocity $\beta = 0.99$ for varying parameters σ/r of the initial (nonexpanding) blob. Axis scales are arbitrary.

proper motion parameter μ (PR and references therein).

Further speculations as to how to extract at least H_0 and q_0 from this kind of proper motion could be done very easily. We will leave this topic for investigators who will deal with real observational data. Let us mention in the end that one may find confirmations of quasi-linear growing of the ejected components diameter in the nucleus of extensively studied superluminal source 3C 345 (Biretta et al., 1986).

4 CONCLUSIONS

It seems that the method discussed here as applied to blazars (or, wider, all possible types of radio-variable AGNs) can be facilitated by the idea proposed by Valtaoja (1991) to monitor variable AGNs, in conjunction with observational programs of space VLBI missions. Continuum monitoring of AGNs can provide information on the best observational "season" for VLBI observations of relativistic expansion of discussed outbursts. Such an observational program appears to be an ideal application of a high orbiting VLBI mission that will have extremely high angular resolution in one dimension through the use of an orbiting telescope and one ground radio telescope.

ACKNOWLEDGMENTS. The authors would like to thank Jacques Roland and Boris Komberg for stimulating discussions on the subject of this paper. One of the authors (LIG) would like to thank the LOC of the conference, and especially Esko and Leena Valtaoja, for their kind hospitality and organizing efforts and Turku University for financial assistance during the conference.

REFERENCES

Begelman, M. C., Blandford, R. D., Rees, M. J. 1984, Rev Mod Phys 56, 256
Biretta, J. A., Moore, R. L., Cohen, M. H. 1986, ApJ 308, 93
Pelletier, G., and Rolland, J. 1989, AA 224, 24 (PR)
Rees, M. J. 1966, Nature 211, 468
Rees, M. J. 1967, MNRAS 135, 345
Shklovsky, I. S. 1963, Sov Astronomy AJ 7, 748
Shklovsky, I. S. 1965, Sov Astronomy AJ 9, 22
Valtaoja, E. 1991, in *Frontiers of VLBI*, eds. H. Hirabayashi, M. Inoue, and H. Kobayashi, Tokyo University Press, in press
von Hoerner, S. 1974, in *Galactic and Extra-galactic Radio Astronomy*, eds. G. L. Verschuur and K. I. Kellermann, Springer-Verlag, p. 353

Blazar Models

C.-I. BJÖRNSSON

Stockholm Observatory, S-133 36 Saltsjöbaden, Sweden

ABSTRACT Temporal and spectral variations in flux and polarization are used to constrain potential models for blazars. It is argued that simple "two-component" models are too crude to account for many of the observed properties; in particular, when applied to blazars as a group. Instead, a one component, "Christmas tree", model is favoured, in which the geometry of the magnetic field plays a central role. The importance of the increasing observational evidence for a close connection between the blazar emission region and at least part of the associated, unresolved VLBI-core is emphasized. The physical relationship between these two emission regions is briefly explored in the context of a dynamically dominant magnetic field containing electrons with an anisotropic pitch-angle distribution.

1 INTRODUCTION

Angel & Stockman (1980) in their review defined blazars as those objects belonging to the subset of active galactic nuclei and quasars which have optical polarization larger than $2-3\%$. This still seems to be a physically relevant definition, although the observational evidence indicates that one thereby selects a process rather than an object; for example, strong line objects (OVV's and HPQ's) tend to have both X-ray properties (Worral 1989; Maraschi, this conference) and a redshift distribution (Burbridge, this conference) different from weak line objects (BL Lac's). In this review it is assumed that in studying the blazar process no distinction needs to be made between the different types of objects in which it occurs. Blandford & Rees (1978) set the frame used in most papers to discuss the blazar phenomenon. In their optically thin synchrotron model, the most severe constraints come from a combination of the observed rapid variability and the lack of appreciable depolarization due to cold (i.e. non-relativistic) particles, expected to be present due to the short synchrotron cooling time needed to avoid excessive Compton losses. Since $\tau_F \approx 2 \cdot 10^{-2} B \tau_C / \nu_{15}^2$, where τ_F is the Faraday rotation in radians, τ_C is the Thomson depth due to cold particles and $\nu_{15} = \nu/10^{15}$Hz. The value of the magnetic field strength (B) determines whether Faraday rotation or Thomson scattering dominates the depolarization in the optical. However, one should note that for plausible values of B ($1 \sim 10^4 G$), the need for relativistic effects and/or reacceleration can not be circumvented by invoking an electron-positron plasma.

The close connection between blazars and compact radio sources (CRS) has only become evident in the last few years. Although it has been known for a long time that blazars always are associated with CRS, it was the work of Fugmann (1988) and Impey & Tapia (1988) which first gave strong indications that also the reverse is true; in complete samples of CRS it was found that at least 2/3 of all sources are blazars. The direct link between blazars and CRS is emphasized by the observations of Landau et al. (1986) and Impey & Neugebauer (1988), which show that these sources have smooth spectra from radio/mm to optical; this is in marked contrast to non-blazar/CRS, which reveal evidence for spectral breaks or several components. In fact, there are indications that in some objects or at some occasions, at least part of the radio and optical emission regions coincide; Kikuchi et al. (1988) found that in OJ 287 a large swing in position angle (PA) occurred simultaneously in radio and optical and at this conference Krichbaum and Wagner have presented observations which suggest a large degree of simultaneity between micro-variability (hours-days) in these same frequency ranges in 0716+714. Furthermore, Gabuzda, Wardle, & Roberts (1989) detected intraday variations in the unresolved VLBI-core of 0735-178. Using the 100m Effelberg telescope Witzel & Quirrenbach (1990) found that similar small amplitude ($\leq 10-20\%$) intraday variations are quite common in CRS. This time scale is much shorter than the one usually associated with major outbursts in CRS but typical for blazars.

Circular polarization (CP) in CRS has a clear signature; for a given source, CP shows a preferred sign both as a function of time (Komesaroff et al. 1984) and frequency (Weiler & de Pater 1983). CP most likely comes from the optically thick core. This is corroborated by the fact that CP exhibits more rapid and larger relative variations than do either the degree of linear polarization (P) or the flux (Komesaroff et al. 1984). Björnsson (1990) has used these properties of CP to argue that although the magnitude of CP is probably due to radiative transfer, the preferred sign is most easily understood as due to the magnetic field geometry, i.e. the magnetic field in the unresolved VLBI-core is not turbulent but has a large scale coherence. Since a dynamically dominant magnetic field is expected for the optical emission region (Blandford & Rees 1978), the CP-properties further emphasize the affinity between the optical and radio emission regions. Incidentally, if the whole radio-to-optical spectral range is produced within the same region, the case for a coherent emission mechanism is weakened; most mechanisms are narrow-band ($\Delta\nu/\nu < 1$) and the existence of a needed radius-to-frequency mapping is made less likely if the emission region is independent of frequency.

The increasing observational support for the view that blazars and CRS are different aspects of the same phenomenon makes it inappropriate to leave out CRS in a discussion of blazar models (or vice-versa). However, since this evidence has become available only recently, its implications for blazar/CRS models have not been explored in any detail in the literature. Hence, in this review I have chosen to focus on blazars. A basically phenomenological discussion of variations in blazar polarization with time and frequency is given in section 2; the aim being to determine to what extent the properties of these variations can add to the Blandford&Rees-picture of blazars. Some dynamical implications are mentioned in section 3 in connection with an appraisal of the importance of shocks both for the optical and radio regimes. The physical relationship between blazars and CRS is tentatively explored; a model is presented for temporal variations in CRS, which draws its main features from the deduced properties of blazars.

2 BLAZAR POLARIZATION

2.1 Time variations

It was clear early on that stochasticity is an important property of blazars; variations in the Stokes parameters revealed no apparent correlations (e.g. Miller 1978), for example, interday variations are not due to the addition or subtraction of a subcomponent (Holmes et al. 1984a). However, as the number of observations grew, evidence for some regular features emerged. The existence of *a dynamically dominant magnetic field*, since: (i) Large values of P are observed. (ii) Large Compton losses are absent (Blandford & Rees 1978). *The value of PA is not random*, since: (i) When the number of observations is large enough a preferred value of PA usually becomes apparent (Hagen-Torn, Marchenko, & Yakovleva 1985, 1986). (ii) Large scale variations of PA & P occur. Jones et al. (1985) modelled large PA-changes as accidental rotations in a random walk process. However, as pointed out by Phinney (1985), it is not only the occurrence of a rotation which is important but also its smoothness. For every "smooth" rotation a large number of "jagged" ones should be observed. Although no quantitative analysis of the observations has been attempted, many rotation events seem surprisingly smooth. (iii) For purely random events one expects $P \sim P_0 \sigma(F)/\langle F \rangle$, where P_0 is the intrinsic degree of polarization of the individual subcomponents, $\sigma(F)$ is the flux-dispersion and $\langle F \rangle$ the average flux. It is often found that $P_0 > 70\%$ (e.g. Kikuchi et al. 1976; Brindle et al. 1985), indicating that the subcomponents are not chosen randomly.

The quantity of data needed to make a more extended analysis meaningful exists for one object only, namely BL Lac itself. Although one should be careful not to over-interpret the conclusions reached for one object, in the spirit of the assumption of a common underlying blazar process, it might be instructive to discuss some of the models proposed in the literature in the light of what is known about BL Lac. The time variations considered are those where the average flux changes on a time scale much longer than the flux itself. Typically, these flux variations are smaller than a few tenths of a magnitude and the relevant time scales less than a few weeks. These time variations appear to be different in character from "bursts", where the flux changes monotonically by large amounts ($> 1^m$).

In two global campaigns BL Lac was monitored intensively for approximately one week at a time; during the first campaign the source was in a high state (Moore et al. 1982), while during the second one in a low state (Brindle et al. 1985). On both occasions the power spectrum for the flux variations was roughly consistent with *random walk* (i.e. "f^{-2}"-spectrum) down to time scales of 5 hours where measurement errors started to dominate. Using single night observations, Wiita (this conference) has shown that this behaviour is a common feature of blazars. Furthermore, using polarimetry, Moore et al. (1982) argued that the power spectrum flattens for times longer than 20 days and Moore, Schmidt, & West (1987) observed a steepening for times shorter than 3-5 hours. Even if the subcomponents turn on randomly, the f^{-2}-spectrum is only relevant for time scales larger than the turn-on time and smaller than the decay time. Hence, the observations of BL Lac indicate that the decay time of the individual subcomponents is much longer ($\sim$ a factor 100) than the turn-on time; this sets important constraints on potential acceleration mechanisms. Also, the turn-on time sets an upper limit to the size of an individual subcomponent ($\sim 5 \cdot 10^{14}$ cm).

Although the high state variations both in flux and polarization were consistent with a random walk process ($\sim$ 100 subcomponents each of which $\sim$ 70% polarized), in the low state a preferred value of PA was apparent. Brindle et al. (1985) interpreted this as due to the existence of a constant, polarized component in addition to the randomly varying subcomponents. They showed that such a *superposition of a constant component on a random walk* gave a reasonable description of the observations. In such a model $\sigma(PA)$ and $\sigma(P)/\langle P \rangle$ should increase and $\langle P \rangle$ change with increasing flux. The extensive observations of BL Lac done by Hagen-Torn, Marchenko, & Yakovleva (1985, 1986) show that none of these expectations are born out; except that the preferred value of PA is more pronounced at very low flux levels, $\sigma(PA), \sigma(P)$ and $\langle P \rangle$ are roughly independent of flux. This indicates that the preferred direction is associated with the distribution of subcomponents, i.e. their PA's are not chosen randomly.

The strong evidence for a large scale magnetic field in the emission region makes it a likely cause for the preferred PA. This leads to a *Christmas tree model*, where subcomponents turn on randomly within a given magnetic field configuration. The constancy of $\sigma(PA), \sigma(P)$ and $\langle P \rangle$ follows if flux variations are caused primarily by variations in the flux of the subcomponents and not their number. The occurrence of large values of P mainly when PA is close to its preferred value is also explained in such a model. Although the size of the subcomponents is small, they can be distributed over a much larger region. This is an attractive feature, if the first indications of a close connection between the optical and radio regimes are strengthen in future observations.

2.2 Frequency variations

With the advent of multi-channel polarimeters, it has become clear that frequency dependent polarization (FDP) is an important property of blazars; for example, Ballard et al. (1989) found it in $\sim 40\%$ of their observations. Variations in polarization with time can have a variety of causes; for example, time dependent injection of particles, cooling, a finite source size giving rise to light travel time effects or the properties of the emission region might change. In this sense, FDP is "cleaner" since a necessary condition for it to occur is a non power-law distribution of electrons. Analysis of FDP, therefore, holds out the possibility for a more direct and unambiguous route to an understanding of the blazar process.

The first attempts to explain FDP were done using *two power-laws with different PA's and spectral indices*. Although such a model can account for some of the individual observations, it encounters problems when judged against the properties emerging for blazars as a group: (i) Adding two power-laws, by necessity, gives rise to a convex spectrum ($d\alpha/d\nu < 0$, where $\alpha \equiv -d\ln F(\nu)/d\ln\nu$). With the exception of a few HPQ's where the blue bump becomes noticeable in the B- & U-band (Smith et al. 1988), deviation from a power-law is in the form of a spectral index increasing with frequency ($d\alpha/d\nu > 0$; e.g. Cruz-Gonzalez & Huchra 1984; Ballard et al. 1990). (ii) A minimum in P is expected at the frequency where the power-laws overlap. Hence, $dP/d\nu > 0$ and $dP/d\nu < 0$ should be observed with equal probability, roughly. Excluding those HPQ's with $d\alpha/d\nu < 0$, there is a clear tendency for $dP/d\nu \geq 0$ (e.g. Ballard et al. 1990). (iii) Changes in PA should occur as frequently as changes in P. Although $dPA/d\nu \neq 0$ is not rare, cases with $dP/d\nu \neq 0$ dominate. Also, identifying the two components with either a jet and a disk or a jet/disk and random components implies that the difference in PA between the two components should be considerable. However, this is not substantiated by observations which show that this difference is usually small (e.g. Brindle et al. 1986). Furthermore, to fit internight variations Holmes et al. (1984b) needed the two components to change in unison. Due to the importance of "randomness" in the blazar process, the value of fitting models to individual observations is limited; especially for models for which a considerable amount of finetuning is needed.

The properties of synchrotron radiation imply that for a given magnetic field geometry, P is determined, to a good approximation, by the local spectral index (Björnsson & Blumenthal 1982). Since $P \propto (1+\alpha)/(5/3+\alpha)$ gives qualitatively the correct observed correlation between P and α, Ballard et al. (1990) investigated whether it also gives a reasonable quantitative fit. Although this was often the case, the expected correlation between $\alpha_B - \alpha_H$ and P_B/P_H was not found (B and H refer to the B- and H-band, respectively). Furthermore, such a model can describe neither $dPA/d\nu \neq 0$ nor cases where P changes by a factor of two or more.

Since agencies external to the emission process (e.g. scattering, absorption, internal or external Faraday rotation) can usually be excluded for various reasons as the cause of FDP (e.g. Puschell et al. 1983; Sitko, Stein, & Schmidt 1984), models invoking the geometry of the magnetic field offer an interesting possibility. The basic features of such models are very similar to those of the Christmas tree model: A *one component model* which consists of a continuous and physically connected distribution of polarized subunits, i.e. the emission at different frequencies comes from progressively different regions of the source. Björnsson (1985) pointed out that in addition to the observationally well supported assumptions of a coherent magnetic field structure and a sharp high energy cut-off in the electron distribution, a source model with a pitch-angle distribution skewed towards smaller values is conducive to producing the observed FDP properties. In such a model, FDP is a projection effect and it occurs over the frequency range where the observed frequency depends, mainly, on the pitch-angle ($\nu \propto \sin \mu$, where μ is the pitch-angle). The effect is illustrated by a simple example in Figure 1. Hence, the spectral shape of the radiation is determined by the details of the anisotropic pitch-angle distribution; for example, if the probability of finding an electron with pitch-angle μ is given by $\sin^{-n} \mu$, the resulting spectrum is to a good approximation given by a power-law ($\alpha \simeq n - 2$) over a frequency interval determined by the range in $\sin \mu$. Substantial FDP is possible also in the case of a quasi power-law spectrum and, a priori, no correlation is expected between spectral curvature and the amplitude of FDP in accordance with observations (Ballard et al. 1990).

The symmetry of the magnetic field determines the properties of the observed FDP. Two extreme behaviours can be distinguished: (i) An axially symmetric magnetic field structure can produce large changes in P ($\Delta P \gg P$) but only minor variations in PA, except that PA can flip by 90° when P comes close to zero. This situation corresponds to $\theta \approx \pi$ or 2π in Figure 1. (ii) For a projected magnetic field with no apparent symmetry, P is expected to vary less ($\Delta P \leq P$) while PA in this case can

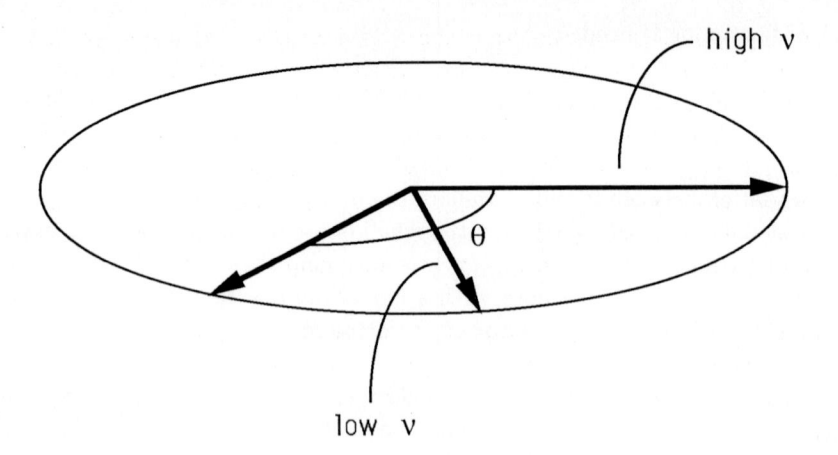

Figure 1. Magnetic field distribution: The magnitude of the magnetic field is constant but its directions are distributed uniformly within an angle θ in a plane which is tilted with respect to the line of sight. The emitted frequency is proportional to the length of the projected magnetic field vector (i.e. the length of the arrows).

exhibit large and smooth variations. This situation corresponds to $\theta \approx \pi/2$ in Figure 1. In general, a behaviour in between these two extremes is expected.

Observations show that PA varies with frequency less often than P and when it does, the total change is normally small ($\ll 90°$). Interpreted in the one component scenario this implies a magnetic field structure with an appreciable symmetry. Since FDP is a projection effect, the highest frequencies come from those parts of the source which have magnetic fields lying closest to the plane of the sky (i.e. largest values of $|\sin\mu|$). For a magnetic field with rotational symmetry one expects the projected field directions to be evenly distributed, roughly, around this (these) special direction(s), smaller $|\sin\mu|$ (i.e. smaller ν) having larger spread. This, then, reproduces the observed increase of P with frequency.

The existence of an anisotropic pitch-angle distribution has consequences which go beyond an explanation of FDP: (i) The salient features of both time and frequency dependent polarization can be explained within the same scenario. (ii) The strong anisotropy needed severely constrains the applicability of stochastic acceleration mechanisms (see also section 3). (iii) Taken together with a dynamically dominant magnetic field, anisotropy implies relativistic streaming, which can increase the observed cooling time to the extent that the need for reacceleration is obviated. (iv) The synchrotron cooling time is $t_s \propto \sin^{-2}\mu$. However, since the radiating particles are streaming towards the observer with a velocity $v \propto \cos\mu$, the observed cooling time t_s^{obs} is shortened; for example, for a purely radial magnetic field distribution, $t_s^{obs} = \sin^2\mu\, t_s$, implying that t_s^{obs} is independent of μ and, hence, ν. If flux variations are due, at least partly, to cooling, the fact that the variational time scale is approximately the same in optical and infrared is readily explained by an anisotropic pitch-angle distribution in a magnetic field with a dominantly radial structure.

3 THE CONNECTION BETWEEN BLAZARS AND COMPACT RADIO SOURCES

3.1 Shocks

It is often assumed that shocks play an important role in accelerating electrons in blazars. However, non-relativistic shocks are beset by a general problem in that they presuppose a dominance of cold (i.e. non-relativistic) particles in the particle energy density; for example, in first order Fermi-acceleration, assuming an electron distribution of the form γ^{-s}, a lower limit to the ratio of cold to relativistic particles is given by γ^{s-1}. Since the density of cold particles estimated by Blandford & Rees (1978) was for particles which had cooled, the actual density has to be increased by a factor γ^{s-1}. With reasonable values of γ and s, the number of reaccelerations becomes so large that a continuous acceleration is indicated. Relativistic shocks do not necessarily suffer from this drawback. An interesting aspect of relativistic shocks is that they can give rise to an anisotropic pitch-angle distribution: (i) Since diffusion across field lines is much slower than along, *Fermi-acceleration at a relativistic shock* is expected to be effective for parallel shocks only. Hence, particles escaping the acceleration region have preferentially small pitch-angles. (ii) Due to the short cooling time, particles running into a *relativistic blast wave* are cold. Thus, the Lorentz factor of the shock must be $\sim \gamma^{1/2}$. With a dynamically dominant magnetic field, the blast wave should move along field lines producing electrons within a pitch-angle $\sim \gamma^{-1/2}$.

The physical relationship between the optical and radio emission regions is important, in particular, in the light of the increasing observational evidence for a close connection between the two (section 1). Although the later stages of outbursts in the radio regime can be explained as due to an adiabatically expanding, optically thick, inhomogeneous synchrotron source (e.g. van der Laan 1966; Blandford & Königl 1979), the initial phases in the mm/infrared can not. In order to fit the observed variations with time of both the flux and the frequency where the flux peaks, rather contrived injection of electrons and/or magnetic field needs to be invoked. An alternative is to explain the initial increase of flux as due to the decreasing importance of Compton-cooling behind a shock which propagates outwards in a jet (Marscher & Gear 1985), i.e. the relative efficiency of synchrotron radiation (radio emission) increases with time. Since the observations in this scenario indicate a weak but relativistic shock, the main increase of electron and magnetic field energy density behind the shock is assumed to be due to compression; hence, the shape of the electron distribution does not change across the shock. The initial acceleration of the particles which determines, for example, the spectral shape, is assumed to occur closer to the central source at the "injection radius".

Due to the cooling behind the shock, the spectral index α of the optically thin radiation in the outburst is steeper by 0.5 than in the quiescent jet. This is indeed the case for the 1983 outburst in 3C273 (Clegg et al. 1983; Robson et al. 1983), where α increased from 0.7 to 1.2. However, this is not generally true; Valtaoja et al. (1988) studied 17 outbursts in 15 different sources and found that in the "cooling phase" all outbursts were consistent with an optically thin spectral index of $\alpha \approx 0.2$, indicating no spectral steepening since $0.2 < \alpha < 0.5$ in the quiescent phase. Furthermore, the relative importance of cooling due to Compton scattering and synchrotron radiation is given by U_{ph}/U_B, where U_{ph} and U_B are the energy densities of photons and magnetic field, respectively. Since the peak flux is assumed to come from where the source becomes optically thick, the increase of U_{ph} due to the shock can not be much larger than the observed increase in flux. Thus, even for a parallel shock (i.e. no increase in B), the relative increase of Compton cooling in the outburst should be no more than a factor of a few. Accordingly, it is expected that if Compton cooling is unimportant in the quiescent jet, it should remain so in an outburst due to a shock.

The problems associated with shock models and the suggested similarities between the properties of the optical and radio emission regions make a different type of models for the outbursts worthy of consideration; namely, pitch-angle scattering of streaming particles. The waves responsible for the scattering could be generated in the same medium as that producing the internal Faraday rotation, suggested by Björnsson (1991) to be the cause of the low polarization in the unresolved VLBI-core.

Let mono-energetic (for simplicity) electrons with small pitch-angles be injected instantaneously into a "jet" containing a dynamically dominant magnetic field. The electrons stream along the field lines and radiate negligible until synchrotron emission is enhanced by pitch-angle scattering. The important time scales are: (i) Synchrotron cooling, $t_s = 3 \cdot 10^8/\gamma B^2 \sin^2 \mu$. (ii) "Source crossing" time, r/c, where r is the scale height of the jet. (iii) Time scale for pitch-angle scattering (e.g. Spangler 1979),

$$t_\mu = \frac{10^3}{w} \left(\frac{\gamma_3 r_{pc}}{B}\right)^{1/2}, \tag{1}$$

where w is the turbulent energy density divided by the total magnetic energy density, $\gamma_3 = \gamma/10^3$ and $r_{pc} = r/3 \cdot 10^{18}$ cm. In deriving equation (1) it was assumed that the energy driving the turbulence is put in on a scale equal to r and that the turbulent spectrum is of the Kraichnan type, which is appropriate for weak turbulence ($w \ll 1$). As the electrons stream along the field lines, the importance of pitch-angle scattering is determined by

$$S \equiv \frac{r}{ct_\mu} = w_{-5} \left(\frac{Br_{pc}}{\gamma_3}\right)^{1/2}, \qquad (2)$$

where $w_{-5} (\equiv w/10^{-5})$ in equation (2) includes turbulence with non-relativistic phase velocities only, since waves with relativistic phase velocities tend to make the streaming electrons loose energy rather than scatter.

It will be assumed that w increases faster with r than $(Br_{pc})^{-1/2}$, so that S increases with time. The properties of the outburst depend critically on the role of cooling.

No cooling ($t_s(\sin\mu = 1) > t_\mu$): The emitted flux is proportional to the number of electrons scattered to large pitch-angles. Thus, the outburst has three phases. (i) $S < 1$, flux increases with time as pitch-angle scattering becomes more efficient. (ii) $S \approx 1$, the flux reaches its maximum. (iii) $S > 1$, the electrons are isotropized and the later stage of the outburst is similar to an adiabatic, van der Laan type expansion. In the streaming scenario this situation corresponds to outbursts.

Cooling efficient ($t_s(\sin\mu = 1) < t_\mu$): The electrons cool before reaching large pitch-angles. Hence, streaming is important and the last adiabatic phase does not occur. Due to the large Doppler factors the observed time scale for flux variations is much shorter than for the "no cooling" case. This situation is conducive to producing the observed rapid, small amplitude variations (e.g. Witzel & Quirrenbach 1990). Since most of the energy is radiated away at $S \approx 1$, equating t_s with t_μ gives the typical frequency of the radiation

$$\nu_9 = 50 \frac{\gamma_3^{3/2}}{r_{pc}^{1/2}}, \qquad (3)$$

where $\nu_9 \equiv \nu/10^9$ Hz. The lowest frequency at which streaming can be important, ν_9^{min}, is obtained from $t_s(\sin\mu = 1) = t_\mu$, yielding

$$\nu_9^{min} = \frac{10^{-2}}{B^3 r_{pc}^2}. \qquad (4)$$

In an electron-proton plasma, streaming enhances CP,

$$CP(\%) \approx \sqrt{\frac{B}{\nu_9 \sin\mu}}. \qquad (5)$$

If the rapid variations with an amplitude $\sim 10\%$ observed by Witzel & Quirrenbach (1990) are due to streaming, $CP \leq 1\%$ is needed since typical values for the total flux is $CP \sim 0.1\%$ (Komesaroff et al. 1984). From

$$\nu_9 \sin\mu = 3\left(\frac{\gamma_3}{Br_{pc}}\right) = 3 \cdot w_{-5}^2(S=1), \qquad (6)$$

it is see that $w_{-5}(S=1) \geq \sqrt{B}$ is required.

With streaming large Doppler boosting factors ($D \sim \sin^{-1}\mu$) can be obtained. The upper limit is set by the Lorentz factor of the emitting electrons. Witzel & Quirrenbach (1990) pointed out that if standard synchrotron theory is to explain their observations, large Doppler factors are implied, $D \sim 10^2$. The need for "geometrical effects" to account for some of the temporal properties of the mm/radio flux has been emphasized by Robson (this conference). Such effects are readily obtained in models where particles stream along field lines of a dynamically dominant magnetic field. One should note that for streaming, the Doppler factor deduced from super-luminal motion (Γ) can be much smaller than D; for example, assume that the magnetic field has a helical structure with opening angle β, then, maximally, $\Gamma \approx \sin^{-1}\beta$ if $\beta > \mu$.

4 CONCLUSIONS

The importance of "randomness" in the blazar process makes it very time consuming to study its properties and underlying physical mechanism(s). The tentative conclusions, which can be drawn at present, are likely to need revision in the future as the observational material increases. It does seem, however, that "two-component" models are too simple to account for many aspects of both the temporal and spectral behaviour of blazars; in particular if one assumes that the properties of the blazar process are independent of the type of object in which it occurs. The type of model favoured in this review is one where individual subcomponents turn on at random within a given magnetic field structure. In such a Christmas tree model, the size of the individual subcomponents is much smaller than the region over which they are distributed. This offers the possibility to accommodate the close connection indicated by observations between the optical emission region and at least part of the unresolved VLBI-core. It is suggested that an anisotropic pitch-angle distribution is a central ingredient both in FDP and radio outburst; in particular in the rapid, small amplitude variations.

REFERENCES

Angel, J. R. P., & Stockman, H. S. 1980, ARA&A,18, 321

Ballard, K. R., Mead, A. R. G., Brand, P. W. J. L., Hough, J. H., Bailey, J. A., & Brindle, C. 1989, in BL Lac Objects, eds. L. Maraschi, T. Maccacaro, & M.-H. Ulrich (Heidelberg: Springer), p. 181

Ballard, K. R., Mead, A. R. G., Brand, P. W. J. L., & Hough, J. H. 1990, MNRAS, 243, 640

Björnsson, C.-I. 1985, MNRAS, 216, 241

_____. 1990, MNRAS, 242, 158

_____. 1991, ApJ, submitted

Björnsson, C.-I., & Blumenthal, G. R. 1982, ApJ, 259, 805

Blandford, R. D., & Rees, M. J. 1978, in Pittsburgh Conference on BL Lac Objects, ed. A. M. Wolfe (University of Pittsburgh), p. 328

Blandford, R. D. & Königl, A. 1979, ApJ, 232, 34

Brindle, C. et al. 1985, MNRAS, 214, 619

Brindle, C., Hough, J. H., Bailey, J. A., Axon, D. J., & Hyland, A. R. 1986, MNRAS, 221, 739

Clegg, P. E. et al. 1983, ApJ, 273, 58

Cruz-Gonzalez, I., & Huchra, J. P. 1984, AJ, 89, 441

Fugmann, W. 1988, A&A, 205, 86

Gabuzda, D. C., Wardle, J. F. C., & Roberts, D. H. 1989, ApJ, 338, 743

Hagen-Torn, V. A., Marchenko, S. G., & Yakovleva, V.A. 1985, Astrophys., 22, 1

_____.1986, Astrophys., 25, 634

Holmes, P. A., Brand, P. W. J. L., Impey, C. D., & Williams, P. M. 1984a, MNRAS, 210, 961

Holmes, D. C. et al. 1984b, MNRAS, 211, 497

Impey, C. D., & Neugebauer, G. 1988, AJ, 95, 307

Impey, C. D., & Tapia, S. 1988, ApJ, 333, 666

Jones, T. W., Rudnick, L., Aller, H. D., Aller, M. F., Hodge, P. E., & Fiedler, R. L. 1985, ApJ, 290, 627

Kikuchi, S., Mikami, Y., Konno, M., & Inoue, M. 1976, PASJ, 28, 117

Kikuchi, S., Inoue, M., Mikami, Y., Tabara, H., & Kato, T. 1988, A&A, 190, L8

Komesaroff, M. M., Roberts, J. A., Milne, D. K., Rayner, P. T., & Cooke, D. J. 1984, MNRAS, 208, 409

Landau, R. et al. 1986, ApJ, 308, 78

Marscher, A. P., & Gear, W. K. 1985, ApJ, 298, 114

Miller, J. S. 1978, Com. on Astrophys., 7, 175

Moore, R. L. et al. 1982, ApJ, 260, 415

Moore, R. L., Schmidt, G. D., & West, S. C. 1987, ApJ, 314, 176

Phinney, E. S. 1985, in Astrophysics of Active Galaxies and Quasi-Stellar Objects, ed. J. S. Miller (University Science Books), p. 453

Puschell, J. J. et al. 1983, ApJ, 265, 625

Robson, E. I. et al. 1983, Nat, 305, 194

Sitko, M. L., Stein, W. A., & Schmidt, G. D. 1984, ApJ, 282, 29

Smith, P. S., Elston, R., Berriman, G., & Allen R. G. 1988, Apj, 326, L39

Spangler, R. S. 1979, ApJ, 232, L7

Valtaoja, E. et al. 1988, A&A, 203, 1

van der Laan, H. 1966, Nat, 211, 1131

Weiler, K. W., & de Pater, I. 1983, ApJS, 52, 293

Witzel, A., & Quirrenbach, A., 1990, to appear in Propagation Effects in Space VLBI

Worrall, D. H. 1989, in BL Lac Objects, eds. L. Maraschi, T. Maccacaro, & M.-H. Ulrich (Heidelberg: Springer), p. 305

Polarized Synchrotron Emission from Shock accelerated Particles

KLAUS-DIETER FRITZ

Max-Planck-Institut für Kernphysik, Heidelberg

1 ABSTRACT
Physical models for the flux outbursts in the nonthermal spectra of Blazar objects have exclusively used power law spectral components in the past. Power laws, however, evolve only in systems which are in steady-state in a certain sense. At least for the highly variable components this is certainly not a proper assumption. The results from time-dependent calculations of first order Fermi acceleration show a significantly different spectral behavior.

2 INTRODUCTION
The highly polarized radiation received from Blazars is commonly interpreted as synchrotron radiation emitted by relativistic electrons. A typical electromagnetic spectrum for those sources is a smooth power law ranging from radio to optical or UV band followed sometimes by a high frequency cutoff. This suggests a universal acceleration process acting in largely different energy regimes. One often proposed mechanism that is capable of producing power law spectra is the first order Fermi process by which electrons are accelerated at shock waves. A shock wave (probably associated with the appearance of a new VLBI knot) converts the ordered kinetic energy of the upstream plasma into the acceleration of particles. If the injection of particles into the process is linked to the upstream plasma density we would expect the injection rate to vary whenever the plasma density changes.

From steady-state first order Fermi theory which has been worked out over the past years for different aspects one expects the spectrum of the accelerated particles (and the associated synchrotron spectrum) to be of power law form with a high frequency cutoff due to synchrotron losses. This is in qualitative agreement with the observations of the smooth "quiescent" component. Superposed on this component one often observes strongly variable flux outbursts with rapidly increasing magnitudes. Usually the degree of polarization increases with frequency (which would be expected for a spectrum with negative curvature). In some cases however the opposite or a more complicated behavior is observed which necessitates more elaborate models. Most of the constructed models (e.g. Björnsson 1985; Marscher and Gear 1985; Ballard,

Mead, and Brand 1990) superpose two or more power law synchrotron components to account for the specific polarization behavior. Some of them have been quite successful to fit at least part of the data.

However, we know at least one thing from such flux outbusts: The physical conditions do *not* seem to be very steady in time (which would be needed to account for power law spectra). This on the other hand casts doubt on the validity of applying steady-state theory to variable sources and needs to be investigated. The objective of this paper is thus to illuminate the applicability of the classical steady-state first order Fermi theory to flux variable sources and to present results from time-dependent calculations. We will focus especially on the polarized part of the synchrotron radiation emitted by the accelerated electrons.

3 THE STEADY-STATE ASSUMPTION

In first order Fermi theory particles are injected at low energy into the process (for example at the shock) and become accelerated to higher and higher energies. The particles have a certain probability to get lost from the system. Reaching steady-state in this process means enough time has been elapsed since the beginning of the acceleration so that the system had been able to achieve equilibrium. Steady-state also means that the parameters entering the acceleration process (e.g. the rate of injected particles) have no significant time dependence. A power law ranging over many decades in momentum thus implies a certain degree of steadiness in particle injection.

We have to keep in mind, though, that the first order Fermi process accelerates particles only at a distinct location in the system namely the shock surface. Having steady-state at the shock surface therefore does not necessarily mean that we have steady-state in any other region in the system. Particles accelerated at the shock need a certain time to be transported (by diffusion or convection) to somewhere else. In other words, we can have steady-state in one region and be far from steady-state in another one.

The term "steady-state" must be confined even further: As we will see below it is well possible to find one part of the spectrum of the accelerated particles to be in steady-state while another is still variable. In addition, synchrotron losses (especially at high frequencies as in the infrared or optical band) strongly affect the particle spectrum and temporal changes of it.

An observation of an emitting region in general yields an integral over a large volume of radiating particles (according to the beam size), probably even over a number of different acceleration sites. This applies particularly to the compact BL Lac objects. The fact that there are periods where Blazar spectra are "quiescent", i.e. are nonvariable, tells us that we cannot have significant particle acceleration going on in these phases as this would steadily increase the total number of energetic particles which would be observable. There are, however, two exceptions: Either particle losses from the system (the observed region) balance the "freshly" accelerated particles or energy losses are strong enough that particles "escape" from the observed momentum range towards low energies. These two exceptions can cause the system to appear steady-state.

4 TIME DEPENDENT PARTICLE ACCELERATION

The basic equation governing the transport in time-dependent geometry is (for details see Fritz and Webb 1990):

$$\frac{\partial F}{\partial \tau} + \overline{U}\frac{\partial F}{\partial \xi} - \frac{\partial}{\partial \xi}(\kappa \frac{\partial F}{\partial \xi}) + \frac{r-1}{3r}(4F + z\frac{\partial F}{\partial z}) + \frac{1}{4a}\frac{\partial F}{\partial z} = T(\tau)\delta(\xi)\delta(z-1) \quad (1)$$

where all quantities have been made dimensionless. $F \propto p^4 f(t,x,p)$ where f is the phase space denity of the relativistic particles; τ and ξ are dimensionless time and distance to the shock, respectively; $z = p_o/p$; $\overline{U}$ is the scaled plasma velocity; r is the compression ratio at the shock and a is a constant parameter specifying the relative contribution of energy gains and synchrotron losses. $T(\tau)$ on the right hand side of eq.(1) denotes the temporal part of the particle source. We assume the source to be located at the shock ($\delta(\xi)$) and being monoenergetic ($\delta(z-1)$) at momentum p_o. The spatial diffusion coefficient κ is held constant for simplicity. Although there is evidence that κ is both depending on p and on ξ we believe that the the principal results are not affected by a variable κ.

Two basic types of variable particle injection were investigated: i) Continuous injection starting at $\tau = 0$, ii) impulsive injection at $\tau = 0$ ($\delta(\tau)$). The fact that we use monoenergetic injection in both cases does not have a significant effect on the results. We could instead use a power law distribution for the injected particles. But the characteristics of the temporal development of the particle distribution become clearer in case of a monoenergetic injection.

Eq.(1) can now be Laplace-transformed with repect to τ and z. The resulting ODE is solved analytically and the solution has to be transformed back into real space. This last part was done numerically. From the momentum distribution of the accelerated particles the synchrotron spectra were calculated.

5 RESULTS AND DISCUSSION

The main results are summarized for the three considered cases:

i) As expected, a continuous constant injection leads to a steady-state spectrum after a certain time (depending mainly on the magnitude of the spatial diffusion coefficient). If we integrate the emitted radiation over a finite volume we obtain a power law for the differential particle density $N(p)$ with spectral index $(r+2)/(r-1)$ for low momenta, a by one steeper power law in a medium momentum range, and a high momentum cutoff due to synchrotron losses. The break by unity is due to the spatial integration over the ξ-dependent high momentum cutoff (particles lose more and more energy as they are advected away from the shock). The steady-state which is determined by the synchrotron losses is first established at the cutoff frequency and moves down in frequency with time. For a detailed description see Fritz and Webb (1990).

ii) The solution for an impulsive injection develops a second cutoff at medium frequencies with a reflattening above. At low freqencies we obtain a power law when integrating over the whole physical space (fig.1) or an increasingly flatter spectrum when integrating over a finite volume across the shock. We never obtain the classical power law streching over the whole frequency range.

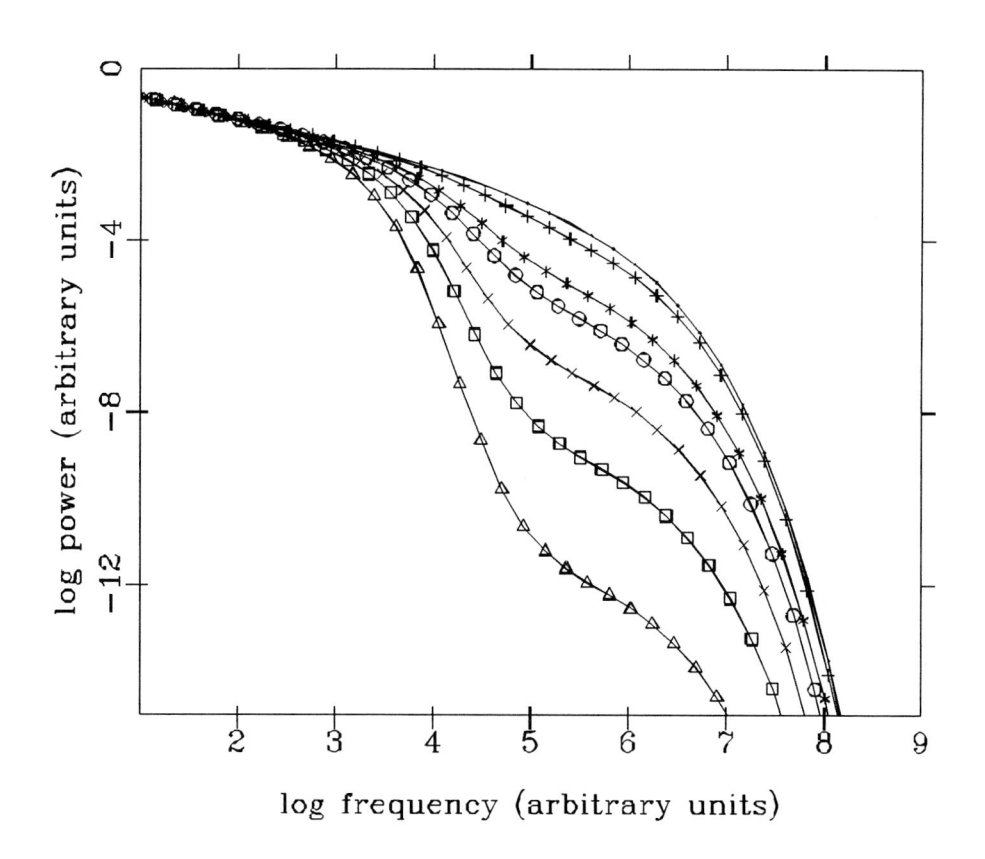

Figure 1. Spectra of polarized synchrotron radiation integrated over the total physical space for impulsive injection at (from right to left) $\tau = 40, 50, 70, 80, 100, 120, 150$. The rising spectra for earlier times have been omitted for clarity of the figure.

Important for our purposes is the fact that the spectrum suffers a positive curvature at some high frequency ν so that $d\alpha/d\nu > 0$, α being the spectral index. This behavior is *always* present whenever there is a *decreasing* rate of injection. Since it takes a while for a particle to gain energy a decreasing rate of injection leads to a relatively large number of higher energetic particles (from earlier stronger injection) which is equivalent to a flatter spectrum or a positive curvature. An increasing rate would have the opposite effect. This positive curvature has the effect of a strong frequency dependent degree of polarization P. Fig.2 shows that P suffers a significant dip with $dP/d\nu < 0$ although the positive curvature is not particularly strong and does not differ very much for different times. Near the local minimum of P the synchrotron spectra at different times can very well be approximated by power laws with similar spectral indices and different amplitudes. Such a feature has actually been observed for the source OJ287 (Holmes et al. 1984).

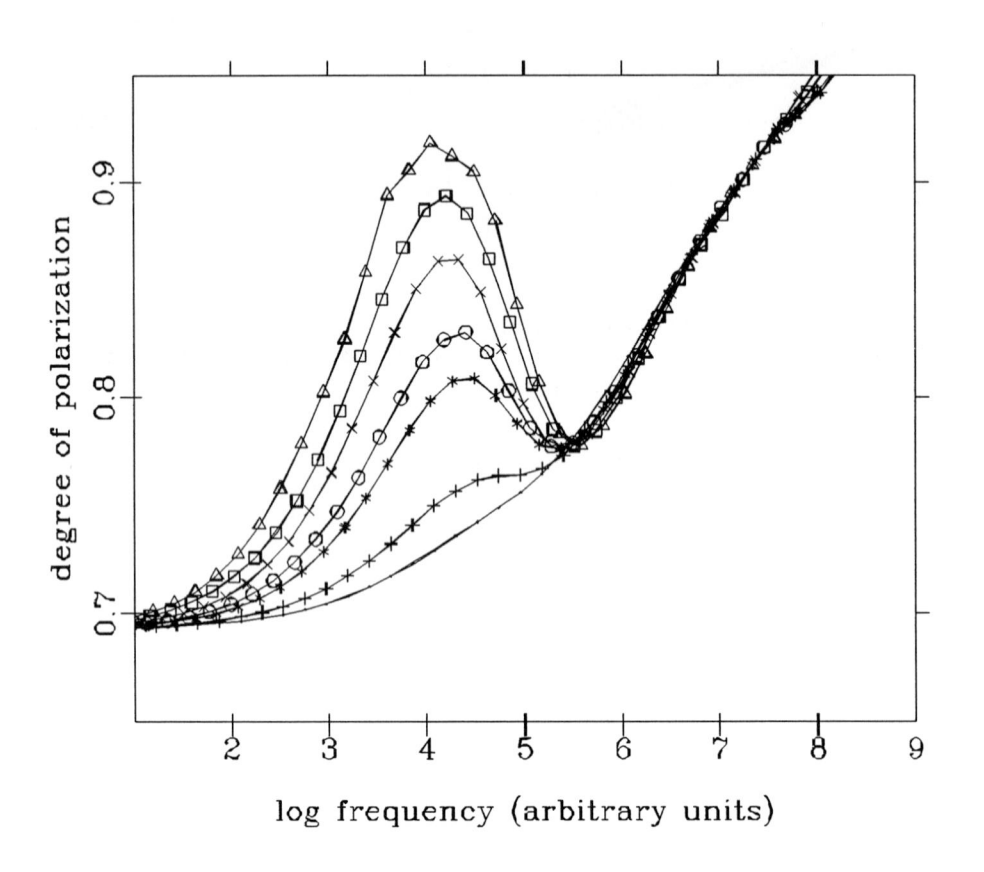

Figure 2. Degree of polarization for the spectra of figure 1.

We know that often also the position angle (PA) of the polarization vector shows strange behavior which cannot be explained by a *homogeneous* magnetic field. Since the high frequency cutoff is a function of position relative to the shock the high energy photons we observe come from different regimes with probably different magnetic field directions. Since the shock is moving through the plasma there is no reason to believe that the magnetic field geometry and with it the position angle should be constant in the synchrotron emitting region . It might therefore well be that indeed there is only one dominant (non power-law) synchrotron component and a complex magnetic field geometry that is responsible for the varying PA. From the presented calculations we see that for relatively simple physical assumptions the complicated spectral observations can at least qualitatively be approached. Future work should investigate the effect of a more complex magnetic field on the temporal behavior of PA.

REFERENCES

Ballard, K.R., Mead, A.R.G., Brand, P.W.J.L., Hough, J.H. 1990, MNRAS, **243**, 640

Björnsson, C.-I. 1985, MNRAS, **216**, 241

Fritz, K.-D., Webb, G.M. 1990, Ap.J., **360**, 387

Holmes, P.A., Brand, P.W.J.L., Impey, C.D., Williams, P.M., Smith, P., Elston, R., Balonek, T., Zeilik, M., Burns, J., Heckert, P., Barvainis, R., Kenny, J., Schmidt, G., Puschell, J. 1984, MNRAS, **211**, 497

Marscher, A.P., Gear, W.K. 1985, Ap.J., **298**, 114

The Connection Between Growing Radio Shocks and Frequency Dependent Optical Polarization in Blazars and Low Polarization Quasars

L.VALTAOJA[1], E.VALTAOJA[2], N.M.SHAKHOVSKOY[3] and Yu.S.EFIMOV[3]

1) Tuorla Observatory, SF-21500 Piikkiö, Finland

2) Metsähovi Radio Research Station, Helsinki University of Technology, SF-02150 Espoo, Finland

3) Crimean Astrophysical Observatory, SU-334413 Nauchny, Union of Soviet Socialist Republics

Blazars have high and variable optical polarization, which often shows various kinds of wavelength dependence with the polarization decreasing or increasing towards the red. In addition, the character of FDP of an object may change in different time scales. Frequency dependent position angles are also observed, but not as often (Smith et al. 1987, Mead et al. 1990, Valtaoja et al. 1991a, Takalo 1991). Because of the rapid variability and the relatively high degree of polarization it is thought that the polarization in blazars is due to synchrotron radiation. But what causes the FDP and the seemingly unsystematic and uncorrelated changes in it, in the flux and in the position angle? Synchrotron radiation in itself produces frequency independent polarization. Dilution by the surrounding galaxy can explain the decrease of the polarization towards the red wavelengths, but it cannot explain the other kinds of observed FDP and the frequency dependence of the position angle. Individual observations have been interpreted successfully with purely numerical models consisting of two separate, variable synchrotron components (Puschell et al. 1983, Holmes et al. 1984, Brindle et al. 1986, Smith et al. 1986, Ballard et al. 1990). Even if the components have not been identified physically there, seems to be something behind these models because the more variable component consistently has flatter spectrum and higher polarization.

Valtaoja et al. (1989,1991) proposed that the high-frequency tail of the shock causing the observed radio variability (Marscher and Gear 1985, Valtaoja et al. 1988, Hughes et al. 1989) could be responsible for the observed behavior of the optical polarization as well. Basically their model consists of two synchrotron components: a growing shock and the underlying jet. This picture agrees with the numerical two-component models if we identify the shock component as the more variable, flatter-spectrum, higher polarization component.

By considering the development of the shock during a radio outburst Valtaoja et al. (1991a) predicted the optical FDP at different stages of the outburst. The exact development of the shock spectrum's optical tail depends on the balance between the growth of the shock and the energy losses, but in general one would expect that when the shock is first seen in the optical region its spectrum is flatter than that of the jet component (stage 1. in figure 1.) since fresh electrons have very flat energy distribution (Valtaoja et al. 1988). Because the compressed shock is also more polarized, one should observe the degree of polarization to decrease to the red. (Similar FDP can be also caused by the dilution of the host galaxy. But in any case when the shock appears the dependence should strengthen.) While the shock is growing the high-frequency tail gets steepened by energy losses. In stage 2. the spectral indexes of the shock and the jet are equal for a moment and no FDP is detected. The spectrum of the shock continues to steepen. At stage 3. the IR bands have an excess of the highly polarized flux and the degree of the polarization increases towards the red. When the shock peaks in the low radio frequencies it is no more visible in the optical wavebands and causes no FDP (stage 4.).

According to this kind of scenario the optical FDP is different depending on in which stage of the radio outburst we happen to observe the object. Variations in the shock parameters and timescales can cause a wide range of optical polarization variations, and because the model has two different components the position angle may also be frequency dependent. Thus the shock moving in the jet can explain many kinds of polarization behavior, but because its time evolution is defined it is still possible to check the validity of the idea with observations. Comparisons between the optical polarization data and the simultaneous radio flares in four BL Lac objects (BL Lac, 3C 279, DA 237 and 0109+224) showed agreement with the general idea (Valtaoja et al. 1991a), but for a detailed testing more optical polarization data with good time coverage are needed.

Most quasars have low polarization (<3%), no detected FDP, and they are not very variable. Are these low polarization quasars (LPQ) physically fundamentally different from the blazars, or could LPQs be the same thing as the HPQs and BL Lacs seen from different viewing angle, as proposed by unified schemes (e.g.,Barthel 1989). Strong evidence in favor of unified models would be the existence of "intermediate" viewing angle quasars having both LPQ and blazar properties. One such example seems to be 3C 273. The detection of a polarization flare in it (Couvoisier et al. 1988) made it the first LPQ with well-documented variable polarization and FDP. After that several observations of 3C 273 (Impey et al. 1989, Valtaoja et al. 1990 and Valtaoja et al. 1991) have shown daily variations both in flux and polarization.

Impey et al. (1989) introduced the term mini-blazar to explain the variations in the polarization properties of 3C 273. In their model 90% of the total flux is coming from the low polarized synchrotron component and the thermal component peaking at short wavelengths, and 10% from the mini-blazar, which is at least 10% polarized and is very variable also in short timescales. The polarization of the mini-blazar would be diluted by the normal LPQ light. They predicted that other LPQs may harbor a blazar component as well, which would strengthen the unified schemes. Valtaoja et al. (1990) also pointed out that the LPQ 3C 273 shows blazar-like activity. They explained the quiescent FDP of the source by one synchrotron component (jet) diluted by a hot accretion disk and the observed daily variability of the polarization with an additional very variable and polarized synchrotron component, a mini-blazar, which they identified as a shock moving in the jet. According to this the only difference between LPQs and blazars would be the strength of the flux and polarization of the two synchrotron components, the jet and the shock, relative to the accretion disk.

Our new optical polarization data (Valtaoja et al. 1991b) strengthened the "boderline case" nature of 3C 273. The data coincided with the beginning of the new outburst monitored at 90 GHz with SEST telescope (Tornikoski et al., these proceedings), and we were able to make a close comparison between the optical and radio data. As figure 2. shows, transient strong FDP with polarization decreasing towards the red was observed in the beginning of the radio outburst, as expected in stage 1. of the shock development. Thus the same radio shock model for the FDP
which we have used for blazars seems to work for 3C 273 also.

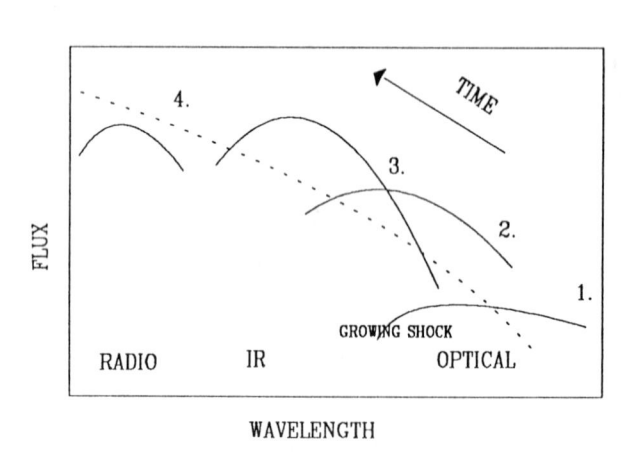

Figure 1. The jet and shock scenario for producing variable frequency dependent polarization in the optical region.

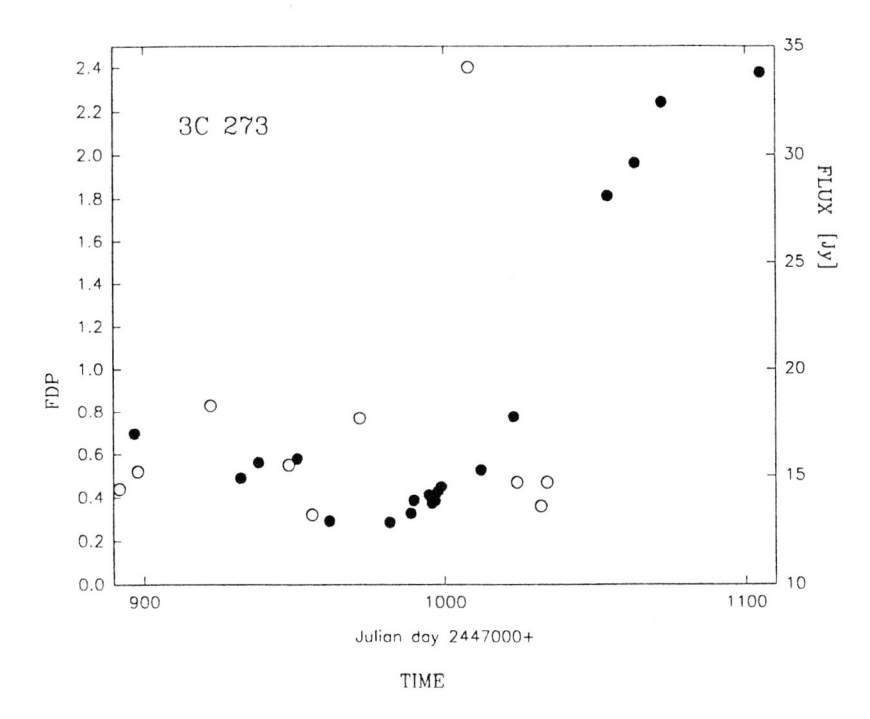

Figure 2. Frequency dependent optical polarization, P(U)/P(I), (open circles) during the 90 GHz radio outburst (filled circles) of 3C 273.

References

Ballard, K.R., Mead, A.R.G., Brand, P.W.J.L., and Hough, J.H. (1989) Edinburgh Astronomy preprint **27**
Barthel,P.D. (1989) ApJ **151**, 606
Brindle, C., Hough, J.H., Bailey,J.A., Axon, D.J., and Hyland, A.R. (1986) MNRAS **221**, 739
Courvoisier, T.J.-L., Robson, E.I., Blecha, A., Bouchet, P., Hughes, D.H., Krisciunas,K., and Schwarz, H.E. (1988) Nat **335**, 330
Holmes, P.A., Brand, P.W.J.L., Impey, C.D., and Williams, P.M. (1984a) MNRAS **210**, 961
Holmes, P.A., Brand,P.W.J.L., Impey, C.D., Williams, P.M., Smith, P., Elston, R., Balonek, T., Zeilik, M., Burns, J., Heckert, P., Barvainis, R., Kenny, J., Schmidt, G., and Puschell, J. (1984b) MNRAS **221**, 497

Hughes, P.A., Aller, H.D., and Aller, M.V. (1989) In BL Lac Objects, edited by L. Maraschi, T. Maccacaro and M.-H. Ulrich (Springer, Berlin), 30
Impey, C.D., Malkan, M.A., and Tapia, S. (1989) ApJ **347**,96
Marscher, A.P., and Gear,W.K. (1985) ApJ **298**,114
Mead,A.R.G., Ballard, K.R., Brand, P.W.J.L., Hough, J.H., Puschell, J.J., Jones, T.W., Phillips, A.C., Rudnik, L., Simpson, E., Sitko, M., Stein, W.A., and Moneti, A. (1983) ApJ **265**, 625
Smith, P.S., Balonek, T.J., Heckert, P.A., and Elston, R. (1986) ApJ **305**, 484
Smith, P.S., Balonek, T.J., Elston, R., and Heckert, P.A. (1987) ApJS **64**, 459
Takalo, L. (1991), A&AS, In press
Valtaoja, E., Haarala, S., Lehto, H., Valtaoja, L., Valtonen, M., Moiseev, I.G., Nesterov, N.S., Salonen, E., Teräsranta, H., Urpo, S., and Tiuri, M. (1988) A&A **203**, 1
Valtaoja, E.,Valtaoja, L., Efimov, Yu. S., and Shakhovskoy, N.M. (1990) AJ **99**, 769
Valtaoja, L., Sillanpää, A, Valtaoja, E, Efimov, Yu.S. and Shakhovskoy, N.M. (1989) In BL Lac Objects, edited by L. Maraschi, T. Maccacaro and M.-H. Ulrich (Springer, Berlin),127
Valtaoja, L., Valtaoja,E., Shakhovskoy,N.M., Efimov, Yu.S., (1991a), AJ **101**, 78
Valtaoja, L., Valtaoja,E., Shakhovskoy,N.M., Efimov, Yu.S., Takalo,L., Sillanpää,A., Kidger,M.R. and de Diego, J.-A., (1991b), In preparation

Photometry and Polarimetry of OJ 287 and Mrk 421 in 1980-1990

S. KIKUCHI

National Astronomical Observatory, Mitaka, Tokyo 181, Japan

1 INTRODUCTION

We have been making simultaneous photometry and polarimetry of several BL Lac objects with the eight channel polarimeter (Kikuchi 1988a) mainly on the 91 cm telescope at the Dodaira Station of the National Astronomical Observatory since late 1980. The observation with the same instrument is the most adequate to study long-term behavior of active sources such as blazars, although some systematic errors may still remain. The VLBI, including polarization and space VLBI, observations give us information on fine structure of active sources. However, limits of angular resolution are present in the VLBI measurements, and only studies on variability are able to show more detailed structure, in other words, structure and dynamics around the central engine. We hope to know what is taking place in blazars on the basis of data with these two kinds of observations in the near future.

OJ287 and Mrk421 are most extensively observed sources at Dodaira, and the results for these sources will be reported briefly. Other blazars observed at Dodaira, but less frequently, are PKS0215+014, 3C66A, PKS0735+178, OI090.4, Mrk501 and BL Lac.

Observations are made mainly with apertures of 13.0" and 18.0". We pay a special attention for subtraction of background sky, since Dodaira is at a distance of about 70 km from the central region of Tokyo.

2 OBSERVATIONAL RESULTS

2.1 OJ287

In Fig.1(a), we give photometric and polarimetric results obtained in 1980-90. As we have shown in the Como meeting, after the outburst in 1983, spiky features in light variations are found on a time scale of less than a few weeks, and color variations on the same time scale are associated (Kikuchi 1989). The change of spectral shape, i.e., the spectrum flattens as the flux increases, is consistent with the explanation by the shock model (Marscher and Gear 1985). We consider that, in the faint stage, an evolution of a single flare or a shock phenomenon has been observed on a time scale of less than a few weeks.

We also note that rapid and large variations in polarization angles occurred in the faint stage of late 1984 – early 86. No preferred direction of polarization was found during this period. If we take into account of an ambiguity by 180° in determining the polarization angles, a range of such variation detected in early 1986 is probably over 180°. It should be noted that the variation has been synchronized with that at 10 GHz (Kikuchi et al. 1988).

2.2 Mrk 421

In Fig.1(b) we show results of photometric and polarimetric observations of Mrk421 corrected for the contribution from the underlying galaxy after Kikuchi and Mikami (1987). In Fig.1(b), we also display the results in 1974-81 obtained by Hagen-Thorn et al. (1983) with an aperture of 24". Subtraction of the galactic light were made similarly. Therefore, in Fig.1(b), we see photometric and polarimetric behavior of the non-thermal component of Mrk421. Both results are connected smoothly, and we find three maxima on 1974, 82 and 90, with an interval of about 8 years. For colors of Mrk421, we find that the source becomes bluer as the brightness increases until early 1987. AT the light maximum on early 1990, however, spectrum has slightly reddened.

We note that photometric behavior is related with that in polarization properties: First, optical fluxes roughly anti-correlate with polarization levels. Second, polarization angles show a large amount of variations, probably more than 180°, in the active states on 1982 and 90, on the contrary for the case of OJ287. In 1984-89, polarization angles were 170-180° in faint states, whereas they were 30-40° in bright phases, although in 1986 a systematic change in polarization angles preceded a flux increase. For the behaviot in 1984-89, the simple explanation by two components or two jets seems to be valid.

Mufson et al. (1990) analyzed the data of the multifrequency observations made on 1980 May, 1984 January and March, and derived different sets of parameters which characterize relativistic jets for different epochs. However, no distinct difference was found in the polarization features for these epochs.

3 DISCUSSION

As we have seen in the previous section, in both sources, rapid and large swings of polarization angles are seen when the sources are weakly polarized. However, the relation between the optical flux densities and polarization angles are different, as we see in Fig.2(a) and (b). It should be noted that, in Mrk421, the polarized flux does not vary appreciably in different phases of activities. Therefore, observed large swings of polarization angles are not explained by some observational errors or other effects which are caused when the polarized flux is very small or the polarization is very weak.

OJ287 and Mrk421 show different behavior in the optical region. If we also take into consideration activities in other wavelength regions, especialy those in the X-ray and radio regions, OJ287 and Mrk421 may belong to different sub-classes.

Rapid variations of polarization angles were observed in other sources. PKS0215+015, which is the most distant one among the known BL Lac objects (z=1.715), showed rapid and large variations in flux and polarization in the optical region in 1984-85 (Kikuchi 1988b). The range of variations in polarization angles exceeded certainly 90°, and possibly 180°, if we take into consideration an ambiguity by 180° in determining polarization angles. Also, in the radio region, Quirrenbach et al. (1989) found a large swing of polarization angles probably over 180° in the quasar 0917+624 on a time scale of les than a day. We consider that a helical magnetic field is playing an important role as pointed out by Konigl and Choudhuri (1985).

As a much larger structure, a helical field structure of a jet of M87 is reported on the basis of the polarization imaging by Fraix-Burnet et al. (1989), and also a bending of a jet is observed in the blazar 1156+295 (McHardy et al. 1990). With the VLBI observation, a helical structure, which is probably the finest structure by a basically imaging technique, is also detected for 3C345 (Krichbaum 1991). It seems natural that a helical structure of magnetic field is also present in a region much closer to a central engine.

References

Fraix-Burnet, D., Le Borgne, J.-F., and Nieto, J.K. 1989, Astron. Astrophys., **224**, 17.
Hagen-Thorn, V.A., Marchenko, S.G., Smehacheva, R.I., and Yakovleva, V.A. 1983, Astrofiz., **19**, 199.
Kikuchi, S. 1988a, Tokyo Astron. Bull., No. **281**, p.3267.
Kikuchi, S. 1988b, Publ. Astron. Soc. Japan, **40**, 547.
Kikuchi, S. 1989, in "BL Lac Objects", ed. L. Maraschi et al., p. 131.
Kikuchi, S., Inoue, M., Mikami, Y., Tabara, H., and Kato, T. 1988, Astron. Astrophys., **187**, L8.
Kikuchi, S., and Mikami, Y. 1987, Publ. Astron. Soc. Japan, **39**, 237.
Konigl, A., and Choudhuri, A.R. 1985, Astrophys. J., **289**, 186.
Krichbaum, T. 1991, in this volume.
Marscher, A.P., and Gear, W.K. 1985, Astrophys. J., **298**, 114.
McHardy, I.M., Marscher, A.P., Gear, W.K., Muxlow, T., Lehto, H.J., and Abraham, R.G. 1990, Monthly Not. Roy. Astron. Soc., **246**, 305.
Mufson, S.L., Hutter, D.J., Kondo, Y., Urry, C.M., and Wisniewski, W.Z. 1990, Astrophys. J., **354**, 116.
Quirrenbach, A., Witzel, A., Qian, S.J., Krichbaum, T., Hummel, C.A., and Alberdi, A. 1989, Astron. Astrophys., **226**, L1.

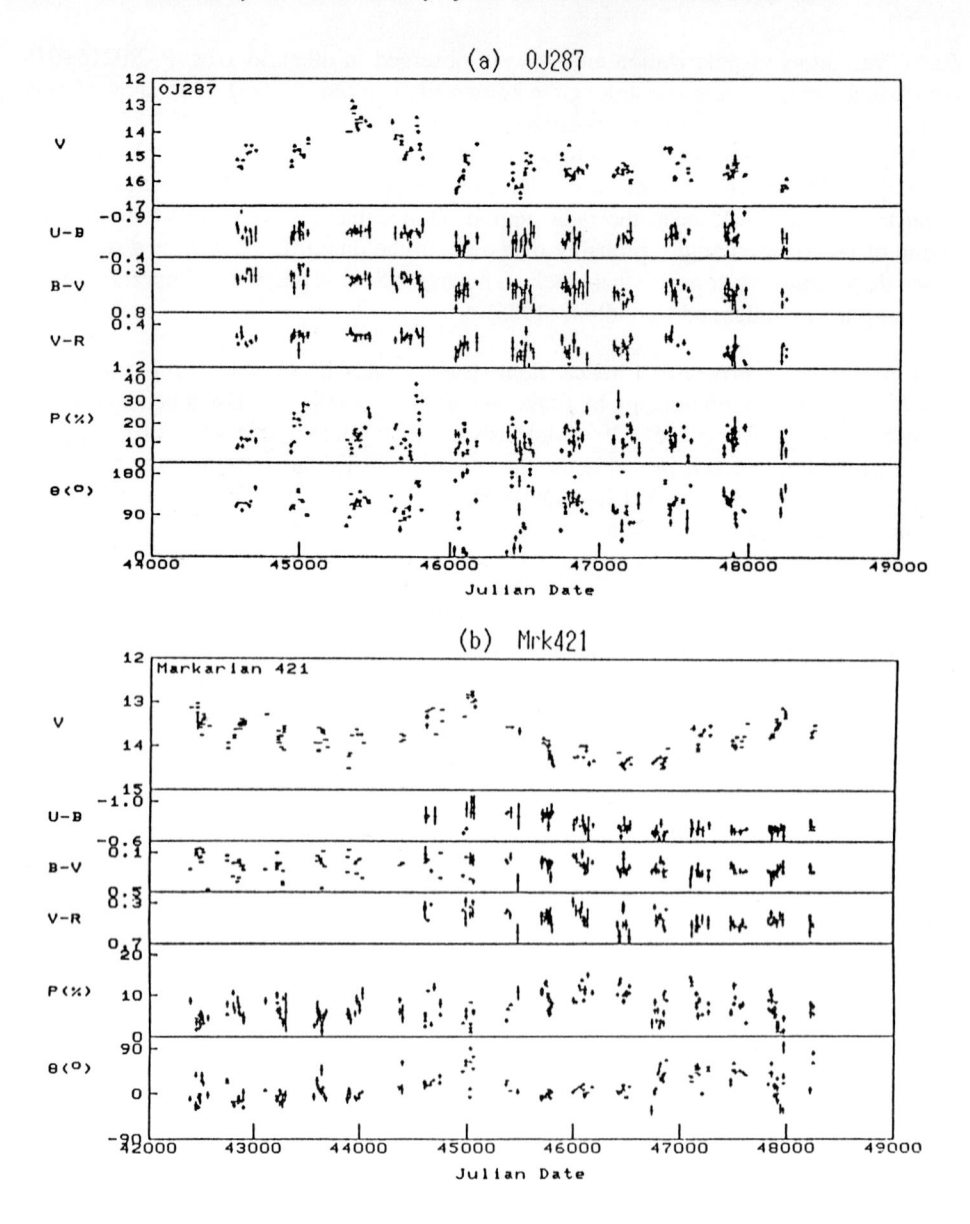

Fig.1. Photometry and polarimetry of (a) OJ287 in 1980-90, and (b) Mrk421 in 1974-90. Data for 1974-81 are after Hagen-Thorn et.al. (1983).

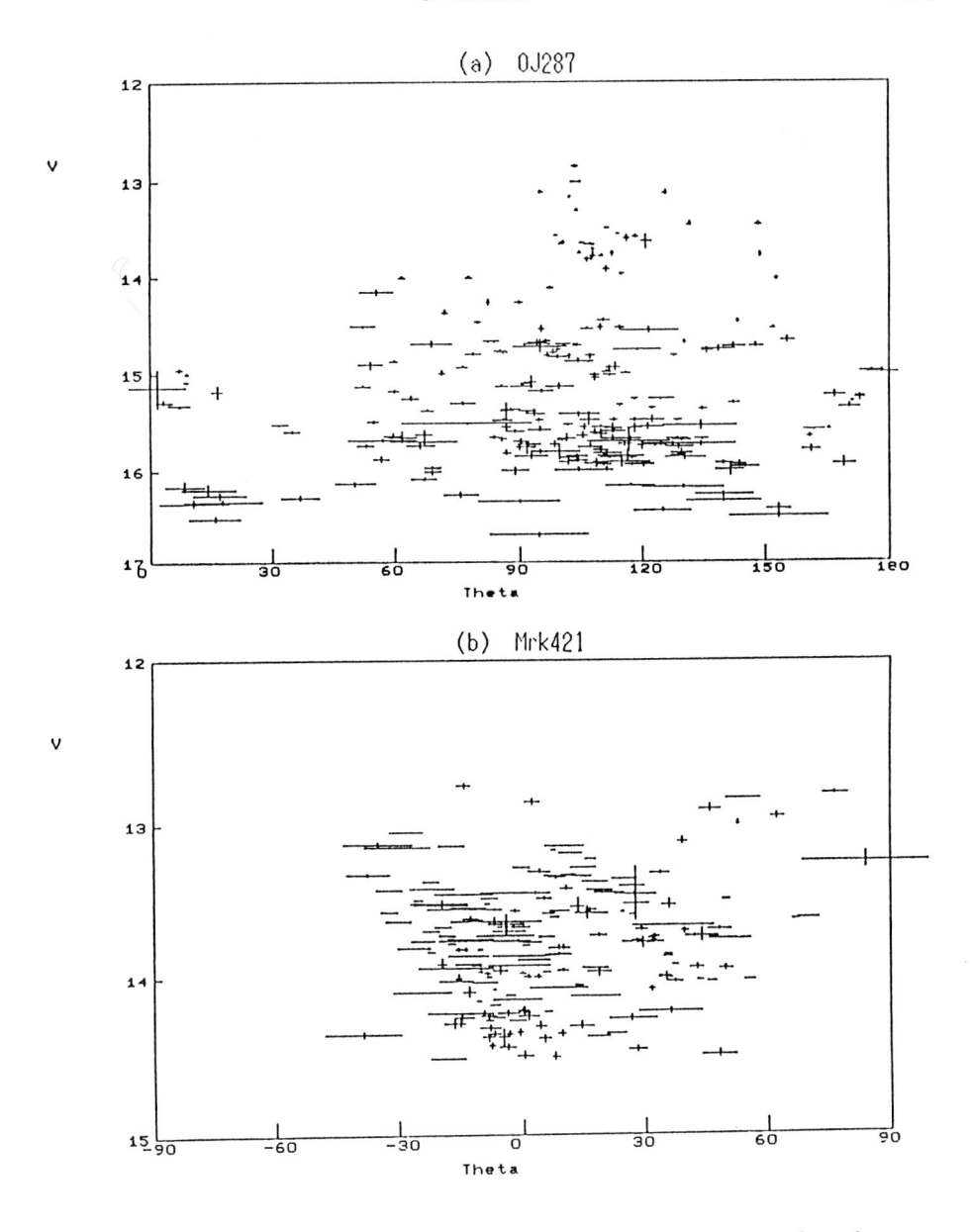

Fig.2. Relations between V magnitudes and polarization angles for (a) OJ287 in 1980-90 and (b) Mrk421 in 1974-90.

First Detection of Nightly Variations in the Frequency Dependent Polarization of OJ 287

L.O. TAKALO[1], A. SILLANPÄÄ[1], M. KIDGER[2] and J.A. de DIEGO[2]

1. Tuorla Observatory
   Tuorla, 21500 Piikkiö,
   Finland

2. Instituto de Astrofisica de Canarias
   La Laguna, Tenerife,
   Spain

Abstract. We present results from a UBVRI photopolarimetric monitoring of blazar OJ 287, during January 1990. For the first time ever we have detected nightly variations in the frequency dependent polarization (FDP) and in the frequency dependent position angle (FDPA) in any blazar. Also we detected a clear rotation in the $p_x p_y$- plane during three nights. The observations can be explained by a model, consisting of a stable jet component and a rotating "clump" in the jet.

1. INTRODUCTION

Frequency dependent polarization (FDP) and/or frequency dependent position angle (FDPA) has been observed in every observed blazar, at least occasionally (e.g. Smith et al. 1987, Mead et al. 1990 and Valtaoja et al. 1991). This FDP and/or FDPA have been observed to change in timescales from days to weeks (Mead et al. 1990, Valtaoja et al. 1991 and references therein). The polarization level has been observed to change in timescales of few hours (e.g. in BL Lac (Moore et al. 1982)). Here we will present the first detection of nightly variations in the FDP and FDPA in OJ 287.

2. OBSERVATIONS

These observation were made on the 2.5 meter Nordic Optical Telescope on La Palma, Canary Islands, using a UBVRI photopolarimeter. The instrument was constructed at the Turku University Observatory. It gives truly simultaneous measurements in all five bands, which is extremely important in our project. A good describtion of the instrument can be found from Korhonen et al. (1984). Using this instrument we monitored OJ 287 during 6 nights in January and March 1990, about 5 hours each night. The nightly data were binned into 3-4 bins; this gave us accurate enough polarization measurements for detecting small variations The observing and data reduction procedures have been described, in detail, by Sillanpää et al. (1991).

3. RESULTS

In Table 1 we list the detected nightly changes in the U and I band polarization and position angle. As can be seen from the table, the polarization changes were larger in the U band than in the I band. On the other hand the position angle behaved the opposite, the I band showed larger nightly variations than the U band. Both the polarization and the position angle varied also from night to night; details can be found from Sillanpää et al. (1991).

Table 1. Observed nightly *changes* in the U and I polarization.

JD date	Pu	σ	Pi	σ	Θu	σ	Θi	σ
7891	0.71	0.25	0.24	0.40	0.7	0.7	6.3	1.5
7892	1.01	0.50	0.35	0.40	2.2	1.5	2.6	1.7
7893	1.12	0.30	1.62	0.50	1.8	1.0	7.6	2.0
7910	0.37	0.25	0.32	0.40	3.8	1.0	2.5	1.5
7952	0.53	0.35	0.67	0.60	4.8	1.5	3.9	2.0
7953	0.81	0.30	0.90	0.50	6.2	1.0	6.7	1.2

σ = the average measurement error during the night.

In Figure 1 we show an example of the detected nightly variations in the FDP and FDPA. The FDP and FDPA showed different kind of behaviour on different nights:

1. During the first three nights of observation at the end of 1989 P(U)/P(I)>1 and PA(U)/PA(I)<1.
2. At the end of January 1990 P(U)/P(I)≈1 and also PA(U)/PA(I)≈1.
3. In the beginning of March 1990 P(U)/P(I)<1 and PA(U)/PA(I)≈1.

This is the first time that nightly changes in the FDP and/or FDPA have been detected in any blazar. For a more detailed discussion of this behaviour see Sillanpää et al. (1991).

If a simple two-component model (e.g. Valtaoja et al. 1991 and references therein) holds, the track of variation of the Stokes parameters is expressed by a straight line in IQU space (Hagen-Thorn, 1980). Similar trends should naturally be visible in the $p_x p_y$-plane, which is clearly not the situation in our observations as we see from Figure 2. The even more interesting feature in this figure is the clear rotation of the polarization vector during these three nights.

This makes it obvious that something is rotating in this object and the best explanation could be a highly polarized, small-size "clump" rotating inside the relativistic jet. During these first three nights we also observed a clear anti-correlation between the flux density and the degree of polarization, ie: increasing flux means decreasing polarization, and vice versa.

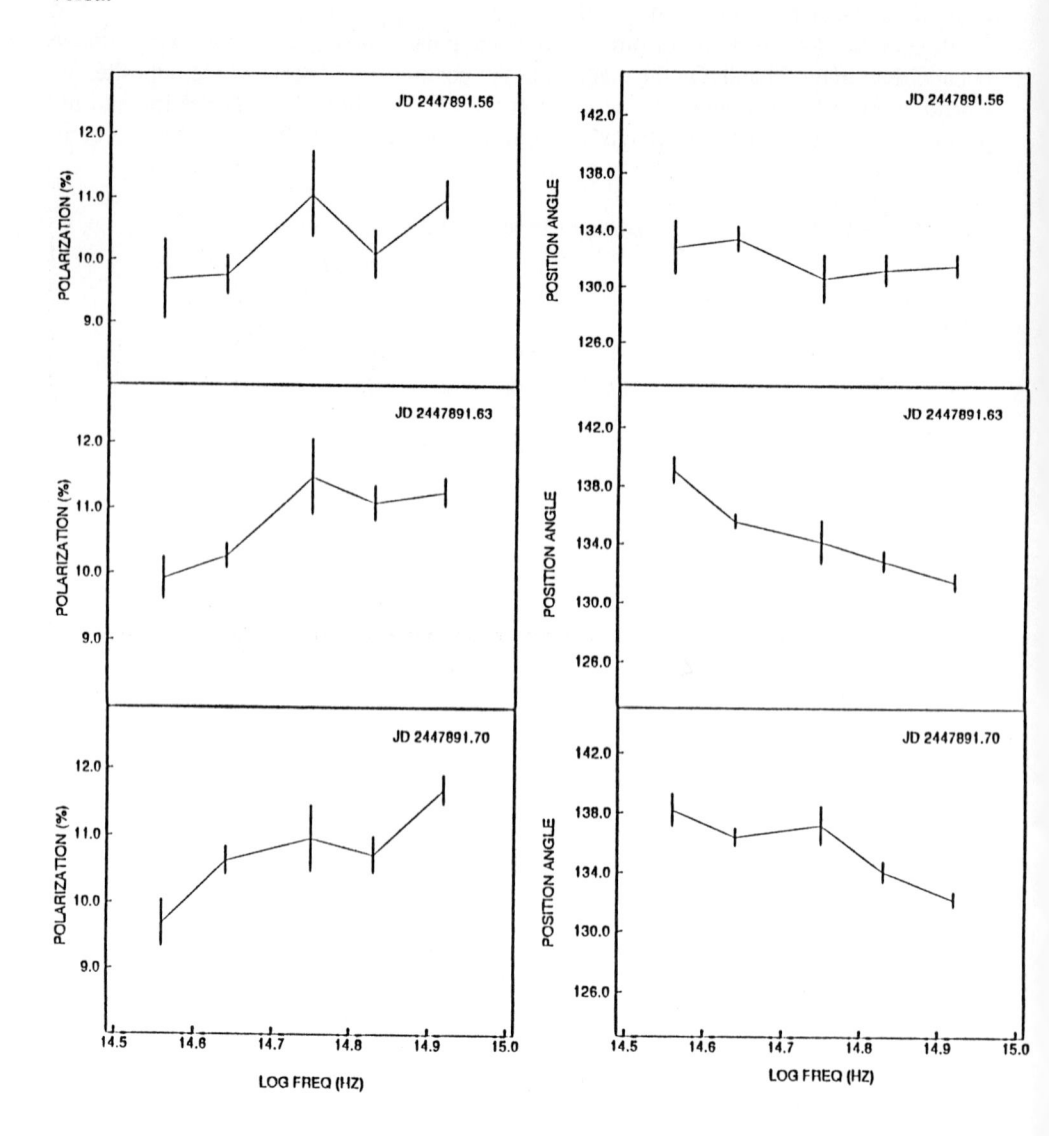

Figure 1. The FDP and FDPA on December 30. 1989. Note the large change in FDPA.

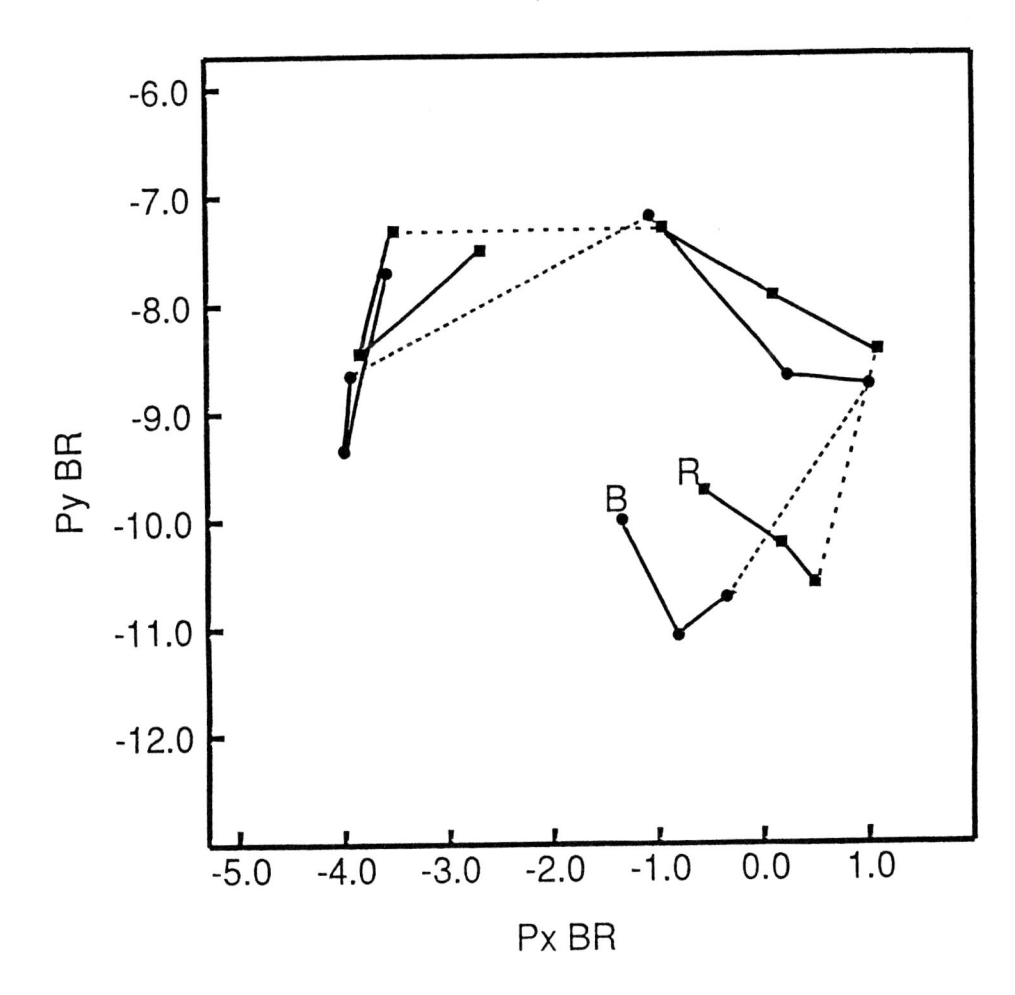

Figure 2. The rotation in the Px Py plane.

This correlation was strongest at the higher frequencies (the correlation coefficient is 0.91, which indicates a significance
of 99 %). We detected, at the same nights, also a correlation between the polarization and the position angle; A decrease in the position angle with increasing polarization. Similar trend is also visible in the observations at the beginning of March (for details see Sillanpää et al. 1991).

We can explain this behaviour by a slightly polarized stable jet and highly polarized, shocked "clumps" rotating inside the jet. The magnetic field of at least one "clump" is almost orthogonal to the jet's magnetic field. If the degree of the polarization in this "clump" is something like 10-15 times the polarization of the jet (Sillanpää, 1989), so even a very slight brightening of this shocked component can weaken the total polarization.

4. MODEL

Many authors have explained the polarization behaviour of blazars more or less succesfully with a two-component model (eg: Kikuchi et al. 1976, Hagen-Thorn 1980, Puschell et al. 1983, Holmes et al. 1984, Brindle et al. 1986, Smith et al. 1987, Sillanpää 1989, Ballard et al. 1990, Valtaoja et al. 1991). A model with more than two components has been used for BL Lac in at least two papers (Moore et al. 1982 and Brindle et al. 1985). A common feature to all these models has been a relativistic jet pointing closely towards us. This is the accepted standard model. It is also quite clear that this jet alone can't explain the polarization behaviour of blazars. But, if we have a relativistic jet emanating from the nucleus of the host galaxy, a natural extension to this theory are shocks in the jet. These shocks could be caused by collisions of relativistic electrons with density perturbations in the interstellar medium. If these highly polarized shocks are rotating around the jet (or if the cloud of relativistic electrons is moving through a helical magnetic field (Königl and Choudhuri 1985)), it is easy to understand the rotation of the polarization vector. If we have many shocks rotating in the jet at the same time there should only occasionally be a clear correlation between flux and polarization as well as the position angle of polarization. This model can also easily explain the quasi-periodicity of OJ 287 seen during the second night of observation (this periodicity and the model will be presented in detail by Kidger et al. 1991) because if a strong "clump" is rotating around the jet, its Doppler factor is changing all the time during the revolution and the flux is allways highest when the "clump" is nearest to the line of sight.

5. CONCLUSIONS

We have presented results from a photopolarimetric monitoring of OJ 287, showing for the first time, that FDP and FDPA can change during one night in this object. We can summarise our polarimetric results as follows:

1. Both the degree and position angle have shown large variations between and even *during* the nights. Variations are also seen in the FDP and/or FDPA.

2. During the first three nights the polarization vector shows a clear rotation.

3. During the same nights we saw a clear anti-correlation between the flux and the degree of polarization; when the flux decreased the polarization increased.

The observed variable polarization and FDP and FDPA in blazars has been explained by a two-component model (e.g. Ballard et al. 1990 and Valtaoja et al. 1991). In this model the polarization behaviour is caused by shocks in the jet. These shocks could be caused by collisions of relativistic electrons with density perturbations in the interstellar medium. If, for some reason, these shocks are rotating in the jet, this could explain the rotation of the polarization vectors.

These observations have clearly shown that when talking about FDP or FDPA in Blazars one MUST have TRULY SIMULTANEOUS observations!

References

Ballard K.R., Mead, A.R.G., Brand, P.W.J.L., and Hough, J.H.: 1990, Mon.Not.R.Astr.Soc.243, 640.
Brindle, C., et al.: 1985, Mon.Not.R.Astr.Soc. 214, 619.
Brindle, C., Hough, J.H., Bailey, J.A., Axon, D.J., and Hyland, A.R.: 1986, Mon.Not.R.Astr.Soc. 221, 739.
Hagen-Thorn, V.A.: 1980, Astrophys.Spa.Sci. 73, 263.
Holmes, P.A., et at.: 1984, Mon.Not.R.Astr.Soc. 221, 497.
Kidger, M., Takalo, L.O., de Diego, J.A., and Sillanpää, A.: 1991 (in preparation)
Kikuchi, S., Mikami, Y., Konno, M., and Inoue, M.: 1976, Publ.Astron.Soc.Japan. 28, 118.
Korhonen, T., Piirola, V., and Reiz, A.: 1984, ESO Messenger
Königl, A., and Choudhuri,A.R.:1985, Astrophys.J. 289, 188.
Mead A.R.G., Blallard, K.R., Brand, P.W.J.L., Hough, J.H., Bailey, J.A., and Brindle, C.: 1990, Astron.Astrophys.Suppl. 83, 183.
Moore, et al.: 1982, Astrophys.J. 260, 415.
Puschell, J.J., Jones, T.W., Phillips, A:C., Rudnik, L., Simpsom, E., Sitko, M., Stein, W.A:, and Moneti, A.: 1983, Astrophys.J. 265, 625.
Sillanpää, A.: 1989, Ph.D. Thesis, University of Turku.
Sillanpää, A., Takalo, L.O., Kikuchi, S., Kidger, M., and de Diego, J.A.: 1991, Astron. J. (in press)
Smith P.S., Balonek, T.J., Elston, R., and Heckert, P.A.: 1987, Astrophys.J.Suppl. 64, 459.
Valtaoja, L., Valtaoja, E., Shakhovskoy, N.M., Efimov, Y.S., and Sillanpää, A.: 1991, Astron.J. 101, 78.

Polarimetric Monitoring of Blazars at the Nordic Optical Telescope

LEO O. TAKALO

Tuorla Observatory
Tuorla, 21500 Piikkiö
Finland

Abstract: We present the first results from a monitoring program of BL Lac objects conducted at the Nordic Optical Telescope on La Palma, Canary Islands. Twelve objects have been observed. All of them show variable polarization, and also, occasionally, frequency dependent polarization (FDP) and/or frequency dependent position angle (FDPA). The FDP is seen to occur more frequently than FDPA.

Several large monitoring programs on blazar polarization have been conducted (e.g. Smith et al. 1987; Mead et al. 1990 and Valtaoja et al. 1991). The polarization of most blazars have been observed to be variable, the shortest variability timescales being few hours (e.g. OJ 287; Takalo et al. 1991). When the observations have been simultaneous in several wavelengths (Mead et al. and Valtaoja et al.), several of the objects have, at least occasionally, shown frequency dependent polarization (FDP) and/or frequency dependent position angle (FDPA). Here we present the first results from a polarimetric monitoring of blazars conducted at the Nordic Optical Telescope during 1989-1990.

The observations were made at the 2.5 meter Nordic Optical Telescope (NOT) on La Palma, using the Turku UBVRI photopolarimeter. This instrument gives *truly* simultaneous measurements in all five bands. A good description of the instrument can be found from Korhonen et al. (1984). The observing procedure has been described in detail by Takalo (1991).

A summary of the results is presented in Table 1, where we have listed the observed range in the V band polarization, the number of times we observed FDP and FDPA and the number of observations. As can be seen from Table 1 all objects show variable polarization, and also FDP and/or FDPA. Examples of these are shown in Figures 1 and 2, respectively. Also from Table 1 one can see that in most objects FDP occurs much more frequently than FDPA. In most of the observed blazars the FDP does not show any prefered "sign" in the FDP; the polarization can be larger either in the I band or the U band, and in most objects this can change from night to night (e.g. PKS 0109+224; Figure 1). In some nearby blazars (e.g. Markarian 421 and Markarian 501), where the host galaxy is prominent and dilutes the synchrotron emission from the nucleus, the FDP has always the same "sign"; the U band polarization being always larger than the I band polarization.

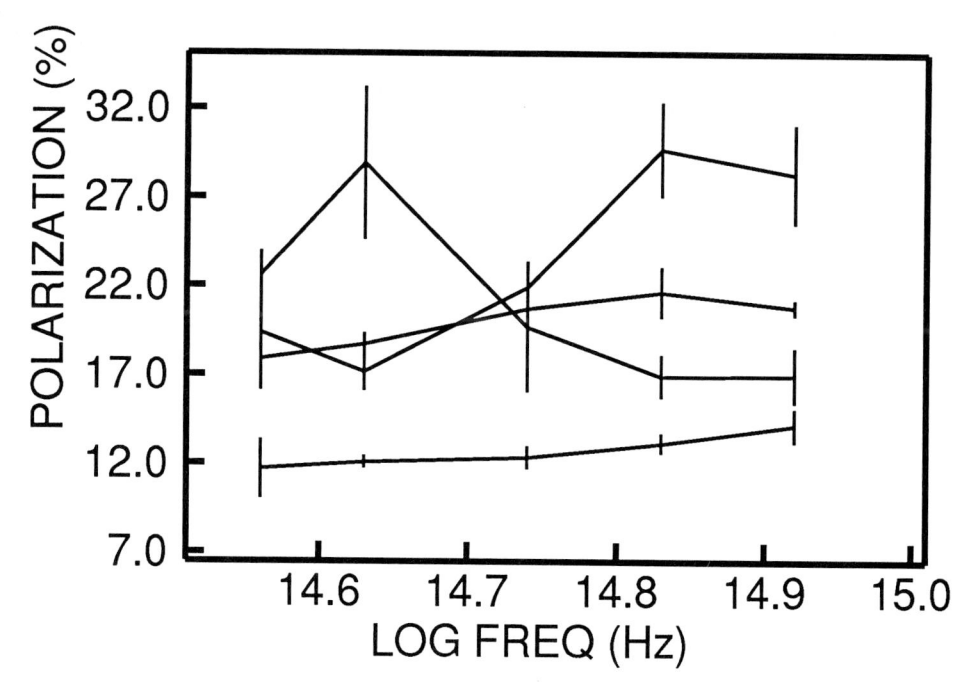

Figure 1. Examples of the observed frequency dependent polarization in PKS 0109+224.

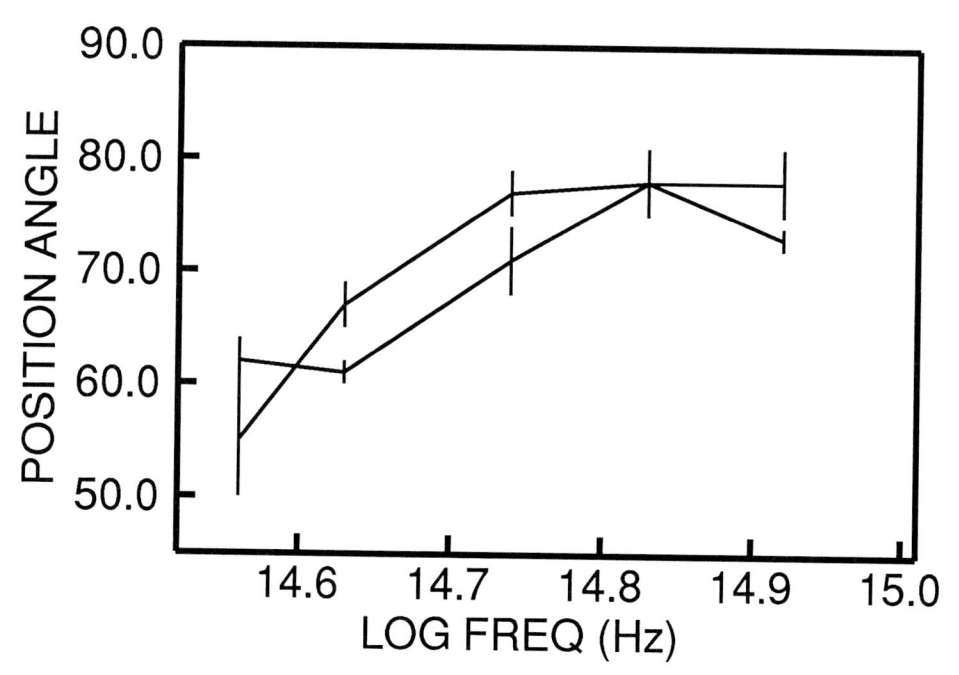

Figure 2. Examples of the observed frequency dependent position angle in PKS 0109+224.

Similar situation is seen in the FDPA, but here the type of variations are less dramatic (Figure 2). Also here we see no difference between the nearby and the other objects.
There seems to be no systematic rule between the occurence of FDP and FDPA; one or the other can be present recardless, if the other one is also present. No correlation can be found between the polarization level and the brighness of an object, nor between the polarization and the position angle. A more detailed statistical analysis of this data is presented in the accompanying paper by Nilsson et al. (1990).

Table 1. Summary of the monitoring results.

Object	p(V)(%)	FDP	FDPA	Number of observations
PKS 0109+224	12-26	9	3	13
3C 66A	11-17	2	1	6
AO 0235+164	13	no	1	1
OF 038	8-15	4	2	8
PKS 0735+178	2-18	5	2	9
OI 090.4	4-7	no	1	2
OJ 287	7-17	4	3	19
Mk 421	1-7	3	1	4
ON 325	8-9	2	1	2
OQ 530	1-7	4	2	4
Mk 501	2-3	8	4	8
BL Lac	1-16	11	10	23

References

Korhonen, T., Piirola, V., and Reiz, A.: 1984, ESO Messenger.
Mead, A.R.G., Ballard, K.R., Brand, P.W.J.L., Hough, J.H., Bailey, J.A., and Brindle, C.: 1990, Astron.Astrophys.Suppl. **83**, 183.
Nilsson, K., Takalo, L.O., and Sillanpää, A.: 1991, This meeting
Smith, P.S., Balonek, T.J., Elston, R., and Heckert, P.A.: 1987, Astrophys. J. Suppl. **64**, 459.
Takalo, L. O.: 1991, Astron.Astrophys.Suppl. (submitted)
Takalo, L.O., Sillanpää, A., Kidger, M., and de Diego, J.A.: 1991, This meeting.
Valtaoja, L., Valtaoja, E., Shakhovskoy, N.M., Efomov, Y.S., and Sillanpää, A.: 1991, Astron. J. (in press)

Statistical Analysis on the Frequency Dependent Polarization in Blazars.

K. NILSSON, L.O. TAKALO, and A. SILLANPÄÄ

Turku University Observatory,
Tuorla, 21500 Piikkiö,
Finland

Abstract: We report on a preliminary statistical analysis on a set of UBVRIphotopolarimetric observations of blazars. No clear correlation can be found between the distributions of the I band polarization level and the frequency dependence of the polarization (FDP) or the frequency dependence of the position angle (FDPA). The occurence of FDP and/or FDPA do not show any correlation either. A linear correlation can be found between the ratio of P(U)/P(I) and the P(I) with 99.9% confidence, but no correlation is found between P(U)/P(I) and P(U).

In recent years new observations of polarization in blazars have clearly shown that in all blazars the polarization depends on frequency (FDP) at least sometimes, also the position angle shows frequency dependence (FDPA). This new data comes mainly from two studies, in which the observations were simultaneous in several wavelengths (Mead et al. 1990 and Valtaoja et al. 1991). It is crucial that the observations are simultaneous in all observed wavelenghts, if one wants to study either FDP or FDPA. This is clearly shown by Takalo et al. (1991), in their study of OJ 287, in which they observed changes in the FDP and FDPA in timescales of hours. Ballard et al. (1990) did a statistical analysis on the Mead et al. (1990) data, trying to find correlations between the occurence of FDP and the polarization level or FDPA and between the polarization level and FDPA. They found no clear correlations. Valtaoja et al performed a similar analysis, and found a correlation between the ratio of P(U)/P(I) and P(I). Here we present preliminary results from a statistical analysis of a new set of polarization observations of blazars.

The data used in this study are from the accompanying paper by Takalo (1991). In this study we have excluded the data for the closeby objects Markarian 421 and 501, because in these objects the FDP is mostly caused by the dilution by the host galaxy.

As can be seen from Figure 1, the distributions between the I band porization with no FDP and with FDP are similar. The same can be said about the distributions with and without FDPA.

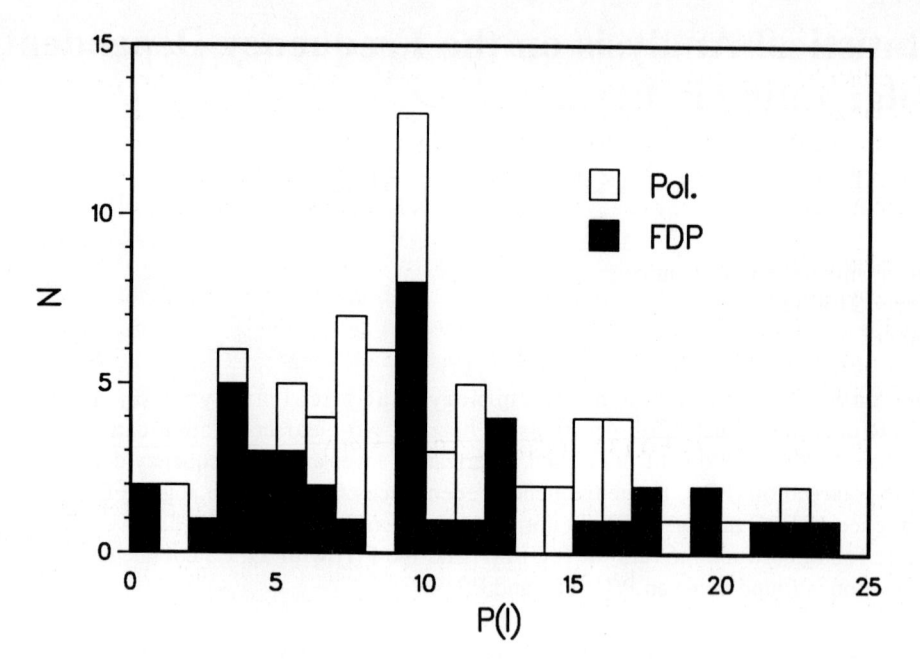

Figure 1. The observed distribution of the I band polarization. The black areas indicate the occurence of FDP.

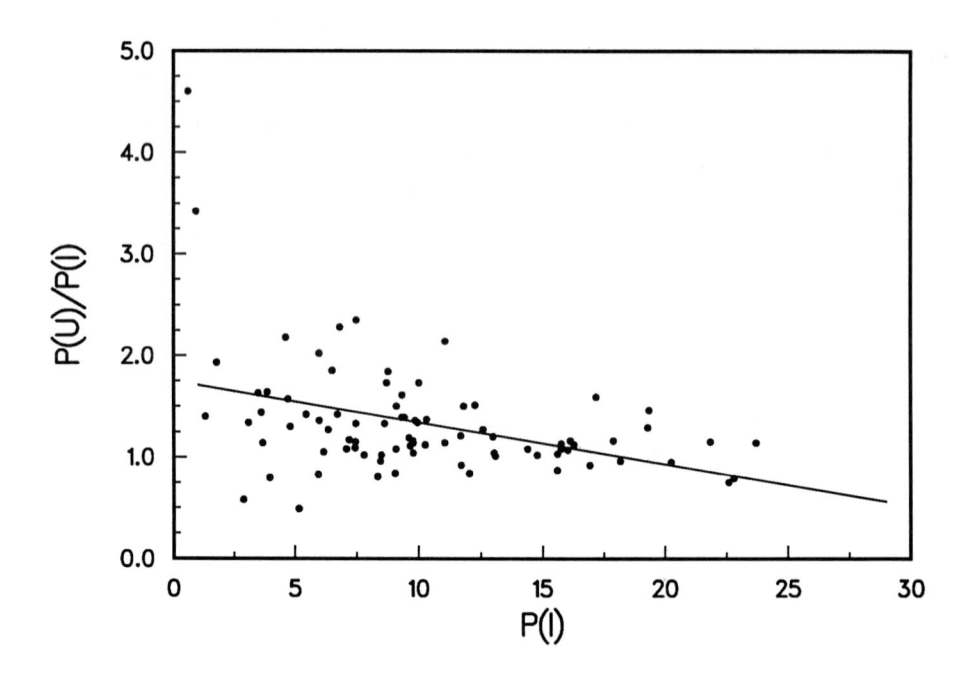

Figure 2. Correlation between the ratio P(U)/P(I) and P(I).

Also we found that no difference exits between the FDP and FDPA distributions. Testing these distributions using the Kolmogorov-Smirnov test, under the null hypothesis that the distributions are the same gave the following significance levels: FDP- polarization 58 pre cent; FDPA- polarization 46 per cent; FDP-FDPA 63.5 per cent.
These significance levels cleary show that there are no differencies between the observed distributions. The same was observed by Ballard et al. (1990) in their study of 44 blazars.

In order to look for a possible correlations between the strenght of the FDP and the polarization levels we have calculated the ratio P(U)/P(I) and plotted this against P(U) and P(I); in Figure 2 we show the P(U)/P(I) versus P(I) plot. The line in the figure is a least square fit to the data. In the figure we see a clear correlation at the >99.9% level. This correlation has the form:

$$P(U)/P(I) = (1.75 \pm 0.13) - (0.041 \pm 0.011) * P(I)$$

This is very similar to the relation found by Valtaoja et al. (1991), using another data sets. On the other hand we found no correlation between P(U)/P(I) and P(U). Valtaoja et al (1991) found a correlation also between these variables. The fact that we see no correlation is most likely to be due to the differences in the data sets.

References

Ballard K.R., Mead, A.R.G., Brand, P.W.J.L., and Hough, J.H.: 1990,
 Mon.Not.R.Astr.Soc. **243**, 640.
Mead A.R.G., Ballard, K.R., Brand, P.W.J.L., hough, J.H., Bailey, J.A., and
 Brindle, C.: 1990, Astron.Astrophys. Suppl. **83**, 183.
Takalo, L. O. 1991, This meeting
Takalo, L.O., Sillanpää, A:, Kidger, M., and de Diego, J.A.: 1991, This meeting.
Valtaoja, L., Valtaoja, E., Shakhovskoy, N.M., Efomov, Y.S., and Sillanpää, A.: 1991,
 Astron. J. (in press)

Rotation of Position Angles in Blazars

A. SILLANPÄÄ and L.O. TAKALO

Tuorla Observatory
Tuorla, SF-21500 Piikkiö
Finland

Abstract. Simultaneous rotation in the polarization position angles at centimeter and optical wavelengths at the rate $\approx 2°\text{-}2.5°$ year^{-1} is found between two major outbursts in 1971-72 and 1983 in OJ 287. A similar trend is visible in the optical region after the year 1983 and also in the radio region in the new observations during the autumn 1989. One new explanation for the apparent rotation is a precessing jet pointing almost directly towards us. The estimated period of this precession is as short as ≈ 150 years. Another explanation for this phenomenon would be that the proportion of the shocked synchrotron radiation to the total flux has diminished between these years. A clear rotation has been observed in the optical polarization position angle in BL Lac during an outburst in autum 1989.

1. INTRODUCTION

The apparent rotation of the Stokes vector around the origin for several blazars (but not for OJ 287) have been reported in many papers (e.g. Ledden and Aller 1978; Aller, Hodge, and Aller 1981; Altschuler 1982;). The rotation rates have differed from $7°$ day^{-1} to $15°$ year^{-1} (Saikia and Salter 1988). Many sources have shown similar rotation at two or three separate radio frequencies but only once before simultaneously in the optical and radio region (Kikuchi et al. 1988) in a few day timescale. A straightforward explanation of the phenomenon is, of course, a physical rotation of the magnetic field within a compact object (Ledden and Aller 1978). Another clear rotation model has been considered by Pineault (1980), who presents a model in which a hotspot or giant flare in an accretion disk around a supermassive black hole produces a rotator event by its orbital motion within the disk. Blandford and Königl (1979) were able to make a quantitative fit of the data with a model invoking an aberration effect in a relativistic jet directed near the line of sight. Königl and Choudhuri (1985) considered the case where the shock illuminated radiation in the non-axially symmetric configuration causes large polarization angle (PA) swings and steplike PA changes, as well as more linear changes of PA.

This paper describes a collected polarization data set of OJ 287 between 1974-1990 in the optical region and 1974-1985 in the radio region (because there exist not published data after this year) in order to consider the apparent rotation of the position angle in the polarization.

A totally different type of rotation has been observed in BL Lac in the autum 1989, when the position angle rotated ~200 degrees in ten days. This rotation was very nicely correlated with the increasing flux.

2. DATA

OJ 287

The radio data used in this paper are published in Aller et al. (1985). We have used only 8.0 GHz data which have the best time coverage. The behavior at different frequencies has been, however, very similar. The optical data have been collected from various sources (see Sillanpää 1991 and references therein).

BL Lac

The BL Lac observations were made at the Nordic Optical Telescope during autum 1989, using the UBVRI photopolarimeter. A good describtion of the instrument can be found from Korhonen et al. 1984. The observing and data reduction procedures have been described by Takalo (1991).

3. RESULTS

In Fig.1 we see the optical and the radio position angles versus time in years for OJ 287. We have rejected the position angle values 0°-30° and 150°-180° because of the 180° ambiguity problem in the polarization position angle. The overall trend in both region has been quite similar. The position angle rose from 75° to $\approx$ 110° between 1974-1990 in the optical region. This gives a slope $\approx 2.0°$ year^{-1} with a standard deviation to the slope of 0.27. The Pearson's correlation coefficient is 0.70 and the significance of the slope is >99.99 %. In the radio region the slope has been $\approx 2.5°$ year^{-1} with the standard deviation 0.45 between 1974-1985. Pearson's correlation coefficient is 0.46 and the significance of the slope is also more than 99.99 %. The rising trend of the position angle has been very similar after the year 1985 in the radio region because Θ was $\approx 120°$ at 8 GHz in autumn 1989 (Hughes et al., 1990, private communication). This position angle (marked x) is exactly the same as predicted by the trend in Fig.1.

In Fig. 2. we show the observed polarization and position angle behaviour for BL Lac during the optical outburst in 1989. As can be seen from the figure the polarization showed large daily variations and at the same time the position angle rotated 200 degrees. Similar behaviour has been observed in OJ 287 by Kikuchi et al. (1988). The optical

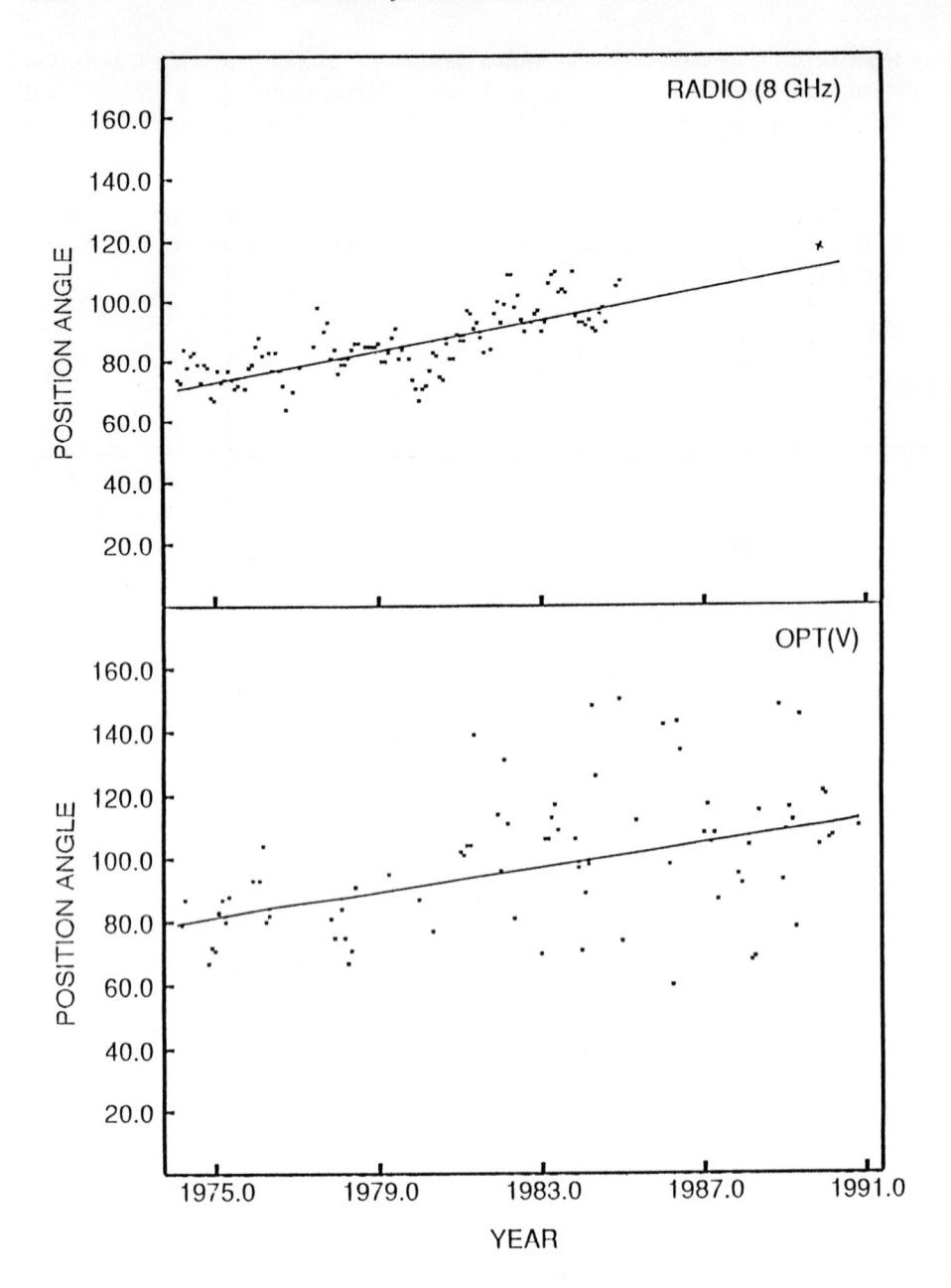

Figure 1. The rotation of the position angle of the polarization in the radio and optical wavelengths between the years 1974 and 1990.

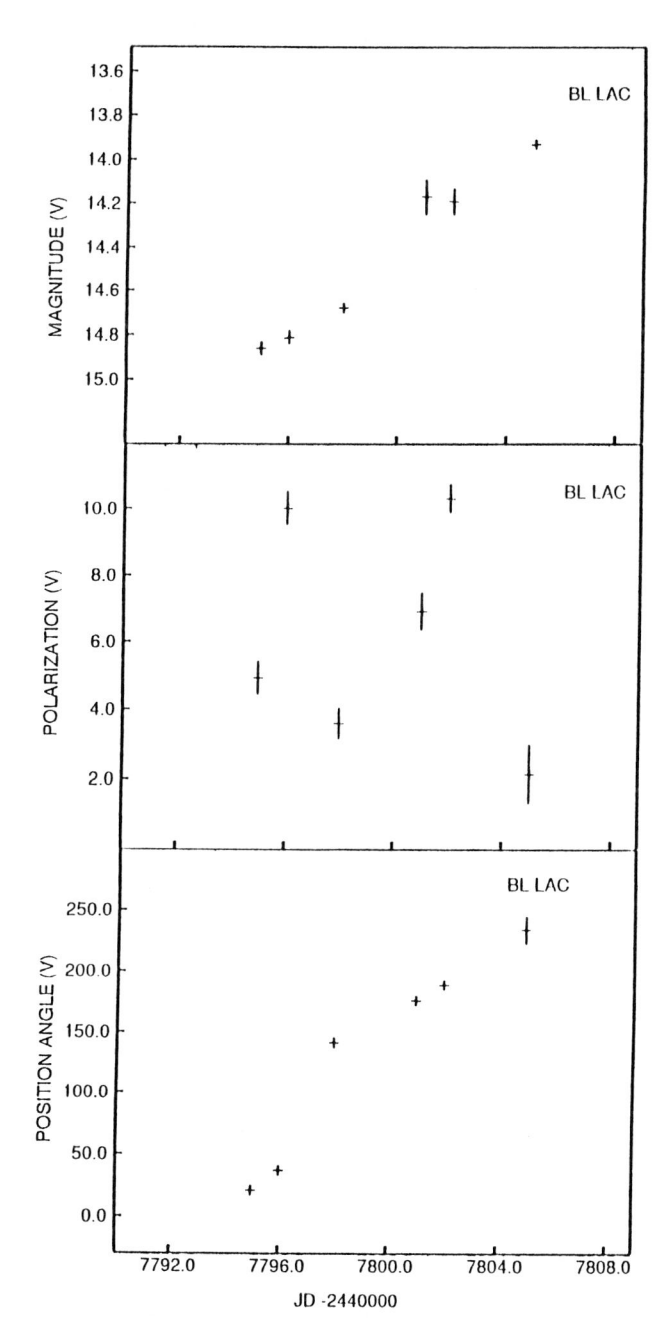

Figure 2. The observed V-band magnitude, polarization and position angle in BL Lac

outburst reached its peak one month later, when the V- magnitude was 13.6 and the polarization was ~0% (Sillanpää et al. 1991).

4. DISCUSSION

The observed position angle rotation in OJ 287 can be explained by two possible models: A precessing jet model and a model in which the share of the shocked flux from the total flux (in the jet) has diminished quite linearly during these years. Sillanpää (1991) has discussed these models and compared them in detail. We have good reason to favour the second model, since besides the rotation of the polarization, also the colour of OJ 287 has changed during the same time (Takalo and Sillanpää, 1989). The colour index B-V has changed from $\approx$ 0.35 to 0.60 between the years 1971-1987 and during the winter 1989-1990 it has been still bigger ($\approx$ 0.62; Kikuchi, 1990, private communication). This behavior can be explained by the steeper spectrum of the initial jet which has became more dominant in the total flux density between these years. The precessing jet model has difficulties to explain this fact. On the other hand the linear trend in the change of the position angle can be more easily explained by the precessing jet model.

The position angle rotation and the polarization variability seen in BL Lac can be explained by the model proposed by Königl and Choudhuri (1985), in which the shock illuminated radiation in non-axially symmetric jet causes the position angle swings. This rotation will be discussed in more detail by Sillanpää et al. (1991).

We have showed that two different position angle rotations can occur in blazars, indicating the existence of at least two different physical reasons for these rotations.

References
Aller, H., Aller, M., Latimer, G., Hodge, P.: 1985, *Astrophys. J. Suppl.* **59**, 513
Aller, H., Hodge, P., Aller, M.: *Astrophys. J. (Letters)* **248**, L5
Altschuler, D.: 1980, *Astron. J.* **85**, 1559
Blandford, R., Königl, A.: 1979, *Astrophys. J.* **232**, 34
Kikuchi, S., Inoue, M., Mikami, Y., Tabara, H., Kato, T.: 1988, *Astron. Astrophys. Letters* **190**, L8
Korhonen, T., Piirola, V., and Reiz, A.: 1984, *ESO Messenger*
Königl, A., Choudhuri, A.: 1985, *Astrophys. J.* **289**, 188
Ledden, J., Aller, H.: 1979, *Astrophys. J. (Letters)* **229**, L1
Pineault, S.: 1980, *Astrophys. J.* **241**, 528
Saikia, D., Salter, C.: 1988, *Monthly Notices Roy. Astron. Soc.* **26**, 93
Sillanpää, A.: 1991, *Astron.Astrophys.* (in press)
Takalo, L.O., Sillanpää, A.: 1989, *Astron. Astrophys.* **218**, 45
Sillanpää, A., Takalo, L.O., and Nilsson, K.: 1991, (in preparation)
Takalo, L.O.: 1991, *Astron.Astrophys.Suppl.* (in press)

Accretion Disk Models for Microvariability

PAUL J. WIITA, H. RICHARD MILLER, NAVARUN GUPTA AND SANDIP K. CHAKRABARTI[*]

Georgia State University, Atlanta and *Tata Institute of Fundamental Research, Bombay and International Centre for Theoretical Physics, Trieste

1 INTRODUCTION

Careful observations over the past few years of a substantial number of BL Lacertae objects and OVV quasars have been able to conclusively demonstrate the reality of intraday changes or microvariability in the optical (e.g., Miller 1988, Miller, Carini & Goodrich 1989, Carini 1990, Miller et al. 1991, Wagner 1991) and radio (e.g., Quirrenbach et al., 1989, Krichbaum, 1991). In this paper we present models for the optical microvariability discussed in the companion paper of Miller et al. (1991). Our models involve the random excess emission produced by flares or "hot-spots" on the accretion disks around supermassive black holes that are believed to provide the ultimate powerhouse for AGN. Preliminary results were reported in Wiita et al. (1991). Related models, concentrating on rapid X-ray variability, have been independently proposed by Abramowicz et al. (1991), Zhang & Gang (1991) and Zhang (1991).

As discussed in Carini (1990) and Miller et al. (1991), the use of CCD cameras in N-star photometric modes allows for extremely accurate and rapid differential photometry; the extragalactic object can be directly compared with several stars in the same field of view and any changes due to atmospheric effects can be essentially eliminated. For a few sources these optical data are now sufficiently numerous that power density spectra (PDS) can be computed. The best data on OJ 287 give $s \equiv d\log(PDS)/d\log f$ ranging between -1.3 and -1.9, while for BL Lacertae, the best data train yielded $s \approx -1.4$. While some X-ray measurements of AGN yield PDS slopes near -1, or flicker noise (Lawrence et al. 1987), others give somewhat steeper slopes (Treves et al. 1989), more in accord with these optical observations and those of optical polarization in BL Lac (Moore et al. 1982). Unfortunately, the data trains are not yet long enough, nor the signal-to-noise ratio high enough, to detect clear breaks in the PDS. Such breaks might indicate the inner edge (at the high temporal frequency end) or outer edge (at the low frequency end) of an accretion disk.

No periodicities have been detected in this optical microvariability data. Only on one night of the Georgia State University observations of 18 sources was even a hint of periodicity seen. On 1986 Nov. 9, OJ 287 exhibited a weak peak in the power spectrum at about 32 minutes, which was not statistically significant. Thus the earlier claims for periodic variations on timescales of tens of minutes (e.g., Visvanathan & Elliot 1974; Carrasco et al. 1985) have not been confirmed. Equivalently good data for 0716+714 over a three week period indicates possible quasi-periodicities of ~ 1 day and ~ 7 days (Wagner 1991).

2 ACCRETION DISK MODELS

2.1 Stationary β-disks

The lack of clear periodicities, which could be nicely identified with single 'hot-spots' on an accretion disk, or with a dominant global oscillation of a disk, is a bit disappointing for disk afficionados. The power spectra have slopes which are neither white noise nor flicker noise. It thus becomes likely that if accretion disks are fundamentally responsible for the observed rapid variations, more than one region of the disk is involved at any one time. Thus the questions we have set out to answer are: can an ensemble of 'hot-spots' or flares on the surface of an accretion disk produce light-curves and PDS similar to those observed?

The absence of large 'big blue bumps' in BL Lac spectra does not provide the strong evidence in favor of accretion disk components that exists in quasars (e.g., Sun & Malkan 1989). But, it has been shown that in some of the best studied cases, accretion disks seem to contribute to blazar spectra (Wandel & Urry 1991). We employ simple β accretion disk models, which are not subject to the large scale instabilities expected of the more standard α disks (Lightman & Eardley 1974) and whose spectra are easier to calculate because they are denser (Wandel & Petrosian 1988) to yield the non-variable portion of the disk emission. These disks are described by M_8, the mass of the central black hole divided by $10^8 M_\odot$; the accretion rate, $\dot{m} = \dot{M}c^2/L_{Edd} = 4.4(\dot{M}/M_\odot yr^{-1})M_8^{-1}$; $\beta = t_{\phi r}/P_{gas}$; and $a = J/M$, the angular momentum parameter. For simplicity, we treat Schwarzschild black holes and set $a = 0$.

The temperature, azimuthal velocity (in units of c), and half-thickness as functions of radius for such disks are given by

$$T(r) = (6.4 \times 10^7 K)(\beta M_8 \dot{m}^2)^{-0.2} r^{-0.9},$$

$$v(r) = (r - 2)^{-1/2}, \quad \text{and}$$

$$h(r) = (1.4 \times 10^{13} \text{cm}) M_8 \dot{m} \phi,$$

where the relativistic correction, $\phi = 1 + (r/3 - 1)^{-1} - 2(r/3 - 1)^{-1/2}$ (Novikov and Thorne 1973), and $r \equiv R/r_g$, with $r_g \equiv GM/c^2$, so $r_{in} = 6$.

Since these disks are optically thick throughout, the emergent spectrum is quite simple, at least for $\nu \leq 10^{17}$ Hz, the bands we care about, and is essentially the sum of black bodies. The spectrum is modified by both gravitational redshifts and Doppler shifts, and these contributions depend upon both the radius and the angle at which each portion of the disk is observed. If the angle between the local disk normal and the line of sight is ψ (we ignore the curvature of the paths the emitted photons follow in the vicinity of the black hole, which is valid for the regions from which most of the optical photons are emitted), the angle between the direction of motion of that patch of disk and the line of sight is θ, while γ is the special relativistic Doppler factor, and $G = \nu_{obs}/\nu_{em} = (1 - 2/r)^{1/2}$ is the gravitational redshift, we have

$$dI_{\nu,obs}d\nu_{obs} = \frac{G^4 \cos\psi}{\gamma^4(1 + v\cos\theta)^4}dI_{\nu,em}d\nu_{em}.$$

The quiescent disk spectrum is then produced by adding up the contributions from many zones (typically 200 in radius and 50 in azimuthal angle, ϕ) as observed at an inclination angle i to the normal to the equatorial plane of the disk. For these models the disk outer radii were taken to range between 1000 and 2000, $1 \leq M_8 \leq 30$, $0.1 \leq \dot{m} \leq 10$, and $0.1 \leq \beta \leq 1.0$. At this stage in these calculations effects due to inner parts of these geometrically thin disks being eclipsed by the outer portions of the disk are not incorporated; this is equivalent to our assuming $i \leq 75°$.

2.2 Phenomenological Hot-spots and Flares

To date, these models are phenomenological in that we allow for putative disk instabilities by computing the additional flux due to a group of 'hot-spots' (or flares) which are randomly distributed in both r and ϕ. The time at which they (instantaneously) turn on is also randomly distributed over the duration of the simulation. The basic time unit is taken as the period for an orbit at the inner edge of the accretion disk, which is $P_{in} = 3.71 \times 10^4 \, M_8$ s. All of the phenomenological simulations discussed here went on for 1000 timesteps, with $\Delta t = 0.01$ or 0.1, corresponding to measurements lasting 10 and $100 P_{in}$, respectively. The flares were turned on for a fixed multiple of P_{in}, which ranged between 0.5 and 10.

At each timestep a new azimuthal angle for each of the currently turned on flares is computed. These contributions to the observed flux as seen from the assumed inclination angle vary since new values of ϕ_i change θ_i and ψ_i.

So far we have considered two heuristic models for the variable components. The first is analogous to a 'hot-spot', where one could imagine a denser region, caused by density waves, or an extra pulse of mass moving through the disk, brightening that region but not much raising its temperature. For this situation we multiply each

$dI_{\nu,i,em}$ by a constant factor, taken as 10 or 100. The second model, more similar to a stellar flare, assumes that the temperature of each randomly selected zone is raised by a constant factor, taken as 5 or 10. In §3 below we sketch out the likely physics behind these heuristic models.

2.3 Results

The additional contributions to the observed flux produced by the varying regions were added to the quiescent disk spectrum, yielding light curves. Then the PDS was computed from these variability data, after the mean and the linear trend (which was always positive, due to the assumption about the way in which the flares were turned on and off) were subtracted off.

A total of 44 simulations with the parameters mentioned above have been analyzed. Between 12 and 150 flares were turned on (and usually off) during each run. Typical multi-color light curves are shown in Figure 1. Simulated power spectra are shown in Figures 2 and 3, with a direct comparison with observations illustrated in Figure 2.

The results of these phenomenological simulations can be summarized as follows: (1) The general form of observed light curves are reproduced well with plausible heuristic parameters. (2) Power law PDS are produced, with $\langle s \rangle = 1.76 \pm 0.25$. (3) More flares $\longrightarrow$ steeper PDS. (4) 'Hot-spots', where $[dI'_\nu = K_1 dI_\nu]$ $\longrightarrow$ shallower slopes than 'flares', where $[T' = K_2 T]$. (5) For short-lived flares ($\Delta t_f < P_{in}$), the inclination hardly affects the PDS, but the magnitude of the variations declines as the inclination rises. (6) For long-lived flares ($\Delta t_f > P_{out}$), eclipses are important and better power-laws are produced as i increases (e.g. Abramowicz et al. 1991). (7) Color variations differ for 'hot-spot' or 'flare' models, with the latter being bluer. (8) Higher frequency optical/UV bands usually emit more flux than lower frequency bands, but at low brightness states this can reverse. (9) The PDS of different colors in the optical and UV bands are very similar.

3 PHYSICS OF FLARES AND HOT-SPOTS

3.1 Unstable Spiral Shocks

We have performed hydrodynamical simulations of an accretion disk in a binary system (Chakrabarti & Wiita 1991). The resultant tidal forces can induce spiral shocks which break into pieces, and then dissipate. New shock fragments then appear, and they provide a natural physical mechanism to yield the random 'hot-spots' required by our phenomenological models.

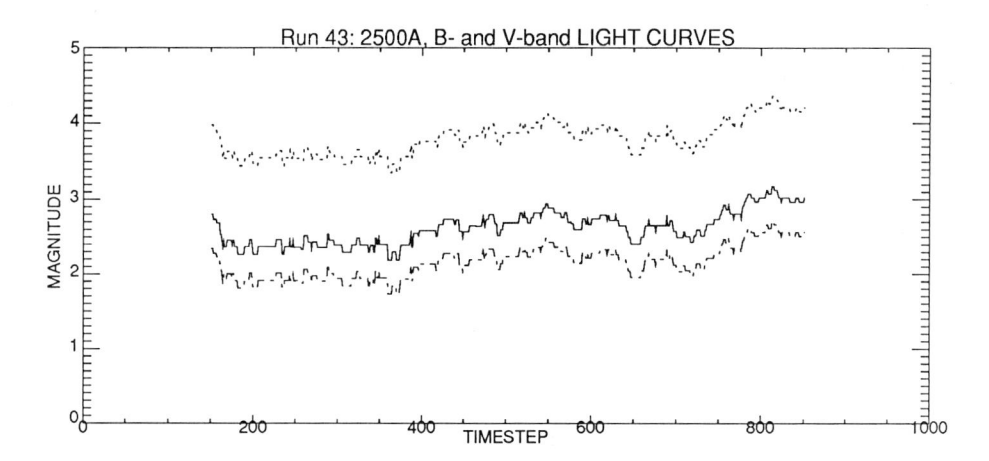

Fig. 1. Light curves for a heuristic model with $M_8 = 10$, $\dot{m} = 0.1$, $\beta = 0.1$, $i = 0°$, $t_{max} = 10(P_{in})$, $\Delta t = 0.01$, $t_{on} = 1.0$, for 150 'flares'. One 'timestep' corresponds to 1.04 hours. The solid curve corresponds to the 'magnitude' of the B band, the dotted to that of the IUE long wave (≈ 2500 Å) band, and the dashed to the V band, plotted so that the UV band is brightest with an arbitrary zero point. All bands have PDS of slope $s \approx -1.9$.

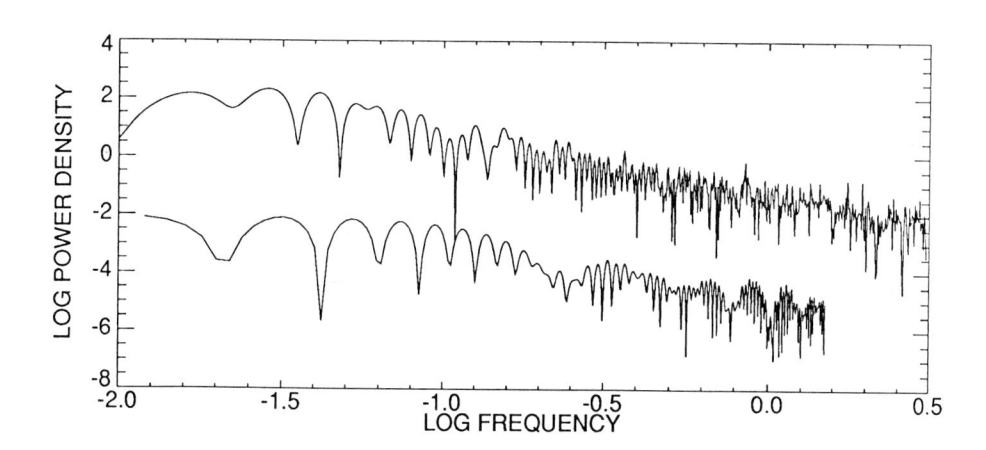

Fig. 2. V-band PDS of: (lower) observations of OJ 287 from 1-4 April 1988; (upper) model with $M_8 = 30$, $\dot{m} = 1$, $\beta = 1.0$, $i = 60°$, $t_{max} = 10(P_{in})$, $\Delta t = 0.01$, $t_{on} = 1.0$, for 100 'flares'. For both, with arbitrary vertical offset, $s \approx -1.9$.

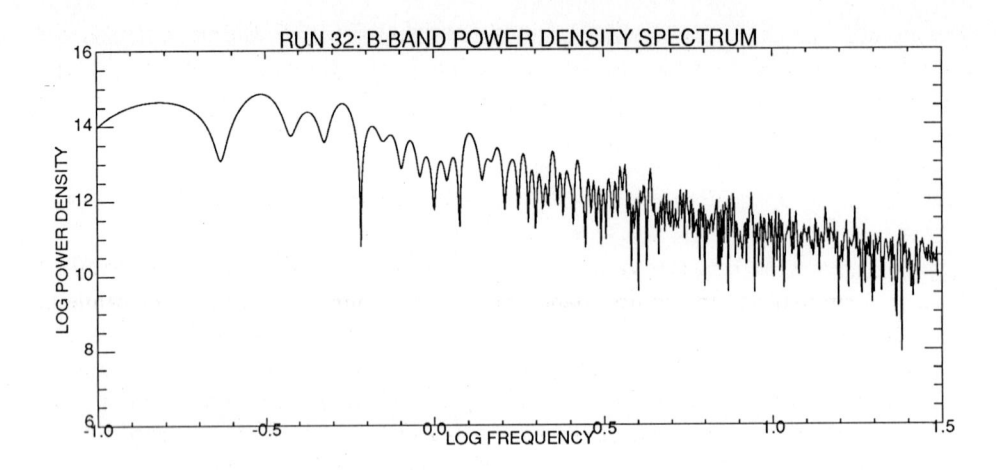

Fig. 3. B-band PDS for a model with $M_8 = 10$, $\dot{m} = 1$, $\beta = 1.0$, $i = 0°$, $t_{max} = 10(P_{in})$, $\Delta t = 0.01$, $t_{on} = 1.0$, for 150 'hot-spots'; the PDS has a slope of $s \approx -1.75$.

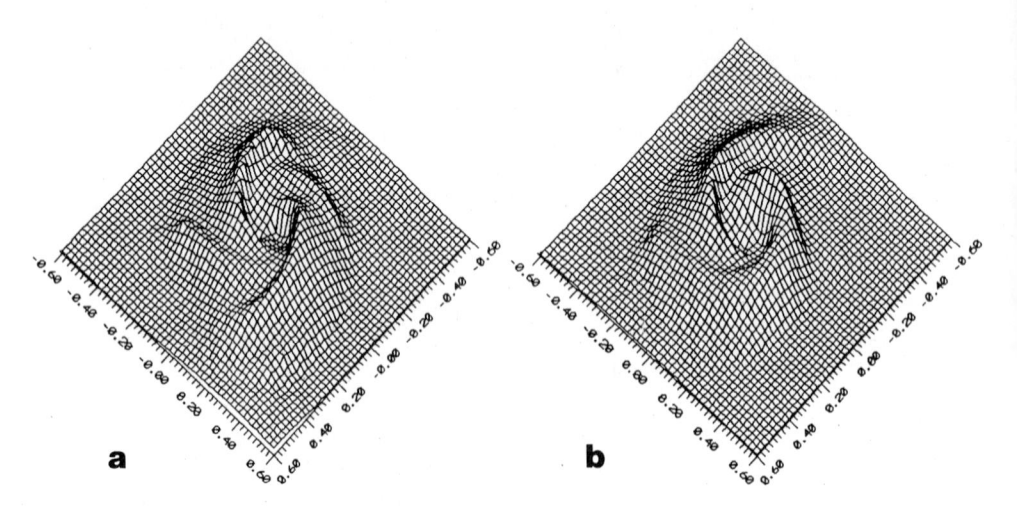

Fig. 4. Density variations for the spiral shock simulation discussed in the text, as viewed at an 80° angle. The central black hole is at (0,0), and the binary companion is at a radius of 1.0 and orbits with a period of 2π. Instabilities are clear, for in (a) a four-armed perturbation is visible at $t = 4.08$, while in (b) a three-armed perturbation is illustrated at $t = 5.37$.

The simulations use the well tested code of T. Matsuda and E. Shima, choosing the orbital frequency of the binary as $\Omega = 1$, the binary separation as a unit length, and the binary mass ratio as unity (cf. Sawada et al. 1986). For the illustrated case, the initial disk has a uniform density and constant angular momentum ($\ell = 0.2$ in our units). Matter is supplied at the outer Lagrangean point with an adiabatic index, $\gamma = 1.4$. The grid has 21 radial zones (from 0.05 to 0.65) and 51 azimuthal zones.

Figure 4 illustrates the density as a function of X,Y coordinates at two different times in a simulation. In the first panel, at $t = 4.08$, four spiral arms exist, while at $t = 5.37$ (Fig. 4b) there are three spiral arms (one *binary* orbital period is $t = 2\pi$). The density enhancements, or 'hot-spots' in the spiral arms appear and disappear with timescales of $\Delta t \approx 1$, and do not reside at fixed radial distances. For example, note the density enhancement visible at $r \approx 0.4$ in Fig. 4b. They are produced at the inner edge of the disk and propagate outwards before dissipating.

In the inner regions the 'hot-spots' live substantially longer than one *local* orbital period, $P(r)$; blobs with lifetimes of ~ 1.0–1.5 time units arise near $r \sim 0.1$ where $P(r) \approx 0.2$. But in the outer regions, where the Keplerian periods are longer, the blobs last less than one local orbital period. Thus these naturally occurring non-axisymmetric instabilities appear to satisfy the requirement that the X-ray emitting blobs be long-lived with respect to $P(r)$ (e.g. Abramowicz et al. 1991), since X-rays will be emitted by the innermost disk regions. This picture is also in accord with the idea that optical variability producing 'hot-spots', inhabiting the outer part of the disks, give the best fits to observations if they are short-lived with respect to the local orbital period (Wiita et al. 1991). Hence, these transient blobs of excess matter may be responsible for *both* X-ray flickering and optical microvariability in AGN.

3.2 Magnetic Instabilities

Another distinct way of producing disk irregularities would be in analogy with solar flares, where magnetic energy storage and release would be important. This would most likely occur in the expected corona sandwiching the thin accretion disk (e.g. Wandel and Urry 1991). Under these conditions, small regions would have substantially elevated temperatures for relatively short periods. Unfortunately, simple approaches involving the amplification of large scale fields and their emergence through magnetic buoyancy, seem to occur over timescales too long to explain microvariability.

The recent discovery by Balbus and Hawley (1991; also Hawley and Balbus, 1991) of a powerful local shear instability in *weakly* magnetized disks may be of great relevance here. This instability, analogous to a classical interchange instability, grows on an orbital time scale and is independent of the magnetic field strength as long as it is below equipartition. They argue that this instability can interchange angular momentum and thus may well be of importance to the understanding of turbulent viscosity in accretion disks.

We note that the magnetic reconnection expected when this instability saturates should yield many small flares with very fast rise-times and regeneration times comparable to a few local orbital periods, just as we require for fitting the light curves and PDS. However, this is no more than speculation at this time and we plan on performing detailed calculations to check this exciting possibility.

4 CONCLUSIONS AND FUTURE WORK

We have demonstrated that simple models, based upon multiple regions of excess emission present on, or just above, accretion disks around supermassive black holes yield light curves and power density spectra very similar to those of high quality observations. Non-axisymmetric shocks appear to provide a plausible explanation for such regions, but other models involving magnetic reconnection also deserve study.

While the general agreement between the data and the models is heartening, it should be stressed that both remain preliminary. Much longer trains of optical observations are needed, preferably with evenly spaced observations, in order to more fully characterize the PDS and search for breaks in the spectra that could yield solid estimates for parameters such as M_8. In order to obtain this type of information at least six observatories, quasi-evenly spaced in latitude, would have to dedicate a substantial portion of their observing time to monitoring a small group of sources for many months. These observations would be most useful if they could routinely cover at least two bands, so that color changes were also well defined.

The modelling can easily be improved, and we expect to do so in the near future. It will be necessary to incorporate the underlying continuum power-laws in all future analyses. Eclipses by the outer part of the disk and light bending near the black hole can be included rather straightforwardly, and would allow investigations of high inclination angles and higher frequency emission (dominated by the region close to the BH in disk models) than we have considered to date. Kerr black holes, being more realistic, should also be investigated. Instead of the simpler β-disk models, more complicated, but "standard" α-disks could be employed, and the differences in predictions could be examined. Both 'hot-spot' and 'flare' models must be developed quantitatively. Then multicolor predictions can be compared with improved observations and should allow us to distinguish between various possibilities.

This work was supported in part by NSF grants AST-8717912 and AST-9102106 at GSU. SKC is grateful to Profs. T. Matsuda and E. Shima for allowing their code to be used and to Prof. A. Salam for hospitality at the International Center for Theoretical Physics.

REFERENCES

Abramowicz, M.A., Bao, G., Lanza, A. & Zhang, X.-H. 1991, A&A, in press
Balbus, S.A. & Hawley, J.F. 1991, ApJ, in press
Carini, M.T. 1990, Ph.D. thesis, Georgia State University
Carrasco, L., Dultzin-Hacyan, D. & Cruz-Gonzalez, I. 1985, Nature, 314, 146
Chakrabarti, S.K. & Wiita, P.J. 1990, presented at 15th Texas Symposium on Relativistic Astrophysics, Brighton 1990, ICTP Preprint IC/90-476
Hawley, J.F. & Balbus, S.A. 1991, ApJ, in press
Krichbaum, T. 1991, this volume
Lawrence, A., Watson, M.G., Pounds, K.A. & Elvis, M. 1987, Nature, 325, 694.
Lightman, A.P. & Eardley, D.M. 1974, ApJ Lett, 187, L1
McHardy, I. & Czerny, B. 1987, Nature, 325, 696
Miller, H.R. 1988, in *Active Galactic Nuclei*, eds. H.R. Miller & P.J. Wiita (Berlin: Springer Verlag), p. 146
Miller, H.R., Carini, M.T. & Goodrich, B.D. 1989, Nature, 337, 627
Miller, H.R., Carini, M.T., Noble, J.C., Webb, J.R. & Wiita, P.J. 1991, this volume
Moore, R.L. et al. 1982, ApJ, 260, 415
Novikov, I.D. & Thorne, K.S., 1973, in *Black Holes*, ed. C. DeWitt & B.S. DeWitt, (New York: Gordon and Breach), p. 343
Quirrenbach, A., Witzel, A., Krichbaum, T., Hummel, C.A., Alberdi, A. & Schalinski, C. 1989, Nature, 337, 442
Sawada, K., Matsuda, T. & Hachisu, I. 1986, MNRAS, 219, 75
Sun, W.-H. & Malkan, M.A. 1989, ApJ, 346, 68
Treves, A. et al. 1982, ApJ, 341, 733
Visvanathan, N. & Elliot, J.L. 1973, ApJ, 179, 721
Wagner, S.J. 1991, this volume
Wandel, A. & Petrosian, V. 1988, Stanford CSSA-Astro-87-21 preprint
Wandel, A. & Urry, C.M. 1991, ApJ, 367, 78
Wiita, P.J., Miller, H.R., Carini, M.T. & Rosen, A. 1991, in *Structure and Emission Properties of Accretion Disks, 6th I.A.P. Meeting / IAU Colloq. 129*, eds. C. Bertout, et al. (Gif-sur-Yvette: Editions Frontieres) in press
Zhang, X.-H. 1991, in *Variability of Active Galactic Nuclei*, eds. H.R. Miller and P.J. Wiita (Cambridge: Cambridge Univ. Press), p. 256
Zhang, X.-H. & Gang, B. 1991, A&A, in press

Blazar Microvariability: A Case Study of PKS 2155–304

H. R. MILLER, M. T. CARINI, J. C. NOBLE, J. R. WEBB* AND P. J. WIITA

Georgia State University and *Florida International University

1 INTRODUCTION

The presence of microvariability has been clearly demonstrated for a number of blazars by the observations reported recently by several investigators (e.g. Miller, 1988, Miller, Carini and Goodrich 1989, Carini 1990, Miller and Carini 1991, Carini, Miller and Goodrich 1990, Miller and Carini 1991, Wagner 1991). These results suggest that high signal-to-noise, high time-resolution multifrequency observations provide the most stringent constraints for physical models which attempt to explain the phenomenon of microvariability. In order to achieve the highest quality data, one must select a source to be investigated which is both bright and variable. PKS 2155-304 turns out to be an excellent choice for this study.

PKS 2155-304 has exhibited significant variability in both its optical flux (Miller and McAlister 1983, Carini and Miller 1991) and polarization (Griffiths et al. 1979). In addition, significant ultraviolet (Maraschi et al. 1986 and Urry et al. 1988) and X-ray (Urry and Mushotzky 1982, Agrawal, Singh and Reigler 1987, Tagliaferri et al. 1989, and Treves et al. 1989) variations have also been detected. All of these properties indicate that PKS 2155-304 is an ideal choice to add to our sample. The purpose of this study is to determine if microvariability is present at both optical and ultraviolet wavelengths for this source.

2 OBSERVATIONS

The high time resolution optical observations of PKS 2155-304 reported here were obtained with the 0.9 meter telescope at Cerro Tololo Interamerican Observatory equipped with a direct CCD camera and an autoguider. Observations were also obtained with the 42-inch telescope at Lowell Observatory similarly equipped but without an autoguider. The observations were made through a V filter with an RCA CCD. Repeated exposures of 90 seconds were obtained for the star field containing PKS 2155-304 and several comparison stars from Miller (unpublished). These standard stars, located on the same CCD frame as PKS 2155-304 provide comparison stars for use in the data reduction process. The observations were reduced using the method of Howell and Jacoby (1986). Each exposure is processed through an aperture photometry routine which reduces data as if it were produced by a multi-star photometer. Differential magnitudes can then be computed for any pair of 'stars' on the frame. These simultaneous observations of PKS 2155-304, several comparison stars, and the sky background allow one to remove variations which may be due to fluctuations in atmospheric transparancy and extinction. The aperture photometry routine used for these observations is the Apphot routine in IRAF.

An analysis of the error in a given CCD observation must consider many possible sources of error associated with that particular observation. The sky background, the read noise of the CCD chip and the dark count are all possible sources of error. If their contributions are found to be negligible, then simple photon statistics, i.e. $1/\sqrt{N_*}$, where N_* is the total number of sky subtracted counts in the object, adequately describes the error in the observations. If the contributions of the aforementioned sources of error are not negligible, then simple photon statistics underestimates the error associated with an observation. In this case, the errors can be calculated via the CCD equation (Howell 1989):

$$S/N = \frac{N_*}{\sqrt{N_* + n_{pix}(N_{sky} + N_d + N_r^2)}}$$

where N_* = the number of sky subtracted counts in the object; n_{pix} = the number of pixels in the measuring aperture; N_{sky} = the number of counts in the sky; N_d = the dark count of the CCD chip and N_r = the read out noise of the CCD chip. Howell (1989) has shown that for the RCA CCD chips used in this investigation, simple Poisson statistics will not adequately describe the errors for the objects with V $\geq$ 18.0. The primary reason for this is that the RCA CCDs used in this investigation have high read out noise, which enters into the CCD equation squared, and thus can contribute significantly to the noise.

In the process of differential photometry, one observes a variable object, V, a comparison star, C and a check star, K. Ideally, one would like to have V, C and K nearly equal in brightness. In this case, the σ of the C-K light curve would describe the errors in the V-C light curve, since all three objects are nearly equal in brightness. An alternative to this situation, which has been shown to be acceptable (Howell, Mitchell and Warnock 1988) is to have V equal to C or K and have the remaining comparison somewhat brighter than V. Unfortunately, neither of these situations is easily obtainable in practice. Usually, V, C and K are at different magnitudes and one must find a way to scale the σ in the C-K data to the σ in the V-C data. Howell et al. (1988) describes a method by which one can calculate a scale factor Γ^2 which allows one to scale the σ^2_{C-K} to the σ^2_{V-C} via knowledge of the characteristics of the CCD, the counts in V, C, K and the sky, and the size of the measuring aperture. One assumes that C and K have no intrinsic variability and that the process of differential photometry eliminates the common sources of error between the objects on a CCD frame, *i.e.* a common sky background and extinction effects. This allows one to assume that the CCD noise variations and photon statistics are independent in the three objects. In addition, Howell et al. (1988) report that even random errors not explicitly considered in the CCD equation are scaled from C-K into V-C. This method has been applied to observations of cataclysmic variable stars obtained with one meter class telescopes and with accuracies as good as 0.05 magnitudes for objects at V = 19.0 (Howell et al. 1988, Howell and Szkody 1988).

Following Howell et al. (1988), the variance in the V-C data can be thought of as containing two components. One of these, $\sigma^2_{V-C}(\text{INS})$ is the variance in V-C resulting from the noise sources in the data and the other, $\sigma^2_{V-C}(\text{VAR})$ is the variance resulting from the intrinsic variability in V-C. Thus

$$\sigma^2_{V-C} = \sigma^2_{V-C}(INS) + \sigma^2_{V-C}(VAR).$$

If V were not variable, then the second term on the right hand side of the equation would equal zero, and the variance in V-C would lead to a description of the errors in the V-C data. This also says that if one could somehow remove the variations in V-C that were intrinsic to the variable, then the variance would again lead to a description of the error in V-C.

The errors for V-C have been calculated assuming that the variations intrinsic to V can be identified and removed. If the observed variations covering several hours can be characterized as a linear trend, then these trends can be fit with a straight line via a least squares analysis. The deviation of each point from this line is calculated, summed and normalized to the number of data points and the square root of this normalized sum is then taken and used as the error for the V-C data. This method has also been used on the standard stars and the results compared to the standard deviations calculated in the usual fashion. No significant difference was found between the σs calculated by either method.

Ultraviolet observations were obtained in October 1989 with the International Ultraviolet Explorer (IUE) spacecraft. The Observations were made with the large (10"×20") aperture and the source acquisition was by blind offset from a nearby SAO star. The spectral resolution was ~ 7Å. Integration times were 60 minutes in both the short-wavelength (SWP) and long-wavelength (LWP) cameras. The data reduction was performed at the IUE Regional Data Analysis Facility at NASA/Goddard Space Flight Center using the GEX algorithm for extracting the spectra.

3 DISCUSSION

A concerted effort was made on 1988 September 25-28 to detect and study the character of optical variations on timescales as short as several minutes for PKS 2155-304. The results of this study are presented in Figure 1. Variations of ~ 0.01 mag/hour are observed on each of the four nights, and suggest that the fluctuations which are observed during each night are part of a smoothly changing longer-term variation that is associated with this source. Large amplitude (~ 0.25 mag) events on the timescale of hours are not present during the time when these observations were made, although variations of this amplitude are clearly present on the timescale of a few days.

Simultaneous ultraviolet and optical observations of PKS 2155-304 were obtained 1989 October 2-6. The ultraviolet flux was observed to change by $\sim 10\%$ in one day (see Figure 2). Optical observations obtained on 1989 October 2 and 6 yield $V = 13.46$ and 13.59 respectively suggesting a similar change for both the UV and optical fluxes, consistent with the results of Edelson et al. (1991), which combines our data with those of other observers. Thus this blazar is the first for which one is able to demonstrate that significant variability is occurring with similar timescales and amplitudes at both UV and optical wavelengths. This suggests that the emission in both bands arise as a result of the same, or two closely related physical processes.

One model (Wiita et al. 1991a,b) would explain the observed microvariability as the result of excess emission produced by *flares* or *hot spots* randomly appearing and disappearing on the accretion disks around supermassive black holes. A similar model, stressing the X-ray fluctuations in AGN, has also been proposed by Zhang (1991) and Abramowicz et al. (1991). The lack of observed periodicities is in accord with this class of models, which assume contributions from typically 10–100 flaring regions at any given time. Preliminary results reported by Wiita et al. (1991b) yield good matches to the power density spectra (PDS) observed for sources where the data has been taken with high sampling frequencies over several consecutive nights (e.g. Carini 1990).

It would be extremely useful to continue the observations of PKS 2155-304 as well as many other members of our sample. Broadening the temporal coverage in the optical, by involving a string of longitudinally spaced observatories with one meter class telescopes to provide continuous coverage of individual sources over periods of weeks to months could really test for the presence of periodicities and provide far better defined PDS. The utility of the simultaneous observations in the V and

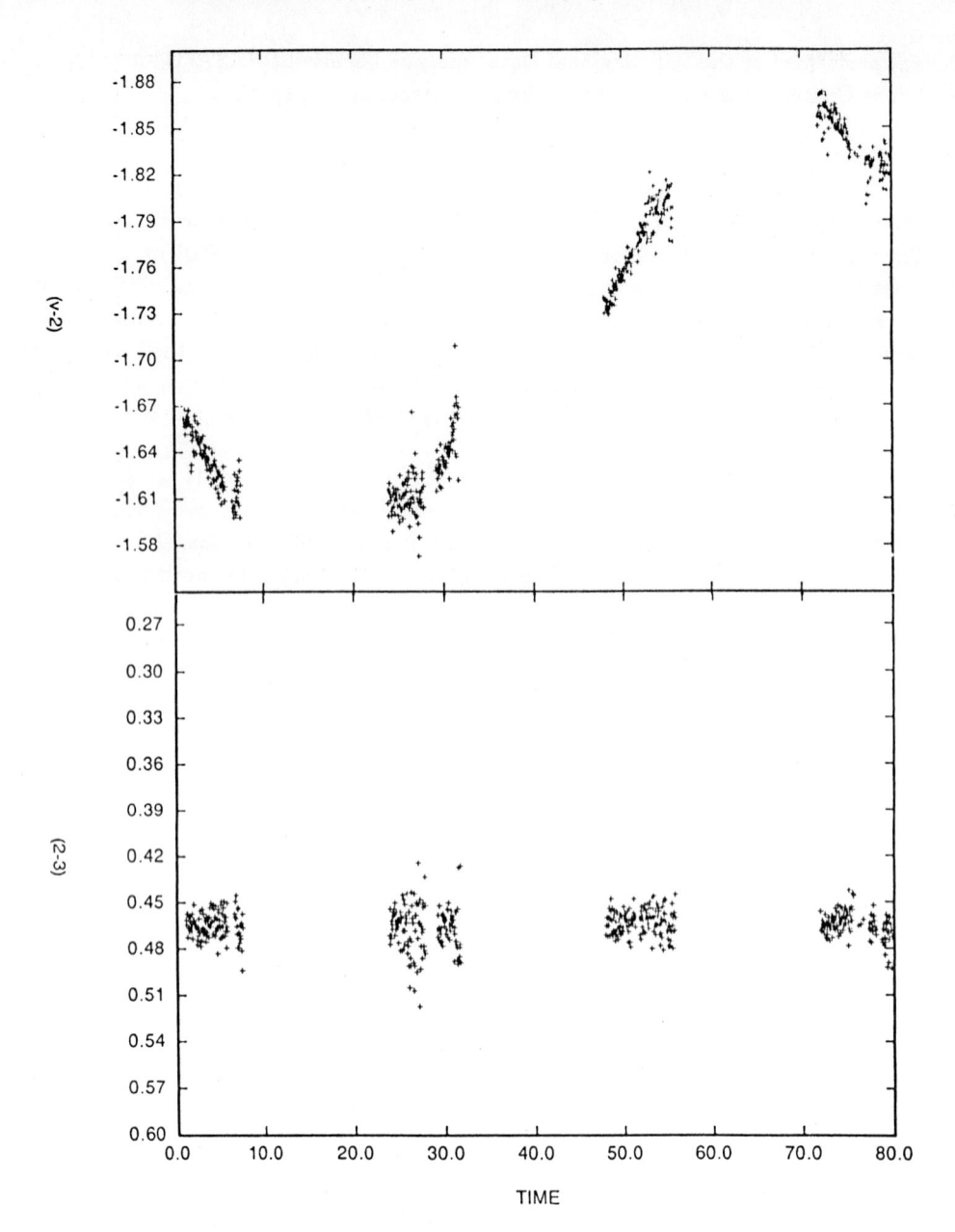

Figure 1. The optical variations observed for PKS 2155-304 through a V-filter for the period 1988 September 25-28 are shown in the upper panel. The scatter of the two comparison stars are shown in the lower panel.

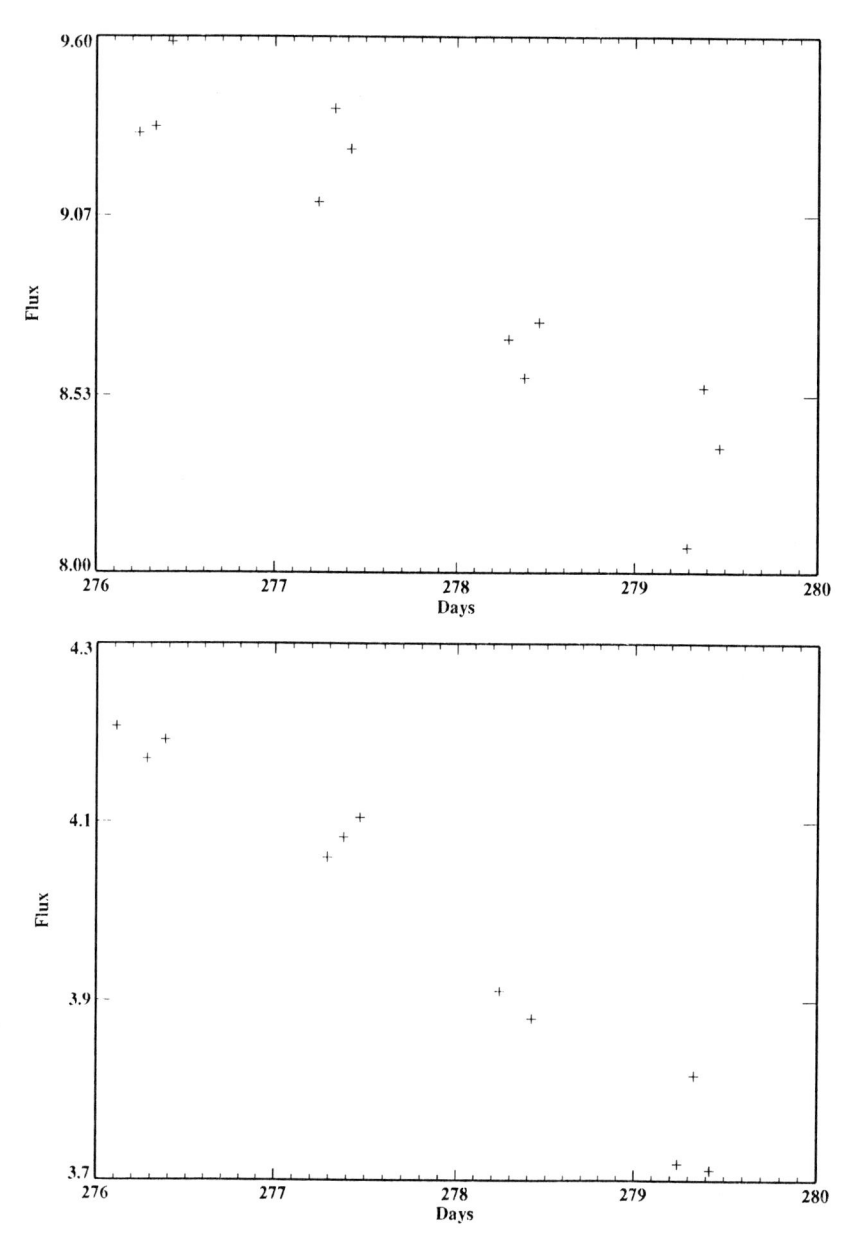

Figure 2. The ultraviolet variations observed for PKS 2155-304 1989 October 2–6 utilizing the SWR camera on IUE are shown in the upper panel. Similar observations using the LWP camera on IUE are shown in the lower panel.

IUE UV reported here encourages us to try for additional campaigns of this type. Theoretical models could be tested and constrained when substantial amounts of multi-color data for a group of sources becomes available.

This work was supported in part by NASA grant NAG5-1141 and by NSF grants AST-8717912 and AST-9102106. HRM, MTC and JCN wish to thank Lowell observatory for generous allocations of observing time.

4 REFERENCES

Abramowicz, M.A., Bao, G., Lanza, A. & Zhang, X.-H. 1991, A&A, in press
Agrawal, P.C., Singh, K.P., & Reigler, G.R. 1987, MNRAS, 227, 525
Carini, M.T. 1990, Ph.D. thesis, Georgia State University
Carini, M.T. & Miller, H.R. 1991, ApJ, in press
Carini, M.T., Miller, H.R., & Goodrich, B.E. 1990, AJ, 100, 347
Edelson, R.A., et al. 1991, ApJ(Letters), in press.
Griffiths, R.E., Tapia, S., Briel, U. & Chaisson, L. 1979, ApJ, 234, 810.
Howell, S.B. 1989, PASP, 101, 616.
Howell, S.B., Mitchell, K.J., & Warnock, A. 1988, AJ, 95, 247
Howell, S.B. & Szkody, P. 1988, PASP, 100, 224
Maraschi, L., Tagliaferri, G., Janzi, E.G. & Treves, A. 1986, ApJ, 304, 637.
Miller, H.R. 1988, in *Active Galactic Nuclei*, eds. H.R. Miller & P.J. Wiita (Berlin: Springer Verlag), p. 146
Miller, H.R. & Carini, M.T. 1991, in *Variability of Active Galactic Nuclei*, eds. H.R. Miller and P.J. Wiita (Cambridge: Cambridge Univ. Press), p. 256
Miller, H.R., Carini, M.T. & Goodrich, B.D. 1989, Nature, 337, 627
Miller, H.R. & McAlister, H.A., 1983, ApJ, 272, 26
Tagliaferri, G., Stella, L., Maraschi, L., Treves, A., & Morini, M. 1989, in *BL Lac Objects*, eds. L. Maraschi, T. Maccacaro & M.-H. Ulrich (Berlin: Springer Verlag), p. 143
Treves, A., et al. 1989, ApJ, 341, 733
Urry, C.M., Kondo, Y., Hackney, K.R.H., & Hackney, R.L. 1988, ApJ, 330, 791
Urry, C.M. & Mushotzky, R.F. 1982, ApJ, 252, 38
Wagner, S.J. 1991, in *Variability of Active Galactic Nuclei*, eds. H.R. Miller and P.J. Wiita (Cambridge: Cambridge Univ. Press), p. 256
Wiita, P. J., Miller, H.R., Carini, M.T., Rosen, A. 1991a, in *Structure and Emission Properties of Accretion Disks*, I.A.U. Colloquium No. 129, in press.
Wiita, P. J., Miller, H.R., Gupta, N., & Chakrabarti, S.K. 1991b, this volume
Zhang, X.-H. 1991, in *Variability of Active Galactic Nuclei*, eds. H.R. Miller and P.J. Wiita (Cambridge: Cambridge Univ. Press), p. 256

On the possible cause of rapid flux variability in active galactic nuclei

V.G. GORBATSKY

Leningrad University Observatory

ABSTRACT

The origin of rapid AGN radiation flares is considered. It is supposed that they are the results of synchrotron emission from clouds of relativistic electrons arising suddenly. It is shown that the data on the connection between the duration of the flare Δt and the radiation frequency ν ($\Delta t \alpha \nu^{-1/2}$) may be explained in this way. The strength of the magnetic field in the flare region is estimated.

1 INTRODUCTION

The flares of radiation on different frequencies ($10^8 \leq \nu \leq 10^{18}$ Hz) are observed in active galactic nuclei (AGN) of any type. Characteristic timescale of variability $t_{char} = dt/d \ln F_\nu$ (F_ν is the radiation flux) was introduced (Bregman et al., 1988). On the assumption that $d \ln F_\nu \approx \Delta F_\nu/F_\nu$ ($\Delta F\nu$ is the amplitude of the flux variation) the correlation between t_{char} and ν was established (Bregman, 1990):

$$t_{char} \alpha \ \nu^{-1/2} \ . \tag{1a}$$

The duration of a flare Δt must be close to t_{char} because the value of ΔF_ν averaged over Δt, is of the same order as ΔF_ν. Therefore

$$\Delta t \ \alpha \ \nu^{-1/2} \ . \tag{1b}$$

The purpose of this note is to show that Equation 1b may be obtained in the framework of the hypothesis of an abrupt appearance of a cloud of relativistic electrons (r.e.) as the cause of flares in all spectral regions.

2 DURATION OF A FLARE

If r.e. lose their energy (ε) due to synchrotron radiation, the timescale of energy loss may be determined by means of equation (Pacholczyk, 1970)

$$\frac{d\varepsilon}{dt} = -\xi_g \varepsilon^2 \ . \tag{2}$$

Here

$$\xi_g = 2.37 \cdot 10^{-3} \, H^2_\perp \tag{3}$$

and $H_\perp$ is the perpendicular magnetic field component. The influence of the inverse Compton effect on the energy losses is not taken into account in Equation 2.

Well-known expression for the frequency υ_c of the maximum of the synchrotron emission (from an electron having energy ε) is

$$\upsilon_c = 6.27 \cdot 10^{18} \, \varepsilon^2 \, H_\perp \ . \tag{4}$$

It follows from Equations 2 – 4.

$$\frac{d\upsilon_c^{1/2}}{dt} = -K\upsilon_c \ , \tag{5}$$

where

$$K = 0.95 \cdot 10^{-12} \, H_\perp^{3/2} \ .$$

The solution of Equation 5 for the initial conditions

$$t = t_0, \quad \upsilon_c = \upsilon_c^{(0)}$$

has the form

$$\frac{1}{\upsilon_c^{1/2}} - \frac{1}{(\upsilon_c^{(0)})^{1/2}} = K(t-t^0) \tag{6}$$

On the assumption that the radiation in the frequency band near $\upsilon_c^{(0)}$ will be decreased to $\upsilon_c^{(0)}/m$ ($m > 1$) and therefore taking $\upsilon_c = \upsilon_c^{(0)}/m$, $t - t_0 = \Delta t$, the expression for Δt may be obtained

$$\Delta t = \frac{\sqrt{m-1}}{K} \, [\upsilon_c^{(0)}]^{-1/2} \tag{7}$$

which coincides with Equation 1b.

3 MAGNETIC FIELD STRENGTH

It is known from observations that the values of t_{char} lie in the interval from 10^3 s (hard X-rays, $\upsilon \approx 10^8$ Hz) for variety of different objects and also for individual objects (Bregman, 1990). The magnetic field strength as determined from Equation 7 (m = 4) is

$$H_\perp \approx \frac{10^8}{(\upsilon_c^{(0)})^{1/3} (\Delta t)^{2/3}} \; G \qquad (8)$$

Taking $\Delta t \approx t_{char}$ one can find that

$$0.1G \leq H_\perp \leq 1G.$$

These values are common for AGN of all types. They correspond to the field strength at distances 10^{17} cm to 10^{18} cm from the "central machine" (e.c. $\gtrsim 10$ G).

The question arises does abrupt releasing of r.e. occur only at such distances and not at smaller ones? Obviously for values of $H_\perp$ much higher (e.g. $\geq 10G$) the flare timescale must be too small and the detection of very rapid flares is a very difficult problem. The detection of superfast flares (with $\Delta t \approx 1^s$ to 10^s in optical region and corresponding values of Δt on other frequencies) may serve as testing for the proposed explanation of correlations (1) and (1b).

4 ENERGIES

Using the observational data on Δt_υ and surplus energy ΔQ_υ that was radiated in the course of a flare it is possible to estimate the number of emitting r.e.. Total energy emitted from the cloud containing N_s r.e. (each having the same energy ε) in the frequency band $\Delta \upsilon_c$ and for time interval Δt may be found from the expression (Gorbatsky, 1987)

$$\Delta Q_{\upsilon_c} = N_{\varepsilon 0} \, \varepsilon_0 \approx 0.2 \cdot 10^{23} \, N_{\varepsilon 0} \, H_\perp \Delta \upsilon_c \Delta t_{\upsilon_c} \qquad (9)$$

In the case that the r.e. energy spectrum has the form $N(\varepsilon) = N(\varepsilon_0)(\varepsilon_0/\varepsilon)^\gamma$ one can determine the values of γ and $N(\varepsilon_0)$ using the data on Δt_υ and ΔQ_υ for different frequencies. Preliminary estimates made for the flares in quasar 3C345 led to $\gamma \approx 2$ and $N(1-10 erg) \approx 10^{48} - 10^{49}$ (Gorbatsky, 1987). When $\gamma \geq 2$ the number of r.e. providing X-ray radiation is small in comparison with the number of "optical" and "IR" r.e.. Due to this circumstance one can consider the evolution of r.e. radiating in a given spectral region apart from r.e. which emit on other frequencies.

Relativistic electrons lose their energy also due to the expansion of a plasma cloud. At the present time it is difficult to take into account such an expansion in the calculations of energy losses. Presumably the effects of expansion must not play predominating role because of a strong magnetic field frozen in thermal plasma. However, it is necessary to take into account the influence of the cloud expansion as well as of the inverse Compton effect in more exact calculations.

The crucial problem concerning of the hypothesis considered above is the origin of r.e.. This problem is still poorly understood not only for r.e. presumably causing the flares but also for r.e. that provide the constant component of the AGN radiation.

REFERENCES

Bregman J.N., Glassgold, A.E., Huggins, P.J., Kinney, A.L., McHardy, I., Webb, J.R., Pollock, J.T., Leacock, R.J., Smith, A.G., Pica, A.J., Aller, H.D., Aller, Hodge, P.E., Miller, J.S., Stephens, S.A., Dent, W.A., Balonek, T.J., Barvainis, R., Neugebauer, G., Impey, C.D., Soifer, B.T., Matthews, K., Elias, J.H. and Wisniewski, W.Z. 1988, *Astrophys. J.* **331**, 746.

Bregman J.N. 1990, *Astron. Astrophys. Rev.* **2**, 125.

Gorbatsky V.G. 1987 *IAU Symp. 121, Observ. Evidence of Activity in Galaxies*, Ed. E.Ye. Khachikian, p. 313.

Pacholczyk A.G. 1970, *Radio Astrophysics*, W.H. Freeman, San Francisco.

Intraday Variability of Compact Extragalactic Radio Sources

T. P. Krichbaum, A. Quirrenbach[*] and A. Witzel

Max-Planck-Institut für Radioastronomie, Bonn, Germany (F.R.G.)
[*] now at Naval Research Laboratory, Washington, D.C., USA

Abstract

Repeated multifrequency flux density observations of a sample of extragalactic flat spectrum radio sources indicate that intraday variability with amplitudes $\gtrsim 2.5\%$ occurs in at least 30% of these objects. Intensity variations with amplitudes of up to 30% percent occur on timescales of less than a day. The variability of the degree of the polarization detected in the quasar 0917+62 is anti-correlated to the observed intensity fluctuations. Three weeks of joint radio and optical observations of the BL Lac-object 0716+71 show pronounced variability in both domains and a change of the typical variability timescale after the first week. This transition appeared simultaneously in the radio and optical band. We discuss our observational results in the framework of various models and conclude that the dominant contribution to the observed strong variations is most likely to be intrinsic to the sources. In addition underlying fluctuations less pronounced and on longer timescales may be attributed to refractive interstellar scattering.

1. Introduction

The first detections of flux density variability in extragalactic radio sources were reported by Sholomitskii (1965) and by Dent (1965). The general properties of such variations which generally occur on the time scale of a few months in many compact objects could successfully be explained by adiabatically expanding synchrotron "clouds" in a number of radio sources (Pauliny-Toth and Kellermann, 1966; and van der Laan, 1966). It was necessary, however, to incorporate the effects of relativistic motions to explain these "rapid" variations without violating the inverse Compton limit of 10^{12} K (e.g. Kellermann and Pauliny-Toth, 1969). The existence of compact structure in extragalactic radio sources inferred from these models led to the development of Very Long Baseline Interferometry. This technique gave direct confirmation of the earlier concepts by the detection of milliarcsecond scale jets and apparent su-

Table 1: The Experiments

Date	Duration [hrs]	No. of sources	100 m-telescope u_0(6 cm)	u_0(11 cm)	Pol.	VLA	optical
May 85	50	17	–	0.27%	–	–	–
Aug 85	87	31	–	0.26%	–	–	–
Dec 85	100	31	–	0.27%	–	–	–
Mar 88	112	17	0.39%	0.27%	–	–	–
Jun 88	104	19	0.45%	–	+	–	–
Dec 88	147	14	0.29%	0.18%	+	–	–
May 89	127	14	0.38%	0.46%	+	+	+
Feb 90	449	4	–	–	–	+	+
Apr 90	108	48	0.71%	–	–	–	–

perluminal motions of features within these jets (e.g. Cohen et al., 1971).

First suggestions that propagation effects could contribute to the observed variability of compact radio sources were made by Shapirovskaya (1978). At decimeter wavelengths, refractive interstellar scintillation (RISS) dominates the variability in numerous sources (see e.g. Spangler et al., 1989). The importance of RISS for the interpretation of radio light curves can also be seen in the unusual "events" in the BL Lac object 0954+65 reported by Fiedler et al., 1987, which were explained as high amplification events due to refraction by localized clouds or filaments in the solar neighbourhood (Romani et al., 1987). Low-amplitude variations on time scales of a few days to weeks – the so called flickering – can be attributed to RISS caused by the "normal" component of the interstellar medium in our galaxy (Heeschen 1984, Heeschen and Rickett, 1987).

Variations on even shorter time scales, reaching amplitudes of more than 20% in less than 24 hours, have been reported by Witzel et al., 1986, Heeschen et al., 1987, and Quirrenbach et al., 1989a. In at least one case, the variations of the total flux density are accompanied by even more dramatic variability of the polarized flux density (Quirrenbach et al., 1989b). The origin of these variations has not yet been unambiguously determined. It is immediately clear that any intrinsic explanation of intraday radio variability has to address the problem of extremely high apparent brightness temperatures; on the other hand, very small and dense scatterers are required to reproduce the short time scales in RISS scenarios. Furthermore, it seems difficult to explain the strong polarization variations in RISS models. A decisive test can be provided by simultaneous radio and optical observations.

2. Observations of intraday variability

Prior to 1985, little was known about short time scale variability of extragalactic

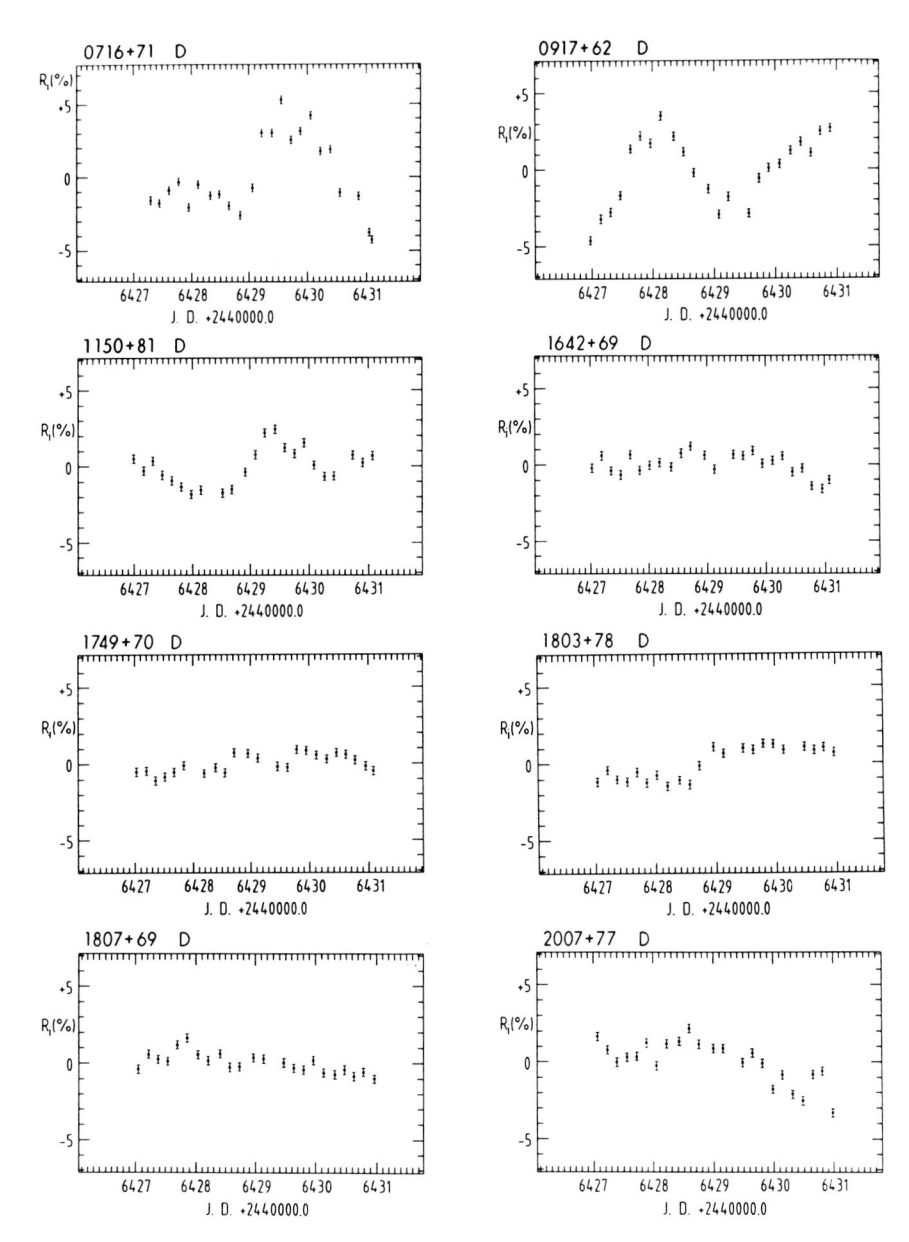

Fig. 1: Relative intensity variations of selected flat spectrum radio sources plotted versus time. The observations were made in December 1985 with the 100 m-telescope at Effelsberg (Heeschen et al., 1987).

sources in the radio range. A large sample of extragalactic sources had been investigated with daily sampling using the NRAO 300 foot and Arecibo telescopes (Heeschen 1984, Simonetti et al., 1985). The observing programs in Effelsberg were originally intended to continue the statistical approach of these observations with a faster sampling rate (e.g. 4 hours). Therefore, a well-defined flux density limited sample of flat spectrum sources ($\alpha \geq -0.5$, $S_\nu \propto \nu^\alpha$) and a matched comparison-sample of steep spectrum sources were chosen from the 1 Jy-catalogue (Kühr et al., 1981). All sources are located at declinations $\delta \geq 60°$ to allow for observations "around the clock", and at galactic latitudes $b^{II} \geq 10°$ to avoid confusion in the galactic plane. In 1985, the measurements were carried out at 11 cm, a wavelength close to that of the previous observations on longer time scales. In addition, it was possible to make use of the stability of the atmosphere and the supreme quality of the 100 m-telescope at this wavelength. Table 1 summarizes the experiments carried out so far. The individual observing runs typically lasted for about 100 hours. Since each observing schedule included about equal numbers of flat and steep spectrum sources with similar distributions of flux densities and positions, a detailed comparison of the two classes was easily possible. Moreover, the steep spectrum sources could be used for calibration purposes, since they were not expected to display significant variations on the time scales of hours to days (and, in fact, did never do so). The apparent average variations of all steep spectrum sources were used to remove the effects of elevation dependent gain of the telescope, atmospheric opacity, and slight changes of the receiver gain from the data. We take the average r.m.s. variations of the corrected data of all steep spectrum sources – denoted by u_0, see Table 1 – as a measure of the remaining r.m.s. accuracy of our measurements. With $u_0 \sim 0.3\%$ fluctuations of more than $\sim 1\%$ could be reliably detected.

To our surprise, some of the fluctuations observed in the first experiments were by no means small: in several sources, including the BL Lac object 0716+71 and the quasar 0917+62 (see Fig. 1), variations with amplitudes of $\sim 10\%$ were found. Therefore, we continued our investigation of intraday variability in 1988, with special emphasis on a more detailed study of individual sources, including quasi-simultaneous multi-frequency observations, polarization measurements and coordinated experiments with radio and optical telescopes (see Table 1). In April 1990 a second complete sample of flat spectrum sources with declinations $35° \leq \delta \leq 50°$ was observed to obtain more information about the frequency of occurrence and the sky distribution of intraday variations. Simultaneous observations with the 100 m-telescope and the VLA at 6 cm wavelength in May 1989 enabled us to verify our data reduction procedures and error estimates through direct comparison of the independent data sets of the two telescopes (see Fig. 2). From this comparison it is evident that even details like the "splitting" of the maximum at J.D. 2 447 655.5 are real.

Table 2: The sample of flat sources and their types of variability observed so far.

No	Source	Id	Type	b^{II}	5/85 11 cm	8/85 11 cm	12/85 11 cm	3/88 6/11 cm	6/88 6 cm	12/88 6/11 cm	5/89 6/11 cm	4/90 6 cm
1	0016+73	QSO	vc	11°	0	I	I	I/0	-	-	-	-
2	0153+74	QSO	dble	12°	I	0	0	-	-	-	-	-
3	0212+73	QSO	jet	12°	0	0	0	-	-	-	-	-
4	0615+82	QSO	vc	26°	0	0	I	-	-	-	-	-
5	0716+71	BLL	vc	28°	-	II	II	II/II	II	II/II	II/II	-
6	0836+71	QSO	jet	34°	0	0	0	-	0	0/0	0/0	-
7	0917+62	QSO	c ?	45°	-	II	II	II/II	II	II/II	II/II	-
8	1150+81	QSO	c	36°	II	I	II	II/II	II	II/II	-	-
9	1642+69	QSO	c	37°	-	II	I	II/II	II	II/I	0/0	-
10	1749+70	BLL	c	31°	II	I	?	II/I	-	-	II/II	-
11	1803+78	BLL	c	29°	0	I	II	0/II	II	I/I	0/I	-
12	1807+69	Gal	jet	29°	-	0	I	-	-	-	-	-
13	1928+73	QSO	jet	24°	0	I	0	0/0	I	I/I	0/0	-
14	2007+77	BLL	c	23°	-	I	I	II/II	II	I/I	II/II	-
15	2021+61	Gal	dble	14°	-	0	0	-	-	-	-	-
16	0108+38	Gal	dble	−24°	-	-	-	-	-	-	-	0
17	0133+47	QSO	c	−14°	-	-	-	-	-	-	-	II
18	0316+41	Gal	irr	−13°	-	-	-	-	-	-	-	0
19	0710+43	Gal	dble	22°	-	-	-	-	-	-	-	0
20	0711+35	QSO	dble	20°	-	-	-	-	-	-	-	?
21	0804+49	QSO	vc	33°	-	-	-	-	-	-	-	II
22	0814+42	BLL	c	33°	-	-	-	-	-	-	-	II
23	0859+47	QSO	jet	42°	-	-	-	-	-	-	-	0
24	0906+43	QSO	jet	43°	-	-	-	-	-	-	-	0
25	0923+39	QSO	dble	47°	-	-	-	-	-	-	-	0
26	0945+40	QSO	jet	50°	-	-	-	-	-	-	-	I
27	1624+41	QSO	jet	44°	-	-	-	-	-	-	-	0
28	1633+38	QSO	jet	42°	-	-	-	-	-	-	-	0
29	1641+39	QSO	jet	41°	-	-	-	-	-	-	-	0
30	1652+39	Gal	jet	39°	-	-	-	-	-	-	-	?
31	2200+42	BLL	jet	−10°	-	-	-	-	-	-	-	?
32	2351+45	QSO	jet	−16°	-	-	-	-	-	-	-	0
33	2352+49	Gal	dble	−12°	-	-	-	-	-	-	-	0
34	0235+16	BLL	vc	−39°	-	-	-	-	-	-	-	II
35	0300+47	BLL	c ?	−10°	-	-	-	-	-	-	-	I
36	0735+17	BLL	jet	18°	-	-	-	-	-	-	-	II
37	0851+20	BLL	c	36°	-	-	-	-	-	-	-	I
38	0954+65	BLL	c ?	43°	-	-	-	-	I	I/II	II/II	-
39	1226+02	QSO	jet	64°	-	-	-	-	-	-	-	0
40	1253−05	QSO	jet	57°	-	-	-	-	-	-	-	0
41	1308+32	BLL	vc	83°	-	-	-	-	-	-	-	I
42	1404+28	Gal	c	73°	-	-	-	-	-	-	-	0
43	1502+10	QSO	c ?	55°	-	-	-	-	-	-	-	0
44	1611+34	QSO	c ?	46°	-	-	-	-	-	-	-	?
45	1741−03	QSO	c ?	13°	-	-	-	-	-	-	-	?
46	1749+09	QSO	c	18°	-	-	-	-	-	-	-	I
47	1821+10	QSO		11°	-	-	-	-	-	-	-	0
48	2214+35	QSO		−18°	-	-	-	-	-	-	-	II
49	2251+15	QSO	jet	−38°	-	-	-	-	-	-	-	0

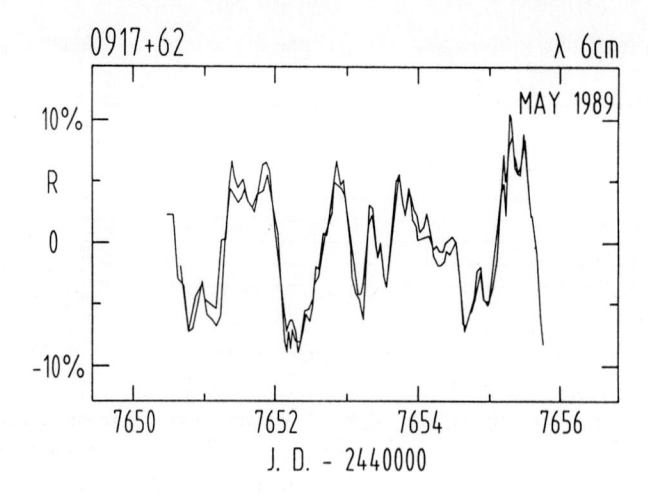

Fig. 2: Light curves of the quasar 0917+62 observed in May 1989 at 6 cm wavelength with the 100 m-telescope and the VLA.

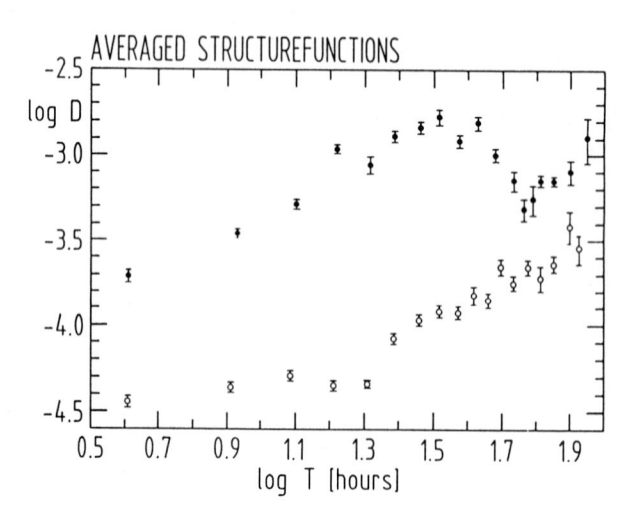

Fig. 3: Sample-averaged structure functions of type I- (open circles) and type II-variables (filled circles). Individual structure functions of each source were averaged over the three observing sessions at 11 cm wavelength in 1985 (see Heeschen et al., 1987).

3. Results

To give an overview about the typical behaviour of flat spectrum sources, Fig. 1 shows a selection of light curves observed at 11 cm in December 1985 (see Heeschen et al., 1987). Whereas some sources (0716+71, 0917+62, 1150+81, and 1803+78) show pronounced variations with a characteristic time scale shorter than the total length of the observations (100 hours), others (1642+69, 1807+69, 2007+77) exhibit weaker, but still significant variations with most of the power in longer time scales. The latter class (called "type I") is characterized by monotonically increasing structure functions; the former group ("type II") shows structure functions with minima at a lag that corresponds to the typical timescale in the light curve. (The structure function D is defined by $D(\tau) = \langle (R(t+\tau) - R(t))^2 \rangle$, where $R(t) = (S(t) - \langle S \rangle)/\langle S \rangle$ is the residual flux density at time t.) In Figure 3 we show averaged structure functions of all sources classified as type I and II, respectively. The prevalence of a preferred time scale of $\sim$ 1 to 2 days even in the averaged structure function of all type II sources seems to be a real effect that is not caused by the finite length of the observing run.

A summary of the results obtained with the Effelsberg telescope is given in Table 2. The first part of this table lists the sources in the complete high declination sample, the second part consists of the sources in the complete intermediate declination sample, the third section list several miscellaneous "interesting" objects. We give source names, optical identifications, VLBI structures (vc = very compact, c = compact, jet = jet-like morphology, dble = double structure, irr = irregular), galactic latitudes, and classification of the variability type at each epoch (types I, II, and 0, 0 = not significantly variable, − = not observed). From this table it can be seen that 14 out of 49 sources – i.e. $\sim$ 30% of all observed flat spectrum sources – have shown type II variability at least once. In those sources that have been observed several times, the results of the single epochs are remarkably consistent with each other; objects like 0917+62 and 0716+71 showed marked variations each time they were observed. It is also evident that type II variability occurs almost exclusively in sources that do not show extended VLBI structures. A detailed statistical analysis of the variability amplitudes of all objects indicates that virtually all sources with compact or very compact VLBI structures show detectable variations on the $\geq$ 2% level (see Witzel and Quirrenbach, 1991).

Simultaneous observations with the Effelsberg 100 m-telescope (at 6 and 11 cm wavelength) and the VLA (at 2, 3.5, 6, and 20 cm wavelength) in May 1989 revealed the multifrequency characteristics of the variations in 0917+62 (see Fig. 4). It is evident that the variability amplitudes are largest at 11 cm wavelength. The main peaks seen at this wavelength can be traced down to 2 cm with decreasing amplitudes without any significant time delays. Additional peaks that are barely visible at 11 cm (e.g.

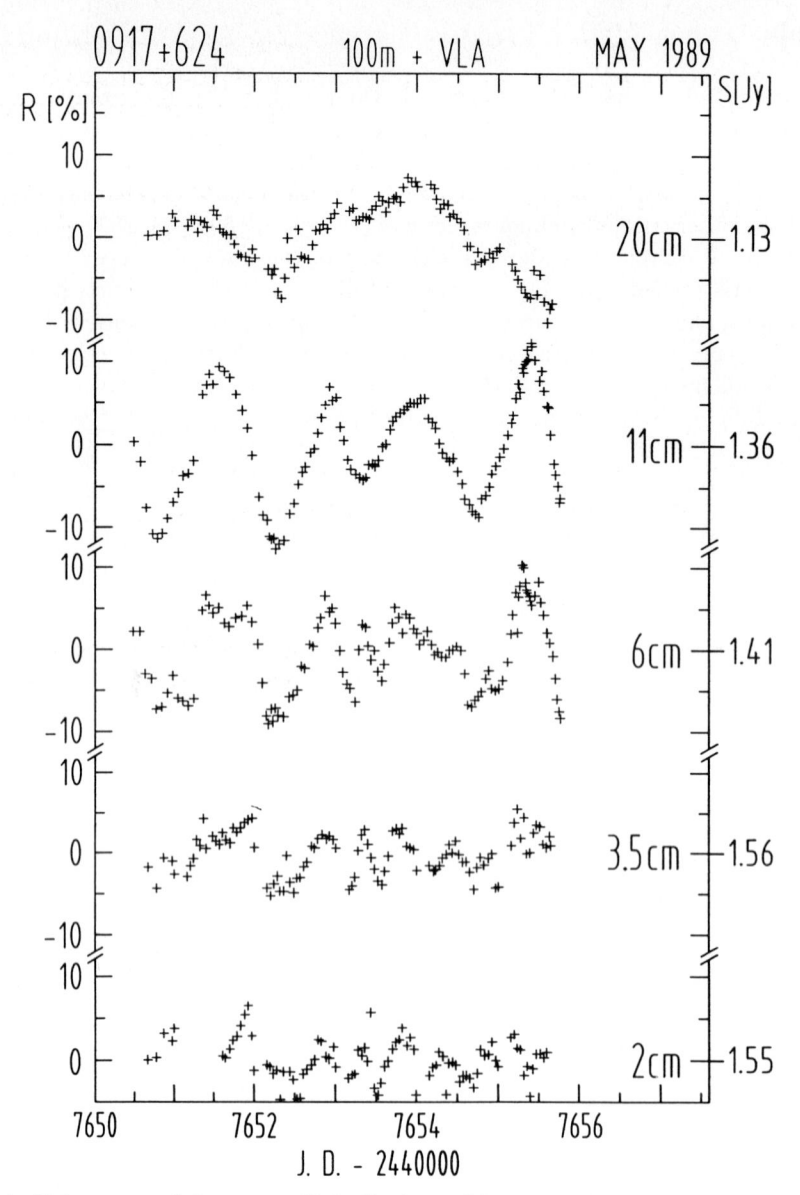

Fig. 4: Light curves of the quasar 0917+62 observed in May 1989 at 2, 3.6, 6, 11, and 20 cm wavelengths.

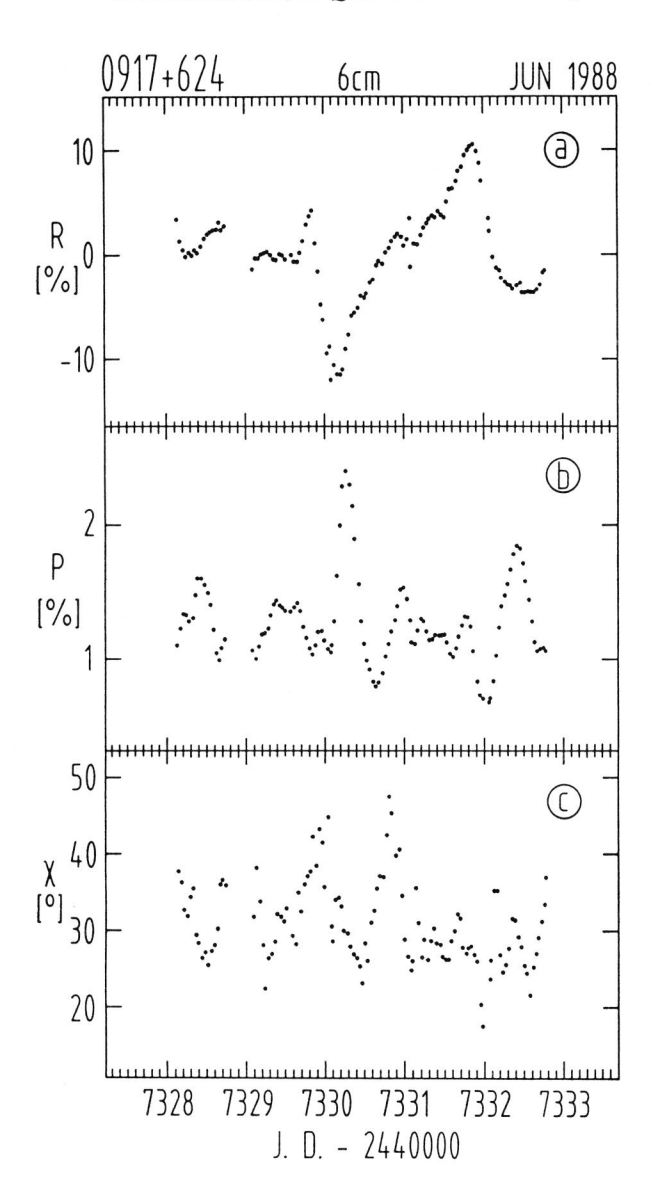

Fig. 5: Relative amplitude R (a), polarized flux density P (b), and polarization angle χ (c) of 0917+62 at 6 cm wavelength in June 1988.

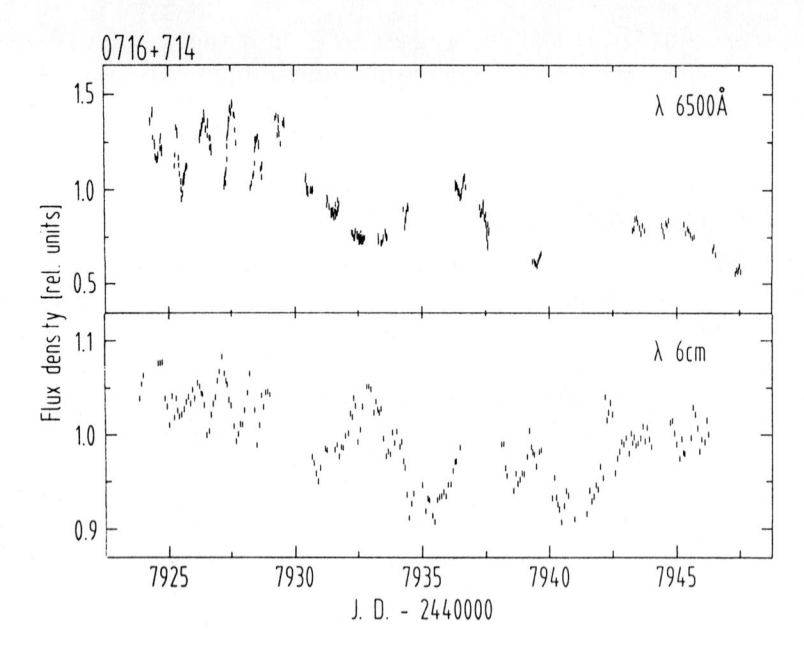

Fig. 6: Optical (top) and radio (bottom) light curves of the BL Lac object 0716+71 in February 1990. Fractional deviations from the normalized mean flux density are plotted versus Julian date. The r.m.s. accuracy for both data sets is typically 1 %.

at J.D. 2 447 653.4) form obvious features with increasing amplitudes at the shorter wavelengths. At 20 cm, the variations seem to be quenched; less fine structure is visible in the light curve. It should be pointed out that this behaviour, especially the decrease of the amplitudes towards shorter wavelengths and the lack of a clear time delay is quite different from the "canonical" outbursts observed on the time scale of months to years in many quasars and BL Lac objects; it is more reminiscent of "low-peaking flares" identified in high-frequency monitoring data (see E. Valtaoja, *this workshop*).

A lightcurve of 0917+62 at another epoch, June 1988, is shown in Figure 5 a. Again, brightness changes of $\sim 15\%$ within 0.3 days are seen, and very fast variations that are not adequately traced with our ~ 1 hour sampling are detected at J.D. 2 447 331.1. At the same time, the polarization degree changes by a factor ~ 3 (see Fig. 5 b); the variations of total intensity S and polarization P are anti-correlated: maxima of P occur near times of minimum S (e.g. at J.D. 2 447 330.2). The polarization angle χ did not show dramatic variations; the deviations of $\pm 15°$ from the mean value of $\chi = 35°$ seen in Figure 5 c are an upper limit to real variations of χ, since some spurious variations due to incomplete correction of the instrumental polarization cannot be excluded. This behaviour seems to be typical for 0917+62; it has been observed in at least four observing runs. Only once did the polarization angle show a large rapid change, which can best be described by swing of 180°, or equivalently by a path of the polarization vector in the QU-plane that encircles the origin (Quirrenbach et al., 1989b).

For more than three weeks in February 1990, a few selected sources which had shown variability of type II were monitored simultaneously in the radio and optical domain. The radio observations were carried out at the VLA, using four antennas during reconfiguration time. The optical data were taken with the Calar Alto 2.2 m and a 0.7 m telescope of the Landessternwarte on Königstuhl in Heidelberg. CCDs were used for relative photometry of the program source and several comparison stars in the field; for details of these observations and the data reduction see S. Wagner, *this workshop*. Figure 6 shows the light curves of the BL Lac object 0716+71 at Johnson R-band (650 nm) and at 6 cm (see also Quirrenbach et al., 1991). The optical light curve shows fast variations (typical time scale ~ 1 day) during the first week of the observations; at J.D. 2447930 there is a transition to a somewhat fainter state; in this state slower variations (time scale ~ 1 week) are observed. The same behaviour can be found in the radio data, although the total amplitude of the variations is much lower here ($\sim 15\%$ as compared to a factor ~ 2 in the optical). It should be noted that there is no clear one-to-one correspondence between peaks in the optical and radio regimes, but the different time scales with a transition around J.D. 2447930 is clearly visible in both regimes. This finding is confirmed by a correlation function analysis. In Figure 7 we plot the structure functions of the optical and radio light curves of 0716+71 separately for the first and second parts of the observing time. The correspondence of the time scales in both wavelength regimes is evident from the minima in these functions. More details and further results are given by S. Wagner (*this workshop*).

4. Interpretation

The astrophysical interpretation of intraday radio variability has first to address the question whether the observed fluctuations are intrinsic to the sources or due to propagation effects. Gravitational microlensing as a possible source of rapid variability of quasars has been discussed by Gopal-Krishna and Subramanian (1991). They conclude that microlensing may provide a viable explanation for intraday variations, although even under optimistic assumptions very high Lorentz factors ($\gamma \gtrsim 30$) are required. It seems difficult, however, to reconcile the predictions of microlensing with the multifrequency data of 0917+62 described above: since the source sizes increase with wavelength and the lensing effect itself is achromatic, the variability amplitudes should continuously decrease with wavelength due to "smearing" effects. Such be-

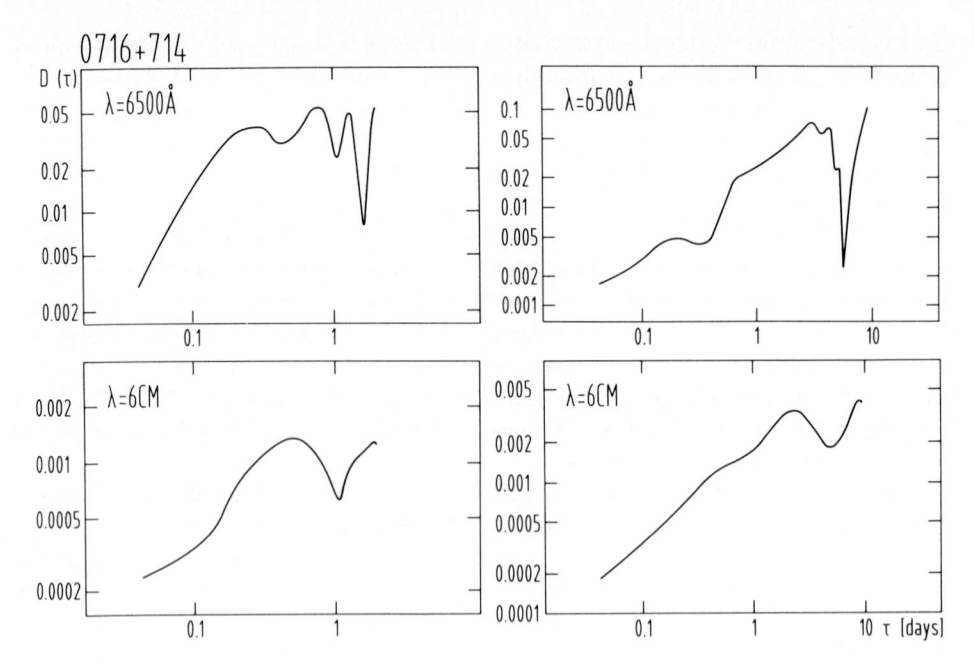

Fig. 7: Structure functions of 0716+71 derived from the light curves in Fig. 6. Part 1 (left) corresponds to $J.D. \leq 2\,447\,930$, part 2 (right) corresponds to $J.D. > 2\,447\,930$. Similar timescales in the optical (top) and in the radio domain (bottom) are obvious.

haviour is obviously not observed.

Refractive interstellar scintillation is a more plausible cause of rapid radio variations. Although this mechanism seems to be excluded by the correlations between radio and optical variability in 0716+71, RISS could still contribute significantly to the fluctuations in other sources. In fact, taking Rickett's (1986) parameters for the "standard" interstellar medium, a source e.g. located at $b^{II} = 45°$ is expected to show variations with r.m.s. fluctuations of $6\% \cdot \theta_{mas}^{-1}$ at 11 cm wavelength, with a typical time scale of 14 days·θ_{mas} (here θ_{mas} denotes the source size in milliarcseconds). This suggests that the presence of low-level fluctuations in virtually all sources with compact VLBI structure is due to scintillation in the interstellar medium (Witzel and Quirrenbach, 1991).

In addition to this "normal" component of the ISM, there is evidence that clouds or filaments in the solar neighbourhood can lead to enhanced scattering. A good example for probable scattering in the north galactic spur (galactic loop I) is given by the quasar 1741-03 (Hjellming and Narayan, 1986). The high amplification events observed e.g. in 0954+65 (Fiedler et al., 1987) also could be due to similar structures. Since the sources of our high declination sample happen to be located behind the galactic loop III, it is tempting to attribute the intraday variations in these sources at least partly to scattering in this old supernova remnant, a possibility which was discussed by Heeschen et al., 1987. Wambsganss et al., 1989 have indeed shown that scattering by a two-dimensional distribution of identical scatterers with Gaussian electron density profiles (central electron density $n_e \sim 300\,\mathrm{cm}^{-3}$, cloud size ~ 0.1 AU) can lead to light curves similar to those observed in 0917+62 at 6 and 11 cm wavelength. In this scenario, the polarization variations could be explained by a two component model for the source: if the source consists of a jet with dominant magnetic field parallel to its axis and a shock component with magnetic field perpendicular to the jet axis, scattering could preferentially amplify the smaller shock component and therefore lead to the observed anti-correlation between S and P by cancellation of the polarizations of the two components during maximum amplification.

It is clear, however, that propagation effects in the ISM cannot produce variations in the optical range. Therefore, the correlated radio and optical variations in 0716+71 provide strong evidence for intrinsic variability (see also Quirrenbach et al., 1991). A detailed model based on shocks propagating in a preexisting jet has been elaborated by Qian et al., 1991. The model is an extension of the models developed to explain long-term variations and VLBI structures of radio sources (see e.g. Hughes et al., 1989, and the contributions of P.A. Hughes, H.D. Aller, and M.F. Aller, *this workshop*). Qian et al. assume that the observed polarized emission is the superposition of the quiet emission of the underlying relativistic jet and a variable shock component. The magnetic field in the jet is probably mostly random, but shows a weak alignment with the jet axis. Due to field compression in the shock region, the variable component contains a more ordered magnetic field, which is perpendicular to the jet axis. Thus the polarizations of the two components tend to cancel each other, and the cancellation is more complete during the maxima of the variations. To illustrate this model, assume that the quiescent and shock components have polarization degrees of 2.5% and 10%, respectively, and that the shock component varies between 0 and 20% of the quiescent flux density. Then variations of S by 20% are observed; they are anti-correlated with changes of P by a factor of 6; χ remains constant at a right angle to the jet axis. This simple example reproduces the variations in 0917+62 remarkably well; it is also compatible with VLBI observations of this source (Quirrenbach 1990). Within this framework, the wavelength dependence of the variations can be naturally understood as an optical depth effect; the spectral turnover of the variable component would then be close to 11 cm. Alternative models of intrinsic variability are based on helical magnetic field structures (e.g. Königl and Choudhuri, 1985) or on "hot spots" in rotating jets that inherit angular momentum extracted from the accretion disk (e.g. Camenzind, 1990). Marscher et al. (*this workshop*) discuss thin shocks in a turbulent jet as a source of the observed rapid variations the flux density, polarization and polarization angle.

The most severe problem for the intrinsic models of intraday radio variations is posed by the extremely high apparent brightness temperatures $T_B \gtrsim 10^{19}$ K implied from the short time scale through the light travel time argument. Brightness temperatures in excess of 10^{12} K should not be observed in synchrotron sources, since they would cool rapidly to this temperature due to second order inverse Compton scattering (e.g. Kellermann and Pauliny-Toth, 1969). This problem could be resolved by postulating Lorentz factors $\gamma \gtrsim 100$ for the bulk plasma flow and applying the usual D^{-3} correction to the observed brightness temperature, where D is the Doppler factor of the plasma flow. Evidence from several independent indicators including apparent superluminal motions and X-ray flux densities shows that $\gamma \lesssim 10$ is a more likely range for most flat spectrum sources (e.g. Witzel et al., 1988). Other possibilities to explain brightness temperatures violating the inverse Compton limit invoke coherent emission mechanism or anisotropic electron populations. But also special geometric configurations could lead to instantaneous interactions between propagating shocks and preexisting perturbations in the jet, so that the light travel time argument could not be applied to derive the size of the emitting region. For example, if irregularities in the jet have a typical separation that is comparable to the jet radius *in the rest frame of the source*, the brightness temperature has to be transformed by a factor $D^{-3} \cdot \gamma^{-2}$, so that $\gamma \approx 10$ is sufficient to explain the short time scale variations (Qian et al., 1991).

In summary, we propose that the observed intraday variations consist of a superposition of low-amplitude fluctuations due to refractive interstellar scintillation and much more dramatic variability with an origin intrinsic to the sources. The latter variations are – at least in some sources – correlated with variability of the polarization and of the optical emission. The full diagnostic potential of these effects for the innermost parts of the jets in compact radio sources has yet to be explored; especially the radio-optical correlations should contain a wealth of information about the physical relation of the regions emitting in the two wavelength ranges and about particle acceleration mechanisms.

References

Camenzind, M., 1990, in: *Reviews in Modern Astronomy 3, Accretion and Winds*, ed. G. Klare, Springer (Berlin), p. 3.
Cohen, M.H., et al., 1971, *Astrophys. J.*, **170**, 207.
Dent, W.A., 1965, *Science*, **148**, 1458.
Fiedler, R.L., Dennison, B., Johnston, K.J., Hewish, A., 1987, *Nature*, **326**, 675.

Gopal-Krishna, and Subramanian, K., 1991, *Nature*, **349**, 766.
Heeschen, D.S., 1984, *Astron. J.*, **89**, 1111.
Heeschen, D.S., and Rickett, B.J., 1987, *Astron. J.*, **93**, 589.
Heeschen, D.S., Krichbaum, T., Schalinski, C., Witzel, A., 1987, *Astron. J.*, **94**, 1493.
Hughes, P.A., Aller, A.D., Aller, M.F., 1989, *Astrophys. J.*, **341**, 54.
Hjellming, R.M., and Narayan, R., 1986, *Astrophys. J.*, **310**, 768.
Kellermann, K.I., and Pauliny-Toth, I.I.K., 1969, *Astrophys. J. (Letters)*, **155**, L71.
Königl, A., and Choudhuri, A.R., 1985, *Astrophys. J.*, **289**, 188.
Kühr, H., Pauliny-Toth, I.I.K., Witzel, A., Nauber, U., 1981, *Astron. Astrophys. Suppl.*, **45**, 367.
Pauliny-Toth, I.I.K., and Kellermann, K.I., 1966, *Astrophys. J.*, **146**, 643.
Qian, S.J., Quirrenbach, A., Witzel, A., Krichbaum, T.P., Hummel, C.A., Zensus, J.A., 1991, *Astron. Astrophys.*, **241**, 15.
Quirrenbach, A., Witzel, A., Krichbaum, T., Hummel, C.A., Alberdi, A., Schalinski, C., 1989a, *Nature*, **337**, 442.
Quirrenbach, A., Witzel, A., Qian, S.J., Krichbaum, T., Hummel, C., Alberdi, A., 1989b, *Astron. Astrophys. (Letters)*, **226**, L1.
Quirrenbach, A., 1990, Ph.D. thesis, University of Bonn.
Quirrenbach, A., et al., 1991 *Astrophys. J. (Letters)*, in press.
Rickett, B.J., 1986, *Astrophys. J.*, **307**, 564.
Romani, R.W., Blandford, R.D., Cordes, J.M., 1987, *Nature*, **328**, 324.
Shapirovskaya, N.Y., 1978, *Astron. Zh.*, **55**, 953.
Sholomitskii, G.B., 1965, *Astron. Zh.*, **42**, 673.
Simonetti, J.H., Cordes, J.M., Heeschen D.S., 1985, *Astrophys. J.*, **296**, 46.
Spangler, S., Fanti, R., Gregorini, L., Padrielli, L., 1989, *Astron. Astrophys.*, **209**, 315.
van der Laan, H., 1966, *Nature*, **211**, 1131.
Wambsganss, J., Schneider, P., Quirrenbach, A., Witzel, A., 1989, *Astron. Astrophys. (Letters)*, **224**, L9.
Witzel, A., Heeschen, D.S., Schalinski, C., Krichbaum, T., 1986, *Mitt. Astron. Ges.*, **65**, 239.
Witzel, et al., 1988, *Astron. Astrophys.*, **206**, 245.
Witzel, A., and Quirrenbach, A., 1991, in: *Proceedings of the conference "Propagation Effects in Space VLBI" held in Leningrad, June 1990*, ed. N. Kardashev and L. Gurvits, in press.

Rapid Variability in the BL Lac object S5 0716+714

Stefan J. Wagner

Landessternwarte, Königstuhl, 6900 Heidelberg, Germany

ABSTRACT

Simultaneous multifrequency light-curves of a densely sampled four-week monitoring campaign of the flatspectrum BL Lac candidate 0716+714 is presented. The object exhibits quasiperiodic variability on time-scales of days to weeks and short-term flickering. Several independent pieces of evidence strongly suggest that the radio emission and optical emission emerge from volumes of very similar dimensions which are closely connected or even identical. This imposes severe restrictions on any microphysical acceleration mechanisms.

1 INTRODUCTION

Variability on time-scales from minutes to years is a criterion used to define BL Lac objects. Since this variability is observed in all wavelength regimes, coordinated multifrequency studies may constrain models of the origin and propagation of the variability in physical and wavelength space. While flux changes on time-scales of weeks to years were found to be correlated in the optical-IR-Radio bands with significant time-lags (months) (see e.g. Bregman, 1990) few studies of this kind have been carried out on shorter time-scales.

This was basically caused by the conception that the volumes emitting at radio frequencies are larger than a lightweek (about 10^{16} cm) and could not show intrinsic variability on time-scales as short as those undoubtedly present at shorter wavelengths. Any observed rapid variability at low frequencies was generally attributed to propagation-induced effects, predominantly refractive interstellar scintillation (e.g. Heeschen, 1987).

Quirrenbach et al., 1989 confirmed intraday variability for a larger sample of compact, flat-spectrum radio-sources. They found several pieces of evidence indicating that an intrinsic origin of the observed flickering had to be considered seriously. An unambiguous confirmation of this intrinsic nature could only be given by finding correlations of the rapid variability observed in the radio regime with similar behaviour at shorter wavelengths where RISS is of no importance. The rivaling explanation for propagation induced variability - the microlensing effect (see e.g. Nottale, 1986) -

can be excluded for intraday flickering. The microlensing theory requires relativistic motions perpendicular to the line-of-sight by the source, the lens or the observer if variability on such short time-scales is to be explained. Since such motion can be ruled out for all realistic lenses and the observer, microlensing could only enhance intrinsic variability.

In a statistic investigation Wagner et al. (1990) demonstrated that the same three out of six sources observed simultaneously in optical and radio regimes showed variability on time-scales of hours to days in both wavebands with very similar characteristics. Here the results of a much longer and better sampled multifrequency monitoring campaign are presented for one of our sources (0716+714) finally confirming the correlated intraday variability and thus its intrinsic nature.

2 OBSERVATIONS

0716+714 was observed with four antennae of the VLA at four different frequencies every 2 hours during February 1990 (J.D. 2447923 - 2447949). Parallel to this, observations in three optical bands we carried out with the 2.2-m-telescope on Calar Alto and one of the 0.7 m telescopes of the Landessternwarte in Heidelberg. Details of the observations and reduction strategies are published elsewhere. For a comparison of the results from the two optical telescopes see Wagner, 1991. The measurement accuracies are about 1% at 6 cm and 0.8% (Calar Alto) or 1.2% (Heidelberg) in the optical light-curves. In all cases the duration of an individual measurement was much shorter than the time-span between succeeding observations.

Fig. 1a *Light-curve of 0716+714 at 650 nm during February 1990*
Fig. 1b *Simultaneous observations at 420 nm (Johnson B)*

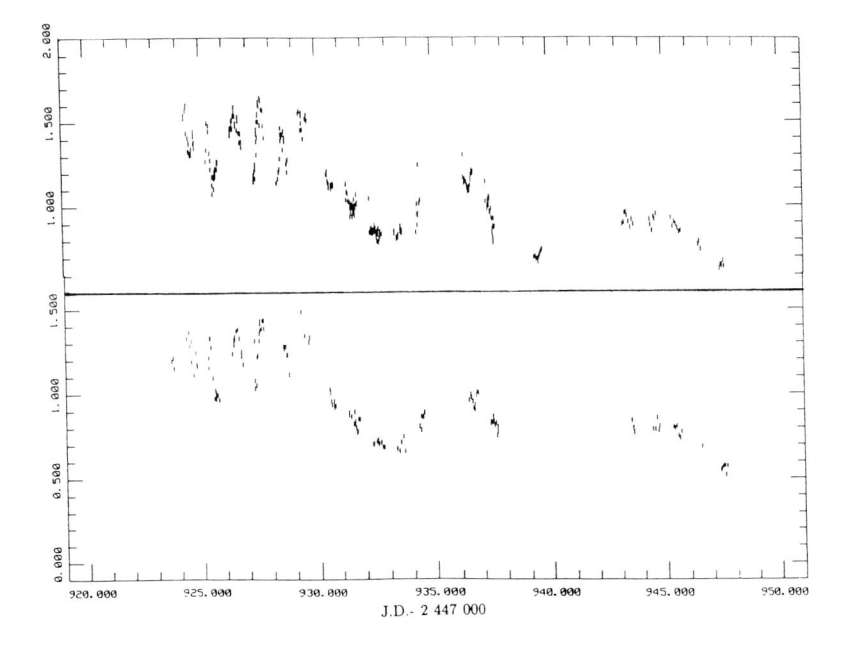

3 THE OVERALL CHARACTERISTICS OF THE LIGHT-CURVES

The combination of the two optical telescopes gave an almost permanent coverage of the light-curve for a total period of 27 days. During nighttime single measurements were obtained twice per hour on average. The complete light-curve at 650 nm is shown in Figure 1a. Figure 1b displays the light-curve at 420 nm (i.e. Johnson B, slightly modified by the wavelength dependent sensitivities of the CCD-chips involved).

Although the light-curve measured at 420 nm is not sampled as densely as the light-curve at 650 nm, it is obvious that the two light-curves measured in the optical regime are almost identical. The same is true for the light-curve measured at 800 nm, in the I band. Throughout the optical regime the spectral index remains constant during the rapid, quasiperiodic state of the first week as well as during the change in modes and during the slow mode of variability.

The good match also indicates that extinction and reddening effects which could in principle affect even relative photometry can be reduced to a negligable level by using using field stars of different colours as calibrators.

The overall light-curve (Fig. 1a) displays a very interesting pattern. During the first week a quasiperiodic modulation with time-scales slightly larger than one day is observed. With an expanded time-scale the almost "saw-tooth" shaped light-curve is clearly visible. The amplitude of the variations amount to 40%. At J.D. 2447930 the behaviour changed abruptly into a state with significantly larger time-scales (only two "cycles" with a mean periodicity of 6 days could be observed). Although the mean brightness during the "slow state" is 40% lower than during the high state of the first week of the campaign, the peak-to-peak - amplitudes are similar to those observed during the "fast state", i.e. about 40%.

Fig. 2 650 nm (errorbars) and 420 nm (Δ) observations during a single night.

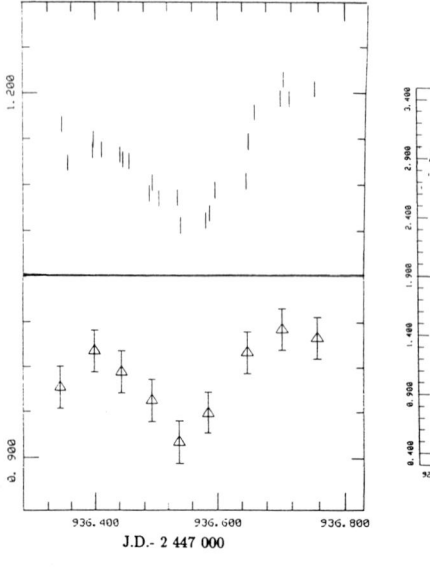

Fig. 3 Optical (dots) and radio observations in February 1990.

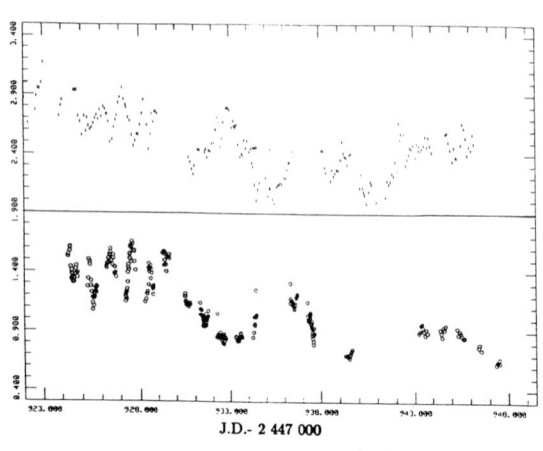

The optical spectrum remains constant within the errors during the whole campaign. This is demonstrated by the perfect coincidence of the 650 nm (R-Band) and 420 nm (B-Band) observations even during individual turnovers, displayed in Figure 2.

Figure 3 compares the optical light-curve (650 nm, R band) with the radio measurements (6 cm, VLA). The close correspondence is clearly visible. During both parts of the campaign the time-scales of the variabilities in the two regimes are similar. Although the time-lags between 6cm and 650 nm data are different in the fast and in the slow state, the change between the two modes occurs within the same 24 h period in both wavelength regimes.

4 STATISTICAL PROPERTIES

We examined the light-curves at all frequencies separately for the rapidly variable first week and for the second part of the campaign independently as well as for the total campaign. The power spectra are described in Wagner, 1991, the structure functions are shown in Quirrenbach et al., 1991. Here we just compare the power-spectra of the first week in the radio and optical regimes (Figure 4). In either band a strong signal related to the specific sampling pattern can be identified. In the optical regime the day-night modulation introduces a spurious peak at 12 h (the average length of the night), at the VLA a 24 h signal – probably due to technical interferences – appeared. If those erroneous periodicities are ignored, two interesting aspects become apparent. Firstly there are two dominant frequencies of 0.8 and 1.3 days, secondly these two frequencies are identical in the optical and radio light-curves. The same result is derived from a detailed analysis of the structure functions (Wagner et al., 1991).

Fig. 4 *Power spectra of the 6 cm (left) and 650 nm (right) data during the first week of the February 1990 campaign.*

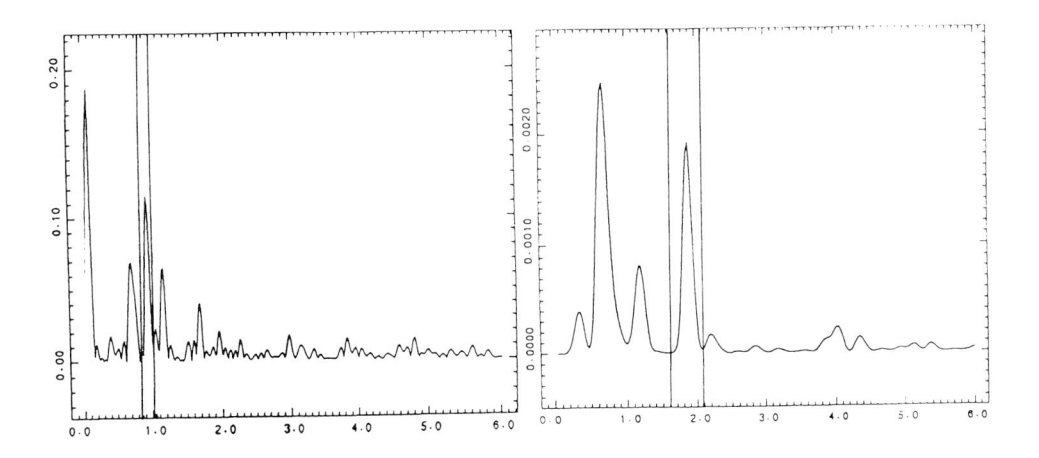

5 THE RELATIONSHIP BETWEEN OPTICAL AND RADIO VARIATIONS

A short description of the optical light-curve can be given as follows:
1.) The light-curve is bimodal.
2.) The change between the model occurs at JD 2447930.
3.) In the first part the light-curve is quasiperiodic.
   A Fourier analysis reveals two frequencies of 0.8 and 1.3 days.
4.) In the second part the light-curve has a characteristic time-scale of 6 days.
5.) The amplitudes of the rapidly and slowly variable parts are identical.
6.) During the change of modes the average level dropped by an amount similar to the amplitudes of the variability.

Surprisingly these descriptions are fully valid for the radio light-curve as well. Although it is possible that the similarity of one of the characteristics of the light-curves in the two frequency regimes is by chance, it is unlikely that the similarity of all of the characteristics is a coincidence.

A common suggestion for rapid variability of compact sources at radio wavelengths is interstellar scintillation in the form of RISS. This scattering is only working at low frequencies and has no effect at 10^{14} Hz. If the optical light-curve is not influenced by RISS, the close similarity of all characteristics of the light-curve between radio and optical rules out that the radio variations are dominantly influenced by RISS. Minor contributions of interstellar szintillation have to occur for such compact sources and are observed (with lower amplitudes) see Krichbaum et al., this volume.

6 CONSTRAINTS ON MODELS OF THE EMISSION PROCESS

Because of the similarities described above we assume a common origin of the fluctuations seen in the two frequency regimes. Cross-correlation functions for the total period of simultaneous observations as well as for each of the the two modes separately are given by Wagner, 1991. The quasi-periodicity causes multiple maxima of similar height in the cross-correlation functions and does not allow an unambiguous identification of the lag. If the lag is less than 50 hours (as implied by the fact that the change in modes occurs at both frequencies simultaneously within less than 50 hours), it is not surprising that no spectral variations are seen within the optical regime.

Variability on time-scales of a day may be explained by flares occuring on a face - on disk surrounding the central SMBH (e.g. Wiita, this volume). While it may be possible to account for the acceleration of particles by magnetic reconnection and produce coherent synchrotron radiation (Lesch et al., in preparation), it is not clear whether the observed periodicities can be explained naturally. Multiple events might lead to cancellation of the circular polarisation expected in sources producing coherent synchrotron radiation, but no simple szenario seems to be able to account for the optical depth problems introduced by the near-simultanity of optical and radio radiation.

An attractive alternative is given in jet-models. Relativistic beaming can partly account for the enormous brightness-temperatures derived from the radio-variability. Shocks propagating down the relativistic jets (see e.g. Marscher, this volume) have been studied in some detail. While turbulence may well explain the short timescales, it does not account for the observed quasi-periodicity. Furthermore the simple shock model predicts that the radio-emitting volume is larger than the one where the electrons emitting the optical radiation are situated. This is in conflict with the observation that the auto-correlation function of the radio variability is as narrow as that of the optical data. If turbulence is avoided, additional effects are required if the jets are not super-relativistic. The shape of the lightcurve indicates that geometric considerations are important. If helical magnetic fields are included in the shock-models, the field geometry will force the density perturbations on helical paths thus introducing changes of the directions and the axis of maximum beaming. This would lead to the observed quasiperiodic variability. Both large-scale shocks (e.g. Königl and Choudhuri) or small-scale bullets (Camenzind et al., in preparation) are possible. In all of these scenarios it still has to be tested whether the suggested acceleration models can explain the energetics and whether the observed multi-frequency data can be accomodated without facing the obvious optical-depth-problems.

ACKNOWLEDGEMENTS:
I thank all my collaborators for their help with the observations and reductions and the DFG (Sonderforschungsbereich 328) for support.

REFERENCES:
Bregman, J.: 1990, *Astron. Astrophys. Rev.* **2**, 125
Heeschen, D. and Rickett, B.: 1987, *Astron. J.* **93**, 589
Königl, A. and Choudhuri, A.: 1985, *Astrophys. J.* **289**, 173
Nottale, L.: 1986, *Astron. Astrophys.* **177**, 383
Quirrenbach, A., Witzel, A., Krichbaum, T., Hummel, C., Alberdi A. and Schalinski, C.: 1989, *Nature* **337**, 442
Quirrenbach, A., Witzel, A., Wagner, S., Sanchez-Pons, F., Krichbaum, T., Wegner, R., Anton, K., Erkens, U., Haehnelt, M., Zensus, J., Johnston, K. et al.: 1991, *Astrophys. J.* in press
Wagner, S., Sanchez-Pons, F., Quirrenbach, A. and Witzel, A.: 1990, *Astron. Astrophys.* **235**, L1
Wagner, S.: 1991, in "Variability of Active Galaxies", Lecture Notes on Physics, Vol. 377, W. Duschl, S. Wagner, M. Camenzind (Eds.)

Fast radio variability of Blazars in observations with RATAN-600

S.A. PUSTIL'NIK, K.D. ALIAKBEROV

Special Astrophysical Observatory, Nizhinij Arkhyz, USSR, 357147

ABSTRACT

The results of search for rapid variability of 8.2 cm radio flux (time scales of 1 to 10 days) in 13 BL Lac type objects are reported. Significant variability (with confidence level more than 0.999) was detected for 5 of them: 0235 + 164, 0300 + 470, 0754 + 100, 0851 + 202, 1034 − 293. The indexes of fast variability of these objects (r.m.s. deviation/average flux) were 0.03 to 0.08. Relative amplitudes of variations at time scales of few days sometimes reached values of 0.1 to 0.2. No significant correlations between fast variations of optical and radio fluxes were found during coordinated observations of some objects.

1 INTRODUCTION

BL Lac objects as a class of extragalactic objects are well known as very active ones. Extremely rapid radio variations on interday scale of BL Lac and OJ 287 were first detected in the early 70's by Andrew et al. (1971) and McLeod et al. (1971). Since that time up to early 80's there were no systematic studies of this remarkable phenomenon.

Starting our program in 1982 we aimed to understand how common is fast radio variability among BL Lac objects and what are the parameters of such variability (see Pustil'nik, Aliakberov, 1987 for details).

2 OBSERVATIONS

We observed 13 program BL Lac objects and 11 reference sources with steep and flat spectra with RATAN-600 radio telescope during December 9 – 31, 1982. We used the main RATAN-600 mode of observations namely registration of a source crossing the diagram during passage the meridian. To search for variations we used the most sensitive receiver at 8.2 cm (f = 3.65 GHz). 10 to 17 flux measurements per program source were obtained with time resolution of 1 day (sometimes 2 days). The list of program objects is in Table 1.

Table 1. Program BL Lacertae objects

Object	Other name	Object	Other name
0109+224	OC 215.7	0851+202	OJ 287
0138-097	OC-065	1034-293	OL-259
0235+164	OD 160	1147+245	OM 280
0300+470	OE 400	1219+285	W Com
0754+100	OI 090.4	1308+326	OP 313
0829+046	OJ 049	1418+546	OQ 530
		2200+429	BL Lac

3 ANALYSIS OF THE DATA

To search for the variability of program objects we compared observed variance of antenna temperature T_A of the object with the expected one for a nonvariable source with the same value of T_A. The variance of T_A for a nonvariable source S^2 has to go as usual: $S^2 = S^2_N + a^2 \cdot T^2_A$, where S^2_N is the additive noise component, and the second term is a multiplicative component caused by the instability of the system gain. The dependence $S(T_A)$ was fitted from observations of 11 reference sources during the same period. It is shown in Figure 1 by the line through filled circles.

We checked the H_0–hypothesis on nonvariability of a program object with the help of the Fisher criterion for comparison of variances. We rejected H_0 at the confidence level of 0.999 or more in 5 of the 13 observed BL Lac objects. They are marked by triangles in Figure 1. In Figures 2 and 3 we show light curves for these 5 variable BL Lacs. In Figure 4 we show the behaviour of some reference sources.

4 RESULTS

The analysis of the data obtained abled us to draw the following conclusions:

- indexes of rapidly varying BL Lac objects were $\mu = 0.03 - 0.08$, $\mu =$ (r.m.s. deviation / average flux density).
- minimum characteristic times of variations were 2 to 4 days.
- relative amplitudes of interday variations sometimes reached values of 0.1 to 0.2.
- detection rate of fast cm-wavelength variability in a representative sample of BL Lac objects was near 40 % at the epoch of observations.

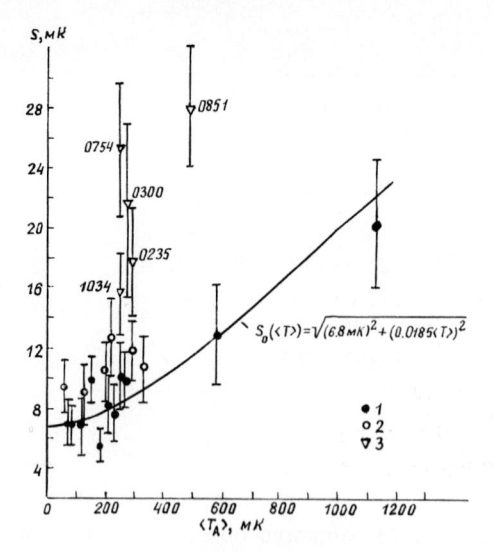

Figure 1. Standard deviations of antenna temperatures S(mK) at 8.2 cm wavelength versus average antenna temperature < T > for all observed sources. Filled circles are reference sources, open circles are program Lacertids without pronounced variability, triangles are program Lacertids with pronounced variability during Dec. 1982 set. Empirical dependence S = S(<T>) is determined with the help of least squares method only on the date of reference sources.

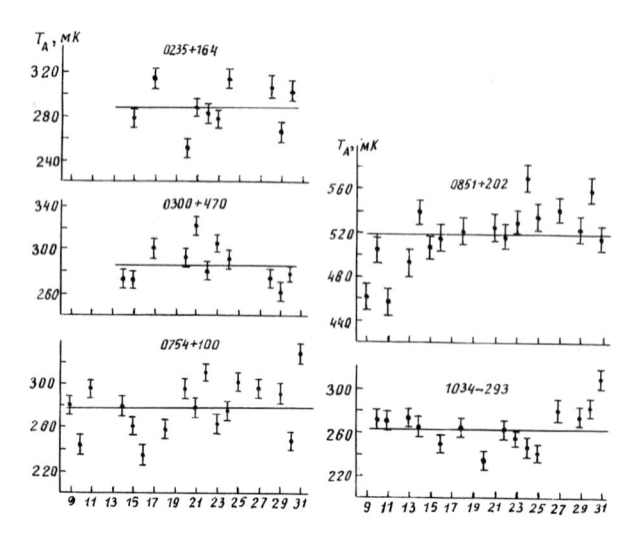

Figure 2a. Light curves at 8.2 cm wavelength for three variable in this set Lacertids – OE 400 (0300+470), OI 090.4 (0754+100) and PKS 1034-293 – on observations with RATAN-600 in December 1982. X-axis is time (Days of December), Y-axis is antenna temperature T (mK). Horizontal lines are average values of T.

Figure 2b. Light curves at 8.2 cm wavelenght for two other variable in this set Lacertids – AO 0235+164 and OJ 287 (0851+202) on observations with RATAN-600 in December 1982.

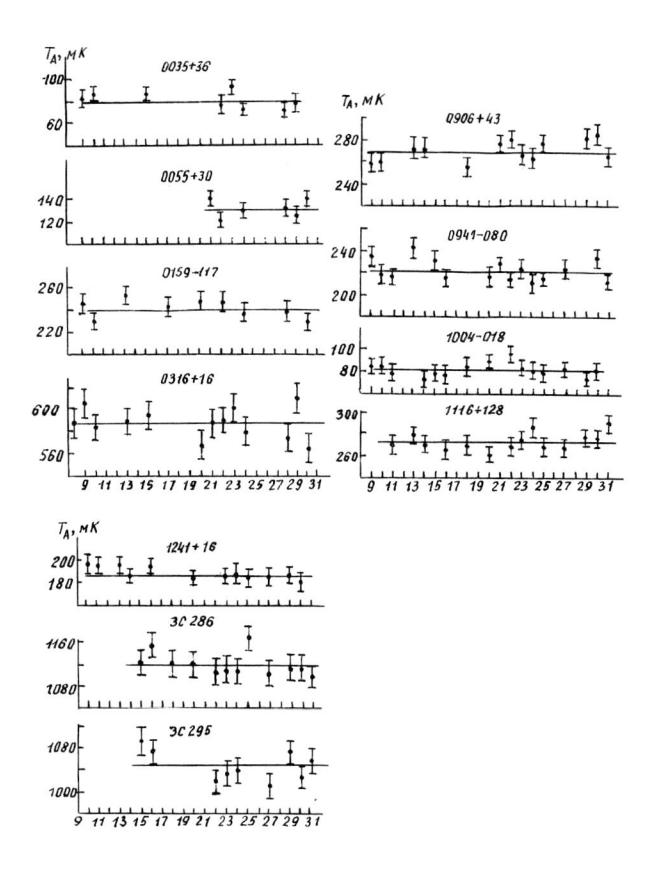

Figure 3. The behaviour of antenna temperatures T at 8.2 cm of 11 reference sources on observations with RATAN-600 in Dec. 1982. Horizontal lines are average values of T.

- these findings do not contradict the earlier suggestions about positive correlation of short term cm-mm variations of flux with long term activity of radio sources (in comparison with the long term light curves by Aller et al., 1991).
- coordinated optical photometry of some of these BL Lac objects (see Pustil'nik et al., 1986) was carried out. No significant correlations were detected between variations in optical and radio regimes. The only indication of such correlation was a case of OI 090.4 (Pustil'nik et al., 1981). To be confident we need better sampled data in both regimes, as was recently demonstrated (Krichbaum et al., 1991, Wagner 1991, this volume).
- the estimates of brightness temperature for the fast variable components lead to values of T_B (var) of about 10^{16} to 10^{17} K.

5 POSSIBLE INTERPRETATIONS

In our view an internal explanation of the nature of the observed variations is more likely than an external one. Relativistic jet models (Blandford, Konigl, 1979) seem to be most natural.

A. In straightforward application of light travel time arguments T_B (var) scales as $D^3 T_B$ (rest). From the inverse Compton limit T_B (rest) $< 10^{12}$ K (Kellermann and Paulini-Toth, 1969) we can derive estimates for the bulk Doppler factor of the jet: $D > 26 - 49$, which seems to be too large. Corresponding estimates of the Lorentz factors $\Gamma > 13 - 25$ (Table 2) are significantly larger than what is usually derived from VLBI data.

Table 2. Derived minimum Lorentz-factors in two models.

Object	Redshift	Γ(light trav.)	Γ(shock)
0235+164	0.852	25	14.5
0300+470	> 0.2	> 16	9.3
0754+100	> 0.2	> 17	9.7
0851+202	0.306	18	10
1034-293	0.312	13	8.4

B. There is also another possibility. In spite of the fact that our light curves of rapidly variable BL Lac objects are obviously undersampled, they resemble remarkably the light curve of quasars 0917+624, obtained with good resolution by Quirrenbach et al., 1989. This quasar displayed quasiperiodic variations of total and polarized radio flux and large swings in P.A., as was predicted in the model by Königl and Choudhuri, 1985 (shocks in the jet with a helical mode of magnetic field). In this model subsequent peaks on light curve are due to a passage of a shock (with Γ_s) through an appropriate phase of a wave of helical mode. In such a model T_B (var) scales as $\Gamma_s^2 D_s^3 T_B$ (rest). For our variable sources we can derive Γ_s of 8.5 to 14.5, which seem to be quite natural for strong shocks in relativistic jets.

References

Aller M., Aller H., Hodge P. 1991, this volume.
Andrew B.H., Harvey G.A., Medd W.J. 1971, *Ap.Lett.* **9**, 151.
Blandford R.D., Königl A. 1979, *Ap.J.* **232**, 34.
Kellermann K.I., Pauliny-Toth I.I.K. 1969, *Ap.J.* **155**, L71.
Königl A., Choudhuri A.R. 1985, *Ap.J.* **289**, 188.
Krichbaum T. et al. 1991, this volume.
McLeod J.M., Andrew B.H., Medd M.J. et al. 1971, *Ap.Lett.* **9**, 19.
Pustil'nik S.A., Lyutyj V.M., Neizvestnyj S.I. 1981, *Sov. Astronomy Lett.*, **7**, 305
Pustil'nik S.A., Pustil'nik L.A., Neizvestnyj S.I., Lyutyj V.M. 1985, *Commun.Spec.Astroph.Obs.*, **48**, 27.
Quirrenbach A., Witzel A., Krichbaum T., Hummel C.A., Alberdi A., Schalinski C. 1989, *Nature* **337**, 442.
Wagner S.J. 1991, this volume.

Rapid Light Variations of BL Lacertae

V.T. Doroshenko[1], V.M. Lyuty[1], A. Sillanpää[2], E. Valtaoja[3]

1) Sternberg Astronomical Institute, Crimean Laboratory, P/O Nauchny, SU-334413 Crimea, USSR
2) Tuorla Observatory, University of Turku, SF-21500 Piikkiö, Finland
3) Metsähovi Radio Research Station, Helsinki University of Technology, Otakaati 5A, SF-02150 Espoo, Finland

Introduction

Investigation of rapid variations in BL Lac objects gives unique information about the sizes and structures of their nuclear regions and of the nature of the emission mechanisms. The prototype, BL Lac itself, shows optical variability on all timescales (Racine 1970; Miller and McGimsey 1978; Miller et al. 1989). In general, the mean amplitude of variability increases with time, but sometimes systematic changes up to 1.5 mag/hour have been observed (Weistrop 1973). No significant periodicities have been found (Moore et al. 1982; Brindle et al. 1985; Doroshenko et al. 1986).

The characteristics of color changes are uncertain. A good correlation between color and brightness variability in BL Lac has been reported (e.g., Racine 1970), but fairly large changes in U-B may occur without significant changes in brightness (Miller and McGimsey 1978), or color variability may be uncorrelated with brightness (Moore et al. 1982).

We have made simultaneous UBVRI observations of BL Lac with the 1.25-m reflector of the Crimean Astrophysical Observatory during 8 nights in 1982-83. In an earlier paper (Doroshenko et al. 1986) we have searched the data for periodic variability. Here we present a reanalysis of our observations with the aim of investigating nonperiodic intranight variations. For details of observations we refer to the previous paper.

Observations and results

We made differential photometry relative to the comparison star "c" (Bertaud et al. 1969), which was observed repeatedly during each night. A chopping UBVRI photopolarimeter was used. The observations consisted of single integrations of 10s in 1982 and 20s in 1983 simultaneously in each color. For better photon statistics, we have combined integrations to obtain total source integration times of 200s. This corresponds to a time resolution of 8 minutes. For each 8-minute segment we have calculated the

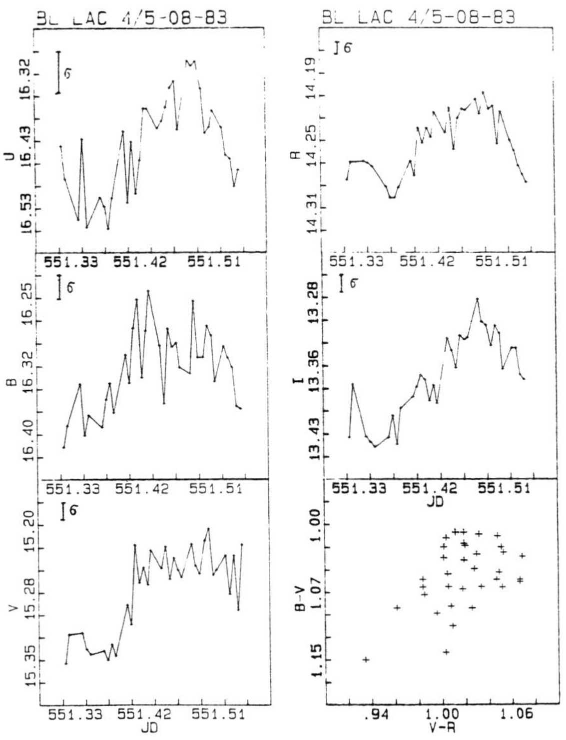

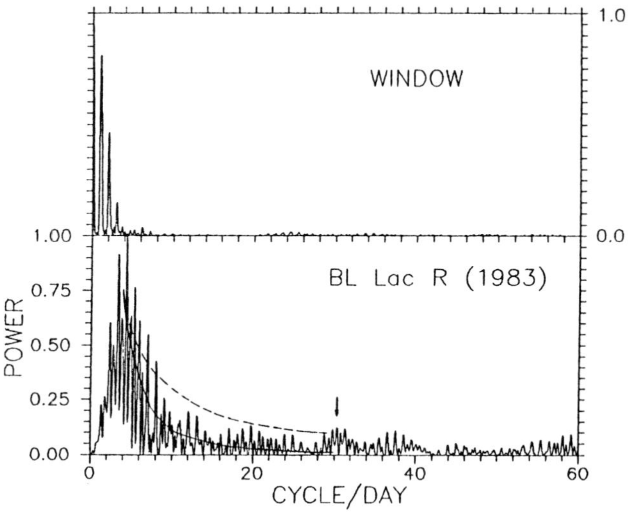

Figure 1. UBVRI lightcurves on 1983.08.04. One sigma error bars are shown. The internight colors B-V vs. V-R are also shown in the lower right corner. The variations are mainly due to errors in the V flux. **Figure 2.** (on the right) The spectral window and the power spectrum for the nights 1983.08.03-06. The solid line represents the spectrum for random walk noise and the dashed line for flicker noise.

magnitudes and colors. Typical errors for one 200s data point, due to photon statistics, are $\sigma_1 \approx 0.05$ mag in the U band and ≈ 0.01 mag in the R band.

To search for variability we have calculated for each night and each color the mean error σ_1 of a single 200s observation and the standard deviation σ_2 of the observations for each night. If $\sigma_2/\sigma_1 > 1.5$, the source has probably varied during the night (99% significance level). According to this criterion, variability in the R band (with the best photon statistics) was observed during all 8 nights. In Figure 1 we show the lightcurves for 4.-5.8.1983, when $\sigma_2/\sigma_1 > 1.5$ in all bands. A variation of 0.1-0.15 mag in ≈ 3 h is seen in all bands.

Figure 2 shows the spectral window and the power spectrum for the R band data for the 4 nights 3.-6.8.1983. The solid line shows the expected power spectrum for random walk noise ($P \propto f^{-2}$), which seems to be a good fit to the observations in the timescales $1-6$ h. A similar result holds for the I band, while for U and B we find $P \propto f^{-1.5}$. Moore et al. (1982) found $P \propto f^{-1.7 \pm 0.3}$ for timescales 5 h $- 5$ d. Our result shows that the random walk noise characteristics of BL Lac continues down to at least 1 h timescales. The peak in the R power spectrum at 30 cycles/day is not visible during any other night, so we do not consider it a sign of a true periodicity.

No definite intranight color variations were found. As an example, in Figure 1 we have plotted the B-V vs. V-R colors for the night 4.-5.8.1983. Both colors are dominated by the noise in the V band. Any possible intranight variations are below the 0.05 mag level. However, for longer time intervals we find a clear correlation between colors and and the V magnitude. When BL Lac is in a bright stage, the dependence is as expected for synchrotron radiation (Beskin et al. 1985). During the low brightness levels, as in our observations, color indices deviate from the predictions, most likely due to the effect of the underlying elliptical galaxy (Kinman, 1975).

References

Bertaud, C. et al.: 1971, *Astron. Astrophys.* **24**, 357
Beskin, G.M. et al.: 1985, *Sov. Astron. J.* **62**, 432
Brindle, C. et al.: 1985, *M.N.R.A.S.* **214**, 619
Doroshenko, V.T. et al.: 1986, *Astron. Astrophys.* **163**, 321
Kinman, T.J.: 1975, *Astrophys. J.* **197**, L49
Miller, H.R., Carini, M.T., Goodrich, B.D.: 1989, *Nature* **337**, 627
Miller, H.R., McGimsey, B.Q.: 1978, *Astrophys. J.* **220**, 19
Moore, R.L. et al.: 1982, *Astrophys. J.* **260**, 415
Racine, R.: 1970, *Astrophys. J.* **159**, L99
Weistrop, D.: 1973, *Nature Phys. Sci.* **241**, 157

Detection of Microvariability in Mrk 421

U.C. JOSHI AND M.R. DESHPANDE

Physical Research Laboratory
Navarangpura
Ahmedabad-380009, India

SUMMARY

Photopolarimetric observations of BL Lacertae object Mrk421 were made on 2.3 meter Vainu Babbu Telescope (at Kavalur) of Indian Institute of Astrophysics, Banglore. Rapid variability in flux in a time scale of about 45 minutes has been detected on January 23, 1990; the flux rapidly decreased by ~ 0.4 mag followed by a slower increase reaching back the normal level in about 45 minutes. The variation in flux is accompanied with a variation in the degree of polarization. The degree of polarization shows more rapid fluctuations than the flux. The upper value of the observed degree of polarization without any filter is 4 % with a maximum decrease of about 1 %. The polarization after correcting for the galaxy contamination is nearly wavelength independent while the angle shows slight wavelength dependence. The photometric observations show that Mrk 421 is in bright phase during the present observing run, the observed flux at visual wavelength of 0.55 micron is 28 mJy and the spectrum is quite steep, the spectral index being $\alpha = 1.1$.

1 INTRODUCTION

Rapid optical variations are expected in some of the BL Lac objects. It is vital to investigate the high speed optical variability of these objects to understand the nature of their central energy sources. In particular, the presence of periodicity in the radiation would support the gravimagnetic rotator model (Lipunov 1987) while in case of the accreting black hole model we expect quasi periodic variation. Mrk 421 is a highly variable BL Lacertae object and has been studied in quite detail by various researchers. Simultaneous multi frequency observations of Mrk 421 in radio, optical, UV and X-ray bands were carried out in the recent past to understand the nature of the enigmatic nucleus (Brodie et al. 1987). Comparison of the polarimetric observations on Mrk 421 made in the past by various authors (Maza et al. 1978; Bailey et al. 1981; Makino et al. 1987; Mead et al. 1988; Kikuchi 1991; Takalo 1991; Valtaoja 1991) show a large variation in both the degree of polarization and position angle. The most extensive data obtained on Mrk 421 is by Kikuchi (1991) covering the period 1980 to 1990. However, there is no attempt to investigate the micro variability (i.e. on a time scale of a few hours).

As a part of a program for seeking the evidence of rapid optical variability in BL Lac objects, we have carried out observations of Mrk 421. Although there is a keen observational interest in a short time scale variability, such data is lacking. In spite of the extensive study of Mrk 421, like multi frequency observations, spectroscopy, photometry and polarimetry etc., only scattered information is available on the optical variability (Xie et al. 1988; Brodie et al. 1987; Makino et al. 1987). In the present paper we have presented the results based on the photopolarimetric observations of Mrk 421 made on January 23, 1990.

2 OBSERVATIONS AND ANALYSIS

Photopolarimetric observations of some BL Lacertae objects were carried out by us on 2.34 m Vainu Bappu Telescope (at Kavalur) of the Indian Institute of Astrophysics, Bangalore. Observations were made between January 18 – 23, 1990 with PRL photopolarimeter; the details of the photopolarimeter are given elsewhere (Deshpande et al., 1985). This photopolarimeter measures the degree of polarization as well as the flux in optical region. The instrument works on the principle of rapid modulation to measure the degree of polarization and the flux is measured by counting the total number of the photons. Polarimetric standards and zero polarization standards were observed to check the performance of the polarimeter and to get the off set in the polarization angle and to measure the instrumental polarization. In fact the performance of the polarimeter was checked by inserting a Glan prism in the beam. The efficiency of the polarimeter was found to be 98 % which is considered to be very good.

3 RESULTS

The results reported here are based on the observations made on January 23, 1990. One of the exciting observation is the detection of the rapid variability in flux as well as in degree of polarization in BL Lacertae Mrk 421.

Observations, without any filter through an aperture of 8.9 arc sec, carried out on January 23, 1990, show rapid variability in flux. In a time scale of about 45 minutes the flux rapidly decreased by about 0.4 mag and then slowly increased to the normal level (Figure 1). In Figure 1 the instrumental magnitude is plotted against the time. Here we have not tried to convert the instrumental magnitude to the photometric standard and also it does not affect the conclusion. In fact this object was observed by us on January 22 in visual wave band; the polarimetric observations, with integration time 100 seconds, indicated short time fluctuations, though the signal to noise ratio for measured polarization was low for an integration time of 100 sec. On January 23 we decided to make observations without any filter to increase signal to noise ratio keeping the

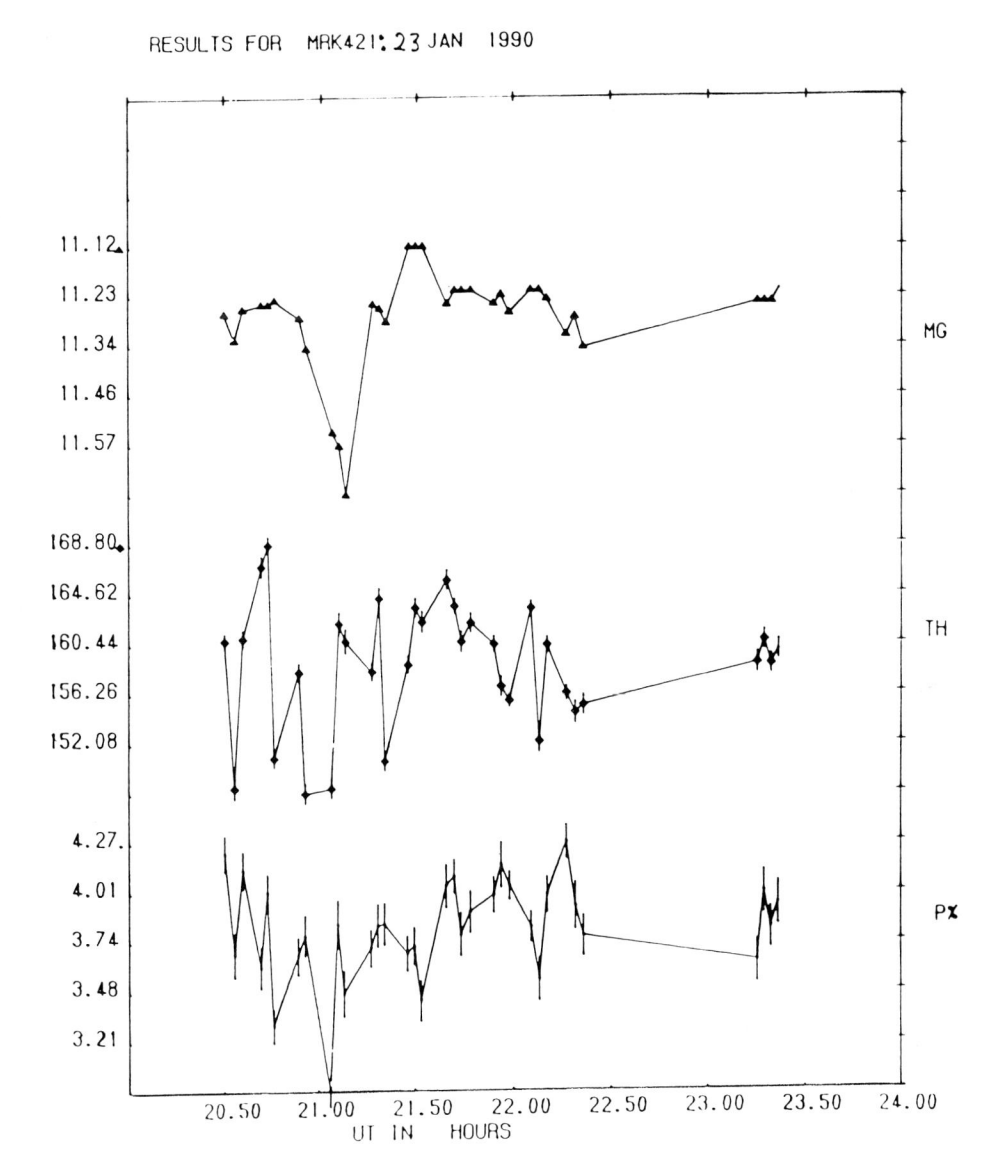

Figure 1. The variability of luminosity, in magnitude (MG), position angle (TH) in degrees and degree of polarization (P%).

integration time 100 sec. Figure 1 shows the variation of polarization and position angle along with the magnitude with time; the X-axis is the time in UT expressed in hours and Y-axis is the degree of polarization in percent. Polarimetric data shows fluctuations in a time scale ~ 20 minutes superposed on a smooth decrease of polarization in phase with the flux decrease. The signal to noise ratio (S/N) for the polarimetric data is always better than 8. The variations thus appear to be very significant. Present polarimetric and photometric observations indicate that short time variation exists and appears to be random in nature. This finding supports the shocked turbulent jet models for blazars by Marscher and Gear (1985).

Mrk 421 has been found to be in bright phase during the present observing run; the observed flux at visual wavelength of 0.55 micron being 28 mJy. The observations made in UBV filters were corrected for the galaxy contamination using the model given by Kikuchi and Mikami (1985) as discussed in Makino et al. (1987) for the underlying galaxy. Present observations show that the spectral index for the non thermal component is much steeper ($\alpha = 1.1$) than the value found ($\alpha = 0.6$) by Makino et al. (1987) in the January – March 1984 observations when the object was in its faint phase. In a shock compressed region the high energy electron is expected to loose energy much faster due to the compression of the magnetic field than in an uncompressed region and the spectrum is expected to be steeper. This further supports the model by Marscher and Gear (1985).

The photometric observations in UBVR band were also made on January 23 to study the wavelength dependence of the polarization. These observations have been made between 22:30 and 23:15 UT. The object, after passing through a dip in flux prior to the above time, has reached the normal flux level and did not show significant fluctuations during these measurements. The observed values of the degree of percent polarization in UBVR bands are respectively 5.31 ± 0.60, 4.21 ± 0.33, 4.39 ± 0.25, 3.81 ± 0.30 and the corresponding angles of polarization are 167, 161, 156, 153 degrees, respectively. The upper level of the degree of polarization observed without filter is typically 4 % with a decrease of about 1 % while the flux level has dropped by about 0.4 mag. The mean value of the polarization through UBVR filters matches with the value measured without filter. The polarization values in UBVR bands corrected for the galaxy contamination are plotted in Figure 2. The degree of polarization for the non thermal component is almost wave length independent. The position angle shows slight wavelength dependence. One explanation for the wavelength dependence of the angle may be in terms of the presence of two or more components having different spectral indices. Makino et al. (1987) have also suggested the presence of two components to explain the X-ray emission from Mrk 421 during the period January – March 1984. The X-ray spectral index changed from ~ 1 at the January epoch to ~ 2 at the March epoch. Also the spectral slope between UV and X-ray was found to be steeper than the UV slope (Makino et al. 1987). The present polarimetric observations and the past simultaneous multi frequency observations are indicative of the presence of two components having different spectral indices, the harder component being strongly variable.

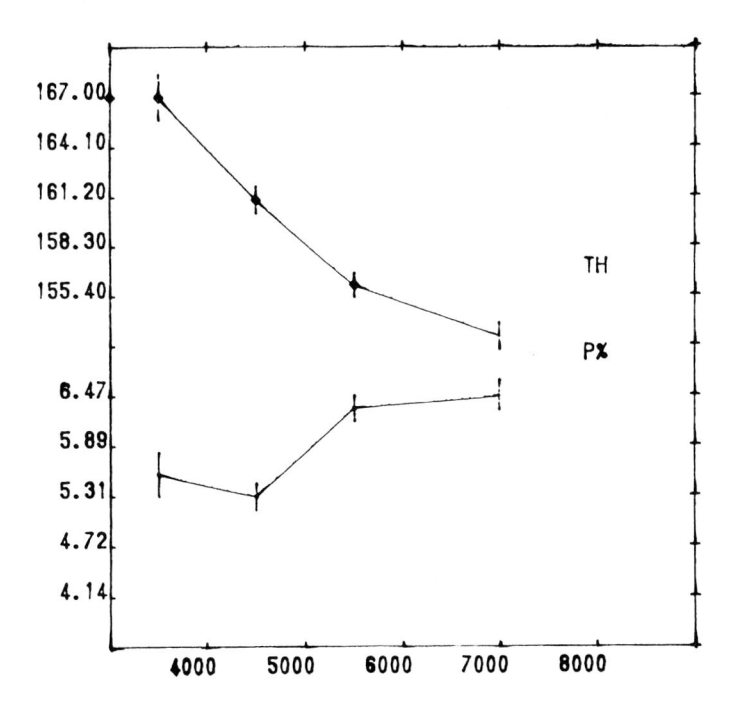

Figure 2. Wavelength dependence of degree of polarization (P%) and position angle (TH) in degrees for non-thermal component in Mrk 421.

ACKNOWLEDGEMENTS

We are grateful to Prof. J.C. Bhattacharya, director of Indian Institute of Astrophysics, Bangalore, for providing telescope time. We are thankful of Prof. R.K. Varma and Prof. P.V. Kulkarni for their encouragement. This work has been supported by the Department of Space, Govt. of India.

REFERENCES

Bailey J., Cunningham E.C., Hough J.H. and Auxon D.J.: 1981, *M.N.R.A.S.* **197**, 627.
Brodie J., Bowyer S. and Tennant A.: 1987, *Ap. J.* **318**, 175.
Deshpande M.R., Joshi U.C., Kulshestha A.K., Bansidhar, Vadher N.M., Mazumdar H.S., Pradhan N.S. and Shah C.R.: 1985, *Bull.Astron.Soc.*, India **13**, 157.
Kikuchi S.: 1991, in this volume.
Lipunov V.M.: 1987, *Astrophys. Space Sci.* **132**, 1.
Makino F., Tanaka Y., Matsuoka M., Koyama K. Inoue H., Makishima K., Hoshi R., Hayakawa S., Kondo Y., Urry C.M., Mufson S.L., Hackney K.R., Hackney R.L., Kikuchi S., Mikami Y., Wisniewski W.Z., Hiromoto N., Nishida M., Burnell J., Brand P., Williams P.M., Smith M.G., Takahara F., Inoue M., Tsuboi M., Tabara H., Kato T., Aller M.F. and Aller H.D.: 1987, *Ap. J.* **313**, 662.
Marscher A.P. and Gear W.K.: 1985, *Ap. J.* **298**, 114.
Maza J., Martin P.G. and Angle J.R.P.: 1987, *Ap.J.* **224**, 368.
Mead A.R.G., Ballard K.R., Brand P.W.J.L., Hough J.H., Brindle C. and Bailey J.A.: 1990, *Astron. Astrophys. Suppl. Ser.* **83**, 183.
Takalo L.O.: 1991, in this volume.
Valtaoja L.: 1991, in this volume.
Xie G.Z., Lu R.W., Zhou Y., Hao P.J., Li X.Y., Liu X.D. and Wu J.X.: 1988, *Astron. Astrophys. Suppl. Ser.* **72**, 163.

Rapid Variability of Mkn 501

J.A. DE DIEGO & M.R. KIDGER

Instituto de Astrofísica de Canarias (Spain)

L.O. TAKALO

Turku University Observatory (Finland)

1. INTRODUCTION

Mkn 501 is a BL Lac object with known rapid variability and has visible and near-Infrared fluxes which are high enough to permit good quality time resolved photometry.

It is centred in the nucleus of a giant elliptical galaxy of redshift 0.0306 (Wills & Wills 1974) which causes problems because of its strong, extended emission.

Here we report optical and infrared observations taken during summer 1989 and 1990, which show variations on timescales less than an hour, possibly indicative of magnetic fields with B≈20G.

2. COORDINATED VISIBLE AND INFRARED MONITORING

2.1. Instrumental description

The 1989 observations were made using the Carlos Sánchez Telescope (CST) for infrared and visible photometry and the Jacobus Kapteyn Telescope (JKT) for CCD photometry.

The CST is a dedicated infrared, 1.52 m, f/13.8 (cassegrain) telescope located at Teide Observatory, on the island of Tenerife and can take infrared and optical photometry simultaneously. The single channel, infrared photometer system, has an In-Sb detector cooled to the N_2 triple point. We used an aperture of 15 arcsec and an integration time of 50 sec per beam. FOVIA (Adapted Visible Photometer), integrates the television image. An aperture of 10 arcsec and an integration time of 10 sec per beam were used. A star is located about 3 arc minutes south-east which serves as a reference for the visible photometry.

The JKT is a 1 m, f/15 (cassegrain) reflector, sited in Roque de los Muchachos

Observatory on the island of La Palma. The observations were made with a GEC type, dye coated, blue sensitive CCD. This allows us to locate the image of the object and reference star in the same exposure. U and B filters were used. The exposures were 200 sec in B and 300 sec in U, with 80 sec of system dead time after every exposure. A pseudoaperture of 6 arcsec was used for data reduction.

2.2. Observations

Simultaneous B and K monitoring was taken on 2 nights whilst monitoring in B only, was done on a three. On the final night of observation, the object was monitored in U.

On the first night (24-07-89) we can see a 0.4 mag flare in B at the end of the monitoring record; in K the flare is less obvious (fig.1). The following night (25-07-89) we observed a steady decline of 0.10 mag per hour in B before interruption by clouds. In K, no such a trend is observed, but the flickering at the start of the observations appears to be real. A χ^2 and a Kolmogorov-Smirnov tests for the data between 22:50 and 00:00 U.T. yield a probability of reality of these variations of about 99% and 95%, respectively.

Three further nights of monitoring in B show only very much lower amplitude variations. The data for 27-07-89 in B show no variations. On August 13/14th (fig. 2) we see a slow fade and rapid flickering, which was repeated the following night (14/15th). There is a suggestion of saw-tooth form in these last variations.

On August 15/16th 1989, we made observations in U. The coated GEC chip used with the JKT, retains a high quantum efficiency and flat response. Although the counts are reduced by a factor of aproximately 5, the galaxy background is very much suppressed. Fig. 3 shows the light curve for this night, which includes a steady rise of 0.046 mag and a rapid fall of 0.05 mag.

3. RAPID INFRARED VARIATIONS

For the infrared observations in summer 1989, we used the same instrumentation and techniques discused above, but the total integration time was 10 to 15 minutes, according to the s/n obtained.

On August 1/2nd, we observed a sharp increase in brightness starting at 01:15 U.T. and rising 0.3 mag in an hour and a half. On August 6/7th, we detected a flare of 0.2 mag in 30 min (fig. 4), shortly after a gap for calibration and beam recentring, hence we are completely certain that this variation was intrinsic to Mkn 501. Neither 3C66A nor III Zw 2, observed in an identical manner, showed any variations in the

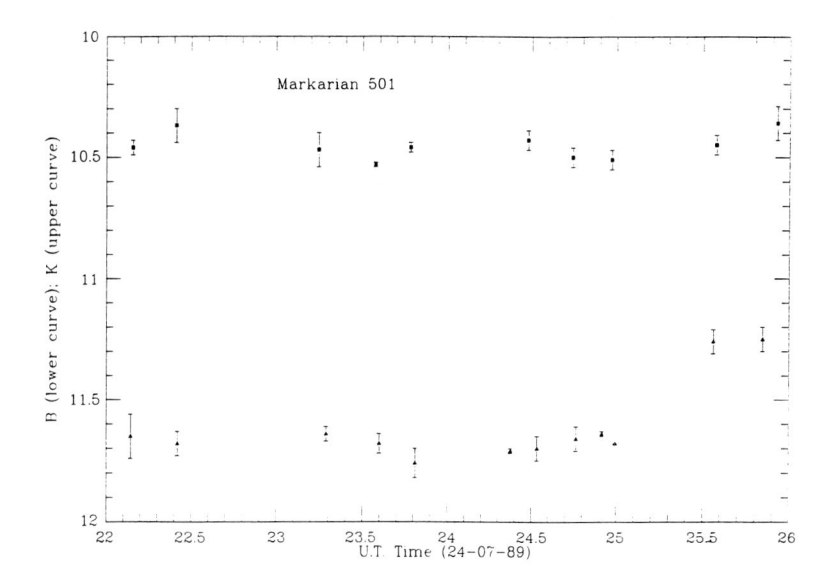

Fig. 1.- Light Curve in B and K on 24-07-89

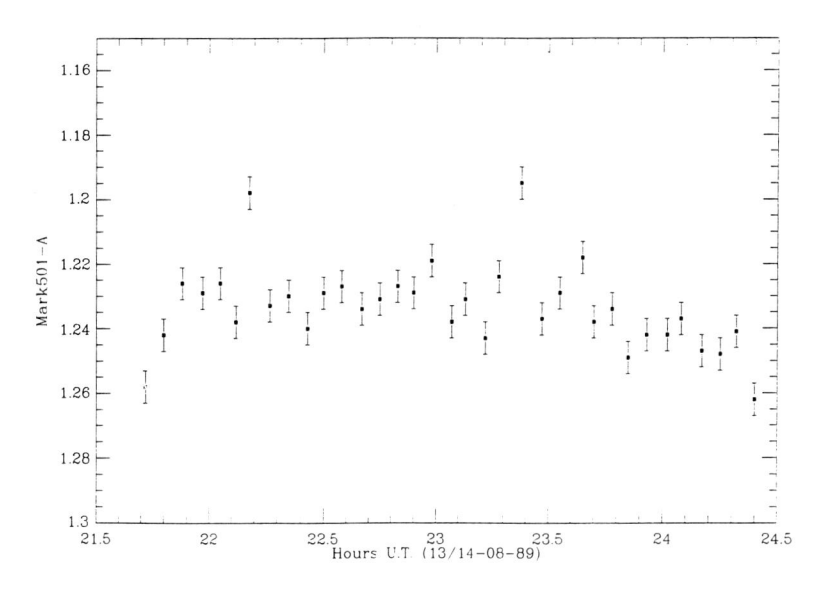

Fig. 2.- Light Curve in B on 13-08-89

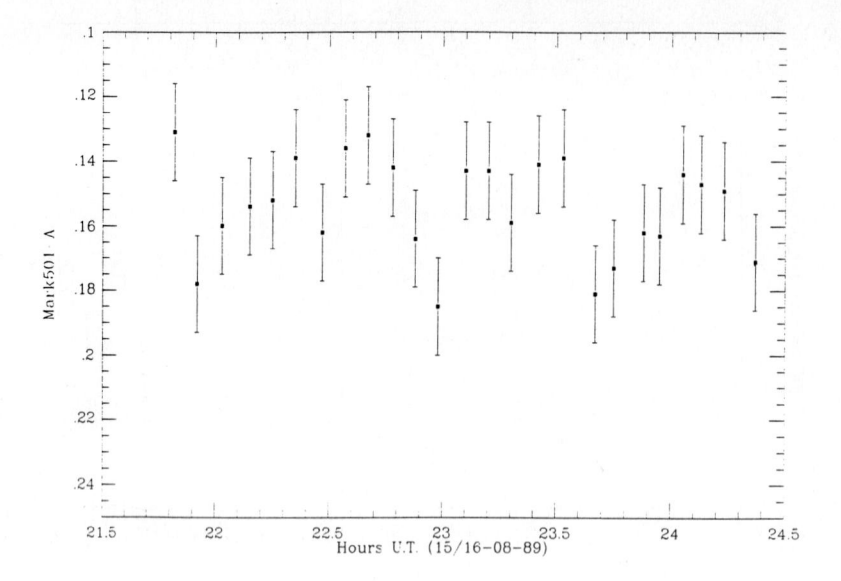

Fig. 3.- Light Curve in U on 15-08-89

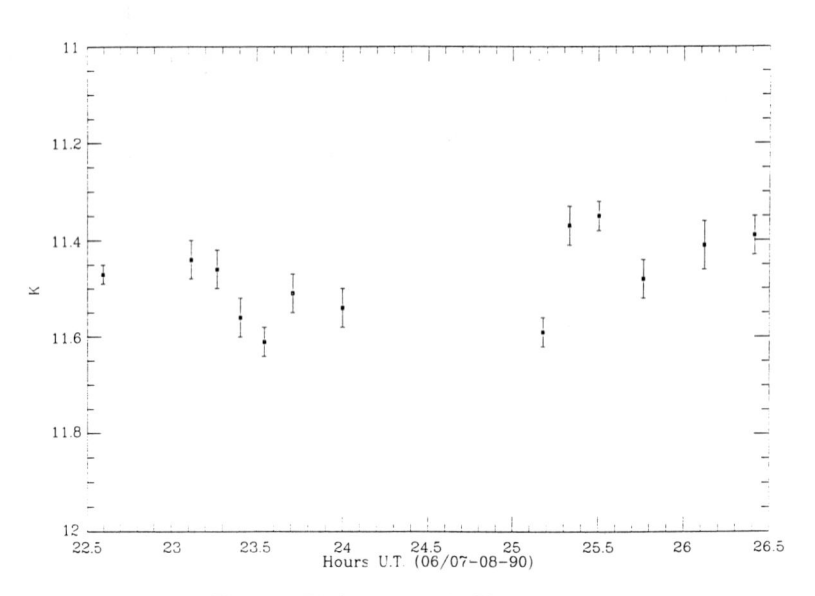

Fig. 4.- Light Curve in K on 06-08-90

same nights.

If we assume that these flares are of synchrotron origin, that the electrons are injected at time t=0, and that the energy spectrum is maintained powerlaw and isotropic, then, as a result of energy losses, the synchrotron spectrum decays exponentially above

$$\nu_c = 1.07 \times 10^{24} B^{-3} t^{-2} Hz$$

with B in gauss and t in seconds. So, for the flare on August 6/7th, when the decay time was approximately 15 min in filter K ($1.35 \times 10^{14} Hz$), a magnetic field of the order of 20 Gauss is inferred.

4. CONCLUSIONS

The B-K simultaneous monitoring on two nights show rapid variations, especially in B, on intraday time scales. Generally, the amplitude of variation in B is greater than in K, which suggests that the source is dominated by thermal emission from the host galaxy in the near infrared. We estimate a ratio of thermal to non-thermal components of about 10 at 2.2 μm. In the visible, the nonthermal source, whilst not dominant, contributes a much higher fraction if the total emitted flux. However, this model is unable to account for the possible flickering seen in K but not in B.

We believe our rapid monitoring in U to be almost unique. Several groups have published the results of photographic monitoring in U but with very large errors and dealing almost exclusively with slow variations. We believe that MKN 501 is almost certainly active the U on rapid time scales as the galactic contribution is considerably reduced. We demonstrate that time resolutions of a few minutes are possible in U with relative photometry accurate to around 0.01 mag (an advance of a factor of 20 on photographic results).

MKN 501 shows significant activity on time scales of a few hours in the near-infrared, although it is not active on all nights. There is evidence of occasional, far more rapid variations in the infrared, although our time resolution is inadequate to determine exactly how rapid they are. Such flares suggest the presence of strong magnetic fields around 20 G.

REFERENCES
Wills, D., Wills, B.J. (1974), Ap. J. 190, 271

Blazar Type Variability in a Seyfert Nucleus

D. DULTZIN-HACYAN[1], W. SCHUSTER[1], J. PEÑA[1,2], L. PARRAO[1], R. PENICHE[1,2], E. BENITEZ[1], R. COSTERO[1]

[1]Instituto de Astronomía, UNAM. México. [2]INAOE, México

ABSTRACT

Differential photoelectric photometry of the Seyfert nucleus NGC 7469 is presented. During the first observing season (15-30 November 1989) night to night measurements were made and we found variability (brightening trend) with a maximum amplitude of $0.^m11$ (5.5σ) in 13 days. Sigmas are not rms errors, they are calculated from photon counting rates and are thus photometric errors (sky measurement errors included) which take into account the difference in brightness of the nucleus and comparison stars (and between the comparison stars).

During the second season (August 1990) all night monitoring was done during 5 photometric nights. Random variability was found during the first four nights with average amplitudes $\sim 0.^m04$ ($\sigma \geq 4.5\sigma$) on average time scales of ~ 13 min. A maximum $\Delta V = 0.^m081$ (9.3σ) was observed in $\Delta t = 2.35$ hrs. During the fifth night the fluctuations are less conspicuous. Nearly simultaneous differential photometry of two comparison stars shows average fluctuations within 1σ during encompassing intervals. The photometric error due to possible misscentering of the nucleus was calculated and is negligible.

1. INTRODUCTION

NGC 7469 is a very well studied Seyfert galaxy with an extremely bright star-like nucleus and a close ($80''$) irregular physical companion: IC 5283 (Burbidge, Burbidge and Prendergast, 1963). It is known to be variable in all wavelengths from IR to X-rays (e.g., Rieke, 1978; Penfold, 1979; Peterson, 1982; Westin, 1984; Marshall, Warwick and Pounds, 1981) and there exists strong evidence of a recent ($\sim 10^8$ yrs) circumnuclear starburst event (at $r \sim 8''$) from observations at different wavelengths (e.g., Ulvestad et al., 1981; Curti et al., 1984; Wilson et al., 1986; Bonatto and Pastoriza, 1990).

Recently Aslanov et al. (1989) reported rapid (10-15 min) random fluctuations in the V bandpass of $\sim 0.^m06$ amplitude for the nucleus of this galaxy, as well as a quasicyclic $0\overset{m}{.}10$ term, with a possible period of $2\overset{d}{.}281$ which disappeared latter on. Except for one night, they obtain a confidence of variability of 99.95% by the χ^2 criterion. Observations were done with a 60 cm telescope and no estimation of the photometric errors is given.

Our own results of the 1989 season and those reported by Aslanov et al. (1989) encouraged us to monitor this galaxy during several nights in August 1990 with a larger telescope.

2. OBSERVATIONS

The observations were done with the 1.5 m telescope of the Observatorio Astronómico Nacional in Baja California, México, fitted with a 6-channel danish photometer (described by Schuster and Nissen 1988) which allows simultaneous observations in two sets of filters. We used filters vb and y, which have the following central λ_{eff} (and bandpasses $\Delta\lambda$ in Å), respectively: $v - \lambda$ 4110 (170), $b - \lambda$ 4685 (183) and $y - \lambda$ 5488 (235).

The data of a fourth filter $u - \lambda$ 3505 ($\Delta\lambda 330$) were discarded because they were extremely noisy for the nucleus and the faintest control star C_2. Filter y is similar in effective wavelength to Johnson's V bandpass. A diaphragm of 20" was used. The integration times were: 2 min for the nucleus and C_2 (6 integrations of 20 sec each) and 1 min for C_1 (3 integrations of 10 sec each). The observing sequence was: 3 (times) C_1, sky, $3C_1$; 3 nucleus, sky, 3 nucleus; $3C_2$, sky, $3C_2$; 3 nucleus, sky, 3 nucleus and repeated. Comparison stars C_1 and C_2 are from Lyutyi (1973).

All the data are given in instrumental magnitudes and are extinction corrected. Photometric error bars were calculated from the counting rates; several adyacent skies were added in order to have the same integration times as for the nearest nucleus or comparison star. In this way we take into account the difference in brightness of the nucleus and comparison stars. The errors for each observation were added quadratically ($\sigma = \sqrt{\sigma_1^2 + \sigma_2^2}$) to give an error bar for each differentail point. Error bars in Figures 1 and 2 are 2σ average values for the season and for each night respectively.

The contribution of the underlying galaxy (which is larger than the diaphragm) was calculated using the luminosity profile given in Doroshenko et al. (1989) for a possible nucleus misscentering error of 5" (which is indeed a large margin), the photometric error turned out to be negligible. The details of this calculation, reduction procedure and photometric error's calculation will be given elsewhere (in preparation).

3. RESULTS

Our results are displayed in Figures 1 and 2. A more detailed analysis (as well as complete data tables and results for all 3 filters) will be given in a forthcoming paper. Fluctuations were considered "significant" on the basis of two criteria: magnitude amplitude variations equal or above 3σ for the nucleus minus C_1 <u>and</u> mean amplitude variations within 1σ for nearly simultaneous observations of C_2 minus C_1 during a time interval larger (and encompassing) the corresponding interval for the nucleus minus C_1 (notice that the above "3σ" and "1σ" refer to different sigmas as shown in Figure 2). The main results of such analysis are the following:

During the first four nights average amplitude variations in the equivalent V bandpass are between $0\overset{m}{.}032 - 0\overset{m}{.}047$ for significant consecutive fluctuations. The average number of such fluctuations per night was 6 with mean "significance" values for each night ranging from 3.8σ to 5σ and mean timescales of $\Delta t = 13.2$ min The highest amplitude variation during one night was $\Delta V = 0\overset{m}{.}081$ (9.3σ) in $\Delta t = 2.35$ hrs on August 17. The shortest mean timescale for $\geq 3\sigma$ amplitude fluctuations, $\Delta t = 9.3$ min, was found for the b filter on August 21. Taking into account the 3 filters, the highest variability (in terms of number of significant fluctuations) was found on August 20. The last night (22 August) the fluctuations are less significant.

No correlation is found between amplitude and wavelength for the filters used.

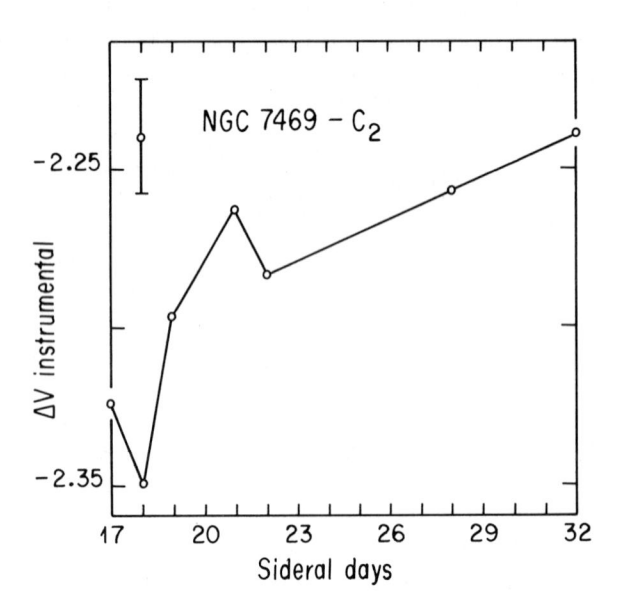

Figure 1. Differential photometry of the nucleus of NGC 7469 from data of November 1989. The error bar represents the mean quadratic sum of photometric errors of the nucleus and C_2 (see text). Instrumental magnitudes are along the ordinate and time (sideral date) is along the abscissa.

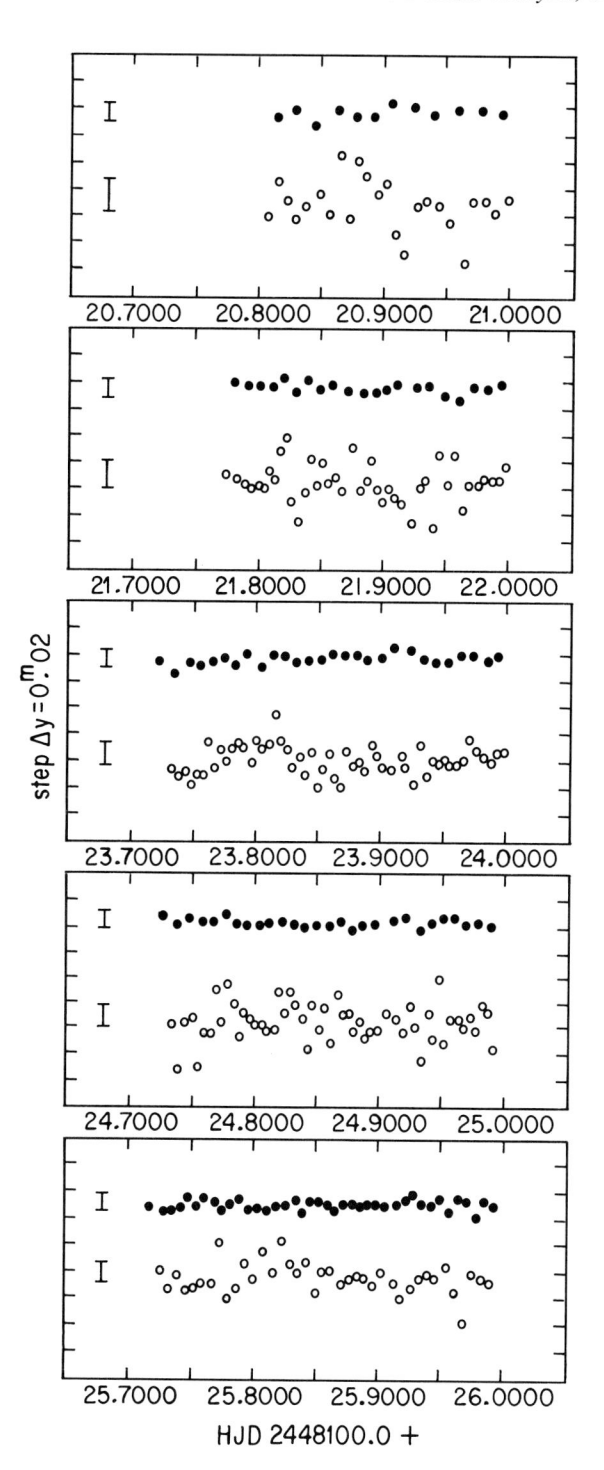

Figure 2. Differential photometry for five nights in August 1990. From top to bottom the dates (and duration) of monitoring are: 17 (4.557 hrs), 18 (5.38 hrs), 20 (6.37 hrs), 21 (6.31 hrs) and 22 (6.33 hrs) August. In the vertical axis we plotted instrumental magnitudes (step is $\Delta y = 0\overset{m}{.}02$). Julian date is along the abscissa. Δy of comparison stars $(C_2 - C_1)$ is on top (filled circles) and Δy of nucleus minus C_1 is on bottom (open circles). Error bars are 2σ mean (for the night) photometric errors added quadratically (see text).

4 CONCLUSIONS

Random variability in timescales of minutes is confirmed. No periodicity was found at such timescales nor in day to day variability.

No correlation is found between amplitude and wavelength.

The only interpretation of this type of variability given until now is, to our knowledge, in terms of phenomena occuring in an accretion disk (*e.g.* Wiita *et al.* this volume) around a supermasive black hole, thus this galaxy provides a neat example where evidence for the black hole accretion and circumnuclear starburst phenomena appear together.

We acknowledge S. Hacyan and A. Sarmiento for mathematical help in calculating the error produced by a possible misscentering of the nucleus, due to the underlying galaxy.

REFERENCES

Aslanov, A.A., Koslov, D.E., Lipunova, N.A. and Lyutyi, V.M. 1989, *Sov. Astron. Lett.*, **15**, 132.

Bonatto, C.J. and Pastoriza, M.G. 1990, *Ap. J.*, **353**, 445.

Burbidge, E.M., Burbidge, G. and Prendergast, K.H. 1963, *Ap. J.*, **173**, 1022.

Curti, R.M., Rudy, R.J., Rieke, G.H., Tokunaga, A.T. and Willner, S.P. 1984, *Ap. J.*, **280**, 521.

Doroshenko, V.T., Lyutyi, V.M. and Rakhimov, V. Yu. 1989, *Sov. Astron. Lett.*, **15**, 207.

Lyutyi, V.M. 1973, *Sov. Astron.*, **16**, 763.

Marshall, N., Warwick, R.S. and Pounds, K.A. 1981, *M.N.R.A.S.*, **194**, 987.

Penfold, J.E. 1979, *M.N.R.A.S.*, **186**, 297.

Peterson, B.M., Foltz, C.B., Byard, P., L. and Wagner, R.M. 1982, *Ap. J. Suppl.*, **49**, 469.

Rieke, G.H. 1978, *Ap. J.*, **226**, 550.

Schuster, W.J. and Nissen, P.E. 1988, *Astr. Ap. Suppl.*, **73**, 225.

Ulvestad, J.S., Wilson, A.S. and Sramek, R.A. 1981, *Ap. J.*, **247**, 419.

Westin, B.A.M. 1984, *Astr. Ap.*, **132**, 136.

Wiita, P.J., Miller, R.H., Carini, M.T. and Rosen, A. 1990, *preprint* and this volume.

Wilson, A.S., Baldwin, J.A., Sun, S.D. and Wright, A.F. 1986, *Ap. J.*, **310**, 121.

Periodicity in OJ287?

M.R. KIDGER, J.A. de DIEGO

Instituto de Astrofisica de Canarias

L.O. TAKALO, A. SILLANPÄÄ, K. NILSSON

Turku University Observatory

ABSTRACT: Nine different groups have published their results of monitoring of OJ 287 at different epochs and in different ranges from the centimetric to the visible and with time resolutions from a few seconds. A bewildering array of possible periods have been found from 6 minutes up to 11.5 years. Various other AGNs have shown possible period medium or long term variations, but never rapid periodicity, hence it is unsurprising that there has been a general scepticism about results. We present the results of a multisite multiwavelength monitoring campaign which provides evidence of at least two possible periodicities active in 1989-90, including what we believe to be the first detection of a possible periodicity in the near infrared.

PERIODICITY IN BLAZARS

There have been many reports of possible periodicity in blazars and related objects. These divide into three different time scales:

i.) Several to more than 10 years (eg: the outbursts of OJ287; 3C345; AO0235+164; and 3C279?) or sinusoidal (eg: 3C273; 3C446; 3C345?; NRAO140?).

ii.) Medium time scale variations, typically a few to 100+ days. Again, these may be outbursts (eg: 3C345; BLLac) or sinusoidal (eg: OJ287, 3C345).

iii.) Very rapid variations, typically saw-tooth or sinusoidal, with periods less than one hour and apparantly unique to OJ287. Until the recent report by Deshpande et al., 1991 of a possible six minute variation in the white light polarisation of OJ287, all reported periodicities were in the range 13.7-43 minutes. Only on one occasion has a period ever been independently confirmed by a second group (Frohlich et al., 1974 reported confirmation of the 39.2 minute period seen previously by Visvanathan and Elliot, 1973). Kinzel et al., 1988 have written an excellent summary of the periodicities observed in this object to 1988.

THE 1989-90 MULTISITE CAMPAIGN

Observations were made using the 2.56m Nordic Optical Telescope (NOT) and the 2.5m Isaac Newton Telescope (INT) of el Observatorio del Roque de los Muchachos, La Palma, Spain and the 1.55m Carlos Sanchez Telescope (TCS) of Teide Observatory, Tenerife, Spain. At each telescope the observations were made by an experienced observer and complementary overlapping observations were made to permit adequate intercomparison of results without unnecessary duplication. Simultaneous UBVRI photopolarimetry with a time resolution of 20 seconds for photometry and 3.5 minutes for polarimetry, were taken with the NOT. B-band CCD photometry with a resolution of 58 seconds was achieved with the INT, whilst K and white light photometry were taken on the TCS with a time resolution of about twenty minutes for both, although beam decomposition permits a maximum resolution of approximately two minutes.

All three telescopes were scheduled simultaneously on December 30th and 31st 1989 and January 1st 1990. Completely simultaneous observations on all three telescopes were only achieved on one night, although some overlap exists on other nights and further data was taken with the TCS on January 21st 1990 and with the NOT on January 18th and 22nd and on March 1st and 3rd 1990. Some data taken with the SEST in La Silla, Chile on March 3rd 1990 has also kindly been made available to us by Merja Tornikoski who undertook and reduced the observations.

OBSERVING METHOD

One of the strengths of these data is that the instrumentation used on the three main telescopes was completely different thus, in the case that periodicity were seen, instrumental effects could easily be ruled out by simple cross-checking. The NOT observations were taken using the Turku photopolarimeter, a twin-beam chopping system, with a 10" aperture mounted at the f/11 Cassegrain focus. The seeing was in the range 1-1.5" for all the observations. Star 10 of the standard sequence was used as the photometric reference, being observed about every 2 hours. Sky measurements were taken every 13 minutes on completion of each set of 4 integrations of 3.5 minutes. The INT observations were taken with a coated GEC-type chip at the f/3.3 prime focus of the telescope using the Prime Focus Camera Unit. The observations were reduced using an aperture photometry routine developed at the IAC which measures the flux in a 8" circular pseudoaperture. Ten exposures were located on the chip, moving the telescope slightly between each one, to reduce dead-time to acceptable levels. Star 10 was used as the prime reference and star 11 as a check star; both were located on the chip at the same time as OJ287, thus providing good relative photometry. The TCS observations were taken with a single channel photometer, in AC, with an InSb detector cooled to the Nitrogen triple point, mounted at the f/13.8 Cassegrain focus. An aperture of 15" was used. The white light observations were taken using FOVIA, a tv integration system. These observations were calibrated using the star BS3176 which is close to OJ287 in the sky.

RESULTS

U and B: Very little variation was seen in the data from either telescope even when the light curve was active in other colours. The amplitude was generally below 0.05 magnitudes. As U is strongly atmosphere sensitive, this serves as and additional check that the conditions were photometric.

V and R: Variations of up to 0.3 magnitudes were seen even when U and B were non-variable. A series of equispaced maxima with separation about 45 minutes are seen in the data, especially on December 31st/32nd although there is some evidence of similar variations on the 30th/31st too. A clear time delay of 15 minutes is seen from V to R.

I: Occasional irregular, high amplitude variations are seen which are, when visible in other filters as well, of highest amplitude in this band. The sinusoidal variations are less pronounced as they appear to be partially hidden by flickering.

K: The light curve was very active indeed showing variations of up to 0.4 magnitudes on all nights and strong sinusoidal variations which follow the visible light curve on December 31st/32nd. The white light observations followed the K light curve closely, but were significantly advanced with respect to the infrared. The observations taken in January 1990 also show an obvious sinusoidal modulation.

ANALYSIS

We have mainly used the Jurkevich $V*m^2$ test (Jurkevich, 1971) as this is less sensitive to windowing effects than traditional Deeming analysis and all more sensitive to non-sinusoidal periods. We folded the data around a range of trial periods in the range 2-120 minutes, searching for the best fit to the light curve where the best fit is defined as the lowest total standard deviation of the folded light curve. The data were previously detrended by subtracting a quadratic term. Potential periods (identified as minima in the $V*m^2$ curve) were used to fold the data permitting inspection of the mean shape. Spurious periods are obvious when folded as their "mean curve" is pure noise.

The most significant observed periods were 21.8 and 40.4 minutes in the TCS data (data string 50 individual integrations over a total of 16.25 hours). The NOT data gave similar periods of 24.0 and 46.5 minutes (2360 points over 26.98 hours) although the visible amplitude was considerably lower than that in the infrared. Data taken with the SEST, at La Silla, Chile on March 3rd 1990 has best periods of 22.5 and 44.9 minutes and amplitudes intermediate between the infrared and visible light curves. These data seem to provide completely independant confirmation of the existence of possible periodicities at about 22.8 ± 1.1 and 43.9 ± 3.2 minutes.

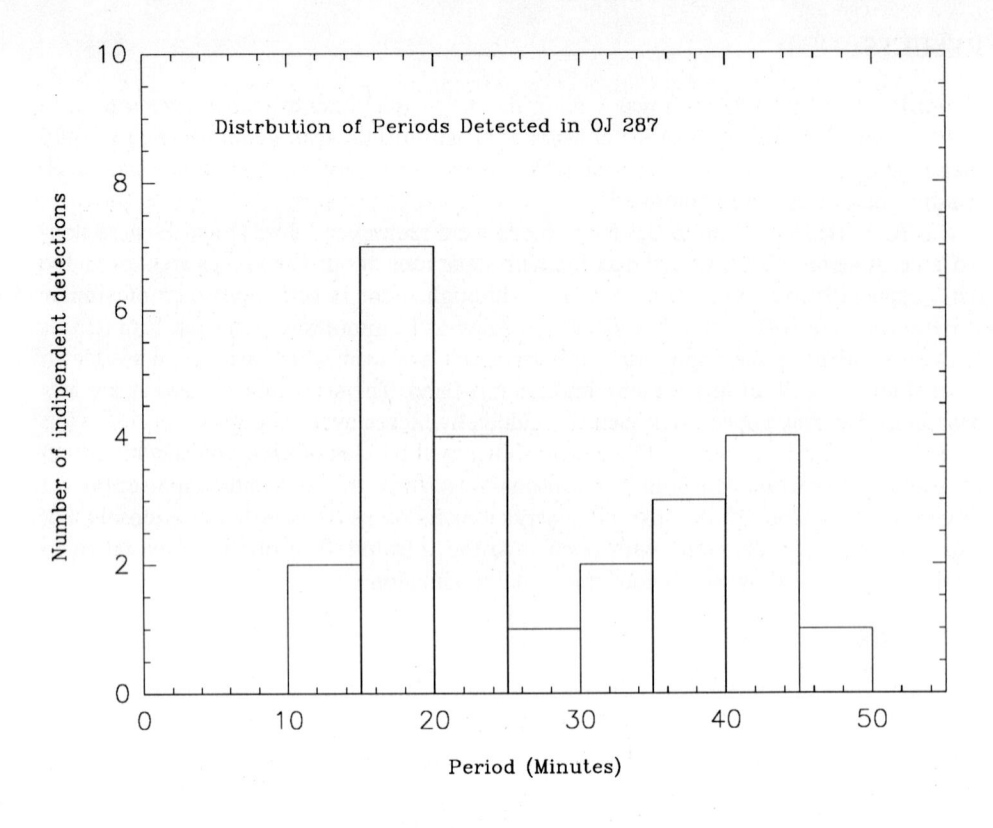

Figure 1. A histogram of the distribution of observed periodicites reported in the literature. The distribution of periods is clearly non-random.

DISCUSSION

Two dominant periods are seen in the data, a result similar to that of Carrasco et al. (1985), where the longer is close to, but slightly less than an exact multiple of the shorter. We note a curious distribution in the periodicities seen by different groups: all are in the range 13-24 or 35-46 minutes, with no intermediate values seen at all. The histogram of the period distribution is non-random and shows peaks around 15-20 and 40-45 minutes. One possible mechanism for the periodicities (although we prefer to regard them as time scales, analogous to semi-regular variable stars rather than periods in the normal sense of the word as they are clearly NOT truly periodic) is orbital mechanics. The fastest observed period will correspond to the innermost stable orbit outside the Schwarzchild radius and will, assuming it to be generated by a hot-spot, be subject to blurring from differential rotation, as well as being unstable and short-lived. Groups observing whilst such a structure is present in the accretion disc will see periodicity, whilst other groups observing at different epochs will not. The second "favoured" period might correspond to a second stable orbit further from the singularity, or to an equivalent closest stable orbit for a second singularity such as has been suggested may exist in OJ287 (Sillanpää et al. 1989).

REFERENCES

Carrasco, L., Dultzin-Hacyan, D., Cruz-Gonzalez, I.; 1985, Nature, 314, 146
Deshpande, M.R., Sen, A.K., Joshi, U.C., Vadher, N.M., Shah, A.B., Chauhan, J.S.; IAUC 5234
Frohlich, A., Goldsmith, S., Weistrop, D.; 1974, MNRAS, 168, 417
Jurkevich, I.; 1971, Ap.Spa.Sci., 13,154
Kinzel, W.M., Dickman, R.L., Predmore, C.R.; 1988, Nature, 331, 48
Sillanpää, A., Haarala, S., Valtonen, M.J., Sundelius, B., Byrd, G.G.; 1988, Ap.J., 325,628
Visvanathan, N., Elliot, J.L.; 1973, Ap.J., 179,721.

Long-Term Periodicity in 3C345?

M.R. KIDGER, J.A. de DIEGO

Instituto de Astrofisica de Canarias

L.O. TAKALO, A. SILLANPÄÄ, K. NILSSON

Turku University Observatory

ABSTRACT: Possible medium or long period variations have been reported several times in 3C345. Only one, a possible 11.4 year period in the mean level of light curve is still a serious candidate to be a true period. We investigate the compatibility of this period with the light curve in the light of recent observations.

INTRODUCTION

Various groups have reported the detection of periods in 3C345, these include 80 and 321 minutes (Kinman et al., 1968); 800 and 1600 days Barbieri et al., 1977) and 11.4 years (Webb et al., 1989). The periods observed by the first two groups disappeared after their publication, whilst the available light curve, composed of 26 years of data, is still rather short to confirm or deny the longest of these periods. The possible 11.4 year cycle however makes a strong prediction (Kidger and Takalo, 1990) of an upturn in the light curve in 1990/91, after an unprecedentedly long monotonic decline. A rise in the level of the light curve did subsequently start in August 1990, up to major outburst level, which was reached in February 1991.

THE LIGHT CURVE

Observations were made in January and February 1991 using the CCD camera of the 2.56m Nordic Optical Telescope of El Observatorio del Roque de los Muchachos, La Palma, Spain, as part of our monitoring programme of this object. Our results showed a large increase in brightness since Spring 1990 (Kidger et al., 1990) which was confirmed almost simultaneously by Schramm and Borgeest (1991). The flare has an amplitude of about 2 magnitudes in B relative to the previous minimum.

ANALYSIS

Kidger and Takalo (1990) noted that the 11.4 year period of Webb et al predicted an upturn in the light curve in late 1990 or early 1991. This has duly occurred, in agreement with the predictions. The 1600 day period of Barbieri et al. (1977) predicts the 1984 maximum approximately, but the 1991 outburst appears to be out of phase with this period. Fourier transform analysis shows that the major peaks in the power spectrum are still around 800 and 1600 days, but are becoming less important as new data is added. The 1600 day peak has split into two separate maxima at period 3.8 and 4.9 years. A prominent maximum is seen at a period of 11.4 years, in agreement with the period of Webb et al. (1989). This peak in the power spectrum now contains almost as much power as the 800 and 1600 day structures.

The 11.4 year modulation is best seen in the baseline to the light curve upon which the rapid flares are superimposed. Although clearly not a pure sinusoid, it is possible that this is a genuine modulation. This period though has only run for about 2.5 cycles since the start of the modern monitoring record, thus it is still not possible to say whether this variation is a genuine period or a "statistical fluctuation".

CONCLUSIONS

It is still too early to decide whether or not the possible 11.4 year period in the light curve is genuine or not. At present we can say that, at least, it does not contradict the evidence and that it did, sucessfully predict the 1991 outburst. This period is most likely though to be analogous to the very strong sinusoidal modulation seen in 3C273 from about 1920-50, which subsequently disappeared very rapidly after only two cycles. We do not believe though that the basis of quasar behaviour in general is periodic, although the occasional exceptions may provide very useful information on the sizes of the emitting regions.

REFERENCES

Barbieri, C., Romano, G., de Serego, S., Zambon, M.; 1977, Astron.Ap., 59, 419
Kidger, M.R., Takalo, L.O.; 1990, Astron.Ap., 239, L9
Kinman, T.D., Lamla, E., Ciurla, T., Harlan, E., Wirtanen, C.A.; 1968, Ap.J., 152, 357
Schramm, J., Borgeest, U.; 1991, IAUC 5191
Webb, J.R., Smith, A.G., Leacock, R.J., Fitzgibbons, G.L., Gombola, P.P.; Shepherd, D.W.; 1989, Astron.J., 95, 374

100 years of observations of 3C 273: New evidence for possible periodic behaviour

M.K. BABADZHANYANTS AND E.T. BELOKON'

Astronomical Observatory of Leningrad University

1 INTRODUCTION

The quasar 3C 273 has one of the most extensive optical data sets beginning at 1887. Thus it is suitable to search for possible periodic variations. The presence of such a behaviour would argue for the existence of a single supermassive object.

Fig. 1 presents the light curve of 3C 273 containing all the data (B-band) published by 1990 (1153 data points for the time interval 1887 – 1991). Different data sets were immediately compared with each other and appropriate corrections for systematic differences were introduced (Belokon' 1991; Babadzhanyants and Belokon' 1991a).

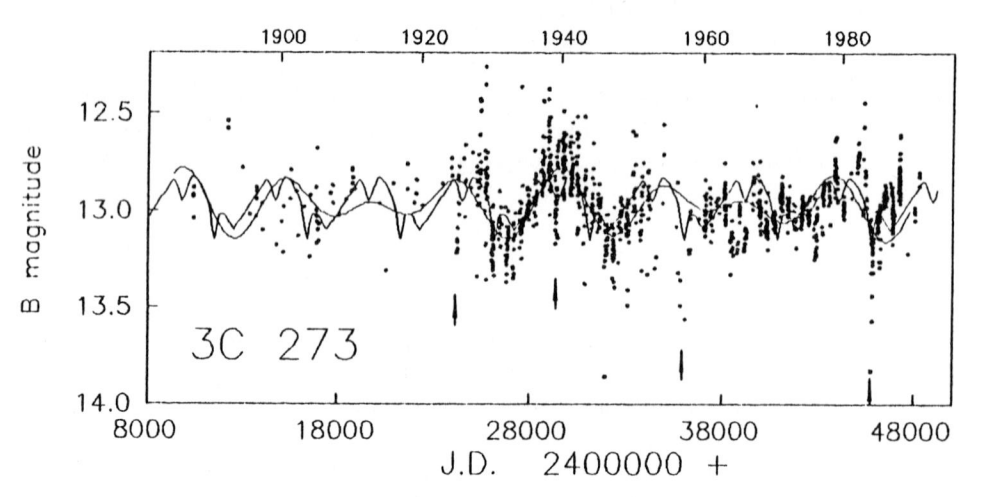

Fig. 1   Light curve of 3C 273. The data set is approximated: 1) by the curve reconstructed by means of the phase diagram for the 13.4-yr period; 2) by the sum of two sinusoids with 13.4-yr and 18.3-yr periods (amplitudes and phases are determined from Fourier analysis). The arrows mark downfalls in the brightness (see the text).

For the search of the periodicity Fourier analysis modified to unevenly spaced data was used (Deeming 1975). There have been many attempts to investigate the optical variability of 3C 273 for periodicity (see Jurkevich 1972; Angione and Smith 1985 and references therein). The main advantage of our analysis is the substantial increase of the data set both by observations in 1938 – 1980 (Kurochkin 1969; Takalo 1982; Lyuty and Metlova 1987) not used earlier and by the data obtained during the last 10 years (see Belokon' 1991 for references). Some distortions of the light curve which arise from the heterogeneity of the data were also eliminated (Belokon' 1991; Babadzhanyants and Belokon' 1991a).

2 THE TIME SCALE CORRESPONDING TO THE MILLIARCSECOND STRUCTURAL VARIATIONS

A connection between optical flares and ejections of superluminal jet components from the core has been found for several superluminal sources: 3C 345, 3C 120, OJ 287 (Babadzhanyants and Belokon' 1985, 1987, 1991b; Belokon' 1987). There is also evidence for possible periodicity of these flares at least on limited time intervals for 3C 120 and OJ 287 (Belokon' 1987; Webb 1990; Sillanpää et al. 1988).

3C 273 is one of the most throughly investigated superluminal sources. A series of high-resolution hybrid maps at $\upsilon = 5$ GHz and $\upsilon = 10.65$ GHz (Unwin et al. 1985; Cohen et al. 1987) and two maps at $\upsilon = 2.3$ GHz and $\upsilon = 8.4$ GHz (Charlot et al. 1988) give a possibility to estimate the frequency of jet component ejection to be $\sim 1 \cdot yr^{-1}$ (Belokon' 1991).

The more extensive VLBI data for 3C 120 in 1978 – 1983, which allows to determine the epochs of components ejection more accurately, suggest a period $P_{VLBI} = 303 \pm 3$[days] of ejection for five jet components. It is in good agreement with the period $P = 302$[days] of optical flares ($\Delta \approx 0\overset{m}{.}6$) revealed in the same time interval (Belokon' 1987).

Therefore, for 3C 273 the Fourier analysis was performed in a frequency range corresponding to periods of hundreds of days. This is the first case of searching 3C 273 for periodicity on this time scale. Because of small amplitude of optical variability of 3C 273 we choose for the periodicity search an interval 1962 – 1983 where half of the data points are photoelectric observations ($\sigma = 0\overset{m}{.}02 - 0\overset{m}{.}03$). Photographic data points with reported $\sigma > 0^m1$ were excluded. To avoid the heavy weight given to extraordinary downfalls in brightness beginning in 1984, the data obtained after 1983 were not used in the analysis. In 1976 –1983 the "slow" light variations with the time-scale of thousands of days were removed.

Fig. 2 shows the power spectrum for the total interval 1962–1983. Power spectra for different segments of this interval and detailed discussion are given by Belokon' (1991). We conclude that there is strong evidence for the existence of 307-day period in light variations of 3C 273 in 1962–1983, i.e. during 26 cycles. The mean light curve has a roughly sinusoidal shape with the total amplitude of $0^{\rm m}15$. The significance level for the peak at P = 307[days] is $\alpha = 10^{-19}$.

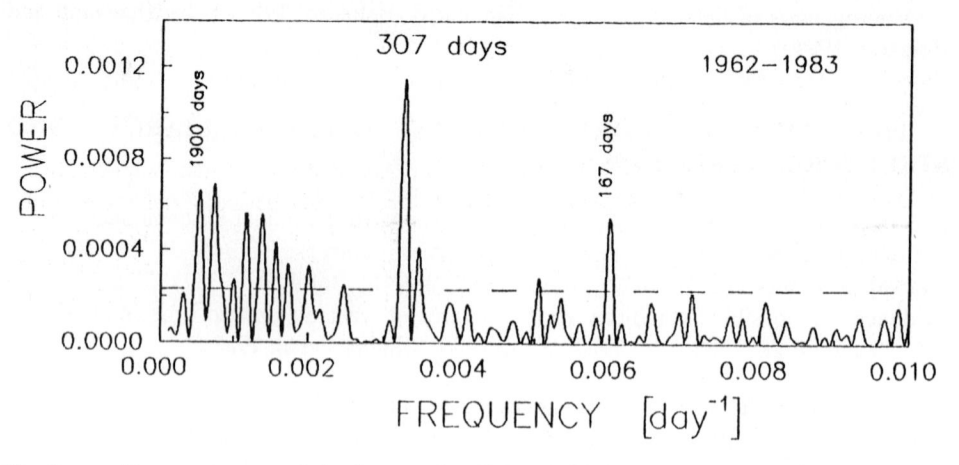

Fig. 2   Power spectrum of the data set for 1962–1983. The number of data points is 370. The "slow" variations of brightness in the 70's were first removed. The supposed period is 307 days ($\alpha \approx 10^{-19}$). Peaks at 167 and 1900 days are aliases (spectral window shows the peak at 365 days). Power corresponding to the significance level $\alpha \approx 0.01$ is marked by a dashed line. For details see Belokon' (1991).

An important argument for reality of the 307-day period is that the corresponding peak was the only constant feature present in the power spectra for all analyzed segments of time interval 1962–1983. Its height is proportional to the amplitude squared of the periodical component and was nearly the same in all cases. But its normalized heights and consequently its significance increases (α decreases) with the extension of the time interval. The phases of the 307-days component determined from the Fourier analysis independently for all segments and for the total time interval agree very well with each other.

The accordance of the 307-day period in optical variations with the frequency of jet component ejection ($\sim 1 \cdot yr^{-1}$) suggests the same connection between the optical flares and the new jet components as in 3C 345 and 3C 120. Note that the straightforward comparison of the epoch of jet component ejection (obtained by linear extrapolation assuming constant velocities) with the optical events turns out to be difficult because of the scanty optical data and the insufficient accuracy of zero-epoch determination.

A powerful (ΔB ≈ 0.^{m}6) and reliably registered optical flare occured in 1983 coinciding with the IR/mm flare reported by Robson et al. (1983). This optical flare is in good agreement with the 307-day period and probably is associated with the jet component C7b.

It is now difficult to understand how the observations after 1984 correspond to the 307-day period. From 1984 up to 1986 the brightness level was below its mean value. A similar minimum occured in the end of 50's and lasted for nearly four years (Fig. 1). The light variations during these minima may be caused by a complicated structure of the minimum itself. OJ 287 has the same deep minima with a pronounced double structure in its optical light curve (Sillanpää et al. 1988).

In 1988, brightness level of 3C 273 was rather high, and the maximum seen in April (Courvoisier et al. 1990, Fig. 4) agrees well with the 307-day period. Krichbaum et al. (1990) associated this flare with the new jet component C9 detected at υ = 43 GHz very close to the core (0.2 mas). The fine structure of the jet (components E1 – E3) found at υ = 100 GHz within 0.3 mas (Bååth et al. 1990) may probably be related to rapid optical flaring like that observed in early 1988 (Courvoisier et al. 1990, Fig. 5).

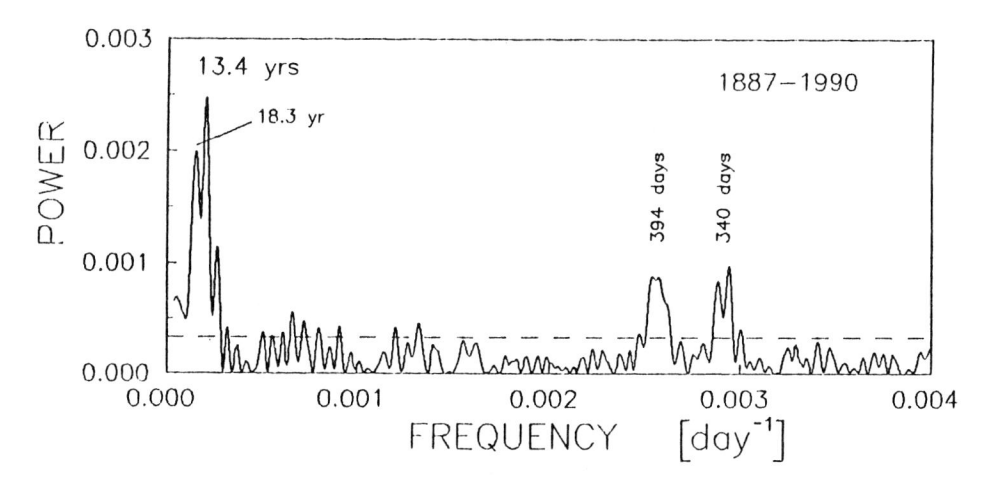

Fig. 3 Power spectrum of the data set for 1887 – 1991. The number of data points is 1153. The supposed period is 13.4 years ($\alpha \approx 10^{-38}$). Peaks at 340 and 394 days are blended aliases (spectral window shows the peak at 365 days). Power corresponding to the significance level α = 0.01 is marked by a dashed line. For detailed discussion see Babadzhanyants and Belokon' (1991a).

3 THE TIME SCALE OF YEARS

We present also a power spectrum for the whole ~100-yr interval for searching periods between about 1 and 70 yr (Fig. 3). The main features are peaks at 13.4 and 18.3 years. Fig. 1 shows two approximations of the observed data: 1) by the mean curve reconstructed from phase diagrams for the 13.4-yr period; 2) by the sum of two sinusoids with 13.4- and 18.3-yr periods (the amplitudes and phases were determined from Fourier analysis). The mean curve obtained using 18.3-yr phase diagram was also compared with the data set. We concluded that a 13.4-yr period is apparently the main one, and a 18.3-yr component describes variations of the 13-yr flares amplitude in the given time interval (see Babadhanyants and Belokon' 1991a for details).

There is also some additional independent evidence for the reality of the 13.4-yr period:
1) The light curve shows several deep downfalls in brightness. Four of them (J.D.~2424200; 2429300; 2436000; 2446000) are especially pronounced and reliably registered. All these brightness downfalls occur at phase 0.0 or 0.6 of the 13.4-yr period, i.e. exactly at minima or maxima of optical flares (Fig. 1). They do not affect the detection of the 13-yr period – their removal from the data set did not change substantially the power spectrum. Some decrease in the brightness level near the phase ~0.6 occurred also during the last two 13-yr cycles. OJ 287 shows similar sharp minima in optical light curve occuring with the same period (~11 years) as that of its flares (Sillanpää et al. 1988).

2) The hight quality VLBI map at $\upsilon = 5$ GHz (Zensus et al. 1988) shows a jet extending over ~70 mas. There are three extended emission regions along the jet separated by distinct gaps where the intensity drops by a factor of 10. These regions must contain the jet components C2-C8 and have superluminal velocities ~$1 \cdot mas \cdot yr^{-1}$. The distances between the gaps are about 12 – 13 mas that corresponds to 13-yr cycles of "slow" optical variations.

A similar association of the 12.5-yr cycle in optical variability with the emerging of large-scale (~tens mas) components was supposed for 3C 120 assuming the same superluminal velocities as at milliarcsecond scale (Belokon' 1987). This assumption was confirmed later by observations (Benson et al. 1988).

3) Krichbaum et al. (1990) noted that there is some evidence for a periodic variation of the apparent superluminal velocities (β_{app}) within the jet components C2 – C9. Fig. 4 shows β_{app} versus ejection epoch (obtained by linear extrapolation). The cycle of β_{app} variations proves to be ~13 years, too. The greatest velocities correspond to the brightness maxima in the 13-yr cycles.

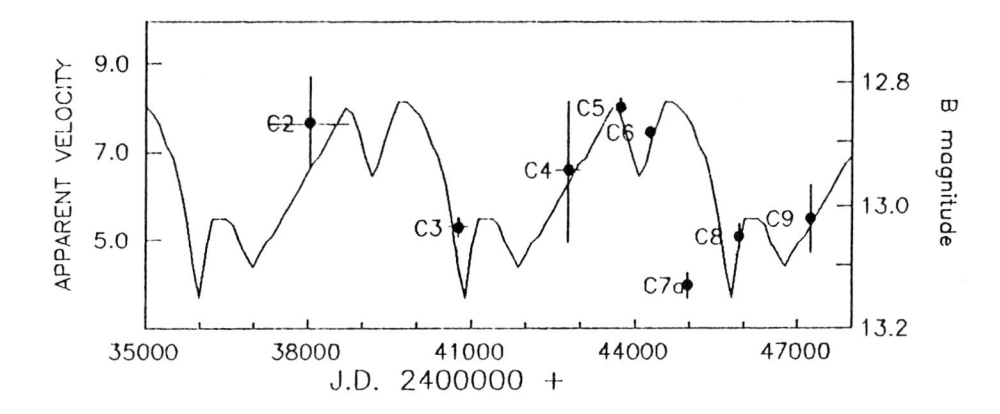

Fig. 4   Apparent superluminal velocity (β_{app}) of jet components C2-C9 (Zensus et al. 1990; Krichbaum et al. 1990) versus their epoch of ejection from the core (obtained by linear extrapolation). β_{app} is in units of light velocity. The mean light curve of 3C 273 with the period of 13.4 years is presented for the same time interval (compare with Fig. 1).

Strongly bent jets in regions close to the core are a common phenomenon for superluminal sources. The position angles of the jet components of 3C 273 with respect to the core are changing with their distance from the core. The gradients of position angle variation are positive for C4, C5, C9 while C7 (C7a) and C8 have negative ones (Krichbaum et al. 1990). Epochs of component ejection for these two groups seem to be concentrated at different phase intervals of the 13-yr period.

Krichbaum et al. (1990) supposed that the variations of β_{app} along with the changes of sign of position angle gradient may be explained by an oscillating jet axis with alternating changes of the mean jet inclination within a few degrees. Some evidence for the correlation of β_{app} and the position angle gradient with the 13-yr period suggest that the jet oscillations may not be chaotic but also include a periodical component.

We thank S.V. Sudakov for his contribution to this work.

REFERENCES

Angione R.J. and Smith H.J. 1985, *Astron. J.* **90**, 2474.
Babadzhanyants M.K. and Belokon' E.T. 1985, *Astrofizika* **23**, 459. English translation: 1986, *Astrophysics* **23**, 639.
Babadzhanyants M.K. and Belokon' E.T. 1987, in *IAU Symp. no 121, Observational Evidence of Activity in Galaxies*, ed. E.Ye. Khachikian et al. (Dordrecht: Reidel), p. 305.
Babadzhanyants M.K. and Belokon' E.T. 1991a, *Pis'ma Astron. Zh.* (submitted).

Babadzhanyants M.K. and Belokon' E.T. 1991b, (in preparation).
Belokon' E.T. 1987, *Astrofizika* **27**, 429. English translation: *Astrophysics* **27**, 588.
Belokon' E.T. 1991, *Astron. Zh.* **68**, 1.
Benson J.M., Walker R.G., Unwin S.C., Muxlow T.W.B., Wilkinson P.N., Booth R.C., Pilbratt G. and Simon R.C. 1988, *Astrophys. J.* **334**, 560.
Bååth L.B., Padin S., Woody D., Rogers A.E.E., Wright M.C.H., Zensus A., Kus A.J., Backer D.C., Booth R.S., Carlstrom J.E., Dickman R.L., Emerson D.T., Hirabayashi H., Hodges M.W., Inoue M., Moran J.M., Morimoto M., Payne J., Plambeck R.L., Predmore C.R., Rönnäng B 1990, preprint.
Charlot P., Lestrade J.-F. and Boucher C. 1987, in *IAU Symp. 129, The Impact of VLBI on Astrophysics and Geophysics*, ed. M.J. Reid and J.M. Moran, (Dordrecht: Kluwer) p. 33.
Cohen M.H., Zensus J.A., Biretta J.A., Comoretto G., Kaufmann P. and Abraham Z. 1987, *Astrophys. J.* **315**, L89.
Courvoisier T.J.-L., Robson E.I., Blecha A., Bouchet P., Falomo R., Maisack M., Staubert R., Teräsranta H., Turner M.J.L., Valtaoja E., Walter R. and Wamsteker W. 1990, *Astron. Astrophys.* **234**, 73.
Deeming T.J. 1975, *Astrophys. and Space Sci.* **36**, 137.
Jurkevich I. 1972, *Astrophys. J.* **172**, L29.
Krichbaum T.P., Booth R.S., Kus A.J., Rönnäng B.O., Witzel A., Graham D.A., Pauliny-Toth I.I.K., Quirrenbach A., Hummel C.A., Alberdi A., Zensus J.A., Johnston K.J., Spencer J.H., Rogers A.E.E., Lawrence C.R., Readhead A.C.S., Hirabayashi H., Inoue M., Morimoto M., Dhawan V., Bartel N., Shapiro I.I., Burke B.F. and Marcaide J.M. 1990, *Astron. Astrophys.* **237**, 3.
Kurochkin N.E. 1969, *Perem.zvezdy* **16**, 568.
Lyuty V.M. and Metlova I.V. 1987, *Soviet Astron. Tsirk.* **1475**, 3.
Robson E.T., Gear W.K., Clegg P.E., Ade P.A.R., Smith M.G., Griffin M.J., Nolt I.G., Radostitz J.V. and Howard R.J. 1983, *Nature* **305**, 194.
Sillanpää A., Haarala S., Valtonen M.J., Sundelius B. and Byrd G.G. 1988, *Astrophys. J.* **325**, 628.
Takalo L. 1982, *Astron.Astrophys.* **109**, 4.
Unwin S.C., Cohen M.H., Biretta J.A., Pearson T.J., Seielstad G.A., Walker R.C., Simon R.S. and Linfield R.P. 1985, *Astrophys. J.* **289**, 109.
Webb J.R. 1990, *Astron. J.* **99**, 49.
Zensus J.A., Bååth L.B., Cohen M.H. and Nicolson G.D. 1988, *Nature* **334**, 410.
Zensus J.A., Unwin S.C., Cohen M.H. and Biretta J.A. 1990, *Astron. J.* **100**, 1777.

Rapid 1.3cm Variability of Blazars

H.J. LEHTO

University of Southampton

ABSTRACT

We have searched for periodic and non-periodic time variability of several radio bright Active Galactic Nuclei. OJ 287 and 4C39.25 showed no periodic variations at a level of 0.08% the total flux density. In other sources no variations were seen at a level of about 1%. The non-periodic variations observed in OJ 287 and 4C39.25 are close to the level expected from interstellar scintillation, implying that the size of the unresolved VLBI component in these sources at 22 GHz has to be $\gtrsim 0.20$ times the observed upper limits.

1. INTRODUCTION

When one is looking for variability at the limit of the capability of the instrument or technique one can easily, and unknowingly, bias one's data or interpretation towards one's wishes, which may be either pro or against variability. In this study all attempts were made to avoid such biases in calibrating and editing the data which was done independently of the actual antenna temperature or flux density measurement. If something had to be *fitted* to the data (eg. a gain curve), a minimal degree of freedom was used.

There are several claims of periodic variations in OJ 287 both in optical and radio on time scales between 12 and 40 minutes (Visvananthan and Elliot 1973, Frolich et al. 1974, Valtaoja et al. 1985, Carrasco et al. 1985, Kinzel et al. 1988, De Diego and Kinzel 1990). However, there are also contradicting or non-confirming observations (Kiplinger 1974, Doroshenko et al. 1986, Dreher et al. 1986, Komensaroff et al. 1988). If the claimed variations are real then the amplitude of the optical periodic variability has to vary by a factor of about 10. In the radio the amplitude of the variable component has to vary on time scales of 2 days, *and* have a cutoff in the spectrum of the variable component between 1 and 6 cm steeper than ($\Delta F \propto \nu^{1.7}$). Our goal is to answer the question does OJ 287 vary on sub-hour time scales, and if so, by how much, and are the variations periodic. We included some other bright radio sources in our study.

2. OBSERVATIONS

We used two instruments to observe OJ 287 simultaneously between December 15, 1986 and February 1987. The Green Bank 140 foot was used at 19.5 GHz, and the VLA at 22.335 and 22.765 GHz. The frequencies were chosen to be slightly different from each other, so that if a periodicity was detected, then we would have a preliminary spectral index for it. However, staying in the same band was necessary in case the variability had a steep spectral index. We obtained a total of 99 hours of useful data at Green Bank and 42.5 hours at the VLA.

2.1. Green Bank

At Green Bank the observing was done in ON-ON mode, i.e. the two receivers were pointed alternatively at the source. The pointing was checked every 1.5–2 hours. A bandwidth of 370MHz and a beam separation of 8 arcminutes were used. The data was corrected for pointing, atmospheric absorption, postional gain variations, and small discontinuous jumps in antenna temperature. The major problem at the 140-foot was the pointing, because of the 3 arcsecond mechanical pointing accuracy, variable beam shape, and non-linearly varying pointing errors.

We analysed all the time series with the one dimensional CLEANed Discrete Fourier Transforms (ClDFT). The square of the euclidian norm of the ClDFT is called the CLEANed periodogram, or in the rest of the paper periodogram for short. Details of the method can be found elsewhere (Lehto 1988, Dreher et al 1986). The shortest time scale studied was typically about 3.0 minutes, and the longest, about 1 to 3 hours, equal to one third of the data set (table 1).

By inserting 8 sinusoidals with different amplitudes we estimated the relative amplitude of the sinusoidal that would have been unambiguously detected had it been present in the signal. For each periodogram we also calculated the level at which white noise would cause a peak at *a preselected frequency* with a probability of 1%. This level, very close to the more subjective detection level, was used as an upper limit.

Except for very low frequencies only two frequencies had statistically significant peaks. The first peak, at one percent level, was seen consistently in OJ 287 at $15.0h^{-1}$ (4min). Seen also in many other sources, this oscillation was due to the much larger than expected error in the inductosyn, the mechanical device that reads out the coordinates at which the telescope is actually pointing.

The second peak that was above our detection line was a peak at $4.7h^{-1}$ (12.8 min) in the periodogram of AO 0235+164 on Jan 1, 1987. We do not believe that this peak is intrinsic to the source. On one occasion (Dec 16, 1986) it is seen clearly

Source	Name	Class	Redshift	Date	S Jy	ΔP %	t_{max} min	t_{min} min
0149+219	Q	Q	1.32	350	1.1	3.7	170	3.2
				018	1.4	0.7	120	3.0
				039	1.4	30	30	4.0
0235+164	AO, OD160	BL	0.94	001	2.4	1.0	110	3.4
0851+201	OJ 287	BL	0.306	349	4.4	0.5	90	3.2
				350	4.4	0.45	150	3.0
				351	4.4	0.9	200	3.2
				365	4.4	1.1	200	3.1
				019	4.7	0.55	120	3.2
				040	5.2	3.0	45	2.9
				050	5.2	0.5	120	3.0
1013+208	–	–	–	349	.60	1.6	40	2.9
1228+126	3C274	G	0.004	365	22.4	0.5	10	2.8
				019	22.4	2.3	80	3.0
1354+195	4C19.44	Q	0.72	365	1.8	0.9	90	3.2
				050	1.7	0.7	110	3.4
2200+420	BL Lac	BL	0.069	349	2.7	1.0	30	2.8
				350	2.7	1.1	60	3.2
				365	3.0	0.7	60	3.8
				039	3.4	5.0	40	3.0

Table 1. Limits of variability for various bright radio sources. The first four columns identify the source. The fifth coulmn is the UT date of 1986 or 1987. The sixth coulmn gives the flux density at 19.5GHz. The seventh coulmn gives the upper limit (see text) for sinusoidal periodic variability in units of percent of the total flux density. The eigth and nineth columns show the time scales between which this upperlimit is valid.

in OJ 287 at $4.5h^{-1}$ (13.3 min). Although statistically less signifiant the peak is present at the same level (0.5%–0.7%) in other OJ 287 periodograms including Dec 15 and Dec 17. It is not seen in the periodogram of OJ 287 obtained at the VLA on Dec 17 at a level of 0.2%. A previous claim of a 13-minute periodicity was made with the same antenna (Valtaoja et al. 1985). Our observation suggests that it was instrumental. As a summary of the Green Bank data, we did not detect any periodicites in any of the sources at levels shown in Table 1.

2.2. VLA

The VLA observations were done on five nights separated by 10-19 days. The array was in C or a hybrid C/D configuration. During the first run the main comparison source for OJ 287 was 0839+187, suitable for its proximity. During other runs 4C39.25 was the primary comparison source. We analysed the comparison sources as independent sources due to their faintness or large angular distance from OJ 287.

To obtain simultaneity between OJ 287 and the comparison source the array was split into two subarrays. This created "double simultaneity": two subarrays and Green Bank. Every other antenna in each arm was in the first subarray, and the

Source	Name	Class	Redshift	Date	S Jy	ΔP %	t_{bend} h
0839+187	DWO	Q	.259	351	.555	0.666	2.0
0851+201	OJ 287	BL	.306	351	4.39	0.224	1.0
				365	4.37	0.106	0.5
				019	4.74	0.160	0.5
				030	4.77	0.138	0.5
				040	5.18	0.083	1.0
0923+392	4C39.25	Q	.699	351	4.10	0.393	3.0
				365	4.16	0.106	0.5
				019	4.21	0.168	0.5
				030	4.10	0.134	0.5
				040	4.14	0.081	1.0

Table 2. Limits of variability for various bright radio sources. The first four columns identify the source. The fifth coulmn is the UT date of 1986 or 1987. The sixth coulmn gives the flux density at 22.5GHz. The seventh coulmn gives the upper limit for sinusoidal periodic variability in units of percent of the total flux density between time scales of 1 min and 1 hour. The eigth column shows the time scale where the $1/f^2$ becomes equal to the white noise component.

remaining antennas in the second subarray. Because of possible instrumental differences between the subarrays, each subarray was used alternately to observe the source and the comparison source. The switch time from one subarray to the other was 5 or 12 minutes. Other technical details can be found elsewhere (Lehto 1988).

A ClDFT was calculated for each source, frequency and subarray. The phase information was retained. By shifting the ClDFT's *to the same phase center*, we averaged the ClDFT's in various combinations to obtain periodograms for each night and source+frequency or source+subarray combination. This allowed us to identify several instrumental oscillations. Next for each source a nightly periodogram was calculated from the four ClDFT's shifted to a common phase center. The last periodogram was the average periodogram of the different nights.

The periodograms were calculated up to a frequency of $60h^{-1}$ (1min). Based on the white noise dominated part of the periodogram an estimate for white noise was calculated. We then calculated a 0.1% probability level for each night. As a reminder, this is the probability that at **a given frequency** the power due to white noise only exceeds this level. We consider that this level, with a power of 13.8 times that of the white noise, is also a reasonable "detection level" for any detectable peak in this VLA data. With 540 frequencies of the full periodogram to choose from, the probability that one of them exceeds this level is only 0.4. However, the large number of spectra from one night allows us to identify the nature of each peak. If the peak in the periodogram is real, it has to satisfy the following conditions: It has to be present in at least one of the two frequencies, and has to show up in the data of both subarrays, and has to be in phase between the two subarray data

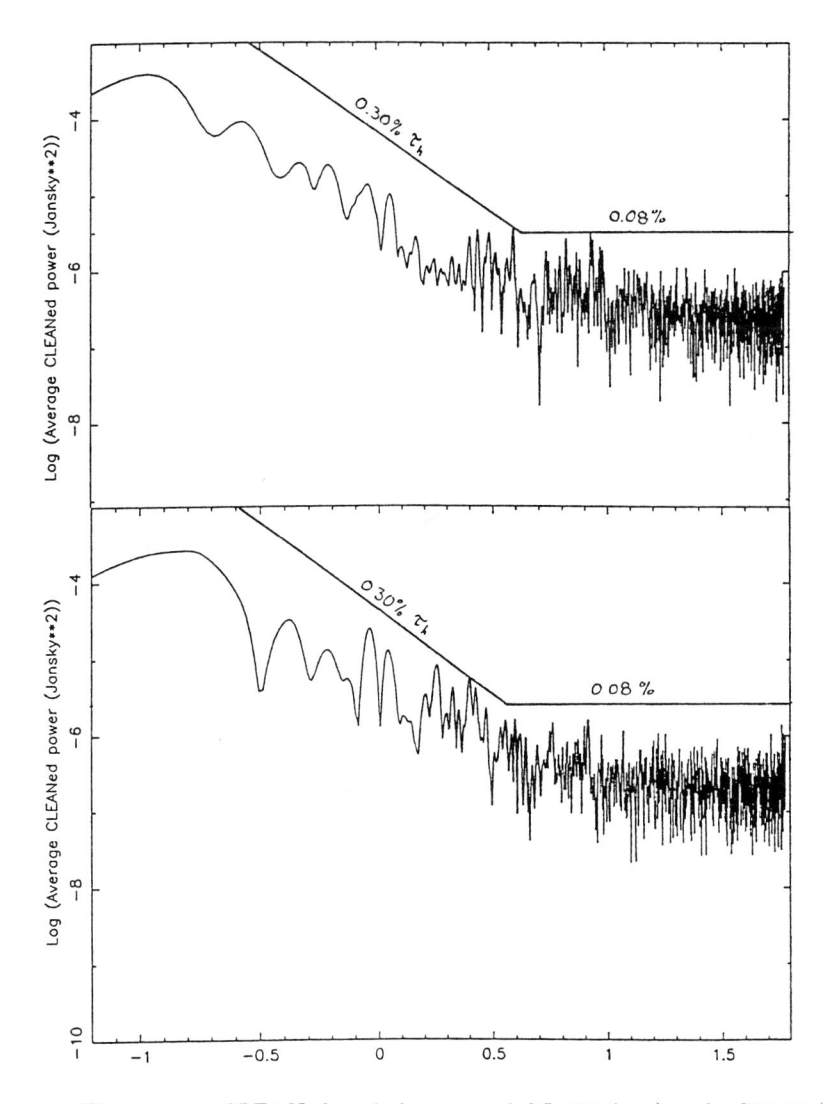

Figure 1. The average CLEANed periodograms of OJ 287 (top) and 4C39.25 (bottom). The abcissa is $log(frequency/h^{-1})$ and the ordinate is the average power in units of Jy^2. A oscillation with a period of 6 min and amplitude of A Jy would have a peak located at 1.0 (=10/h) and a value of about $(A/2)^2$. The upper limits for periodic variations are shown.

sets, implying that it should be relatively more prominent in the phased averaged periodogram of that frequency than in the individual periodograms. For it to be intrinsic to the source it has to be missing from the other source altogether.

All periodograms had a low frequency tail, to which we will return later. At high frequencies ($\geq 1h^{-1}$) there were on average 2 peaks above the 0.1% line. Almost half of the peaks were associated with one subarray only. These peaks had over ten times more power in one subarray than in the other. About one fourth of the peaks were associated with the subarray switch time, and occured at a frequency corresponding to twice that time. About every sixth peak was visible in all spectra of one night in all sources. About every twentieth peak was associated with one frequency only visible in both subarrays with less than one tenth of the power at the other frequency. None of the remaining handful of peaks satisfied the condition of reality, mainly the condition about the phase. We considered the 0.1% percent line as a suitable upperlimit for any nightly periodic variations (Table 2).

The periodograms of OJ 287 and 4C39.25 exhibit a $1/f^2$ low frequency tail. The average spectrum of OJ 287 (Figure 1) is well represented by

$$PG_{OJ\ 287} = 5.202(.023) \cdot 10^{-6} f^{-2} + 0.332(.015) \cdot 10^{-6},$$

where the digits in parantheses give the 1σ error of the fitted parameter. The periodograms have units of Jy^2. Similarly the spectrum of 4C39.25 (Figure 1) is

$$PG_{4C39.25} = 3.265(.031) \cdot 10^{-6} f^{-2} + 0.324(.020) \cdot 10^{-6}.$$

The white noise term is clearly the same for both sources. Besides these two spectral features there are clearly no other obvious features. From these periodograms we obtain the limits for persistent periodic variability

$$\Delta S \leq \begin{cases} 0.08\%, & 1\ \min \leq \tau \leq 16\ \min; \\ 0.30\% \tau_h, & 16\ \min \leq \tau \leq 4\ \text{hours}, \end{cases}$$

where τ_h is the time scale in hours.

3. THE GALACTIC SEEING LIMIT and THE VLBI SIZE

The structure function (Simonetti et al. 1985) corresponding to the $1/f^2$ tail is linear in τ. For each source we calculated the average value of structure function at ($\tau = 4$ h) from all the time series of that source. The linear dependence of the structure function was then scaled to give the same value at 4 hours. For OJ 287 ($\langle S \rangle = 4.4 Jy$) and 4C39.25 ($\langle S \rangle = 4.15 Jy$) the structure functions became $D_N = 1.37 \cdot 10^{-4} \tau_h$ and $D_N = 1.49 \cdot 10^{-4} \tau_h$. Let us now compare these structure functions to the theoretical ones.

If the turbulence of the interstellar medium follows a critical powerlaw (Cordes et al. 1985, Blandford and Narayan 1985)

$$P(q) = C_n^2 q^{-\alpha},$$

then (Lehto 1988) it can be shown that for a relative interstellar screen velocity of 43km/s, $\lambda = 1.33$ cm and $\alpha = 4$, a theoretical structure function expected for the relative brightness scintillation becomes

$$D(\tau) = .482 \frac{csc|b|\langle C_n^2 \rangle}{\theta^3 \xi^2} \tau_h,$$

where θ is the angular size of the compact scintillating component in tenths of milliarcseconds, which contains a ξ proportion of the total flux density.

OJ 287 has about 40% of its flux density at **22GHz** within 0.21±0.10 mas (Lawrence et al. 1985). Comparing this theoretical structure function to the one observed for OJ 287, we obtain a turbulence parameter

$$\langle C_n^2 \rangle \leq 9.65 \cdot 10^{-3} m^{-7}.$$

The central component of the triple source 4C39.25 is unresolved at 0.2 mas and has about 30% of its flux density (Shaffer and Marscher 1987) implying

$$\langle C_n^2 \rangle \leq 1.98 \cdot 10^{-2} m^{-7}.$$

These are in agreement with the values obtained from line of sights to six high latitude ($20° < b < 60°$) pulsars in the longitude range $160° < l < 230°$ ranging from $1.8 \cdot 10^{-4}$ to $1.2 \cdot 10^{-3}$. If the central component containing 40% or 30% of the flux density were much smaller, then one would expect scintillation with a clearly larger amplitude. *We are close to the "Galactic seeing" level.* We also estimate that the saturation time scale (i.e. the time scale were maximum scintillation is achieved) is about 8 hours. Figure 2. shows the various limits we obtain from the observed VLBI size, variability, and C_n^2.

We have not detected any intrinsic periodic variations in a number of BL Lacs and Quasars. If the previous claims of periodic variability in OJ 287 are real, then they have to vary in amplitude by at least a factor of ≈ 10. However, we do not think that any of these claims are firm enough evidence for periodic variability. Most claims are based on 2-4 periods or on misinterpreted periodograms (power spectra).

4. AKNOWLEDGEMENTS

Most of this work was done during my stay at the NRAO* and the University of Virginia. George Seielstad and Ron Maddalena provided considerable support with

---

* National Radio Astronomy Observatory is operated by Associated Universities, Inc., under agreement with the National Science Foundation.

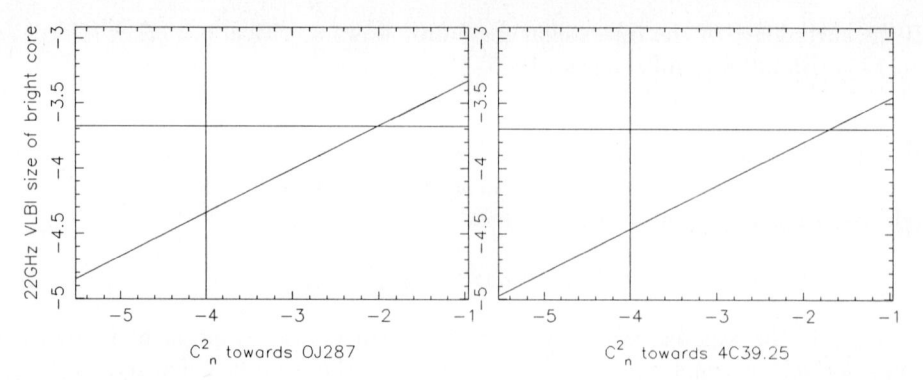

Figure 2. The limits for turbulence parameter C_n^2 and the VLBI size of the 22GHz cores of OJ 287 and 4C39.25. The abscissa is $log(C_n^2)$ and the ordinate log(VLBI size in arcsec). The vertical line indicates the smallest measured value of C_n^2 from pulsar timing measurements. The horisontal line is the upper limit of the size containing respectively 40% and 30% of flux density. The structure functions, as upper limits for variability due to scintillation, exclude the area right of the sloped line. The allowed region of parameters is the triangle in the center of each figure.

the Green Bank observations. Pat Crane, at the VLA, was very helpful. The author is especially indebted to David Heeschen, William Saslaw and James Condon for valuable discussions. Tim Pearson's plotting software PGPLOT was used to obtain the figures.

5. REFERENCES

Blandford, R. and Narayan, R. 1985, Mon. Not. Roy. Astr. Soc., **213**, 591.
Carrasco, L., Dultzin-Hacyan, D. and Cruz-Gonzalez, I. 1985, Nature, **314**, 1446.
Cordes, J.M., Weisberg, J.M. and Bobriakoff, V. 1985, Ap.J., **288**, 221.
De Diego, J.A., and Kinzel, M. 1990, Astrophys. Space Sci., **171**, 97.
Doroshenko, V.T., Lyuty, V.M., Terebizh, V.Yu., Efimov, Yu.S., Shakhovskoy, N.M., Piirola, V., Haarala, S., Korhonen, T., Sillanpää, A. and Valtaoja, E. 1986, Astron. Astrophys., **163**, 321.
Dreher, J.W., Roberts, D.H. and Léhar, J. 1986, Nature, **320**, 239.
Frolich, A., Goldsmith, S. and Weistrop, D. 1974 Mon. Not. Roy. Astr. Soc., **168**, 417.
Kiplinger, A.L. 1974, Ap.J., **191**, L109.
Komensaroff, M.M., Roberts, J.A. and Murray, J.D. 1988, Observatory, **108**, 9.
Lawrence, C.R., Readhead, A.C.S., Linfield, R.P., Payne, D.G., Preston, R.A., Schilizzi, R.T., Porcas, R.W., Booth, R.S. and Bruke, B.F. 1985 Ap.J., **296**, 458.
Lehto, H.J. 1988, Ph.D. Dissertation, University of Virginia.
Shaffer, D.B. and Marscher, A.P. 1987, in Superluminal Radio Sources, ed. J.A. Zensus and T.J. Pearson (Cambridge University Press, UK), p. 193.
Simonetti, J.H., Cordes, J.M. and Heeschen, D.S. 1985, Ap.J., **296**, 46.
Valtaoja, E., Lehto, H., Teerikorpi, P., Korhonen, T., Valtonen, M., Teräsranta, H., Salonen, E., Urpo, S., Tiuri, M., Piirola, V., Saslaw, W.C. 1985, Nature, **314**, 148.
Visvananthan, N. and Elliot, J.L. 1973, Ap.J., **179**, 721.

Multi Frequency Observations and Variability of Blazars

THIERRY J.-L. COURVOISIER

Geneva Observatory

1 INTRODUCTION

Several Authors have expressed some difficulties to use the name blazar to describe a class of objects at this conference. I shared these problems when preparing this contribution. I solved the difficulty following Bregman (1990) who defined blazars as those objects for which the radio to optical and ultraviolet emission is dominated by non thermal processes. This definition is however still so loose that further distinctions are necessary. I distinguish two sub-classes, the smooth continuum objects (BL Lac objects) and the objects with much more structure in their continuum emission. The second sub-class include Optically Violently Variable objects (OVV), probably the highly Polarized Quasars (HPQs) and at least some if not all flat spectrum radio loud objects. I will refer in these notes to the second category as the bumpy objects. The common points between the two categories of blazars sketched here are not evident when looking at the shape of their continuum spectral energy distribution. BL Lac objects have a smooth and regularly steepening continuum emission with maximum output (maximum in $\nu \cdot L_\nu$) in the far infrared, whereas the bumpy objects have their maximum output in the extreme ultraviolet or in the hard X-rays. All these objects show, however, large and rapid variability and various degrees of polarization.

The deep unity of the blazar class of objects lies probably in the physical identity of the non thermal emission process. Observationally, this translates into similar radio to infrared properties such as polarization, existence of jets and variability, although the similitudes are weakened by secondary effects due to orientation of the beamed blazar component to the line of sight. The BL Lac type blazars are largely dominated by the non thermal continuum, while the bumpy objects show in addition several other emission components. In this way, it might be more appropriate to speak of the blazar phenomenon, which is seen alone in BL Lac type blazars and mixed with other emission processes in more complex objects.

I will in this contribution rely on the paradigm that the non thermal emission in the far infrared and radio domains observed in AGN, the blazar phenomenon, is due to synchrotron emission in a relativistic jet. The reasons to believe that this paradigm is reasonable are the high degree of polarization observed in several objects, the superluminal expansion seen in many jets and variability on short timescales. Many of these observations and their interpretations are discussed in these proceedings.

The questions that need be discussed when studying blazars are of two kinds. On one side the nature of the jet needs to be addressed, its structure, its origin and collimation and the parameters yielding the observed emission (electron energy and spatial distributions, magnetic field strength and geometry). On the other side, the connections between the blazar component and the other emission components need be addressed, with the natural follow on question of the relationships between the different classes of AGN. Specifically, one wonders why some blazars are BL Lac like and others show in addition many emission components. We should also investigate whether orientation effects alone are sufficient to explain these differences (see section 4). I will adress here the second set of questions, several contributions relating to the first questions being discussed in other contributions in these proceedings.

Although the BL Lac type blazars have smooth spectral energy distributions from the radio domain to the ultraviolet, they nonetheless show some very complex behaviours, which should be kept in mind when comparing them with the other classes of AGN and in particular the HPQs and OVVs. Let me mention two of these complexities. Multiwavelength studies of some BL Lacs have shown that the optical and infrared variability are very well correlated. They have also shown that the radio light curves at different frequencies are very well correlated. On the other side, these studies have demonstrated that the radio variations and the infrared-optical variations are much more loosely connected. This has been described for example by Bregman et al. (1990) in the case of BL Lac itself and by Valtaoja, Sillanpaa and Valtaoja (1987) in the case of OJ 287. The second property I would like to keep in mind is the complex spectral behaviour of BL Lac objects while their flux varies: Some sources become harder as their flux increases, while others show no changes in their infrared to ultraviolet spectral energy distributions when their flux changes (see e.g. Tanzi et al. 1989). It should also be mentionned that some sources have spectral breaks in the infrared or optical domains, while the energy distributions of others can be well described by a single power law.

2 THE EMISSION COMPONENTS
AGN have in general complex continuum spectral energy distributions. Their total emission is the sum of several components, including the blazar component for radio loud objects. It is very surprising that most components of a given source have roughly the same luminosity, eventhough they are caused by so different processes

as synchrotron emission and dust emission (see below). This is only a more physical way of expressing that the overall energy distribution of AGN is roughly proportional to ν^{-1}. The near equality of the luminosity of the different components should be a clue to the general organization of AGN, it is not understood yet.

The components observed in AGN include synchrotron emission, possibly even two such components with very different variability patterns (see below); emission from dust; the so-called blue bump, most often associated with emission related to the accretion flow; the soft excess component, an unexplained component in excess of the extrapolation of the X-ray power law towards low energies; and the X-ray emission. These components are briefly reviewed in the next paragraphs.

2.1 The Synchrotron Components

The infrared emission of radio loud AGN smoothly joins the radio flux to the near infrared and optical emission. The variability of this component and its polarization strongly suggest that this component is due to synchrotron emission. Since this component smoothly merges into the emission at higher frequencies, it is difficult to estimate its high frequency cut-off. This parameter is important, because it shows at which electron energies the synchrotron cooling of the electrons and positrons is faster than the accelerating timescale. The presence of a near infrared excess emission often complicates the matter. This excess is often described as a bump on top of the synchrotron component, which then extends into the optical and ultraviolet domains (Edelson and Malkan 1986). In at least one case (3C 273), however, it was shown that the near infrared flux remained stable (within the $\simeq 5\%$ measurement uncertainties) while the far infrared (wavelengths larger than 10μ) flux decreased by a factor of two. Since synchrotron emission decreases faster at high frequencies than at low frequencies, this observation could be used to set an upper limit of about 15% (3σ) to the synchrotron contribution to the near infrared flux (Robson et al. 1986). This shows that the synchrotron high frequency cut-off can be as low as few 10^{13} Hz.

A further emission undoubtedly due to synchrotron processes but possibly quite unrelated to the previously discussed component (see section 4) is the very rapidly varying infrared to optical emission observed in 3C 273 in February and March 1988 (Courvoisier et al. 1988). This emission was characterised by daily variations in the optical and near infrared such that $\mid \Delta L/\Delta t \mid \simeq 6 \cdot 10^{40} ergss^{-1}$ ($H_0 = 50km/(s \cdot Mpc)$). The observed cooling time of approximately 2 days could be used to deduce the magnetic field in the emission region. This field was found to be of the order of 0.7 Gauss. A more detailed analysis of the data including spectral information is underway, preliminary results indicate a similar magnetic field. Analysis of the mm data obtained at this epoch shows that the emission must be originating in a medium moving at relativistic velocities towards the observer (Robson et al. in preparation). Subsequent VLBI monitoring showed that a new component had appeared in the jet of 3C 273 precisely at the epoch of the fast optical and near infrared variability (Krichbaum et al. 1990, Baath et al. 1991). This coincidence and the observation that the flare emitting plasma is moving with relativistic velocities strongly suggests that the continum activity is related to the superluminal jet (as had been proposed in Courvoisier 1988).

Daily variations in the polarization level and angle had also been observed during the outburst (Courvoisier et al. 1988). The properties of the rapidly varying synchrotron component of 3C 273 are thus very similar to those of BL Lac objects. This certainly illustrates the unity of the subject and conforts us in identifying the phenomena observed in the radio to infrared domains of 3C 273 (at least but probably of all blazars) with the BL Lac activity. It also illustrates the similarities between the physical processes responsible for the characteristics of different classes of AGN.

The results obtained from continuum optical, infrared, mm and radio observations together with the VLBI monitoring have shown the power of observations covering large parts of the electromagnetic spectrum combined with imaging techniques to understand the physics of blazars (it also illustrates the large efforts necessary to understand these complex objects). There are, however, now several questions that need be addressed. Can we use the spectral evolution of the varying continuum emission to deduce characteristic cooling times for the electrons and to disentangle accelerating and cooling processes, and if so, how do the cooling timescales agree with synchrotron emission? Another outburst took place in 1990, does it mean that the duty cycle of the rapidly varying component is about two years in 3C 273? How similar were the outbursts observed in 1983 (Robson et al. 1983) and the 1988 activity and more generally, are all outbursts similar? (I would be surprised if the answer was yes.) Are all outbursts connected with the appearance of new components in the VLBI jets and conversely, are all new components connected with an optical-infrared outburst? A preliminary analysis of the 1990 data shows that the optical outbursts took place at similar phases of the UV activity, is this a coincidence, and more generally, what is the link between the synchrotron activity and the accretion process? Courvoisier et al. 1990 suggested the presence of a magnetic energy reservoir to store accretion energy (see also Shields and Wheeler 1976), can this idea be firmly established or disproved? Furthermore, How similar are outbursts in different sources? Some of these questions can be studied with the available data, some will require the accumulation of several years of high quality observations covering most of the known electromagnetic spectrum in many sources.

2.2 Near Infrared Emission and Dust Emission

The near infrared emission is characterized by a steep spectrum in many AGN (Edelson and Malkan 1986, Courvoisier et al. 1990, Neugebauer et al. 1987, Courvoisier Robson and Bouchet in preparation). The observations of 3C 273 in early 1986 imply that this component is not dominated by synchrotron emission (see above). This conclusion is strengthened by the striking difference between the far infrared light curve and the near infrared light curve. The latter show very little variations, if any, outside the periods of outbursts, while the far infrared flux is continually changing.

The near infrared emission is probably not due to the thermal emission from the outskirts of an accretion disc. The increasing area of the disc is insufficient to compensate for the decrease in emissivity as the temperature decreases to values giving rise to infrared emission. Accretion disc models therefore tend to produce too small thermal emission in the near infrared to explain the observed flux in this spectral domain.

An alternative explanation is thermal emission from hot ($\simeq 1500K$) dust. A possibility which was discussed by Barvainis (1987). A nice confirmation of this idea is provided in the case of F 9 for which Clavel Wamsteker and Glass (1989) showed that infrared variations follow the ultraviolet variations with lags that increase with increasing wavelengths. This result strongly suggests that dust is heated to temperatures that decrease with increasing distances to the ultraviolet photon source in good agreement with dust emission models.

Barvainis (1990) even suggests that all the infrared to optical emission of radio quiet AGN can be explained by the superposition of dust emission, a galactic disc component and an optical-ultraviolet power law. This model has too many free parameters to be firmly established with existing data, it has nonetheless the virtue of showing that this type of explanation, which is minimal in the sense that it doesn't require any ingredient not present in galaxies to explain the far infrared emission, is possible. In this model, the nuclear activity only appears at wavelengths shorter than about 1μ with the power law component. Dust emission had also been suggested to explain the steep turn on of the infrared emission between mm wavelengths and $\simeq 100\mu$. This turn on had been shown to be steeper than the $\nu^{5/2}$ law expected for synchrotron radiation (Chini, Kreysa and Biermann 1989). Also suggesting that far infrared emission of radio qiet AGN may not be due to the synchrotron process.

Work in progress on the bright southern quasar 0914-62 indicates, however, that using dust distributions to account for the near infrared spectral energy distributions obtained with CVF observations (spectral resolution of $\simeq 100$) may be more difficult than expected. The reason for this difficulty is the fact that the near infrared spectral energy distribution of this object matches closely a power law and shows no curvature, as might be expected from dust close to the sublimation temperature.

2.3 Optical Ultraviolet Domain

In most AGN the optical ultraviolet emission represents an excess compared with the infrared continuum and the infrared-X-ray interpolation. This excess is often called the blue bump. The blue bump is most often associated with thermal emission from an accretion disc (Schields 1978, Malkan and Sargent 1982, Camenzind and Courvoisier 1984 among others). Most studies are based on geometrically thin and optically thick accretion discs, as discussed in Shakura and Sunyaev (1973).

The standard disc model has, however, at least two problems (Courvoisier and Clavel 1991a). These difficulties are related to the dependence of the ultraviolet continuum spectral energy distribution on the luminosity of the source for one and to the lack of delay between ultraviolet and visible variations for the other.

In standard accretion disc models the maximum temperature (the temperature of the inner regions of the disc) is inversely proportional to the mass of the central black hole to the 1/4 power. As a result, more luminous objects, which are expected to be more massive, should show a softer spectrum than less luminous (less massive) objects. This prediction is not met by the observations, neither for selected AGN (Courvoisier and Clavel 1991a) nor for samples of objects (Bechtold et al. 1984, O'Brien, Gondhalekar and Wilson 1988).

The sound travel time in the accretion discs surounding supermassive black holes between the regions emiting the ultraviolet flux and those emiting in the visible domains far exceeds the observed delays between ultraviolet and optical variations (Courvoisier and Clavel 1991a). As a consequence, these models cannot account for the observed variability. Since the amplitude of the variations is large ($\simeq 50\%$), a large portion of the flux cannot be explained by standard thin accretion discs.

A possible way out of these difficulties may be found in thick accretion discs. These discs are radiation pressure dominated and thus less prone to sound travel time difficulties. They are unstable and thus certainly variable. It remains to be seen, however, that they can be stable enough not to be completely disrupted by their instabilities (see Blandford 1990 and references therein).

One difficulty in the study of the blue bump, as for any other emission component, is to be able to isolate the blue bump emission from the other emission components in the infrared and in the soft X-rays. Courvoisier and Clavel (1991b) have shown that this difficulty together with several other problems plague any detailed analysis of the physical processes at the origin of the blue bump.

2.4 The X-Ray Soft Excess

The spectral domain extending from 1200Å to 0.1keV is still very poorly studied. The main difficulty is the strong absorption of any material in this band. This applies to the instruments as well as to the interstellar matter. The satellite ROSAT and in coming years the instrument EUVITA on board the soviet spacecraft Spektrum X-G will hopefully greatly improve this situation at least for a sample of AGN in regions of the sky with very low hydrogen column densities. It is clear, however, when looking at the observations performed by EXOSAT, that a large fraction of the AGN shows an excess emission around 0.1 keV when compared with extrapolations of the spectrum using the 2-10keV power law (Turner and Pounds 1989). The origin of this excess is still poorly understood. One suggestion is that the soft excess is the high frequency tail of the accretion disc emission. Tests of this hypothesis will be to see how the soft excess varies compared with ultraviolet variations. Very few such tests have been done to date, because no instrument had a sufficient soft X-ray sensitivity since the end of the EXOSAT program. In the Seyfert galaxy NGC 5548 work in progress (Walter and Courvoisier 1991) suggests that the temporal behaviour is very different in both domains.

2.5 X-Ray Emission

The X-ray emission of AGN has long been described as a simple power law. This view has changed in recent years, as more sensitive detectors have been flown. It appears now that the X-ray spectrum (between 0.5 and 10 keV) is much richer than anticipated (Pounds et al. 1990). The X-ray emission shows often a soft excess (see above), a curved spectrum, an iron emission line between 6 and 7 keV and an absorption iron edge at 8 keV. Several of these features can be understood if the X-ray emission is generated some distance away from an accretion disc. In this case the primary emission could be a power law and the features summarised here be caused by reflection of the primary flux by the cool (in X-ray terms) and optically thick accretion disc (Pounds et al. 1990).

The previous discussion doesn't imply any specific emission process for the primary X-ray generation. It only presupposes that the emission takes place outside of the accretion disc. One model which satisfies this requirement is the wind and shock model as discussed in the case of 3C 273 by Courvoisier and Camenzind (1989). In this model, a soft photon source, presumably the inner regions of an accretion disc, emits a large number of soft photons. Some of these are then Comptonised in a shock located at some distance from the soft source. Photon number conservation in the Comptonisation process together with the observed numbers of ultraviolet and X-ray photons allows to constrain the geometry of the shock. It can be shown that it covers few percents of the soft source. A further application of the model has been made by Walter and Courvoisier (1990) to the Seyfert galaxy NGC 5548. This galaxy had been observed at numerous occasions by EXOSAT and has shown large amplitude variability. It also has a strong and well defined soft excess component. In our application of the model, the soft component to be Comptonised in a shock at some distance of the accretion disc is taken to be proportional to the count rate observed in the low energy telescope of EXOSAT. Compton theory predicts in this case that the spectral index of the Comptonised flux depends linearly on the log of the ratio of soft photon to Comptonised photons. This ratio is directly observable with EXOSAT which probed both components simultaneously. Figure 1 shows that in the case of NGC 5548 this prediction is very well substantiated by the observations.

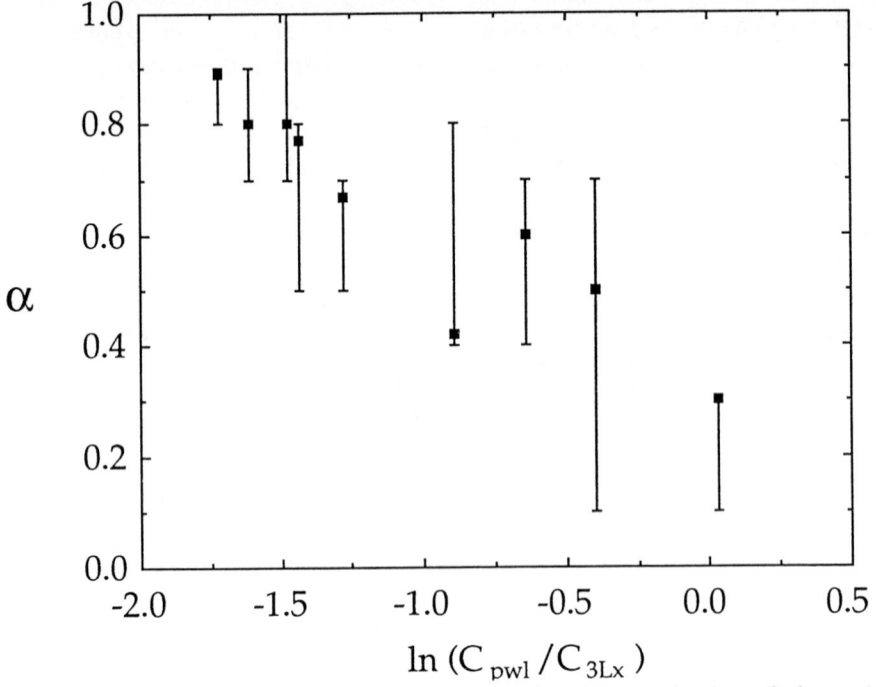

Figure 1: The dependence of the X-ray spectral index on the log of the ratio of Comptonised to soft photons for NGC 5548.

Extension of this formalism to sources for which the soft excess is not well characterised can lead to trivial results insofar as, in this case, the soft flux lies on the same power law as the 2-10 keV flux.

The same excellent correlation can be found in the quasar 3C 273, using the ultraviolet photon flux rather than the EXOSAT LE count rate as a measure for the soft photon source, well in line with the wind and shock model. This measurement is possible for this source, because several more or less simultaneous X-ray and ultraviolet measurements were performed.

It is worth noting that the Comptonisation model described here doesn't necessarily predict a correlation between the soft and Comptonised photon fluxes. When the geometrical properties of the Comptonising medium vary (its optical thickness and covering factor), so does the ratio of Comptonised photons to soft photons. The linear dependence of the slope of the Comptonised component on the log of the soft to hard photon fluxes ratio remains, however, established.

3 CORRELATION ANALYSIS

The multi frequency data we obtained on 3C 273 in the last few years is now of a sufficient quality and spans a sufficient laps of time to be studied in terms of time series analysis. Although the quality of our sampling has significantly improved since the beginning of the campaign, the data are still far from uniform and far too sparsely sampled. We must therefore be very carefull with the choice of the time analysis method. In order not to create any data by interpolation (with highly uncertain results in a badly undersampled environement) and to make a maximal use of the available data, we use the method of Edelson and Krolik (1988). This method calculates an approximation of the correlation function using all available pairs of points. As an additional bonus, the method calculates a measure of the significance of the approximation.

3.1 Ultraviolet and X-Ray Correlation

The cross correlation between the ultraviolet flux as measured by IUE and the 2-10 keV flux as measured by EINSTEIN, EXOSAT and GINGA (and where appropiate extrapolated to the 2-10 keV domain) shows no significant correlation at any of the sampled lags (Courvoisier et al. 1990). This result is possibly due to the fact that the X-ray flux is not generated in close connection with the accretion process (at least if the ultraviolet flux is indeed a signature of the accretion process).

The Comptonisation model discussed above is compatible with the lack of correlation between the ultraviolet and X-ray fluxes if the Comptonisation medium is variable in time. Besides its possible significance for the physics of the emission regions, this fact illustrates the need, when looking for connections between emission components, to look not only for flux corelations but also for more hidden relationships between the parameters describing the emission in the different components.

3.2 Milimeter and X-Ray Correlation
No significant correlation can be found between the fluxes in these two wavelength domains either. This negative result excludes simple synchrotron self-Compton emission processes, as, in this case, the same electron population is emitting in both spectral regions.

3.3 Optical and Ultraviolet Correlation
A close connection is observed at zero lag between the fluxes in these two spectral domains. This connection is true not only in the case of 3C 273, but also in several Seyfert galaxies for which sufficient data has been obtained from the ground and with IUE. The consequence of the short upper limits on the lag for accretion disc models has been discussed above. This close connection is also a clear sign that the emission from $\simeq 1\mu$ to the ultraviolet (i.e. the blue bump) is emitted by one single component.

The correlation decribed here for 3C 273 is also seen in the Seyfert galaxies for which sufficient ultraviolet and visible data is available. This shows that the difficulties described above for the interpretation of the blue bump are generic to quasars and Seyfert galaxies and not limited to high luminosity sources. If luminosity is indicative of the accretion flow, this means that the standard thin disc model has problems not only in the super Eddington accretion regime (where difficulties are expected) but also at much smaller accretion rates.

3.4 Ultraviolet and radio correlation
Quite unexpectedly, the best correlation between the flux of different components is obtained between the ultraviolet and the radio fluxes. This result is surprising because the origin of the two components is thought to be so different, namely thermal optically thick emission related to the accretion flow in the ultraviolet and non thermal synchrotron radiation in the radio domain. This correlation is not due to the flares which occured in 1988 and again in 1990, since they were not observed in the ultraviolet. A confirmation of the correlation is given by the analysis of the visible data with the radio data, excluding the flaring activity (figure 2). The correlation is of similar quality to the ultraviolet correlation with the radio. This is expected, because of the excellent ultraviolet-visible correlation. The sampling of the visible data (Geneva photometry) being considerably better than the ultraviolet sampling, this excludes that the first correlation can be caused by an artifact due to the sampling function. In addition it shows that all the blue bump emission is closely correlated with the radio emission. Similarly, using 22GHz or 37GHz doesn't change the result

in any significant way.

The correlation is such that the blue bump flux precedes the radio emission by approximately 0.4 years. This is contrary to what would be expected if the blue bump emission was due to Comptonisation of the radio photons. It is, however, the sign of the delay expected for a causal connection between the accretion process, the primary source of energy, and the non thermal radiation mechanisms which must ultimately derive their energy from accretion. The numerical value of the delay will provide a useful test of any model relating the two phenomena and is a function of both the geometry and the energy transport mechanisms. At present, the models are still too rough to make firm predictions, as is evident from the fact that the correlation was unexpected by most researchers, including our group. There always remains the possibility that the corelation is a mere statistical fluke, time has, however, firmed our finding since the first detection of the correlation (the data used in fig. 2 are complete until July 1990).

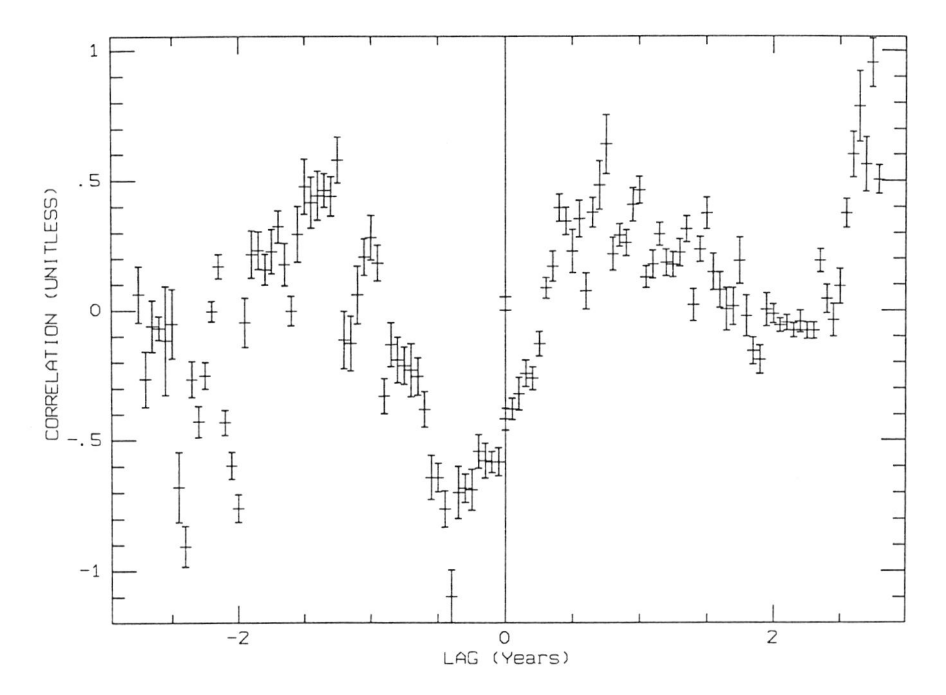

Figure 2: Cross correlation between the V flux and the radio flux in 3C 273.

3.5 Connection with BL Lac Objects
The correlations between the radio and optical emission in BL Lac objects (as op-

posed to bumpy objects) was recalled in the introduction. The similarities between the observations of 3C 273 reported here and the description given for BL Lac by Bregman et al. (1990) are striking. Namely excellent correlation at the high frequencies (ultraviolet-visible in Seyfert and Quasars and visible-infrared in BL Lac), excellent correlations between fluxes at different radio frequencies and a significant but less pronounced correlation betwen the high frequencies (visible) and the low (radio) frequencies. Even the sign and the magnitude of the delay agree roughly.

This striking similarity is very difficult to understand if the blue bump emission of quasars and Seyfert galaxies is of a different nature compared to the emission mechanisms at work in BL Lac objects.

4 THE BLAZAR COMPONENT

We can estimate the contribution of the steady (i.e. the slowly variable) synchrotron component to the continum emission at $6 \cdot 10^{14} Hz$. I assume that the blue bump spectral energy distribution is proportional to ν^0 and use the spectral index of the near infrared component as an upper limit to the index of the synchrotron emission at these frequencies (see above). We know from the variability discussion that the contribution of the synchrotron emission at J is less than or comparable to 10% (2σ) so this contribution is $< 0.1 \cdot (\nu_J/6 \cdot 10^{14})^{-1.5} \simeq 1.2 \cdot 10^{-2}$. If the fast flaring activity in the visible and infrared domains is due to this synchrotron component, its flux must therefore increase by at least a factor 100 in the visible to be comparable with the blue bump emission, whereas a variability amplitude of only a factor 2-3 is observed in the mm and radio domains.

This large difference can be explained by either an electron acceleration mechanism which is considerably more efficient at high energies or, and possibly more convincingly, by assuming that the outburst component, which is so similar to the blazar component, is distinct from the steady synchrotron emission. This latter view has naturally the drawback of adding yet another component to the puzzle of AGN. It might however help in understanding the different categories of objects and their polarization properties.

Following the preceding argument further, one can estimate that the synchrotron emission contributes approximately 10^{-3} of the flux at 1200Å. Since one of the characteristics of BL Lac objects is that the non thermal emission dominates the flux also in the ultraviolet, as is shown by the smooth turn over of the continuum spectral energy distribution, one concludes that in order to observe an object like 3C 273 as a BL Lac, its non thermal emission should be boosted by a factor larger than thausand. The luminosity of 3C 273 is already of the order of $10^{47} ergs \cdot s^{-1}$. Boosting one component by a factor larger than 1000 would therefore lead to a very luminous object indeed. It therefore appears very unlikely that an object like 3C 273 can be viewed as a BL Lac object from any direction. This is all the more so since 3C 273 is already a superluminal source and we therefore know that our viewing angle is close to the jet axis. A large additional Doppler boosting is therefore very improbable. Clearly therefore, orientation cannot be the sole factor governing whether a blazar appears as a BL Lac or a bumpy radio loud quasar or OVV.

One parameter may play an important role in the classification of AGN and help understand the differences between radio quiet sources and radio loud sources. It is the magnetic field. In case there is no magnetic field, there will be no synchrotron emission, the infrared emission can then be dominated by thermal dust emission. If, as is often expected, the magnetic field also plays an important role in the collimation of the jets, so the absence of such a field will also prevent the formation of jets, radio quiet objects are then expected to be jet less. In objects where the B-field is stronger, synchrotron emission plays an increasing role and jets appear, the extreme case is one in which the magnetic field is very high and thus jets and synchrotron emission completely dominates the objects as is the case in BL Lac objects. This is at most a conjecture at this time, it may, however, provide a useful line to pursue. The previous argument implies in this model that very bright objects, like 3C 273 cannot have very strong magnetic fields, since they cannot be turned around to be seen as BL Lac objects.

REFERENCES

Baath L.B. et al. (1991) A&A **241**, L1

Barvainis R. (1987) Ap.J. **350**, 537

Barvainis R. (1990) Ap.J. **353**, 419

Bechtold J. et al. (1984) Ap.J. **281**, 76

Blandford R.D. (1990) in Active Galactic Nuclei, Lecture Notes of the 20th Advanced Course of the Swiss Astrophysical and Astronomical Society, Eds Courvoisier and Mayor, Springer Berlin

Bregman J.D. (1990) A&A Reviews **2**,125

Bregman J.D. et al. (1990) Ap.J. **352**, 574

Camenzind M. and Courvoisier T.J.-L. (1984) A&A **140**, 341

Chini R., Kreysa E. and Biermann P.L. (1989) A&A **219**, 87

Clavel J., Wamsteker W. and Glass I.C. (1989) Ap.J. **337**, 236

Courvoisier T.J.-L. (1988) The Messenger **54**, 37

Courvoisier T.J.-L. and Camenzind M. (1989) A&A **224**, 10

Courvoisier T.J.-L. and Clavel J. (1991a) A&A in press

Courvoisier T.J.-L. and Clavel J. (1991b) Proceedings of the 6th IAP Meeting/IAU Col. 129, Eds Bertout, Collin-Souffrin, Lasota and Tran Than Van, Editions Frontieres, Gif-sur-Yvette

Courvoisier T.J.-L. et al. (1988) Nature **335**, 330

Courvoisier T.J.-L. et al. (1990) A&A **234**, 73

Edelson R.A. and Krolik J.H. (1988) Ap.J. **333**, 646

Edelson R.A. and Malkan M.A. (1986) Ap.J. **308**, 59

Krichbaum T.P. et al. (1990) A&A **237**, 3

Malkan M.A. and Sargent W.L.W. (1982) Ap.J. **254**, 22

O'Brien P.T., Gondhalekar P.M. and Wilson R. (1988) MNRAS **233**, 801

Neugebauer G. et al. (1987) Ap.J.Supp. **63**, 615

Pounds K.A. et al. (1990) Nature **344**, 132

Robson E.I. et al. (1983) Nature **305**, 194

Robson E.I. et al. (1986) Nature **323**, 134

Shakura N.A. and Sunyaev R.A. (1973) A&A **24**, 337

Shields G.A. (1978) Nature **272**, 706

Shields G.A. and Wheeler J.C. (1976) Astrophysical Letters **17**, 69

Tanzi E.G. et al. (1989) in BL Lac Objects, Lecture Notes in Physics **334** Springer Berlin

Turner T.J. and Pounds K.A. (1989) MNRAS **240**, 883

Valtaoja L., SillanpaaA. and Valtaoja E. (1987) A&A **184**, 57

Walter R. and Courvoisier T.J.-L. (1990) A&A **233**, 40

THE ENERGETICS AND MORPHOLOGIES OF BLAZAR OUTBURSTS

James R. Webb
Dept. of Physics
Florida International University
University Park
Miami, Fl 33199

I. ABSTRACT

We discuss the energetics and morphologies of high-amplitude outbursts seen in OVV quasars and BL Lacertae objects. We present long-term light curves of several sources and discuss the types of variability seen, with emphasis on the energetics of the high-amplitude outbursts. We also discuss the application of jet models and accretion disk models to Blazar outbursts.

II. OUTBURSTS AS DIAGNOSTIC TOOLS

Rapid, high-amplitude variations seen in Blazars were quickly recognized as being important in determining the size of the emission regions. If one assumes the entire source participates in the variation, the maximum size of the emission region can be calculated using equation 1 (Terrell 1967).

$$R < \frac{4cL}{|dL/dt|(1+z)} \quad (1)$$

This equation is relaxed and the Doppler boost factor included if we allow for relativistic expansion. Equation 1 can be approximated by equation 2 where D is the Doppler boosting factor (Urry 1987).

$$R < Dct \quad (2)$$

Stringent demands are made on theoretical models to account for the energy observed during an outburst, energy which can either be liberated in "real time" or generated and stored between outbursts. Characteristic timescales inherent in the source can be derived from the duration of the outbursts and the time needed to store energy between successive outbursts.

III. LONG-TERM LIGHT CURVES AND VARIABILITY PROPERTIES

Although Blazars exhibit rapid variability at every observed frequency, the optical variations provide us with a key to understanding the physical processes operating in the central regions of the sources. The University of Florida's Rosemary Hill Observatory (RHO) has contributed to our knowledge of these variations by monitoring a sample of Blazars with intervals of days or weeks between observations. When a major flare is detected the sampling increases, and several exposures per night are obtained. Despite weather, lunar phases, and sidereal motion which prevent regular sampling, these long-term light curves are invaluable in determining the characteristics of optical variations of these sources. Blazar light curves can be classified into four categories originally introduced by Dent et al. (1974).

   I. Rapid flickering w/o significant long-term trends.
  II. Large-amplitude, long-term changes dominate.
 III. Short and long-term trends have similar amplitudes.
  IV. Episodic, large amplitude outbursts with intervening periods of inactivity.

Figure 1 shows some long-term light curves of Blazars which were compiled at RHO and kindly provided before publication to the author by the RHO observers: A. G. Smith, R. J. Leacock, P.P. Gombola, S. Clements, and Damo Nair.

IV. ENERGETICS OF OUTBURSTS

Corrections for line emission and galactic absorption must be made in order to convert the observed magnitudes to continuum flux. The corrected continuum flux is

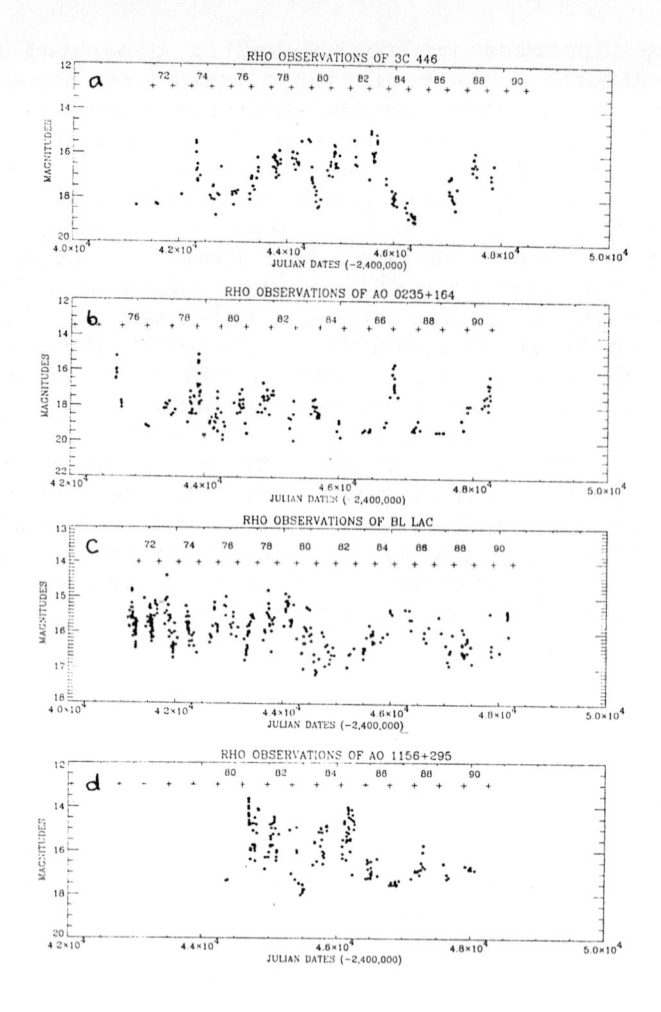

FIGURE 1. Long-term RHO light curves of four Blazars discussed in the text.

then converted to monochromatic luminosity at the source, assuming a cosmological model and values for the redshift of the source, Hubble's constant, and qo. The monochromatic luminosity is converted to bolometric luminosity over the desired spectral range by integrating under the spectral distribution. A power law distribution between 0.3 and 3 microns is usually assumed because this region is observed to vary as a whole during these high-amplitude outbursts. If significant spectral curvature is present, a more complicated model such as a broken power law or a second order polynomial must be used to calculate the bolometric luminosity. Changes in the spectral index are sometimes observed, but do not correlate with either intensity or outburst duration (Webb et al. 1990 and Brown et al. 1989). The mean spectral index is used to approximate the actual spectral distribution when insufficient spectral information is available.

We now integrate under the light curve to get the total energy emitted during the outburst. We should also note that, depending on how well the optical light curve is sampled, this value may be an underestimate or overestimate the actual energy emitted. So that we calculate only the energy involved in the outburst, we subtract off a base luminosity which can only be estimated from the long-term light curve of the source. In the next section we discuss the application of this method to several Blazars.

V. OBSERVATIONS

Pollock (1982) isolated sixty-four discrete outbursts seen in the long-term light curves of 24 separate sources. The optical data consisted primarily of observations made at RHO over a ten-year span; however, additional data and spectral information were taken from the literature to supplement the RHO observations. Pollock found that the maximum luminosities ranged from 2.08×10^{43} ergs/sec (NGC 4151) to 1.23×10^{49} ergs/sec (AO 0235+164) and outburst energies ranged from 3.5×10^{50} (NGC 4151) to 2.8×10^{55} ergs (3C 446), assuming H_o is 100 Km/sec/Mpc and the value of q_o is 0.0. Only four sources in his sample, 3C 279, PKS 1156+295, 3C 446, and AO 0235+164, exhibited outbursts where the maximum luminosity exceeded 1.0×10^{48} ergs/sec.

The 1981 outburst of PKS 1156+295 was one of the most actively observed outbursts to date, as reported in Wills et al. (1981) and Bregman et al. (1981). This source has a measured emission feature attributed to MgII at a redshift of z=0.729 and shows a strong, variable linearly polarized non-thermal continuum. VLBI observations show that most of the radio emission comes from an unresolved core with angular size of less than 2 mas at 13 cm (Wills et al. 1981). Figure 2 shows the 1981 outburst of PKS 1156+295 as observed in the optical B band.

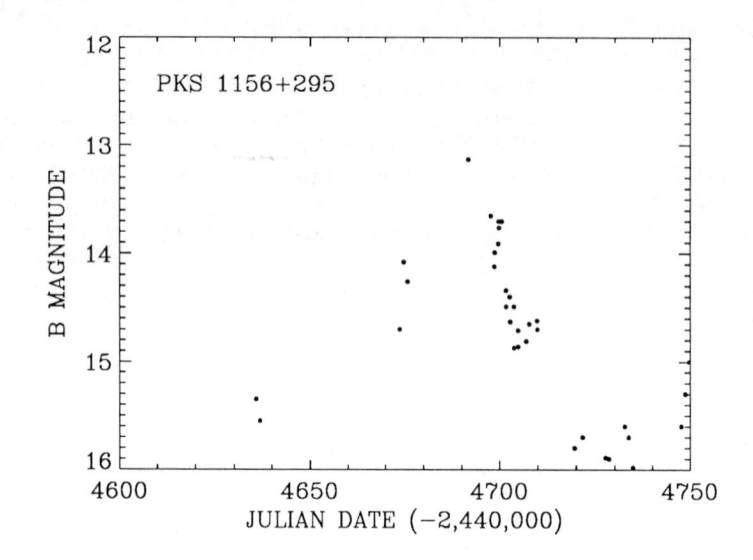

FIGURE 2. The 1981 optical outburst of 1156+295.

In the years before the 1981 outburst, Pollock found that 1156+295 was rarely seen brighter than 16th magnitude on Harvard archive monitoring plates, although this field was observed as many as eight times a year as early as 1926 and more frequently between 1927 and 1951. On March 28, 1981, 1156+295 attained a maximum brightness of B=13.15. Wills et al. found that the outburst, if interpreted in terms of the magnetic storage model (Shields and Wheeler 1976), implied a viscosity parameter less than 1.0×10^{-4} in the disk.

We can significantly reduce the luminosity if we assume the radiation was relativistically beamed in our direction. The approximate dependence of the luminosity and total energy on the amount of Doppler boosting is:

$$L = D^{-(4+a)} L', \qquad (3)$$

and

$$E = D^{-(3+a)} E'. \qquad (4)$$

In equations (3) and (4), a is the spectral index and D is the Doppler boosting factor. The results of luminosity and energy calculations for the 1981 outburst of 1156+295 are listed in Table 1. The maximum observed luminosity in ergs/sec, observed energy in ergs, and the total energy in ergs are listed for three different values of the Doppler parameter. Blandford and McKee (1977) analyzed shock models in which ninety percent of the total energy contained in the shock is emitted as radiation. The total energy listed in Table I was calculated assuming that shock waves are responsible for the flux variation and that only ten percent of the total energy is unobservable.

TABLE I.
ENERGETICS OF THE 1981 FLARE OF PKS 1156+295

$q_o = 0.0$, $H_o = 55$ Km/sec/Mpc

D	1	2	5
L_{max}	1.7×10^{48}	3.7×10^{46}	2.4×10^{44}
E_{burst}	1.3×10^{54}	5.7×10^{52}	9.3×10^{50}
E_{total}	1.4×10^{54}	6.3×10^{52}	1.0×10^{51}

The BL Lacertid AO 0235+164 has shown four separate energetic outbursts since 1975. The long-term light curve of AO 0235 was shown in Figure 1-b. Figure 3 shows the RHO observations of each of the four outbursts. Most of the flaring activity takes place on the timescale of 100 to 200 days. The times between bursts vary: 1,210 days between the

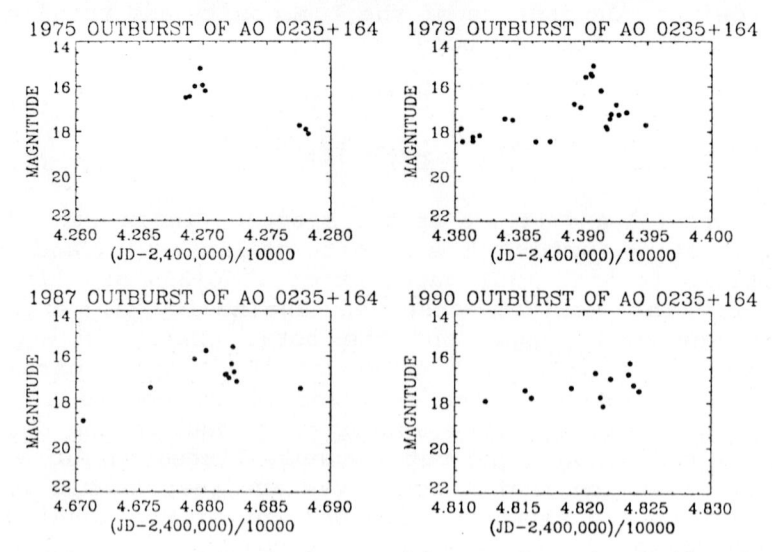

FIGURE 3. Optical observations of each of the four major outbursts of AO 0235+164.

1975 and 1979 bursts, 2,894 days between the 1979 and 1987 bursts, and 1,434 days between the 1987 and 1990 bursts. No strict periodicity is seen, although it is curious that the time between the 1979 burst and the 1987 burst is roughly twice the time between the 1987 and 1990 bursts. The sampling is such that an outburst might have been missed in early 1983, 1,430 days before the 1987 burst.

It is possible that the energy emitted during the outbursts is stored over a timescale of ~1400 days and then released explosively in a much shorter period of time, ~100 to 200 days. The emission region size is constrained by equation 1 to be less than 5.1×10^{15} cm. The energetics of the 1987 outburst are shown in Table 2. The luminosity was calculated by fitting a second order polynomial to the optical-near IR spectrum between 0.36 and 3.5 microns. The duration of the burst in the source rest frame (no Doppler boosting) was 101 days and a base luminosity of 3×10^{46} ergs/sec was assumed.

TABLE II.
ENERGETICS OF THE 1987 FLARE OF AO 0235+164

$q_o = 0.0, \quad H_o = 55$ Km/sec/Mpc

D	1	2	5
L_{max}	6.4×10^{47}	1.0×10^{46}	4.1×10^{43}
E_{burst}	2.5×10^{54}	7.8×10^{52}	8.0×10^{50}
E_{total}	2.8×10^{54}	8.5×10^{52}	8.8×10^{50}

Application of the magnetic storage model to the 1979 outburst indicated that there was too little time between the 1975 and 1979 outbursts to store the needed energy if realistic viscosity parameters were used. Relativistic beaming with Doppler factors of 1.3 or more eliminates the energy storage problem for this outburst (Webb and Smith 1989). Synchrotron-Self-Compton model fits to the x-ray-radio spectrum of AO 0235 yield similar values of the Doppler parameter.

The OVV quasar 3C 279 has also shown well-documented, energetic outbursts in recent years. The 1938 outburst data was gleaned from Harvard patrol plates and an outburst in 1988 was observed by the Florida group along with collaborators at other observatories (Webb *et al.* 1990). Extensive UBVRI observations during the 1988 outburst showed that the spectral index varied between -0.55 and -2.05 with a mean of -1.3 between 0.4 and 0.7 microns. The spectral index did not appear to be correlated with either flux level or time. Tables 3 and 4 show the luminosities and energies for the 1938 and the 1988 outbursts of 3C 279, respectively. Blazar 3C 279 was radiating nearly 25 times its Eddington luminosity in 1938 and 14 times its Eddington luminosity in 1988 if we assume the mass of the central black hole is

8.91×10^8 Solar masses (Padovani 1989). Figure 4 shows the morphology of the 1988 outburst. The dotted line connecting the points outline the curve under which the energy was calculated; it is not intended to delineate the actual brightness change of the source.

TABLE III.
ENERGETICS OF THE 1938 FLARE OF 3C 279

$q_o = 0.0$, $H_o = 55$ Km/sec/Mpc

D	1	2	5
L_{max}	2.8×10^{48}	8.2×10^{46}	7.6×10^{44}
E_{burst}	2.1×10^{55}	1.2×10^{54}	2.8×10^{52}
E_{total}	2.3×10^{55}	1.3×10^{54}	3.1×10^{52}

TABLE IV.
ENERGETICS OF THE 1988 FLARE OF 3C 279

$q_o = 0.0$, $H_o = 55$ Km/sec/Mpc

D	1	2	5
L_{max}	1.6×10^{48}	4.7×10^{46}	4.3×10^{44}
E_{burst}	1.3×10^{55}	7.5×10^{53}	1.7×10^{52}
E_{total}	1.4×10^{55}	8.2×10^{53}	1.87×10^{52}

VI. DISCUSSION

Blazar models must be able to account for the large maximum luminosities seen in outbursts, the large amounts of energy liberated, and the long-term variability characteristics. In sources which seem to vary continuously, such as those in type I or II variability classes, the energy is released with little storage time in between, while sources in variability class IV store energy over a period of time and release it explosively. Continual monitoring suggests that objects such as BL Lac in Figure 1-c evolve and effectively change the variability class one would assign them based on earlier optical observations.

Accretion models are capable of storing energy over a period of time. Magnetic accretion disk models can be

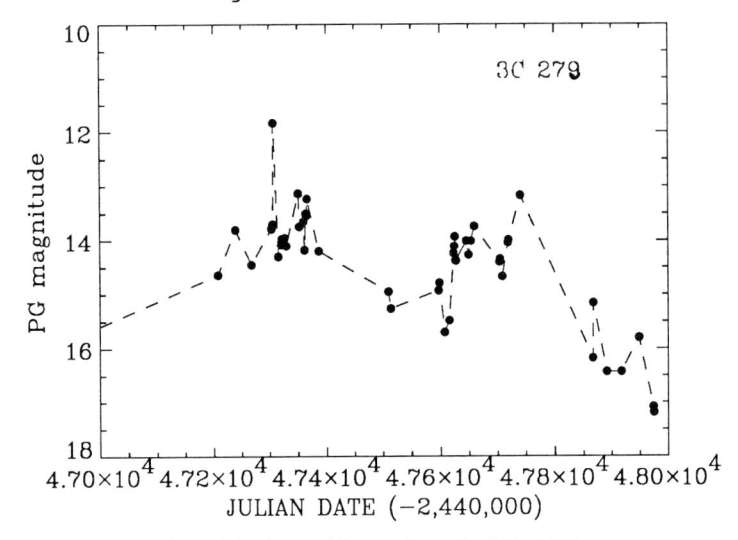

FIGURE 4. The 1988 outburst of 3C 279.

divided into two classes depending on the ratio of the gas pressure to magnetic pressure (Shibata, Tajimi, and Matsumoto 1990). For Solar type disks, $b=p_g/p_m$ is large and the magnetic field lines rise out of the disk creating small flares similar to Solar flares or disk corona. If the ratio p_g/p_m is small, the magnetic field lines are trapped inside the disk and are constantly stretched by differential rotation until they finally erupt from the disk, releasing all of the energy that had been stored in the twisted magnetic field lines. Table 5 shows a possible correspondence between variability types and disk types.

TABLE V.
VARIABILITY TYPE VS. DISK TYPE

VAR. TYPE	DISK TYPE	EXAMPLE
I	High b, strong B field Corona?	1156+295 BL LAC
II	High b, weak B field thermal or viscous instability	3C 120
III	Moderate b, strong B field	3C 446
IV	low b, strong B field cataclysmic disk	0235+164

Magnetic disk models provide a means for energy storage and release, while shocked jet models most easily explain the multifrequency spectra (x-ray through radio). The success of the shocked jet model in explaining radio variability (Hughes, Aller and Aller 1991) makes these models attractive in trying to explain the optical variations as well. Accretion disks store and supply energy, which is subsequently released in the form of shocked jets. An optical flare on the surface of the disk can perhaps inject gas onto the hole at super-Eddington rates, which is subsequently seen as a radio jet expanding with super-luminal velocities. In this picture there could be an appreciable time delay between the optical outburst and the appearance of a radio burst. An alternative is that material is dumped onto the hole and the emerging jet radiates first in the optical and then in the radio region of the spectrum. In the latter case, there should be a more direct correspondence between optical and radio outbursts.

VII. CONCLUSIONS

In explaining Blazar variations, especially high amplitude outbursts, we must account for the release of large amounts of energy in relatively short periods of time. Long-term optical behavior might help us decide which type of accretion disk is present and what the timescales inherent in the disk are.

Further theoretical work should be directed toward coupling disk models which allow for the storage of large amounts of energy and shocked jet models which explain the apparent superluminal motion and the SSC spectrum. Judging from the observational data obtained so far, it is clear that multifrequency monitoring along with optical polarization monitoring at appropriate intervals is necessary to test current models. Multifrequency target of opportunity programs utilizing many observatories are also a high priority. One program utilizing the real-time capabilities of IUE and several ground-based observatories has now been implemented by the authors and some observations of BL Lac during a recent outburst have been obtained.

REFERENCES

Blandford, R. and C. McKee, 1977, M.N.R.A.S., __180__, 343.

Bregman et al., 1981, Ap. J., __274__, 86.

Brown et al., 1989, Ap. J., __340__, 150.

Dent et al., 1974, A. J., __79__, 1232.

Hughes, P., Aller, H. and Aller, M., 1991, This conference.

Padovani, P., 1989, Astron. Astrophys., __209__, 27.

Pollock, J. T., 1982, Phd. Dissertation, Univ. of Florida.

Shibata, K., Tajima, T., and Matsumoto, R., 1990, Ap. J., 350, 295.

Shields, G. and Wheeler, J., 1976, Astrophys. Lett., 17, 69.

Terrell, J., 1967, A. J., 147, 829.

Urry, C. M., 1984, Phd. Dissertation, Univ. of Maryland.

Webb et al., 1990, A. J., 100, 1452.

Webb, J. and Smith, A., Astron. Astrophys., 220, 65.

Wills et al., 1981, Ap. J., 274, 62.2

Optical Spectral Energy Distributions of Variable Sources in Blazars

V. A. HAGEN-THORN, S. G. MARCHENKO, O.V. MIKOLAICHUK

Department of Astronomy, Leningrad State University

ABSTRACT

The results of UBVRI observations of ten blazars (Sitko, Schmidt and Stein 1985; Smith, Balonek, Heckert and Elston 1986; Smith, Balonek, Elston and Heckert 1987) are analyzed. It is shown that in all cases but one, the variability on a time scale of 1 – 2 years can be explained by only one additional source of radiation. All sources have variable fluxes but unchanged spectral distributions. The power-law shape of spectra of these sources is evidence of their synchrotron nature. The change of number of relativistic electrons in the sources is likely to be the cause of variability.

INTRODUCTION

Photometric variability is one of the many manifestations of blazar activity. In order to understand the nature of variable sources it is important to determine their radiation mechanism. The knowledge of continuum spectral energy distributions of the sources is needed for this determination. The data may be obtained from the analysis of multicolor photometric observations of variability.

The method of calculation of color indices of variable sources in a two-component model (galaxy + variable source) has been proposed by Choloniewski [1981]. He noted that for the source with variable flux but unchanged spectral energy distribution the points corresponding to the observed UBV magnitudes must lie in the flux space $\{\phi_U, \phi_B, \phi_V\}$ on straight line, the leading tangents of which determine color indexes U – B and B – V of a variable source. We have developed these ideas and proposed the method of component separation which permits not only to calculate the color indices of variable component but also to find its contribution to the total flux (Hagen-Thorn 1985).

In this paper we shall give the results of the analysis of UBVRI observations of ten blazars. The data are those published by Sitko et al. (1985) and Smith et al. (1986, 1987), the method of analysis is described in Choloniewski (1981) and Hagen-Thorn (1985).

The results for the quasar 3C 345 have been published in Hagen-Thorn and Mikolaichuk (1988) and for nine other objects in Hagen-Thorn, Marchenko and Mikolaichuk (1990). For details, see these papers.

RESULTS

Table 1 contains the list of the objects for which the analysis has been done. In the second and third columns are the adopted values of A_V used for the reduction of the observed fluxes and the time ranges of observations, respectively.

Table 1.
The list of objects and spectral indices of variable components

Object	A_V	The range of observations	$\alpha \pm 2\sigma$
AO 0235+164	0^m26	22.11.82 – 25.01.84	-2.24 ± 0.03 (1982)
			-3.89 ± 0.03 (1983)
PKS 0735+178	0.12	16.02.82 – 28.03.84	-1.50 ± 0.03
OI 090.4	0.07	22.11.82 – 26.03.84	-1.35 ± 0.02
OJ 287	0.06	02.01.82 – 28.03.84	-1.27 ± 0.03
B2 1156+295	0.00	08.01.83 – 14.06.84	-1.87 ± 0.03
B2 1308+32	0.00	16.03.83 – 26.03.84	-1.77 ± 0.08 (-1.45)
OQ 530	0.00	19.04.82 – 14.06.84	-1.50 ± 0.08 (-1.24)
3C 345	0.00	18.02.83 – 14.06.84	-1.71 ± 0.04
BL Lac	1.33	28.05.82 – 13.06.84	-1.94 ± 0.07 (-1.62)
3C 273	0.00	03.01.83 – 13.06.84	-0.61 ± 0.04

The fluxes corrected for interstellar absorption (in millijanskys) are compared in pairs in Figures 1 – 5. In Fig. 1 the errors of individual observations are indicated, in others the values of characteristic errors are given by crosses near the names of spectral bands which are used in the comparison with the fluxes in the V band (in all cases the errors are indicated at the level of 1σ).

One can see that for the objects 3C 345, PKS 0735+178, OI 090.4, OJ 287, B2 1156+29, OQ 530, BL Lac and 3C 273 the symbols corresponding to the observed fluxes are situated along straight lines. Thus the variability on a time scale of 1 – 2 years (Table 1) can be explained in every case by the existence of a single source of variable flux but unchanged spectral energy distribution.

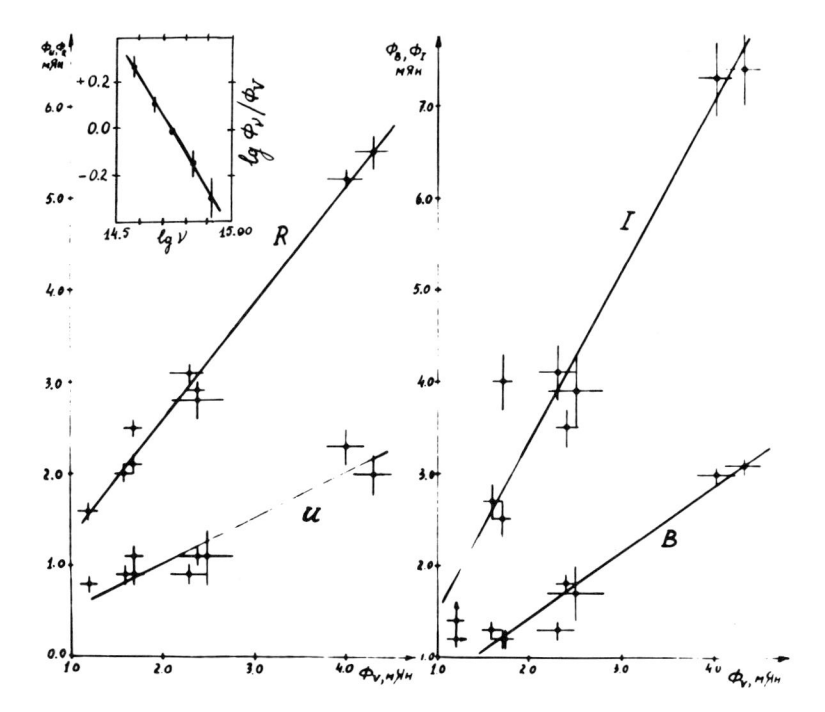

Fig. 1. 3C 345: the comparison of fluxes in the UBVRI bands and the spectrum of the variable source (inset).

In the figure for the object B2 1308+32 numbers 1, 2, 3 are related to consecutive observations showing systematically higher slope than the average for the colors R and especially I (these data correspond to the IR flash described in Sitko, Stein and Schmidt, 1984). Nevertheless in this case one may still represent all data with a single dependence.

But in the case of AO 0235+16 for which dots and triangles represent data for different times (dots for 1983, triangles for 1982) symbols are evidently not situated along a single straight line. Thus the model with a single source of variable flux but unchanged spectral energy distribution is not suitable in all cases of observations. However, at considerable shorter time intervals the model can still be used since the dots and triangles taken separately are situated on straight lines. (The behaviour of AO 0235+16 in January 1983 is analogous to that of B2 1308+32 in the IR flash; thus one may consider that AO 0235+16 also has an IR flash in January 1983).

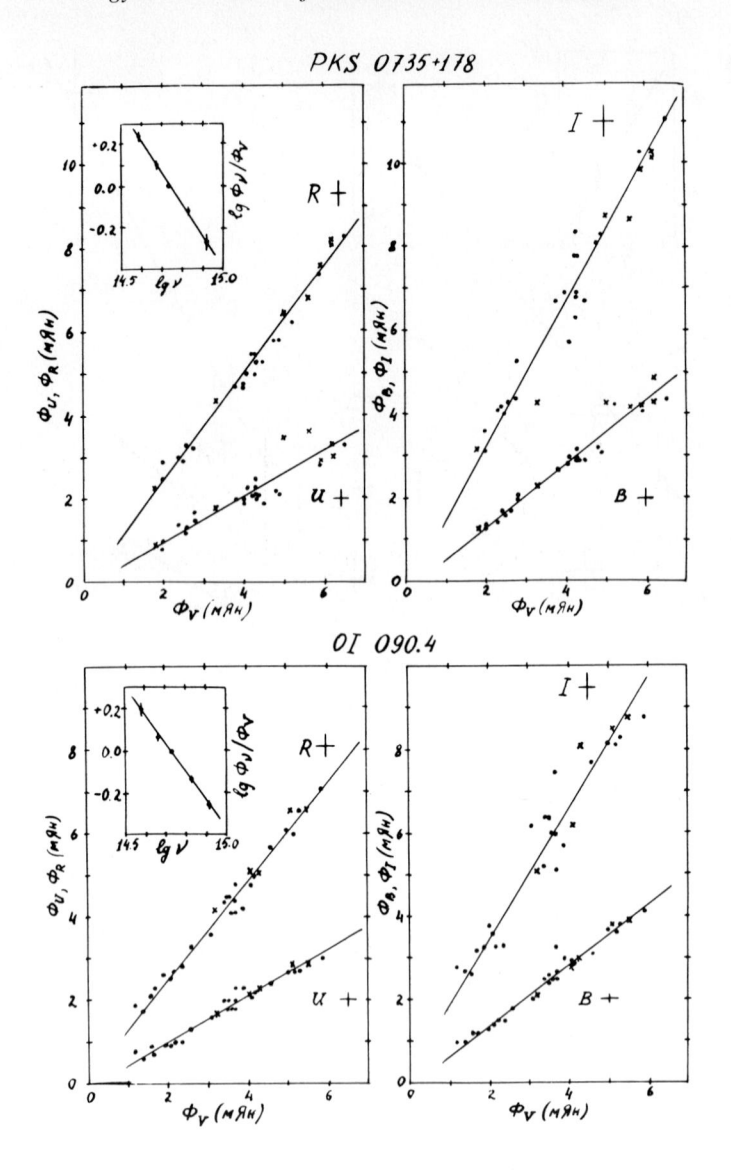

Fig. 2. PKS 0735+178 and OI 090.4: the comparison of fluxes in the UBVRI bands and the spectrum of variable source (inset).

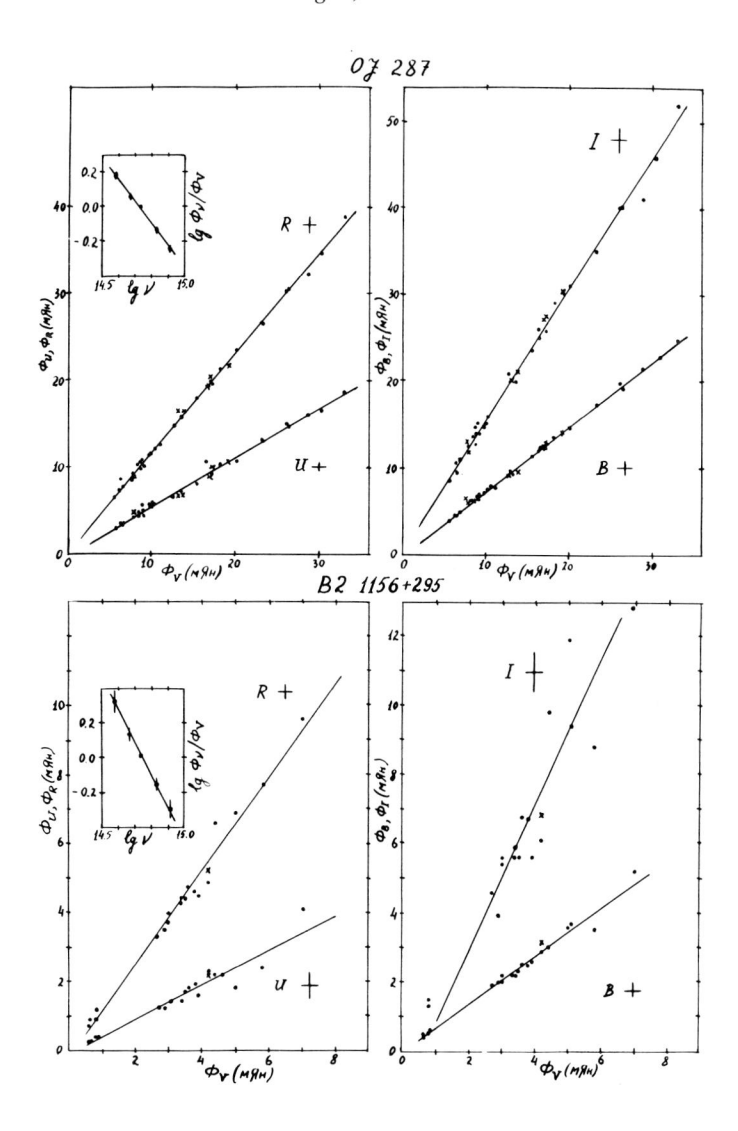

Fig. 3. OJ 287 and B2 1156+295: the comparison of fluxes in the UBVRI bands and the spectrum of variable source (inset).

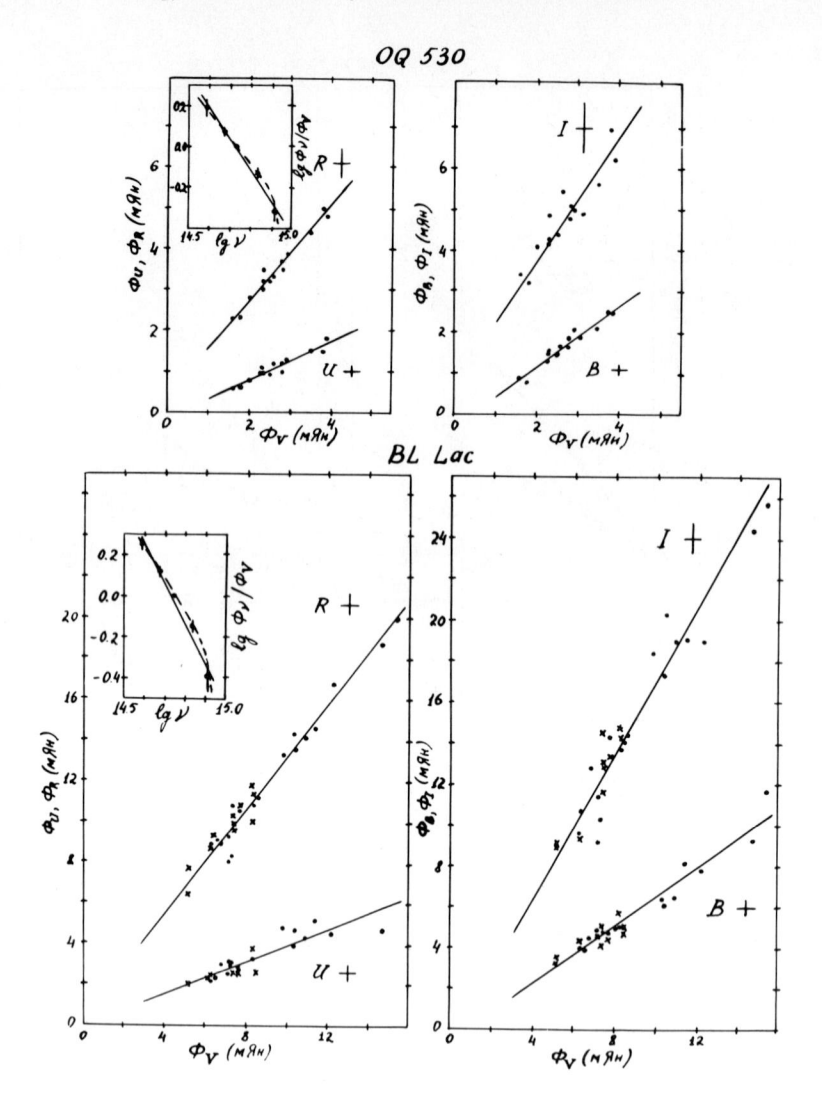

Fig. 4. OQ 530 and BL Lac: the comparison of fluxes in the UBVRI bands and the spectrum of variable source (inset).

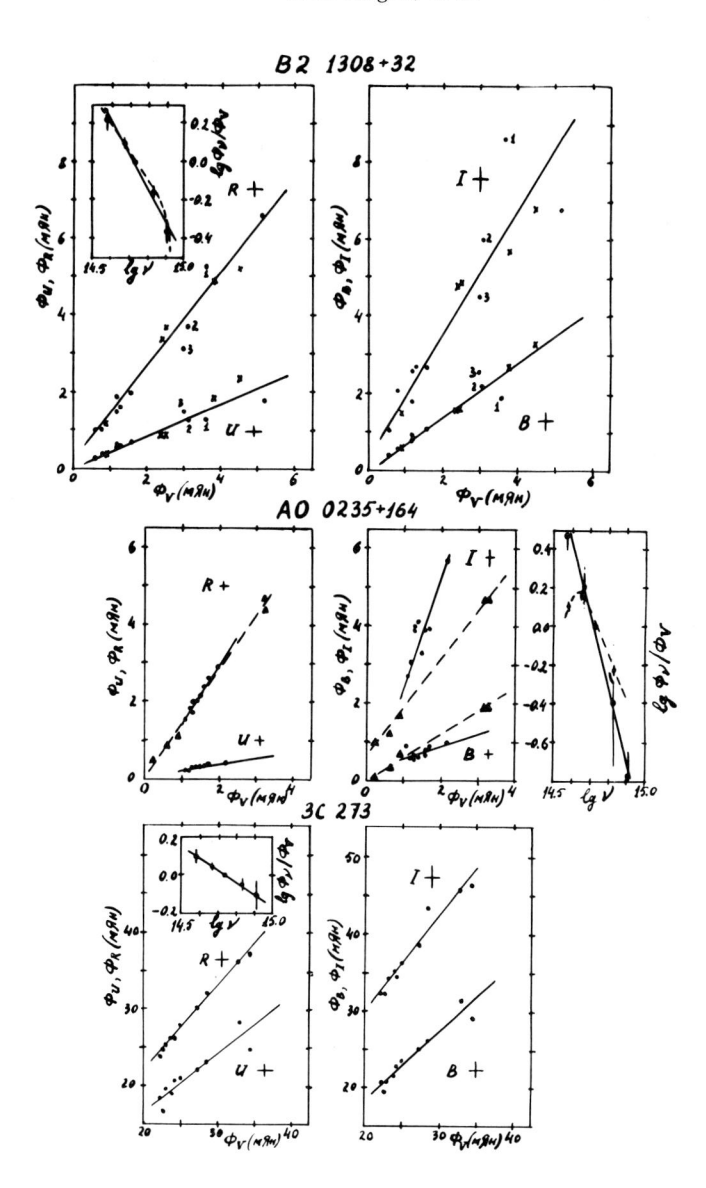

Fig. 5. B2 130: the comparison of fluxes in the UBVRI bands and the spectrum of variable source (inset).

In all cases the straight lines were fitted by the method of orthogonal regression (Hagen-Thorn and Marchenko, 1989) because both ϕ_υ and ϕ_V had accidental errors. The values of slopes give ϕ_υ/ϕ_V for variable sources. Insets in Figures 1 – 5 demonstrate lg ϕ_υ/ϕ_V versus lg υ (the errors are given at the level of 2σ).

Fig. 6 summarizes all the data. One can see that there are practically ideal linear dependencies between lg ϕ_υ/ϕ_V and lg υ for the objects PKS 0735+178, OI 090.4, OJ 287, B2 1156+29, 3C 345 and 3C 273. For OQ 530, BL Lac and B2 1308+326 all points are also situated on the straight lines within the errors of the observations, though there is a tendency for the point of the U band to lie below the line determined by the other points. It is probably an indication of the existence of a high-frequency cutoff. AO 0235+16 has in one case (in the IR flash) a straight spectrum, at the other time the spectrum shows a low-frequency decline.

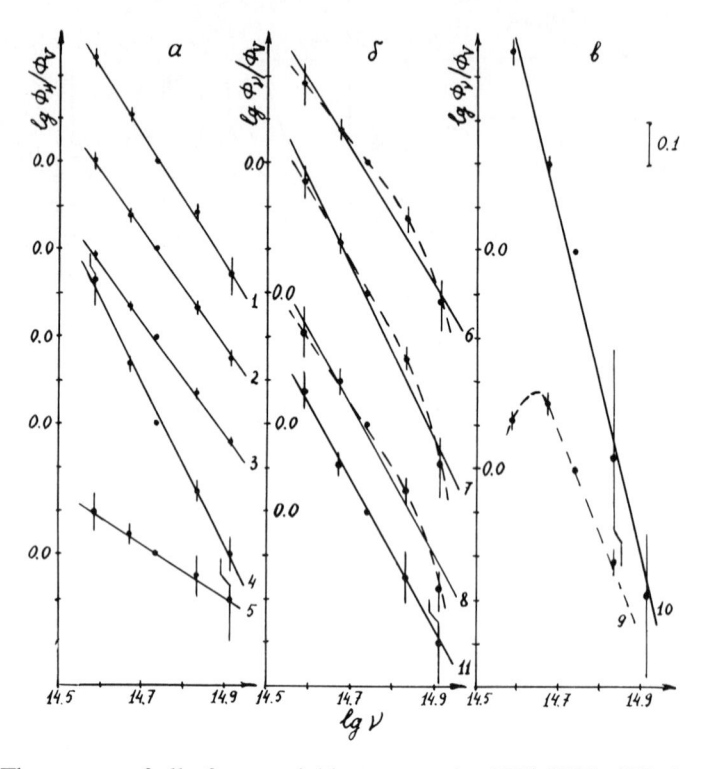

Fig. 6. The spectra of all of our variable sources: 1 – PKS 0735+178; 2 – OI 090.4; 3 – OJ 287; 4 – B2 1156+29; 5 – 3C 273; 6 – OQ 530; 7 – BL Lac; 8 – B2 1308+32; 9 – AO 0235+164 (1982); 10 – AO 0235+164 (1983); 11 – 3C 345.

In all figures solid lines are the results of the fitting procedure based on the least squares method for all five points. Corresponding values of the spectral indices α are given with their errors in the last Column of Table 1. Dashed lines give the spectra which are rectilinear in the BVRI region but have high-frequency cutoff in UV. Spectral indices found from BVRI observations are also placed in last Column of Table 1 (in parentheses). In the case of AO 0235+16 a dashed line gives the spectrum of the variable source observed in November – December 1982. The spectral index is given for BVR region.

DISCUSSION

It can be seen that the spectra are power-law or power-law with high-frequency cutoff practically in all cases. Taking into account the rapid variability and highly variable polarization observed in most cases (Sitko et al. 1985, Smith et al. 1986, Smith et al. 1987) we can conclude that the optical radiation of the variable sources is of synchrotron origin.

As is generally known (see for instance Visvanathan, 1973) the flux density of a homogeneous synchrotron source containing an assembly of relativistic electrons with the energy spectrum $N(E)dE = kE^{-\beta}dE$ is given by

$$\phi_\upsilon = CkH_\perp^{\frac{\beta+1}{2}} \upsilon^{\frac{-\beta-1}{2}} \int_{\upsilon/\upsilon_c}^{\infty} x^{\frac{\beta-3}{2}} F(x) \, dx. \qquad (1)$$

Here C is a constant which depends on the source distance, $H_\perp$ is the orthogonal component of the magnetic field, $F(x)$ is a known function, $\upsilon_c \equiv \text{const.} \; H_\perp E_{max}^2$ is the critical frequency (E_{max} is the energy cutoff in the energy spectrum of electrons). Spectral index α for rectilinear part of spectrum (where $\upsilon << \upsilon_c$) permits one to find the index $\beta = 1-2\alpha$. The region in which $\upsilon \sim \upsilon_c$ is that of the high-frequency cutoff.

On the basis of the above considerations, we may state that the characteristics of a variable source are as follows: a) ϕ_υ is variable in time; b) the dependence $\phi(\upsilon)$ is unchanged; c) the radiation of the source is synchrotron in origin, i.e. Eq. (1) is valid. Let us consider how the observed variability can be explained under these circumstances.

The right hand side of Eq. (1) contains some quantities which in general may vary in time. First it may be the index in the energy spectrum of electrons β. However, in our case the spectral index α does not vary in time, thus β does not vary either. (It should be noted that variable sources have steep spectra; therefore nearly always we have $\beta \geq 4$ meaning that the electron spectra are very steep too). Next in Eq. (1) there are the values of k, $H_\perp$ and υ_c, the latter depending on $H_\perp$ and E_{max}. When the high-frequency cutoff exists the unchanged energy distribution in this region indicates the constancy of ν_c (and most likely the constancy of $H_\perp$ and E_{max} taken separately). Consequently the the only parameter on the right hand side of Eq. (1) which can vary in time is k, that is the flux variability is due to a change in the number of relativistic electrons in the source.

If the spectrum is rectilinear, the flux density in the frequency interval under consideration is not sensitive to the variations of ν_c and consequently of E_{max}. In this case the variability may be connected with changes of both k and $H_\perp$. It should be noted that a change of $H_\perp$ may be caused by both variations of the magnetic field parameters and the motion of electrons through space which contains regions which are unchanged in time but have different magnetic fields. In the latter case the variations of $H_\perp$ may be sufficiently rapid.

It is clear from the above considerations why simultaneous observations of variability in optical as well as, for instance, in far UV are important since in this case one can shift to the region of high-frequency cutoff to rule out (or to confirm) the change of $H_\perp$ as the cause of flux variability. Unfortunately these data are very sparse. A preliminary comparison of the present data for OJ 287 with the results of its quasi-simultaneous observations with IUE (Maraschi, Taglieferri, Tanzi and Treves 1986) shows that at λ 1500 Å there is a hint that the high-frequency cutoff does not depend on time. This fact is evidence in favour of the constancy of $H_\perp$ and indicates that the change of the number of relativistic electrons is the cause of flux variability.

CONCLUDING REMARKS

In conclusion we emphasize the advantage of our technique of variability analysis as compared with the technique usually used in which <u>observed</u> spectral slopes (color indices) are compared with various characteristics.

Sometimes this can lead to invalid conclusions. Smith et al. (1987) concluded that the active component of OQ 530 becomes bluer with increasing flux. Our considerations show that the variable source does not change its colors; the observed decrease of the color indices with increasing flux is due to the reduction of the relative contribution of the redder constant component (underlying galaxy).

It is especially important to use the correct technique of variability analysis in studying the nature of the UV bump in the spectra of active galaxies – an important problem which will be solved in the near future by observations above the Earth's atmosphere.

REFERENCES

Choloniewski, J. 1981, *Acta Astron.* **31**, 293.
Hagen-Thorn, V.A. 1985, *Astrophysics* **22**, 449.
Hagen-Thorn, V.A., Marchenko, S.G. 1989, *Astrophysics* **31**, 231.
Hagen-Thorn, V.A., Marchenko, S.G., Mikolaichuk, O.V. 1990, *Astrophysics* **32** (in press).
Hagen-Thorn, V.A., Mikolaichuk, O.V. 1988, *Astrophysics* **29**, 322.
Maraschi, L., Taglieferri, G., Tanzi, E.G., Treves, A. 1986, *Astrophys. J.* **304**, 637.

Sitko, M.L., Schmidt, G.D., Stein, W.A. 1985, *Astrophys. J. Suppl. Ser.* **59**, 323.
Sitko, M.L., Stein, W.A., Schmidt, G.D. 1984, *Astrophys. J.* **282**, 29.
Smith, P.S., Balonek, T.J., Elston, R., Heckert, P.A. 1987, *Astrophys. J. Suppl. Ser.* **64**, 459.
Smith, P.S., Balonek, T.J., Heckert, P.A., Elston, R. 1986, *Astrophys. J.* **305**, 484.
Visvanathan, N. 1973, *Astrophys. J.* **185**, 145.

Redshift-dependent Optical Variability of Quasars: an Interpretation in Terms of Spectral Changes

E. GIALLONGO[1], D. TREVESE[2], F. VAGNETTI[3]

[1] Osservatorio Astronomico di Roma, I-00040 Monteporzio, Italy
[2] Istituto Astronomico, Università di Roma "La Sapienza", I-00161 Roma, Italy
[3] Astrofisica, Dipartimento di Fisica, II Università di Roma, I-00173 Roma, Italy

1. CORRELATION ANALYSIS OF QSO SAMPLES

The analysis of long term optical variability of AGNs provides important information on their structure and evolution. However, the correlation of variability with absolute magnitude or redshift is affected by the criteria adopted for the selection of the sample. In flux limited samples the absolute magnitude is strongly correlated with redshift due to the steepness of the number count vs. flux relation. As a consequence any correlation of variability with absolute magnitude may be a reflection of a correlation with redshift or viceversa. Combining different samples with different magnitude limits it is possible to reduce this effect. We have combined the early bright data of the Braccesi AB sample (Bonoli et al. 1979) ($B < 18.25$) with the variability data (Trevese et al. 1989) of the faint sample of Koo, Kron and Cudworth (1986). In this way the resulting correlation coefficient between absolute magnitude and redshift is reduced to $\lesssim 0.5$.

According to Bonoli et al. (1979) and Trevese and Kron (1990), the behavior of variability can be described by the $r.m.s.$ magnitude differences as a function of the time lag τ between the observations: $\sigma(\tau) = \sqrt{\langle [m(t+\tau) - m(t)]^2 \rangle - \sigma_n^2}$, where σ_n is the variance of the noise and the average is taken over the whole sample. $\sigma(\tau)$ can be parametrized as $\sigma_{max}[1 - \exp(-\tau/T)]$, with $\sigma_{max} \simeq 0.3$ and $T \simeq 0.5$ yr rest-frame. For $z \lesssim 2$, $\sigma(\tau)$ can be considered nearly constant for rest-frame time lags $\gtrsim 2$ yr. Thus we adopted (Giallongo,Trevese and Vagnetti 1990, GTV) as variability index the absolute value of the magnitude difference at a fixed observed time lag = 4 yr. In this way we can neglect the reduction of variability of high redshift objects caused by cosmological time dilation.

We find a positive correlation of variability with redshift: $\rho = 0.25$ significant at the 95% confidence level. No significant correlation with absolute magnitude is found.

Our result appears in contrast with previous analyses of Pica and Smith 1983 and Cristiani et al. 1990, who find higher variability at lower redshifts. However we have shown (GTV) that this is due to the different kinds of variability indicators adopted and to the different time sampling.

2. INTERPRETATION AND CONSEQUENCES

The absence of correlation with absolute magnitude favors models with a single coherent source rather than randomly flaring subunits (see Pica and Smith 1983).

The positive correlation of variability with redshift could reflect a real cosmological evolution of the quasar activity, in the sense of a stronger instability at early epochs. However we suggest a possible explanation in terms of an observational effect. In fact, all quasars are observed in the same optical band, thus we sample higher redshift objects at shorter rest-frame wavelengths. On the other hand, there is evidence (Edelson, Krolik, and Pike 1990) that Seyfert galaxies are more variable in the UV than in the visual, which corresponds to a steepening of the spectral energy distribution during the less luminous phase. A similar evidence has been found for quasars by Kinney, Bohlin and Blades 1990. A positive correlation of variability with z follows. A rough estimate of the effect can be obtained under the simple assumption of a power-law spectrum (log $f_\nu = -\alpha \log \nu + cost$), whose spectral index α decreases when the object becomes brighter in the blue, leaving the flux unchanged at a fixed rest-frame frequency ν^* somewhere in the red part of the spectrum. In this way the amplitude of variability is

$$\delta m = 2.5 \log[(1+z)\nu_{obs}/\nu^*]\delta\alpha. \qquad (1)$$

Using archival IUE data of low z active galactic nuclei, Edelson et al. 1990 studied the change of the spectral index as a function of the UV flux. From their Fig. 2 we estimate $\delta\alpha/\delta m_{UV}$ and from Eq. 1 we obtain that $\lambda^* = c/\nu^*$ is in the range 6000 − 8000 Å. For our average $\delta B \simeq 0.25$ at the mean redshift $z = 1$, a typical $\delta\alpha \simeq 0.2$ is found using again Eq. 1. Differentiating respect to the redshift we obtain $\partial(\delta B)/\partial z = 1.086\,\delta\alpha/(1+z)$, corresponding to a slope $\simeq 0.1$ for the linear regression of δB versus redshift, consistent with our observed value of 0.07. Our result is also consistent with the color variations found by Hawkins and Woltjer 1985.

The picture emerging from the above analysis is that optically-selected quasars have the same average intrinsic variability at all redshifts and luminosities. Within this framework the quasar-to-quasar dispersion of the spectral index represents an upper limit to the average spectral variability of individual objects. An estimate of the r.m.s. dispersion of the spectral index can be found e.g. in Sargent, Steidel and Boksemberg 1989, who report a value of 0.3 for a sample of 59 high-z quasars. This value is only slightly higher than the one associated with variability. It appears attractive to consider the extreme possibility that the light curves of all the optically-selected quasars are statistically the same on long enough time scales (e.g., > 100 yr); of course, this does not prevent the occurrence of different patterns of variability (e.g. flares, flickering or slow trends), which may represent different phases of quasar life, rather than "genetically" different quasar types.

Quasar samples selected on the basis of variability could be biased against low z objects. As discussed in GTV, the importance of this bias is a function of the variability threshold adopted, thus depends on the signal-to-noise ratio of the data, and becomes unimportant in low noise surveys. Furthermore, the correlation could also affect the luminosity function of quasars, in the sense of a larger uncertainty of the apparent magnitude of high redshift objects which may cause an overestimate of the evolution (details in GTV).

REFERENCES

Bonoli, F., Braccesi, A., Federici, L., Zitelli, V., Formiggini, L. 1979, A&AS, 35, 391

Cristiani, S., Vio, R., and Andreani, P. 1990, AJ, 100, 56

Giallongo, E., Trevese, D., and Vagnetti, F. 1990, ApJ, in press (GTV)

Edelson, R. A., Krolik, J. H., and Pike, G. F. 1990, ApJ, in press

Hawkins, M. R. S., and Woltjer, L. 1985, MNRAS, 214, 241

Kinney, A. L., Bohlin, R. C., Blades, J. C., York, D. G. 1990, ApJS, in press

Koo, D. C., Kron, R. G., and Cudworth, K. M. 1986, PASP, 98, 285

Pica, A. J., and Smith, A. G. 1983, ApJ, 272, 11

Sargent, W. L. W., Steidel, C. C. and Boksemberg, A. 1989, ApJS, 69, 703

Trevese, D., and Kron, R. G. 1990, in 'Variability of Active Galactic Nuclei', Atlanta, H. R. Miller and P. J. Wiita Eds., in press

Trevese, D., Pittella, G., Kron, R. G., Koo, D.C., Bershady, M. 1989, AJ, 98, 108

Emission–Line Variability in the Seyfert 1 Galaxy NGC 3516

IGNAZ WANDERS AND ERNST VAN GRONINGEN

Astronomiska observatoriet, Box 515, 75120 Uppsala, Sweden

1 INTRODUCTION

During the first five months of 1990 six Seyfert 1 galaxies and 2 QSOs were monitored spectroscopically with the 2.5m INT and the 4.2m WHT as part of the LAG programme (Lovers of Active Galaxies) at the Observatorio del Roque de los Muchachos on La Palma. Exceptionally good weather in January and February gave us spectra of our target objects every few days. However, in the later months time gaps of up to two weeks arose. Simultaneous photometric monitoring was done on the 1 m JKT. One of our target objects is the Seyfert 1 galaxy NGC 3516. High–resolution spectra of both Hα and Hβ were obtained with a high signal–to–noise ratio. In total we collected 19 Hβ epochs and 21 Hα epochs. Difference spectra show not only symmetric profile variations but also asymmetric ones and a clear feature (dip) on the blue wing of the Hβ profile slowly grows in the course of time. The data reduction was only recently completed, this is a preliminary report on the results of the campaign.

2 DATA REDUCTION

Using the MIDAS data reduction package the spectra were reduced in the standard manner: Bias subtraction and flatfielding; interactive cosmic ray clearing; wavelength calibration using the arc frames taken for every spectrum; flux calibration using the standard stars HD 84937 and BD 262606; night sky subtraction; atmospheric B–band correction (only Hα spectra). In this way, 2D spectra were created. Due to the extended narrow–line emission, two blobs on either side of the nucleus, and the underlying galaxy, the extraction of 1D nuclear spectra is not trivial. Nevertheless, by defining very carefully the spatial boundaries (along the slit) of the *nuclear* narrow–line emission (the spatially unresolved NLR) the contribution to the 1D nuclear spectrum of extended narrow–line emission could be minimized to less than a few percent. This step is critical as the narrow lines are assumed to be nonvarying and are used as an internal flux calibrator, necessary to compare spectra of different epochs.

The stellar contribution of the underlying galaxy to the nuclear spectrum is significant and many stellar absorption features adorn the Hβ spectra. By adding many columns of many 2D spectra avoiding the nucleus and the noise dominated outer parts of the galaxy, a galaxy template of NGC 3516 was extracted. This was then scaled and subtracted from all Hβ spectra such that the Mg II b absorption lines did not leave any residuals. Typical errors in the galaxy subtraction are 10 %, but because the galaxy spectrum is flat, they do not affect the emission line profiles at large scales.

3 INTERNAL FLUX CALIBRATION

The last step consisted of putting the spectra on a common scale, by scaling all spectra to a reference spectrum (4th June). To find the scaling factors between the spectra one assumes that the flux in the narrow-lines (e.g. [O III] and [S II]) is constant in time. Because of different telescopes, spectrographs, detectors and different seeing conditions during the observing nights, slight wavelength shifts and resolution adjustments have to be made. The following semi-automated method was used.

After choosing a relative flux factor, wavelength shift and smoothing factor and taking the difference of a spectrum and the reference spectrum, a parabola was fitted to the small wavelength range around the narrow lines in the difference spectrum, and χ^2 was evaluated on the residuals of the difference spectrum minus the fit. By varying the flux factor, wavelength shift and smoothing factor with small steps, a χ^2 data cube was obtained.

The factors and shift defining the minimum of the χ^2 data cube were taken for calibrating the spectrum relative to the reference spectrum. As can be seen from the difference spectra shown here, this objective method works very well and provides much more reliable results than eye-fitting trials.

4 DISCUSSION

In the course of time a dip develops on the blue wing of the Hβ profile around a wavelength of 4880 Å. This feature is not visible in the spectra before 21st February but present in all spectra of later date, see e.g. the shown spectra of 31st March and 4th June. The latter were taken with different telescopes which convinces us of the reality of the feature. In the galaxy spectrum of NGC 3516 there is no absorption feature at this wavelength, and we conclude that this feature is part of the BLR. There is no sign of this dip in any of the Hα spectra. The far blue wing of the Hβ and Hα profiles is stronger in June than in January, see difference spectra of 4th June and 31st January. This is a slow development in time and takes place in both Hβ and Hα on a time scale of months. Line profile variations are seen on time scales of less than a week.

The "inverse" double-peaked profile of the difference spectrum of 31st March and

31st January is not a residual of the narrow Hβ line but is a real feature. The Hα and Hβ difference spectra clearly resemble each other but are not exactly equal. This implies that the Hα and Hβ lines respond at different times to continuum variations or are distributed in a different way in the BLR.

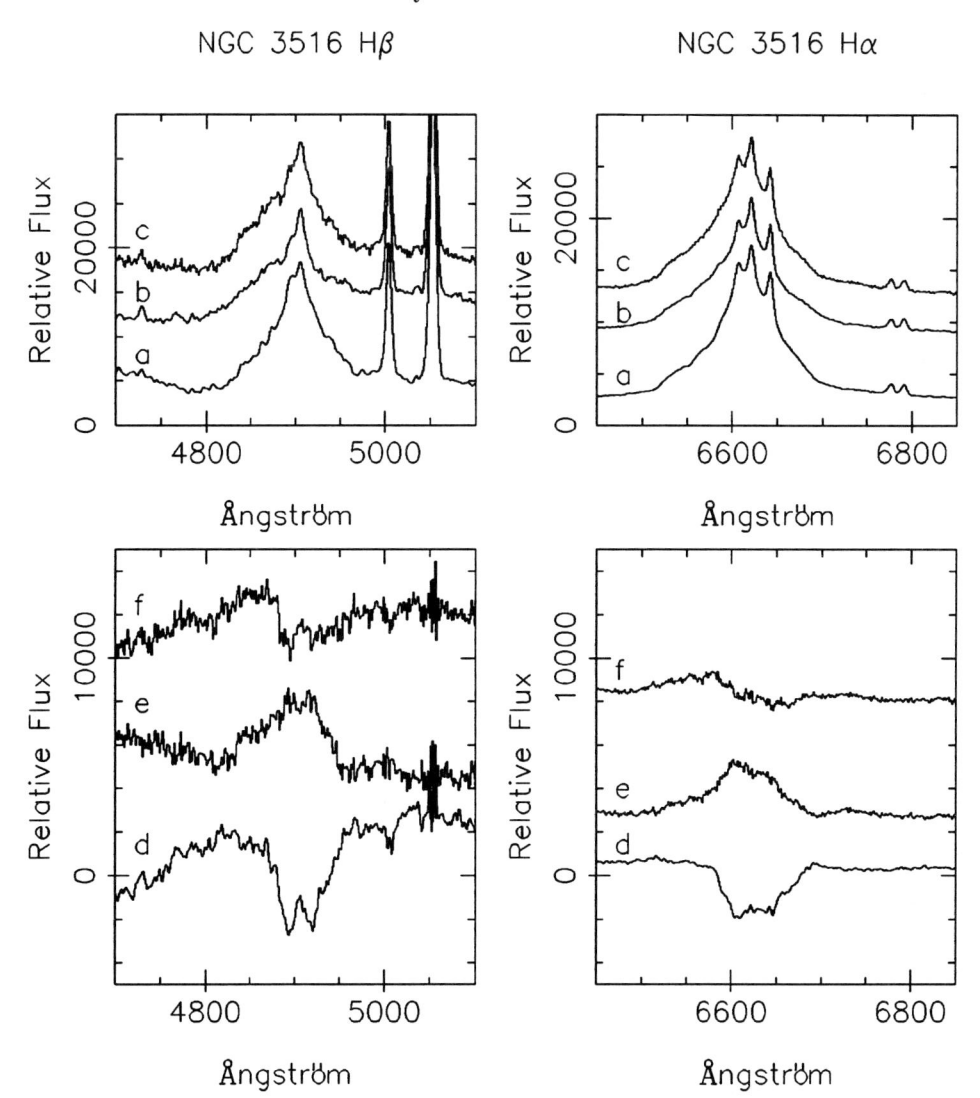

Scaled spectra of a) 31st Jan; b) 31st Mar; c) 4th Jun; and difference spectra d) 31st Mar - 31st Jan; e) 4th Jun - 31st Mar; f) 4th Jun - 31st Jan. Note: all spectra have been given an offset to be able to show them in one plot. Therefore the continuum level shown has no meaning.

Near–infrared observations of hard X–ray selected AGN

Jari Kotilainen

Institute of Astronomy, Cambridge, England

We have undertaken a near–infrared study of a complete hard X–ray (2 – 10 keV) selected sample of AGN (Piccinotti et al. 1982). Most of the 34 sources are Seyfert 1 nuclei, and their near–infrared continuum will be discussed in Kotilainen et al. 1991a,b. The sample includes three blazars (PKS 0548–322, Mrk 501 and PKS 2155–304), which will be considered here. Broadband images of the sample at J (1.25 μm), H (1.65 μm) and K (2.2 μm) filters were obtained between 1987 – 1989 at CTIO, KPNO and UKIRT, with a 62x58 pixel InSb array. The pixel scale is 0.3 – 0.6 "/px. Magnitudes of the three blazars in a 3 " aperture are given in Table 1, columns 2 – 4.

1 VARIABILITY.
PKS2155 was observed twice. The observations are separated by 16 months and it shows clear decrease in brightness in all filters. For the other blazars, the magnitudes are consistent with literature values (Gezari et al. 1987), with PKS0548 being in a bright and Mrk501 in a faint state.

Several of the Seyfert 1 nuclei were observed more than once. Flux variations of more than 30 % between observations separated by 1 – 2 months were detected for 3C120 and the quasar 3C273. For most of the sources, however, the variations were much smaller and on longer timescale. These variations are consistent with being predominantly the time averaged response of the thermal dust component to the varying heating source.

2 DECOMPOSITION.
For nearby (z $\leq$ 0.3) AGN, the emission from the underlying galaxy dilutes the nuclear nonstellar light. This contamination can be very large in the near–infrared, since the energy distribution of a normal galaxy peaks between 1 and 2 μm. To separate the nonstellar and galactic components of the nuclear light, we have extracted radial luminosity profiles of all the sample galaxies in each filter. We have decomposed them using three components:

- an $r^{1/4}$-law bulge
- an exponential disk
- a nuclear (AGN) point source.

These components were then convolved with a gaussian point spread function corresponding to the seeing. The best fitting parameters for the components were determined by minimizing the least–squares fit to the observed profile. Columns 5 – 7 in Table 1 give the resulting nonstellar fractions to the total luminosity in the 3 " aperture. PKS0548 and Mrk501 are galaxy–dominated, whereas PKS2155 (being more luminous and further away) is dominated by the AGN component. Examples of the best fits are given in Figure 1. For the Seyfert 1 nuclei, the average nonstellar fraction increases with wavelength, and for most sources the K band is already dominated by nonstellar light. Also, there is a trend of lower luminosity sources having larger contribution from the host galaxy.

3 SPECTRAL SHAPE

The X–ray–to–near–infrared spectral index and the near–infrared spectral index between J and K bands (after corrcting for the host galaxy) are given in Table 1, columns 8 and 9. PKS0548 has a flat X–IR slope, whereas Mrk501 and PKS2155 have $\alpha \sim -1$. The average slope for the blazars, $\alpha = -0.83 +- 0.19$, is only slightly flatter than the slope for the Sy1s, $\alpha = -0.92 +- 0.11$. This suggests that a single synchrotron powerlaw extends between the two regions, also for Seyfert 1 nuclei. The near–infrared spectral index of Mrk501 and PKS2155 is again close to –1, whereas PKS0548 has a steeper slope. The average index for the blazars is $\alpha = -1.33 +- 0.56$. For the Seyferts, $\alpha = -2.03 +- 0.84$, consistent with blazars being dominated by nonthermal emission and Seyfert 1 nuclei having an additional and probably in many cases dominating thermal dust component to the near–infrared emission. There are, however, large variations between individual Seyferts.

For PKS2155, we have measurements at two brightness levels. At the brighter level, the corrected colours are: J–H = 0.75, H–K = 0.60 and at the fainter level, J–H = 0.83, H–K = 0.59. There is a general tendency of flattening slope with increasing intensity (e.g. Gear *et al.* 1986) in blazars, but only a marginal flattening is seen in PKS2155. On the other hand, for quasars, Neugebauer *et al.* 1989 found that during the flux variations, the near–infrared colours remain roughly constant.

REFERENCES.

Gear,W.K. *et al.*, 1986, *Ap.J.*, **304**, 295.
Gezari,D. *et al.*, 1987, *Catalog of Infrared Observations*, NASA Publications.
Kotilainen,J. *et al.*, 1991a,b, in preparation.
Neugebauer,G. *et al.*, 1989, *A.J.*, **97**, 957.
Piccinotti,G. *et al.*, 1982, *Ap.J.*, **253**, 485.

Table 1.

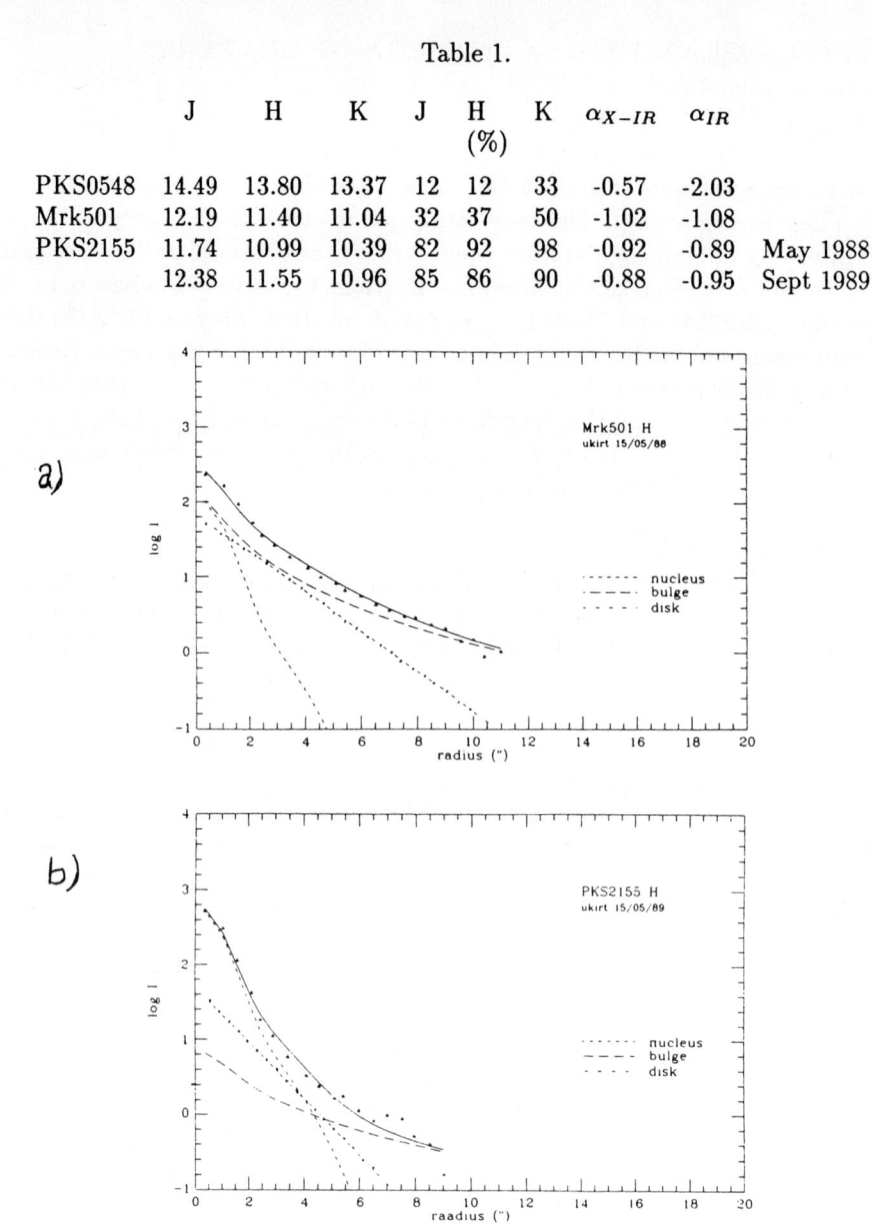

	J	H	K	J	H (%)	K	α_{X-IR}	α_{IR}	
PKS0548	14.49	13.80	13.37	12	12	33	-0.57	-2.03	
Mrk501	12.19	11.40	11.04	32	37	50	-1.02	-1.08	
PKS2155	11.74	10.99	10.39	82	92	98	-0.92	-0.89	May 1988
	12.38	11.55	10.96	85	86	90	-0.88	-0.95	Sept 1989

Figure 1. Examples of the luminosity profile fits in the H band. Triangles represent the observed profile, short dash line the nonstellar nuclear source, long dash line the bulge, dot–dash line the disk and the solid line the total intensity. a) Mrk501. b) PKS2155.

UV AND X-RAY EMISSION OF BLAZARS

Laura Maraschi

Dipartimento di Fisica, Universitá di Milano

Abstract

Observational information on the spectral energy distribution and variability of Blazars in the UV and X-ray bands is reviewed and discussed in the framework of relativistic jet models.

1. Introduction

Despite some recurrent criticism on the term Blazar, used for all objects in which variability and polarization indicate the presence of a non thermal component, I will adopt it in the following. The purpose is that of stressing the likely common nature of the non thermal activity in host objects which may differ in other properties.

BL Lac objects are conventionally defined by the additional requirement, that emission lines are absent or weak. I believe it is the latter specification which confuses the issue: in fact, lines may be weak because they are intrinsically weak in the host object, or because they are washed out by the strong non thermal continuum. Thus a modest power in the non thermal continuum may be sufficient to define a BL Lac in a galaxy with modest nuclear activity, like a Fanaroff-Riley type 1 (FR1) radio galaxy, while a much stronger non thermal power is required to show up as a BL Lac type phaenomenon, in objects with intense emission lines, like FR2 Radio Galaxies and Quasars.

It is important to recall the role of the "search" (or selection) technique in finding the objects. For hystorical reasons the "classical" Blazars mostly derive from radio surveys, with the exception of the Markarian objects, which can be considered optically selected. More recently, the HEAO-1, EINSTEIN and EXOSAT surveys have produced X-ray selected samples of BL Lac objects (Schwartz et al. 1989, Maccacaro et al. 1989, Giommi et al. 1989). It is remarkable that the X-ray selected BL Lacs, though similar in X-ray luminosity to the radio selected ones, are much weaker radio emitters and much closer in distance, (therefore more frequent) than the "classical" radio selected Blazars (Maraschi et al.1986b). Thus X-ray observations have unveiled a population of BL Lac-type objects of which the "classical" radio-selected ones may represent the most extreme examples. The spectral properties of the two classes appear to be substantially different (e.g. Fig 1) and will be reviewed below with emphasis on the UV and X-ray bands (Sections

2 and 3). The few optically selected BL Lac objects known. also appear in X-ray selected samples and share the same spectral properties. Therefore they will be included in the latter group.

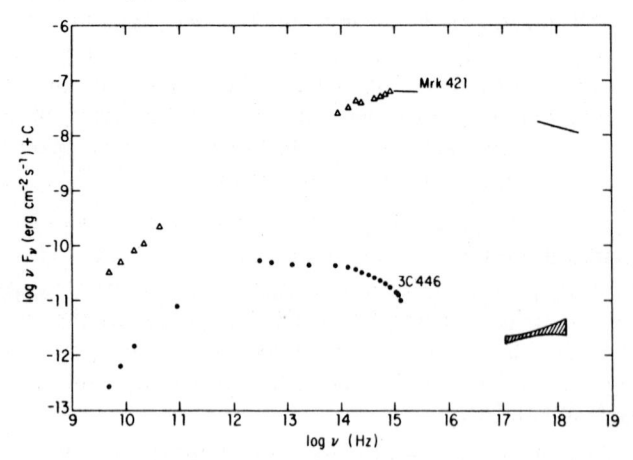

Fig.1 Power per logarithmic bandwidth for MKN 421 (Makino et al. 1987) and 3C446 (Bregman et al. 1986). The first is a typical X-ray selected, the second a typical radio selected Blazar.

The other major role that X-ray observations have played, is that they have shown extremely rapid variability, on timescales of hours in several objects (Feigelson et al. 1986, Morini et al. 1986, Giommi et al. 1990). If one tries to accomodate the radio emission in spatial dimensions compatible with the X-ray variability constraints, allowing for relativistic beaming as originally suggested by Blandford and Rees (1978), the required relativistic corrections become implausibly large. This difficulty led us to develop a relativistic jet model, in which the high frequency emission derives from the innermost region of the jet, while optical and radio frequencies are produced further out (Ghisellini Maraschi and Treves 1985, hereinafter GMT). This model naturally accounts for timescales decreasing with increasing frequencies. The theoretical interpretation discussed in the following will refer to it as a basic scheme. A general review of emission models for BL Lac objects is given by Königl (1989).

The observed spectral and variability properties in the ultraviolet and X-ray bands are summarized in Section 2 and 3 respectively. In Section 4 the properties in the two bands are compared. Section 5 offers a theoretical framework for the understanding of the phaenomenology. Section 6 summarises the results.

2. Spectra and Variability in the Ultraviolet Band.

The ultraviolet observations of Blazars performed with IUE up to 1986 have been reviewed by Bregman Maraschi and Urry (1987). Successive observations have not substantially increased the number of objects observed in the UV, thus the general results reported in that paper remain valid. In particular it is clear that the average spectral index of Blazars in the UV band $<\alpha_{UV}>=1.43$ is steeper than that of Quasars ($<\alpha_{UV}>=0.6$, Richstone and Schmidt 1980).

The UV spectrum of Blazars is on average substantially steeper than that in the IR, $<\alpha_{IR}>=0.94$, but similar to that in the optical, $<\alpha_{OP}>=1.38$. Thus the trend of quasi-continuous steepening of the broad band spectrum from the radio to the IR optical bands seems to be halted in the ultraviolet, though this may in part result from observational selection effects against steep spectrum objects, which tend to be weaker and therefore less observed.

Within the Blazar group, the UV spectrum of X-ray bright objects is strikingly flatter than the group average (Ghisellini et al. 1986). The same is true for the IR and Optical spectra. As a consequence, while the power per logarithmic frequency (νF_ν) of "classical" radioselected BL Lacs peaks in the submillimeter - far IR range ($10^{12} - 10^{13} Hz$), that of X-ray selected BL Lacs peaks in the far UV - soft X-ray range ($\simeq 10^{16} Hz$) (e.g. Fig. 1), which is a remarkable "physical" difference between the two subgroups.

No marked observational distinction in the broad band spectrum characterizes the objects with emission lines with respect to the more properly lineless BL Lac objects. It is worth noting that objects of this type are "rare" and do not appear in X-ray selected samples.

Most of the subsequent observational work was focused on variability and on relations with other wavebands.

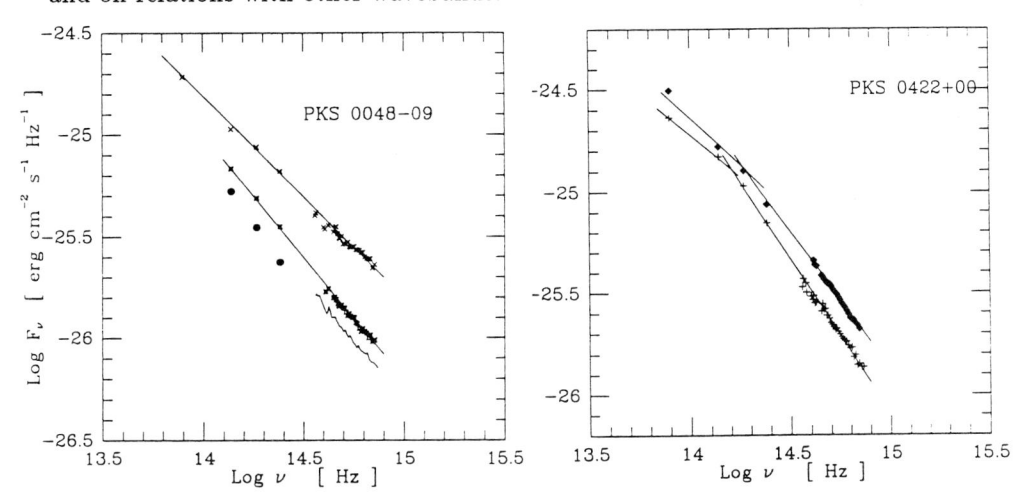

Fig. 2 - Dereddened IR to visible energy distributions of two BL Lac objects in different brightness states (from Tanzi et al. 1989)

An important result derived from the cumulative study of a group of 90 Blazars by Impey and Neugebauer (1988), who showed that the amplitude of variability increases systematically with increasing frequency from the radio to the ultraviolet band. This trend agrees with our findings on specific objects in the IR - optical - UV bands (see Fig. 2, Tanzi et al. 1989) and with the indications derived within the UV band that a flux increase tends to be associated with a hardening of the spectral shape (Maraschi et al. 1986b, Urry et al. 1988).

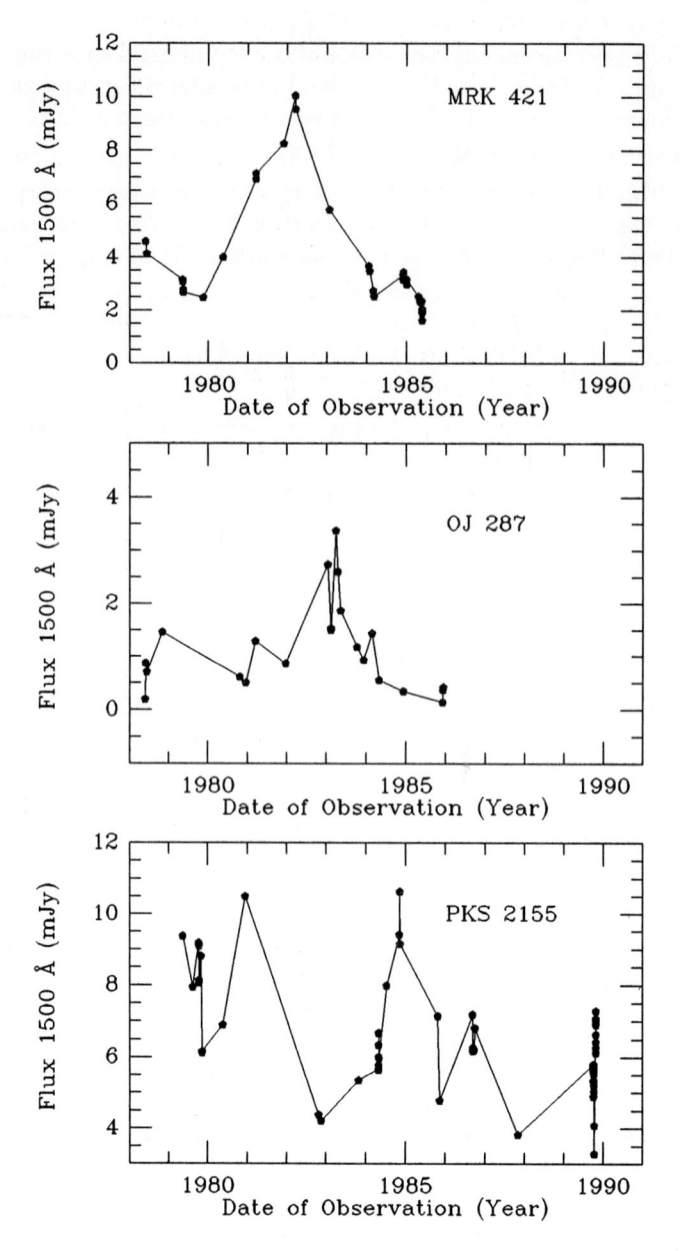

Fig. 3 - UV light curves of the three BL Lac objects most observed with IUE.

A systematic study of variability of Blazars in the UV has been performed by Treves and Girardi (1990) based on the ULDA data bank, complete up to the end of 1986. Three objects have more than 20 UV observations: MK 421 PKS 2155-304 and OJ 287. Their light ccurves are shown in Fig.3. (The data for PKS 2155-304 are complete up to the the end of 1989, Edelson et al. 1991) It is clear that the light curves are still undersampled. Introducing the usual definition of timescale as $\tau = (\Delta t/\Delta F) < F >$ one finds that τ decreases quasi-linearly with the interval between observations Δt.

Much denser sampling is necessary to explore "minimum" timescales and perform Fourier transforms. The recent effort of a group of observers revealed a 12% variation in 5 hours in PKS 2155-304 (Edelson et al 1991) which would yield for this source a minimum observed timescale in the UV of two days.

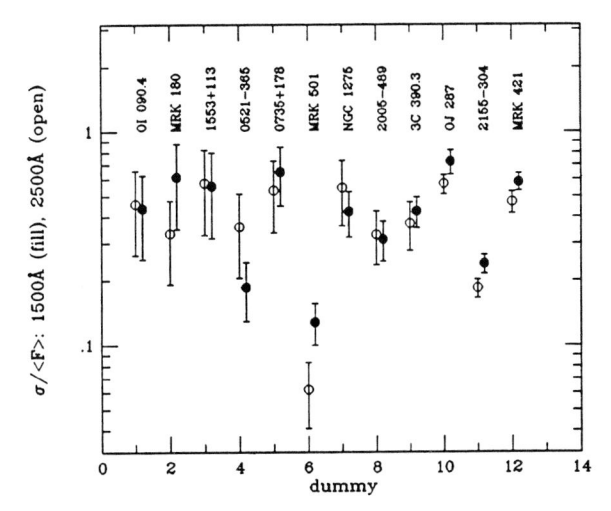

Fig. 4 - Normalized dispersion of UV flux measurements at 1500 A (filled symbols) and 2500 A (open symbols) of Blazars with more than 4 exposures in each IUE cameras.

The normalized dispersion of UV flux measurements, for the 12 objects which have at least 4 exposures in each IUE camera, in the ULDA data base, is reported in Fig. 4. The objects are listed in order of increasing number of observations and the errors have been estimated with simulations and with the bootstrap technique (Treves and Girardi 1990). For the three objects with the largest number of observations there is indication that the average amplitude is larger in the short wavelength band in agreement with the overall trend found by Impey and Neugebauer (1988).

3. X - Ray Observations

Spectral results concerning individual Blazars have been derived from the HEAO-1, EINSTEIN, EXOSAT and GINGA missions. On the whole not more than 20 objects have individual spectra measured by one of the above experiments. It should be kept in mind that the brightest objects in X-rays, for which the best spectral information is available, are radio weak and may not be representative of the entire group of Blazars.

A collection of results on BL Lacs from EXOSAT (12 objects) gives a mean spectral slope $<\alpha_X>= 1.5$ for a single power law fit in the 0.2-8 keV band (Maraschi and Maccagni 1988). while a collection of EINSTEIN IPC spectra (13 objects) gives $<\alpha_X>= 1.1$ in the 0.2-4 keV band (Madejski and Schwartz 1989). In fact from a systematic comparison with the EINSTEIN MPC data for the same objects, the above authors derive evidence that the spectra of BL Lac objects gradually steepen in the 1-3 keV range, as observed in some individual objects (Barr Giommi and Maccagni 1988) and inferred from a cumulative analysis of the X-ray spectra of 17 BL Lac objects measured by EXOSAT (Barr et al. 1989).

Worral and Wilkes (1990) perform an interesting global comparison of the X-ray spectra of three samples: BL Lac objects (24 radio selected + 6 X-ray selected), Highly Polarized Quasars (12) and Flat Radio Spectrum Quasars (19). On the basis of EINSTEIN IPC data the latter two groups have the same average spectral index, $<\alpha_X>= 0.5$, while BL Lac objects, both radio selected and X-ray selcted are significantly steeper, with $<\alpha_X>\geq 1$.

In a higher energy range, 2-20 keV, GINGA results show that MK 421, PKS 2155-304 and H0323+022 have steep spectra $<\alpha_X>\geq 1.7$ in the 2-20 keV range, while 3C 371, BL Lac, 3C 279 and 3C 446 have relatively flat spectra $<\alpha_X>\leq 1$ in the same range (Ohashi 1989). Thus, it appears that there is a significant difference in the X-ray spectral shape of X-ray selected BL Lac objects ($<\alpha_X>\geq 1$) and HPQs and OVVs ($<\alpha_X>\leq 1$, as at lower energies, with radio selected BL Lacs (3C 371, BL Lac) having intermediate properties.

A systematic analysis of the variability properties of BL Lac objects in the EXOSAT data base has been carried out by Giommi et al. (1990). The sample comprises 36 objects, of which 19 are detected in both, the Low Energy (LE, 0.2-1 keV) and Medium Energy (ME, 1-8 keV) experiments. Practically all those which were bright enough to allow detection of 20% variability and were observed repeatedly were found to be variable. Variability within a single observation, that is on time-scales of hours or less, was detected in 8 objects. In 4 of them (ON 325, E1402.3+0416, PKS 2005-489, PKS 2155-304) the amplitude was more than a factor 2 in few hours. The variability function for the Exosat sample indicates that the amplitude tends to be larger in the ME than in the LE band (Maccagni et al. 1989).

Using the ME/LE hardness ratio (averaged over individual observations) to characterise the spectral shape Giommi et al.(1990) find that in all objects which could be studied the hardness ratio increases with increasing ME intensity, that is the spectra harden with increasing intensity. The four best documented cases are shown in Fig. 5. This behaviour was previously noted in individual sources: PKS 2155-304 (Morini et al. 1986), MKN 421 (George, Warwick and Bromage

1988), MKN 501, MKN 180, 1218+304 (George, Warwick and McHardy 1989). Although the detailed spectral behaviour may be complex (Maraschi 1988), the general trend seems well established, at least for X-ray bright (radio-weak) objects.

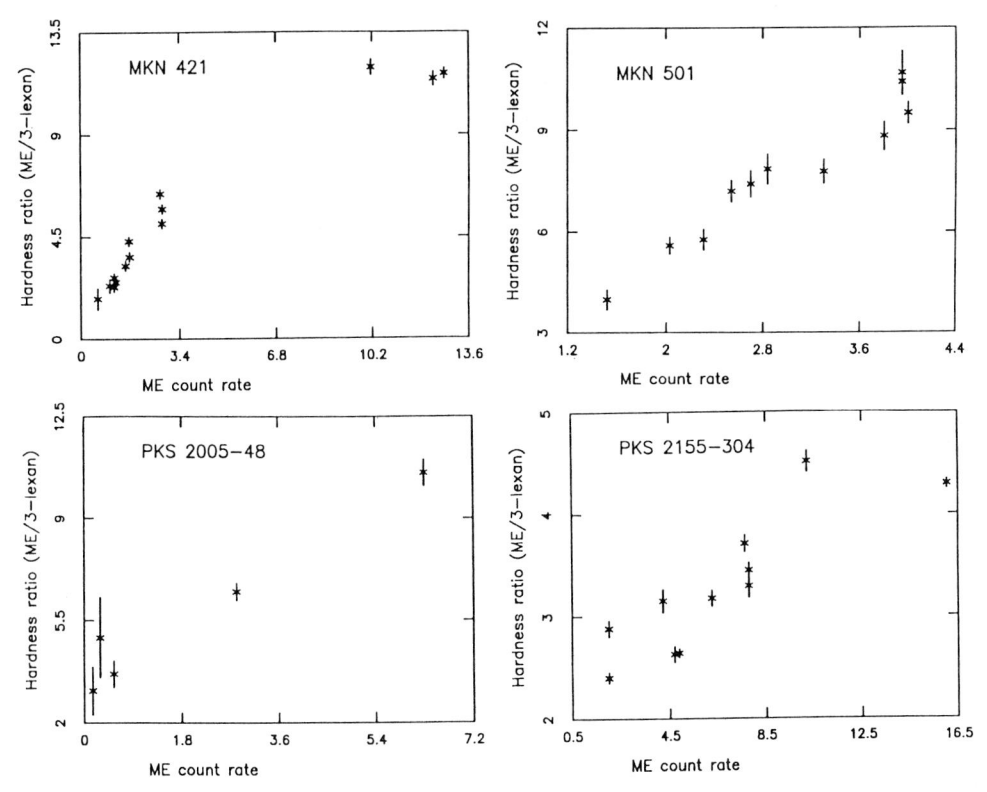

Fig. 5 - Hardness ratios as a function of intensity in the ME EXOSAT band for four BL Lac objects (from Giommi etal. 1990)

The short term X-ray variability of PKS 2155-304 allowed to perform a power spectrum and cross correlation analysis. The power spectra are dominated by long term linear trends. The residual noise is very "red" with slope close to -2. No evidence of lags is found within the 0.1 to 6 keV range (Tagliaferri et al. 1991).

4. Coordinated UV X-ray Observations.

The relation between the UV and X-ray emission is an important difference between Radio selected and X-ray selected Blazars. For the first group, the X-ray to UV flux ratio is smaller than for the second, as measured by the UV to X ray spectral index α_{UX} (Ghisellini et al. 1986) or by other types of analyses (Worral et al. 1989). On the other hand, for radio selected objects the spectral index in the X-ray band is flatter than the extrapolation from the UV, while for X-ray selected objects it is close to, or steeper than the extrapolation. Thus the X-ray emission of the first group stands out as a separate component with respect to the UV, while for the second group it appears as a smooth extension of the UV spectrum to higher energies. This point is illustrated in Fig.1 for two representative objects.

The above conclusions are derived from mostly non simultaneous observations. Quasisimultaneous observations in the UV and X-ray bands of individual objects do not alter this picture (Urry 1988). A great potential of quasisimultaneous observations is the study of correlation between variability in the two bands, but the obvious difficulty of obtaining large sets of coordinated data has up to now allowed only limited results.

Campaigns of coordinated UV and X-ray observations have been carried out for PKS 2155-304 and MKN 421, which belong to the X-ray selected group. About 10 quasisimultaneous observations per object were obtained, with spacings ranging from days to months (Treves et al. 1989, George Warwick and Bromage 1988). The data allow a quantitative comparison between the variability in the two bands, confirming the higher amplitude and shorter timescales in the X-ray band with respect to that in the UV. This result extends to higher frequencies (for two objects), the trend found by Impey and Neugebauer(1988) on a large group of Blazars.

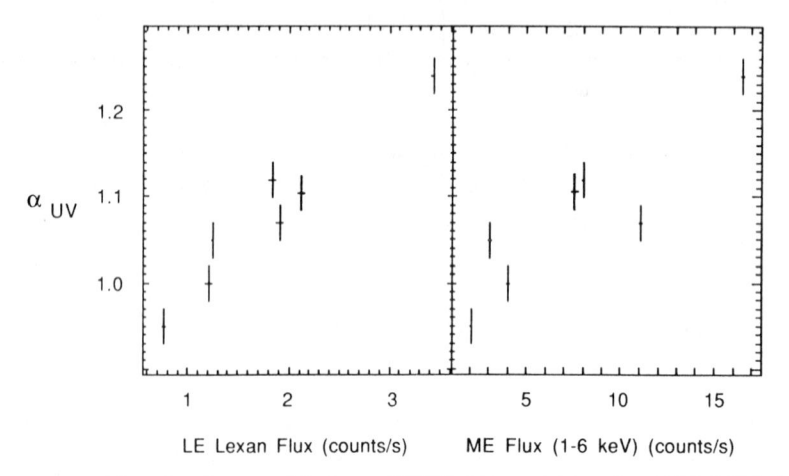

Fig. 6 - Spectral index of PKS 2155-304 in the 1200-3000 A range as vs. X-ray intensity in the LE and ME EXOSAT bands (from Treves et al. 1989).

The correlation in time between X-ray and UV fluxes is rather poor, however, in PKS 2155-304, the slope of the UV spectrum is significantly correlated with the X-ray flux (Fig. 6). This is a remarkable result, indicating that the UV and X-ray emissions are connected, but not by a simple flux correlation. Extensive observational campaigns are needed to better quantify and characterize the connection, using powerful methods of time series analysis. A unique opportunity is offered by the joint operation of IUE and ROSAT. Such a program is being proposed by a wide international collaboration which includes the author.

5. Theoretical framework

The success of the inhomogeneous jet model by GMT is based on the fact that it incorporates in a natural way the decrease of variability timescales with increasing frequency, by attributing the high frequency emission to the most compact region of the jet.

The recognition that Radio selected BL Lac objects can be thought of as a minority of radio strong objects, within the larger population of radio weak X-ray selected objects, led us to a further step i.e. to assume that the X-ray emission is more isotropic (or less beamed) than the radio emission. One can then interpret the different broad band energy distributions as due to different viewing angles to the jet axis and at the same time explain the relative paucity of radio strong objects in X-ray selected samples (Maraschi 1986b). The wider beaming angle of the X-ray emission may result either from a lower value of the bulk velocity near the jet core (accelerating jet) or (and) from a wider spread in the velocity directions (increasing collimation). The first case was discussed in detail by Ghisellini and Maraschi (1989), the second is presently under consideration (Celotti Ghisellini and Maraschi, in preparation).

Irrespectively of the cause, if the beaming cone at the base of the jet is wider than at larger distance, an observer at intermediate angles will receive radiation from the inner regions, but only little from the outer regions whose radiation is concentrated in a narrower cone. In the inner region, X-rays are produced via synchrotron radiation: this component should have a steep spectrum because of the short radiative lifetimes and of high energy cut offs in the particle spectra. The X-ray emission of X-ray selected objects could thus be interpreted as synchrotron emission from the inner region of the jet.

A second X-ray component is expected from the Inverse Compton mechanism in the outer regions of the jet. This component should share the properties of the outer regions, thus should be beamed in a narrower cone and have flat spectrum. In objects seen at small angles, radiation from the outer parts of the jet should dominate, hence the strong power at low frequencies (radio to IR) should be associated with a dominant IC component in the X-ray band, with flat spectrum and a large excess with respect to the UV emission. The variability properties of the IC X-ray component are predicted to be different from those of the synchrotron X-ray component in that they should be closely related to variability at lower frequency. This is a general prediction of the whole scheme.

To be more specific, PKS 2155-304 and MK 421 should correspond to intermediate viewing angles, dominant synchrotron component in X-rays varying on timescales much shorter than the IR to radio emission, which are not relativistically enhanced. At the opposite end, 3C 345, 3C 446, BL Lac, 3C 279 are all radio loud objects with flat spectrum X-rays. In these cases the viewing angle should be within the narrow beaming cone of the outer jet regions and the scheme predicts a strong IC X-ray component, with correlated variations between X-rays and low frequency radiation. The available data for these sources (Bregman et al. 1986, 1988, 1990,

Makino et al. 1989) are consistent with the scheme.

A quantitative discussion of variability within the GMT type jet model can be found in Celotti Maraschi and Treves (1991). A perturbation of fixed amplitude is supposed to travel from the inner to the outer regions of the jet and therefore affects different energy bands at different times. The predicted spectral variations are in agreement with the observational properties established thus far. The correlation of light curves at different frequencies is expected to depend somewhat on the detailed structure of the jet. However our model light curves offer an interesting perspective for interpreting forthcoming well sampled X-ray and UV light curves. Another important result of the model is the finding that, not only the absolute intensity, but also the fractional amplitude of variability, depend on the viewing angle.

6. Summary and Conclusions

Compared to the coverage and sampling obtained in the radio and optical bands, observations of Blazars at high frequencies are still scarce, with only a handful of objects observed more than 10 times in the last decade. Nevertheless some trends in the variability properties are clearly apparent:

a) the average amplitude of variability increases with increasing frequency

b) the minimum observed timescales decrease with increasing frequency

c) the spectra tend to harden with increasing intensity.

The latter effect is definitely present in the X-ray band, but less clear, perhaps limited only to short time scales, in the UV. Very little is known about the correlation of X-ray and UV variability. The available data point to a positive correlation and there are good prospects for substantial advances in this area in the near future.

All what we know about spectra and variability of Blazars can be understood within the following scheme, based on a version of the relativistic jet picture introduced by Blandford and Rees (1978). The highest energy particles and strongest magnetic fields are present near the base of the jet and radiate via synchrotron from the selfabsorption frequency in the infrared up to the X-ray band, with a steep X-ray spectrum. This radiation is emitted in a wide angle cone due to the fact that the bulk velocity is subrelativistic or the collimation is poor or both. Further out, the jet becomes more relativistic and /or more collimated and emits synchrotron radiation from the Radio to the UV band. X-rays are produced in this region via the Inverse Compton effect, with a flat spectrum. Variability can be ascribed to a shock wave travelling from the base to the outer regions of the jet.

The model accounts for the Broad Band Energy Distributions of objects like MKN 421 and PKS 2155-304 for intermediate viewing angles, and for BBEDs of objects like 3C446 and 3C345 if the angle of view is closely aligned with the jet axis. It incorporates "naturally" the observed variability trends mentioned above. It predicts that X-rays with steep spectrum, observed at intermediate viewing angles, are produced close to the jet apex. In this case X-ray variations are expected to occur with shorter timescales and to precede variations in the UV band. Objects with flat X-ray spectra, made through the IC mechanism, correspond to smaller viewing angles and are more strongly beamed. In these objects X-ray variations should occur with small spectral changes and similar timescales as the infrared optical variations. Due to strong beaming, the variability amplitude in the IR optical band may be very large in this case and the timescales shortened with respect to the intrinsic ones.

I look forward to the possibility of testing the scheme further with campaigns of coordinated multifrequency observations on a number of objects representative of the two categories.

Acknowledgement

It is a pleasure to thank the organizers of the Turku Conference and in particular Esko and Lena Valtaoja for a stimulating meeting and for their warm hospitality.

References

Barr, P., Giommi, P., and Maccagni, 1988, *Astrophys. J.*, **324**,, L19.
Barr et al., 1989, in *BL Lac Objects*, eds. L. Maraschi, T. Maccacaro and M-H. Ulrich (Springer–Verlag), p. 290.
Blandford, R.D., and Rees, M.J., 1978, in *Pittsburgh Conference on BL Lac Objects*, ed. A.N. Wolfe (Pittsburgh University Press), p. 328.
Bregman, J.N., et al., 1986, *Astrophys. J.*, **301**, 708.
Bregman, J.N., Maraschi, L., and Urry, C.M., 1987, in "Scientific Accomplishments of the IUE" ed. Y Kondo (Reidel Pub. Co.) p.685.
Bregman, J.N., et al., 1988, *Astrophys. J.*, **331**, 746.
Bregman, J.N., et al., 1990, *Astrophys. J.*, **352**, 574.
Celotti, A., Maraschi, L., and Treves, A., 1991, *Astrophys. J.*, in press.
Edelson, R., et al., 1991, *Astrophys. J.*, **372**, L9.
Feigelson, E.D., et al., 1986, *Astrophys. J.*, **302**, 337.
George, I.M., Warwick, R.S, and Bromage, G.E., 1988, *Mon.Not.R.astr.Soc.*, **232**, 793.
George, I.M., Warwick, R.S., and McHardy, I.M., 1988, *Mon.Not.R.astr.Soc.*, **235**, 787.
Ghisellini, G., and Maraschi, L., 1989, *Astrophys. J.*, **340**, 181.
Ghisellini, G., Maraschi, L., and Treves, A., 1985, *Astron. Astrophys.*, **146**, 204. (GMT)
Ghisellini, G., Maraschi, L., Tanzi, E.G., and Treves, A., 1986, *Astrophys. J.*, **310**, 317.
Giommi, P., et al., 1989, in *BL Lac Objects*, eds. L. Maraschi, T. Maccacaro and M–H. Ulrich (Springer–Verlag), p. 231.
Giommi, P., Barr, P., Garilli, B., Maccagni, D., and Pollock, A.M.T., 1990, *Astrophys. J.*, **356**, 432.

Impey, C.D., and Neugebauer, G., 1988, *Astron. J.*, **95**, 307.
Königl, A., 1989, in *BL Lac Objects*, eds. L. Maraschi, T. Maccacaro and M-H. Ulrich (Springer–Verlag), p. 321.
Maccacaro, T., Gioia, I.M., Schild, R.E., Wolter, A., Morris, S.L., and Stocke, J.T., 1989, in *BL Lac Objects*, eds. L. Maraschi, T. Maccacaro and M-H. Ulrich (Springer–Verlag), p. 222.
Maccagni,D., et al., 1989, in *BL Lac Objects*, eds. L. Maraschi, T. Maccacaro and M-H. Ulrich (Springer–Verlag), p. 281.
Madejski, G., and Schwarz, D., 1989, in *BL Lac Objects*, eds. L. Maraschi, T. Maccacaro and M-H. Ulrich (Springer–Verlag), p. 267.
Makino, F., et al., 1987, *Astrophys. J.*, **313**, 662.
Makino, F., et al., 1989, *Astrophys. J.*, **347**, L9.
Maraschi, L., Tagliaferri, G., Tanzi, E.G. and Treves, A., 1986a, *Astrophys. J.*, **304**, 637.
Maraschi, L., Ghisellini, G., Tanzi, E.G., and Treves, A., 1986b, *Astrophys. J.*, **310**, 325.
Maraschi, L. and Maccagni, D., 1988, Mem.S.A.It., 59, 277.
Maraschi, L., sofia
Morini, M., et al., 1986, *Astrophys. J.*, **306**, L71.
Ohashi, T., 1989, in *BL Lac Objects*, eds. L. Maraschi, T. Maccacaro and M-H. Ulrich (Springer–Verlag), p.296.
Schwartz, D.A., et al., 1989, in *BL Lac Objects*, eds. L. Maraschi, T. Maccacaro and M-H. Ulrich (Springer–Verlag), p.209.
Stocke, J.T., et al., 1985, *Astrophys. J.*, **298**, 619.
Tagliaferri, G., Stella, L., Maraschi, L., Treves, A., and Celotti, A., 1991, *Astrophys. J.*, in press.
Treves, A., et al., 1989, *Astrophys. J.*, **341**, 733.
Treves, A., and Girardi, E., 1990, in "Variability of Active Galaxies" eds. W.J. Duschl, S.J. Wagner and M. Camenzind (Springer–Verlag), p.175.
Urry, C.M., Kondo, Y., Hackney, R.H., and Hackney, R.L., 1988, *Astrophys. J.*, **330**, 791.
Urry, C.M., 1988, in "Multiwavelength Astrophysics", ed. F. Cordova, (Cambridge Univ. Press), p. 279.
Worral, D.M., 1989, in *BL Lac Objects*, eds. L. Maraschi, T. Maccacaro and M-H. Ulrich (Springer–Verlag), p. 305.
Worrall, D.M., and Wilkes, B.J., 1990. *Astrophys. J.*, **360**, 396.

Index of Objects

(References are to first page of papers. Sources in Burbidge and Hewitt, Table 1 (pages 22-38) are not indexed separately.)

0003-067	175	0426-380	151
0007+106	159	0430+052	4, 45, 111, 126, 134, 151,
0016+731	225, 238, 331		159, 384
0048-097	175, 447	0438-436	151
0106+013	159	0440-003	175
0108+388	238, 331	0454+844	187, 225, 238
0109+224	159, 284, 300, 352	0454-234	175
0112-017	175	0458-020	159, 175
0122-005	175	0511-220	175
0133+476	159, 205, 238, 331	0518+165	251
0135-247	175	0521-365	447
0138-097	175, 352	0528+153	151
0148+218	159, 391	0537-441	151, 238
0153+744	225, 238, 331	0538+498	251
0202+149	159	0548-322	444
0202-172	175	0552+398	159
0212+735	187, 205, 225, 238, 331	0605-085	175
0215+014	289	0615+821	225, 331
0219+428	289, 300	0642+449	159
0224+671	159	0710+439	331
0234+285	159, 205	0711+356	238, 331
0235+164	4, 55, 70, 85, 111, 142,	0716+714	85, 159, 187, 205, 225,
	151, 159, 187, 238, 300,		264, 311, 331, 346
	331, 352, 377, 391, 414,	0735+178	111, 159, 170, 187, 196,
	427		234, 238, 264, 284, 289,
0238-084	175		300, 331, 427, 447
0248+430	159	0736+017	111, 159
0300+470	238, 331, 352	0743-006	175
0306+102	238	0754+100	159, 238, 289, 300, 352,
0316+413	4, 111, 159, 205, 331, 447		427, 447
0323+022	447	0804+499	159, 205, 331
0333+321	85, 159, 377	0814+425	159, 331
0336-019	175	0829+046	352
0355+508	159, 205	0834-201	175
0403-132	175	0836+710	225, 238, 331
0414-189	175	0839+187	391
0420-014	111, 159, 175	0846+513	4, 159
0422+004	159, 300, 447	0851+202	39, 45, 70, 111, 126, 134,

Index of Objects

	151, 159, 167, 187, 196, 205, 238, 264, 278, 289, 294, 300, 303, 306, 311, 331, 352, 377, 384, 391, 399, 427, 447	1228+126	427 4, 159, 196, 391
0859+470	331	1243-072	175
0906+015	111	1253-055	85, 111, 134, 142, 151, 159, 175, 187, 196, 205, 284, 331, 377, 414, 447
0906+430	251, 331	1302-102	175
0914-620	399	1308+326	111, 126, 159, 205, 238, 331, 352, 427
0917+624	205, 331	1322-427	170
0923+392	85, 126, 142, 159, 205, 234, 331, 391	1335-127	142, 151, 175
		1354+195	391
0945+408	331	1354-152	175
0953+254	159	1400+162	187
0954+658	159, 238, 331	1404+286	331
1013+208	391	1406-076	175
1032-199	175	1413+135	142, 159
1034-293	352	1416+067	251
1039+812	225	1418+546	159, 300, 352, 427
1045-188	175	1424-418	151
1055+018	159	1442+101	151
1101+384	111, 187, 289, 300, 303, 361, 447	1458+718	251
		1502+106	331
1127-145	175	1504-166	175
1133+704	447	1510-089	111, 151, 159, 175
1144-379	151	1514-241	111, 175
1145-071	175	1519-273	151
1147+245	352	1538+149	159, 238
1148-001	175	1553+113	447
1150+497	251	1611+343	159, 331
1150+812	225, 331	1624+414	238, 331
1156+295	85, 111, 126, 142, 159, 414, 427	1633+382	159, 238, 331
		1637+574	159
1215+303	300, 447	1638+398	205
1218+304	447	1641+399	45, 111, 134, 142, 159, 167, 196, 205, 221, 238, 251, 259, 331, 377, 382
1219+285	126, 159, 238, 352		
1219+755	4	1642+690	251, 331
1223-052	111, 142, 159, 187, 196, 251, 377, 414, 447	1652+398	111, 126, 238, 289, 300, 303, 331, 367, 444
1226+023	4, 45, 102, 111, 134, 142, 159, 170, 175, 196, 205, 221, 229, 238, 264, 284, 331, 377, 382, 384, 399,	1727+502	159
		1739+522	159
		1741-038	159, 175, 331

Index of Objects

1749+096	70, 111, 142, 159, 238, 331	2345-167	151, 175
1749+701	187, 225, 238, 331	2351+456	238, 331
1803+784	159, 187, 205, 225, 238, 331	2352+495	238, 331
1807+698	142, 159, 187, 238, 251, 331, 447	2354-117	175
1821+107	331	E1402.3+0416	447
1823+568	238	IC 5283	372
1828+487	238, 251	NGC 3516	441
1845+797	4, 45, 159, 447	NGC 4319	4
1921-293	85, 111	NGC 5141	414
1928+738	4, 159, 205, 221, 225, 238, 331	NGC 5548	399
2005+403	159	NGC 7469	372
2005-489	447	SS 433	45
2007+776	159, 187, 205, 225, 331	3C 66A	see 0219+428
2008-159	175	3C 84	see 0316+413
2021+614	159, 331	3C 120	see 0430+052
2126-158	175	3C 138	see 0518+165
2128-123	175	3C 147	see 0538+498
2131-021	175	3C 216	see 0906+430
2134+004	126, 159	3C 273	see 1226+023
2141+175	159	3C 274	see 1228+126
2145+067	142, 159	3C 279	see 1253-055
2155-152	175	3C 298	see 1416+067
2155-304	320, 444, 447	3C 309.1	see 1458+718
2200+420	4, 70, 111, 126, 134, 142, 159, 167, 187, 196, 205, 238, 264, 284, 289, 294, 300, 306, 311, 331, 352, 358, 377, 391, 399, 427, 447	3C 345	see 1641+399
		3C 371	see 1807+698
		3C 380	see 1828+487
		3C 390.3	see 1845+797
		3C 446	see 1223-052
		3C 454.3	see 2251+158
2201+315	159	4C 14.60	see 1538+149
2203-188	175	4C 29.45	see 1156+295
2214+350	331	4C 38.41	see 1633+382
2216-038	175	4C 39.25	see 0923+392
2223-052	111, 175	4C 49.22	see 1150+497
2227-088	175	4C 71.07	see 0836+710
2230+114	159, 187, 251	AP Lib	see 1514-241
2243-123	175	BL Lac	see 2200+420
2251+158	111, 151, 159, 170, 205, 238, 331	Cen A	see 1322-427
		CTA 102	see 2230+114
		DA 55	see 0133+476

Index of Objects

DA 193	see 0552+398	OH 471	see 0642+449
DA 237	see 0735+178	OI 090.4	see 0754+100
III Zw 2	see 0007+106	OJ 287	see 0851+202
M 87	see 1228+126	OL 093	see 1055+018
Mkn 180	see 1133+704	ON 231	see 1219+285
Mkn 205	see 1219+755	ON 325	see 1215+303
Mkn 421	see 1101+384	OQ 172	see 1442+101
Mkn 501	see 1652+398	OQ 530	see 1418+546
NGC 1275	see 0316+413	OS 562	see 1637+574
NRAO 140	see 0333+321	OT 081	see 1749+096
NRAO 150	see 0355+508	OV-236	see 1921-293
NRAO 512	see 1638+398	OW 637	see 2021+614
OA 129	see 0420-014	OX 057	see 2134+004
OC 012	see 0106+013	OX 169	see 2141+175
OF 038	see 0422+004		

Index of Subjects

(References are to first page of papers)

Acceleration mechanisms 142, 264, 278, 399
Accretion 39, 45, 311
Accretion disks 311, 320, 331, 346, 377, 399, 414

Black holes
   binary 39, 45, 196, 221, 311
   multiple 39
Blazars 39, 55, 82, 85, 111, 151, 159, 167, 187, 196, 225, 238, 251, 264, 284, 300, 303, 306, 320, 352, 391, 399, 414, 427, 447
   and compact radio sources 264
   classification 4, 55, 70, 82, 447
   definition 4, 55, 111, 264, 399, 447
   models for optical variability 264, 311, 320, 399, 414
   multifrequency characteristics 55, 85, 111, 399, 447
   VLBI 187, 196, 225, 238, 251
BL Lac objects 4, 70, 79, 170, 346, 358
   definition 4, 55, 70, 447
   parent populations 4, 70, 79
   space distribution 4, 70
   spectra 4, 55, 70
   two or more classes 4, 55, 70, 238, 447
   variability 4, 70, 346, 358
Brightness temperatures 70, 111, 256, 331, 352

Central engines 39, 45, 52, 196, 414
Classification of AGN 4, 55, 70, 82, 238, 399, 447
Compact steep-spectrum sources 55, 79, 251
Continuum emission 4, 55, 82, 85, 111, 399, 447
Cosmological evolution 79, 82, 438
Cosmology 167, 259

Doppler boosting 45, 55, 70, 85, 142, 196, 225, 264, 331, 352, 414
Dust emission 399

Faraday rotation 238, 264
Fermi acceleration 85, 264, 278
Flares
   evolution in shock models 85, 102, 111, 142, 399
   optical 45, 111, 311, 320, 327, 414
   radio 70, 111, 126, 134, 142

Gamma rays 85
Gravitational lensing 4, 55, 70, 238, 331, 346
Gravitational radiation 39, 45

Highly polarized quasars 4, 55, 70, 79, 175, 184, 399
Host galaxies 4, 55, 70, 399, 444

Interacting galaxies 39
Interstellar scintillation 85, 151, 331, 391

Jets
   bending of 45, 85, 134, 187, 196, 205, 221, 234, 238, 251
   helical structures 111, 196, 205, 221, 225, 289, 331, 346, 352
   Lorentz factors 187, 221, 225, 229
   magnetic field structure 85, 134, 238
   structure 187, 196, 238, 251
   turbulent 85, 142

Index of Subjects

Line emission 4, 55, 70, 82, 441
Lorentz factors 4, 55, 70, 82, 85, 142, 187, 221, 225, 229, 256, 331, 352

Magnetic fields 134, 142, 238, 264, 278, 327, 399
Mergers 79
Monitoring programs
   optical 289, 300, 414
   radio 111, 126, 134, 142, 151, 159, 167, 170, 175
   VLBI 187, 196, 205, 221, 225
Multifrequency behavior 85, 102, 111, 159, 284, 306, 399, 447

OVV sources 4, 45, 55, 70, 251, 399, 414

Parent populations 4, 55, 70, 79, 82, 111, 238
Periodicity
   infrared 377
   optical 39, 45, 111, 311, 346, 358, 377, 382, 384
   radio 126, 346, 377, 391
   VLBI component ejection 384
Polarization
   circular 264
   frequency dependent optical 264, 278, 284, 294, 300, 303, 361
   mechanisms 134, 142, 264, 278, 306
   optical 55, 111, 264, 278, 284, 289, 294, 300, 303, 306, 361
   radio 126, 134, 142, 331
   radio vs. optical 55, 284, 306
   rotation of position angles 85, 289, 294, 306
   VLBI 238
   vs. other source characteristics 55

QSO-galaxy associations 4
Quasars 4, 79, 126, 134, 170, 175, 184, 205, 256, 259, 284, 327, 438, 444

Radiation mechanisms 4, 264, 278, 399, 427, 447
Radio galaxies 4, 55, 70
Radio-quiet QSOs 79, 399
Rapid variations
   infrared 367, 399
   in Seyferts 372
   mechanisms 4, 85, 311, 320, 327, 331, 346, 352
   optical 45, 289, 320, 331, 346, 358, 361, 367, 372
   optical polarization 111, 264, 289, 294, 361
   radio 111, 126, 151, 331, 346, 352, 391
   radio-optical connections 111, 331, 346, 352
   radio polarization 331
   ultraviolet 320
Redshhifts, noncosmological 4

Samples 4, 55, 151, 175, 251, 256, 444
   optical 289, 300, 303, 414, 427
   radio 55, 70, 111, 126, 134, 142, 151, 159, 167, 170, 175, 184, 331, 352
   VLBI 187, 205, 225, 238
Seyfert galaxies 45, 372, 399, 441, 444
Shock-in-jet models 4, 55, 70, 82, 85, 134, 187, 196, 229, 331, 346, 399, 447
   analytical models 85, 102, 134, 142
   comparison with radio data 70, 85, 102, 111, 134, 142, 151, 256
   optical variability 264, 278, 284
   polarized flux production 134, 278, 294
   VLBI polarization 238
Spectra, broad-band 85, 102, 111, 175, 399, 427, 438, 444, 447
Structure functions 111, 142, 331, 346
Superluminal motion 4, 55, 187, 196, 205, 221, 225, 229, 234, 384
Supermassive stars 45, 52

Time delays 102, 111, 399

Unified schemes 4, 55, 70, 79, 82, 238, 251, 284, 399, 447

Variability
    connection between radio and optical 102, 111, 284, 331, 346
    connection to VLBI 45, 134, 170, 187, 205, 229, 234, 384
    dependence on redshift 70, 438
    from bent jets 45, 85, 111, 134
    from shocks 70, 85, 102, 111, 134, 142, 159, 170, 284
    from turbulent jets 85
    infrared 399
    line emission 441
    mechanisms for 4, 45, 52, 167, 259, 264, 399, 414, 427, 438
    nonthermal continuum 85, 111, 399
    of different classes of AGN 70, 111, 126, 151, 175, 184
    optical 55, 70, 289, 320, 331, 346, 358, 361, 367, 372, 377, 382, 384, 399, 414, 427, 438
    optical polarization 55, 284, 289, 294, 300
    optical vs. radio compactness 55
    radio 70, 85, 102, 111, 126, 151, 159, 167, 170, 175, 184, 331, 346, 352, 391, 399
    radio polarization 134, 142
    timescales 70, 111, 175, 447
    UV 399, 447
    VLBI structural 187, 225
    vs. spectral characteristics 70, 175, 184
    X-rays 85, 447

X-rays 4, 70, 85, 170, 399, 447